Clinical

Hilary Baker LHMC

Hugh L. Moffet, M.D.

*Professor of Pediatrics
Former Chairman, Infectious
Disease Teaching Committees for
the Independent Study Program and
the Second Year Curriculum
University of Wisconsin, Madison*

Clinical Microbiology
SECOND EDITION

J. B. LIPPINCOTT COMPANY Philadelphia • Toronto

SECOND EDITION

Copyright © 1980, by J. B. Lippincott Company. All rights reserved. No part of this book may be used or reproduced in any manner whatsoever without written permission except in the case of brief quotations embodied in critical articles and reviews. Printed in the United States of America. For information address Medical Books, J. B. Lippincott Company, East Washington Square, Philadelphia, Penna. 19105

ISBN O-397-50450-0

Library of Congress Catalog Card Number 79-21674

Printed in the United States of America

1 3 5 6 4 2

Library of Congress Cataloging in Publication Data
Moffet, Hugh L
  Clinical microbiology.
  Bibliography
  Includes index.
  1. Medical microbiology.   I. Title. [DNLM:
1. Microbiology. QW4.3 M695c]
QR46.M7 1980   616.01   79-21674
ISBN 0-397-50450-0

The authors and publisher have exerted every effort to ensure that drug selection and dosage set forth in this text are in accord with current recommendations and practice at the time of publication. However, in view of ongoing research, changes in government regulations, and the constant flow of information relating to drug therapy and drug reactions, the reader is urged to check the package insert for each drug for any change in indications and dosage and for added warnings and precautions. This is particularly important when the recommended agent is a new or infrequently employed drug.

Dedication

At parties, when asked what sort of work you do,
Do you have trouble answering? Do you smile
And wonder if what you do is worthwhile,
Too overspecialized, routine, not new?
Prestige in science goes mostly to the few
Who test and treat by careful, controlled trial.
But does its practice need a flashy style?
Must all compete to prove their value, too?
Of course not. Quiet competence should be,
Like faithful service, honored as the rest,
And people need availability
And kindness, and still deserve to get our best.
If we can do all this and yet feel free
We'll find our work will satisfy this quest.

Preface

The second edition of *Clinical Microbiology* has been extensively revised and enlarged. It continues the approach and philosophy of the first edition. Every section has been updated, using new data from clinically oriented journals. One new chapter has been added, entitled "Miscellaneous Bacteria," which includes Legionnaire's Disease bacillus. The original chapter on Mycobacteria and Fungi has been divided into two chapters.

In spite of extensive revisions, the format and goals of the book have remained the same. The book is arranged by microorganisms, and as one colleague puts it, "It's a parade of bugs." The frequency, importance, clinical manifestations, laboratory approach, biologic characteristics of clinical interest, treatment, and prevention for each microorganism are discussed. Recent references from readily available journals are cited extensively for documentation and for further reading, if desired.

In order to limit the scope of this book to clinically applicable microbiology, a separate discussion of a number of important topics is deliberately omitted. Some topics, such as immunologic principles and tumor viruses, are better covered in other courses of study or by nonmicrobiology textbooks. Familiarity with microbial physiology, metabolism, and genetics is not essential to clinical medicine, so these areas are omitted. Classification of microorganisms and bacteriophage are mentioned only where clinically relevant. While host-parasite relationships and chemotherapy are not discussed in a separate chapter in general terms, they are covered in specific detail for individual microorganisms.

Clinical Microbiology is intended to complement my other book—*Pediatric Infectious Diseases: A Problem-Oriented Approach*—which is useful for care of the patient with a diagnostic or other specialized problem. *Clinical Microbiology* is for the person who wants more clinical or laboratory details about the microorganism believed to be the cause of the illness. Both the beginning student and the experienced person should find a degree of depth suitable to their needs.

Extensive revisions or additions have been made, including the sections on hepatitis viruses, influenza virus, Group B streptococci, diarrheagenic *Escherichia coli*, gonorrhea, *Bacillus cereus* food poisoning, neonatal botulism, *Clostridium difficile* colitis, yersiniae, vibrios, campylobacters, chlamydiae, ureaplasmas, mycoplasmas, and rotaviruses. There are brief sections on rarely recovered microorganisms, and nosocomial microorganisms with new names or new importance. The space devoted to each microorganism is roughly proportional to its clinical importance.

New diagnostic techniques used in the clinical laboratory are integrated into this edition with a series of diagrams which use a new set of symbols for antigen, antibody, labelled antibody, and enzyme substrates. These diagrams illustrate radioimmunoassay (RIA), enzyme-labelled antibody techniques for measuring antigen or antibody (ELISA), direct and indirect fluorescent antibody (FA) techniques, latex fixation, and co-

Contents

1. GRAM-POSITIVE COCCI 1

Staphylococci *1*
Pneumococci (Streptococcus Pneumoniae) *7*
Group A Streptococci *11*
Group B Streptococci *19*
Other Streptococci *21*

2. GRAM-POSITIVE RODS 27

Corynebacteria *27*
Listeria Monocytogenes *30*
Clostridium Species *31*
Actinomyces and Nocardia Species *35*
Bacillus Species *37*

3. ENTERIC GRAM-NEGATIVE RODS 41

General Classification *41*
Salmonella *43*
Shigella *46*
Diarrheagenic E. Coli *50*
Other Enteric Bacteria *55*
Pseudomonas Species *58*
Nonfermenter Group *60*
Vibrio Species *62*
Bacteroides Species *64*

4. SMALL GRAM-NEGATIVE RODS 73

Haemophilus Influenzae *73*
Bordetella Pertussis *78*
Zoonoses *81*
Yersinia Species *81*
Pasteurella Species *83*
Francisella Tularensis *83*
Brucella Species *84*

5. GRAM-NEGATIVE COCCI 90

Meningococcus *90*
Neisseria Gonorrheae (Gonoccus) *95*

6. MISCELLANEOUS BACTERIA   103

Legionnaires' Disease Bacillus *103*
Actinobacillus and Haemophilus Aphrophilus *106*
Aeromonas and Pleismonas *106*
Branhamella *106*
Calymmatobacterium (Donovania) *107*
Cardiobacterium *107*
Chromobacterium *107*
Eikenella *107*
Erysipelothrix *107*
Erwinia *107*
Fusobacterium *108*
Lactobacillus and Bifidobacterium *108*
Peptococcus and Peptostreptococcus *108*
Propionibacterium *108*
Streptobacillus *108*
Veillonella *108*

7. MYCOBACTERIA   111

Mycobacterium Tuberculosis *111*
Other Mycobacteria *117*

8. FUNGI   122

Fungi in General *122*
Histoplasma Capsulatum *122*
Coccidioides Immitis *126*
Blastomyces Dermatiditis *128*
Sporothrix Schenckii *129*
Candida Albicans *131*
Cryptococcus Neoformans *133*
Aspergillus Species *134*
Other Invasive Fungi *135*

9. SPIROCHETES   141

Treponema Pallidum and Other Spirochetes *141*
Leptospira Species *145*

10. CHLAMYDIA, RICKETTSIA, MYCOPLASMA   149

Chlamydia Species *149*
Rickettsia *153*
Mycoplasma Pneumoniae and Other Mycoplasmas *155*

11. MAJOR DNA VIRUSES   162

Herpesvirus Hominis (Herpes Simplex Virus) *162*
Varicella-Zoster Virus *166*
Cytomegalovirus *169*
Epstein-Barr Virus *172*
Adenoviruses *176*
Smallpox and Vaccinia Viruses *179*
Other DNA Viruses *182*

12. HELICAL RNA VIRUSES   188

Influenza Virus *188*
Parainfluenza Virus *196*
Mumps Virus *197*
Measles Virus *202*
Respiratory Syncytial Virus *205*
Rabies Virus *207*

13. OTHER RNA VIRUSES 216

Enteroviruses *216*
Rhinoviruses *222*
Arthropod-borne Encephalitis Viruses (Arboviruses) *224*
Rubella Virus *227*

14. MISCELLANEOUS VIRUSES 234

Hepatitis Viruses *234*
Yellow Fever Virus *243*
Dengue Virus *243*
Colorado Tick Fever Virus *244*
Coronaviruses *244*
Lymphocytic Choriomeningitis Virus *245*
Lassa Virus *245*
Marburg and Ebola Viruses *245*
Tumor Viruses *246*
Slow Virus Diseases *247*
Virus-like Particles in Lupus *249*
Reoviruses *250*
Rotaviruses *250*
Norwalk Virus *251*
Cat Scratch Disease *252*

15. SELECTED PROTOZOA 259

Entamoeba Histolytica *259*
Free-living Amebas *261*
Giardia Lamblia *262*
Toxoplasma Gondii *264*
Malaria *266*
Pneumocystis Carinii *270*
Trichomonas Vaginalis *270*

16. SELECTED NEMATODES 275

Trichinella Spiralis *275*
Toxocara Species *276*

APPENDIX: A REVIEW OF ESSENTIALS 279

Review Questions *280*
Answers to Review Questions *287*

INDEX 289

Clinical Microbiology

Gram-Positive Cocci

STAPHYLOCOCCI

Objectives

1. Describe how *Staph. aureus* is differentiated from *Staph. epidermidis* in the laboratory.
2. Describe the clinical importance of penicillinase, coagulase, staphylococcal enterotoxin, and staphylococcal epidermolytic toxin.
3. Describe how phage typing of *Staph. aureus* is done, how this information can be clinically useful, and under what circumstances it is available.
4. Describe six diseases of which *Staph. aureus* is a very frequent cause.
5. Describe three clinical circumstances in which *Staph. epidermidis* can cause disease.
6. Describe the differences in pathogenesis and frequency between staphylococcal food poisoning and staphylococcal enterocolitis.
7. Describe how teichoic acid antibodies of *Staph. aureus* can be used to recognize severe staphylococcal infections.

Definitions

Staphylococci are gram-positive cocci in clumps, like grapes (staphylo-grapes). *Staph. aureus* was originally named for its golden-colored colony, but now is best defined as a staphylococcus that is either coagulase-positive or ferments mannitol. *Staph. albus* is now called *Staph. epidermidis*, because of its frequent occurrence on normal skin.

Staphylococcus is a genus (plural, genera) in the family *Micrococcaceae*. Some clinical laboratories give a preliminary report of "staphylococcus-microcaccus group," and later the isolate reported as *Staph. aureus*, *Staph. epidermidis*, or *Micrococcus*. Other related genera in this family are *Micrococcus*, *Sarcina*, *Planococcus* and *Peptococcus* (Table 1-1). Aerobic clinical isolates which do not ferment glucose are often reported as *Micrococcus* species, without further identification. These genera rarely cause disease, and there is no apparent clinical value in detailed identification by the laboratory, although antibiotic susceptibility testing is necessary if the isolate appears to be causing disease.

Frequency and Importance

Staph. aureus is an extremely common cause of skin infections, and is an important cause of many serious systemic infections.

Types and Groups

Staph. aureus can be typed using standardized bacteriophage (viruses) which lyse the bacteria.[25] The isolate to be typed is spread over a plate and a

TABLE 1-1. SIMPLIFIED DEFINITIONS OF STAPHYLOCOCCI AND SIMILAR BACTERIA

Microorganism	Major Characteristic
Staph. aureus	Coagulase positive
Staph. epidermidis	Coagulase negative
Staph. saprophyticus	Novobiocin resistant
Micrococcus	Does not ferment glucose
Peptococcus and *Sarcinia*	Anaerobic
Planococcus	Forms tetrads

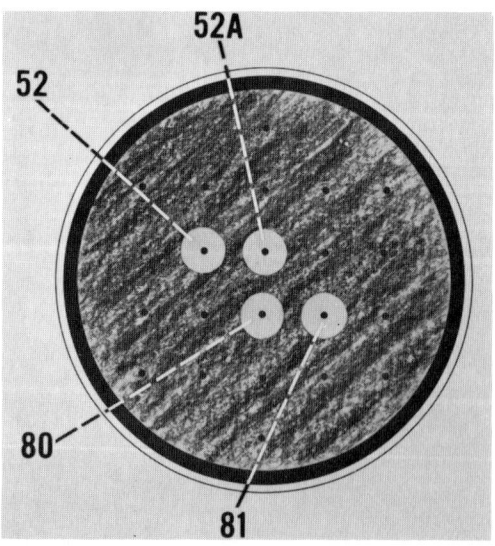

FIG. 1-1. Culture plate inoculated with Staph. aureus phage type 52/52A/80/81, illustrating lysis of this strain by each of these four bacteriophage.

series of 5 to 60 bacteriophage (named by number) are dropped on a particular position on the plate. The type of Staph. aureus is defined by the phage that lyse it; for example, type 52/52A/80/81 is lysed by these four phage (Fig. 1-1). Phage typing of Staph. aureus can be used to trace the spread of an outbreak, but is not useful as an indicator of virulence. Phage typing is usually available only in research or reference laboratories, and will be done by them only for special situations, such as outbreaks.

The hundred or so phage types are grouped into four groups, Group I to IV. For example, Group I contains phage types 29, 52, 52A, and 80 in various combinations, and most members of this group are penicillin-resistant.[1] Group II contains the phage types associated with epidermolytic toxin, and Groups III and IV contain most of the types associated with enterotoxin production.

Non-typable strains are apparently mediated by phage-resistance extrachromosomal genes, which can be "cured" by acridine orange treatment.[18]

Clinical Patterns of Illness

Staph. aureus is the most common cause of the following diseases:

Pyogenic Infections. These include skin infections, particularly wound infections, abscesses, and pustules; osteomyelitis and septic arthritis.

Food poisoning manifested by vomiting, acute bacterial endocarditis, and pneumonia with pneumatoceles or empyema are usually caused by Staph. aureus.[16]

Empyema is purulent pleural fluid. A pneumatocele is an air-filled sac resulting from a necrotizing pneumonia (Fig. 1-2). Meningitis with hemorrhagic skin lesions similar to those seen in meningococcemia has been reported.[15]

Staph. aureus is a rare cause of severe enterocolitis. This disease was observed in the 1960s as a complication of broad-spectrum (especially tetracycline) antibiotic therapy, especially when antibiotics were used to try to suppress bowel bacteria before bowel surgery. This disease was rarely observed in the 1970s. It was characterized by invasion of the bowel mucosa by staphylococci, with resultant diarrhea. Now some authorities believe a *Clostridium* species was the real cause of many cases of presumed staphylococcal pseudomembraneous enterocolitis (see Chap. 2). Staphylococcal enterocolitis should be distinguished from staphylococcal food poisoning, which is manifested by vomiting and retching caused by preformed staphylococcal enterotoxin.

Staph. epidermidis is a *very common* cause of:

False Positive Blood Cultures. Usually, when the laboratory reports the recovery of Staph. epidermidis, it is a contaminant, and is unrelated to the patient's disease. Staph. epidermidis is a common contaminant in blood and spinal fluid cultures because the skin is not properly disinfected before the specimen is taken.

Staph. epidermidis is an *occasional* cause of:

Infections Complicating Plastic Foreign Bodies. In patients with a plastic tubing and a valve system inserted between the ventricle of the brain and the vena cava or peritoneal cavity for relief of the pressure of hydrocephalus, Staph. epidermidis may infect the tubing and produce ventriculitis, meningitis, or bacteremia. Plastic heart prostheses can become infected after heart operations in which an artificial valve or patch has been inserted.[22, 28]

Wound infections are *rarely* caused by Staph. epidermidis,[28] although it can be recovered from wound cultures because it is a normal flora of the skin. Staph. epidermidis and Staph. saprophyticus are also rare causes of urinary tract infections.[26]

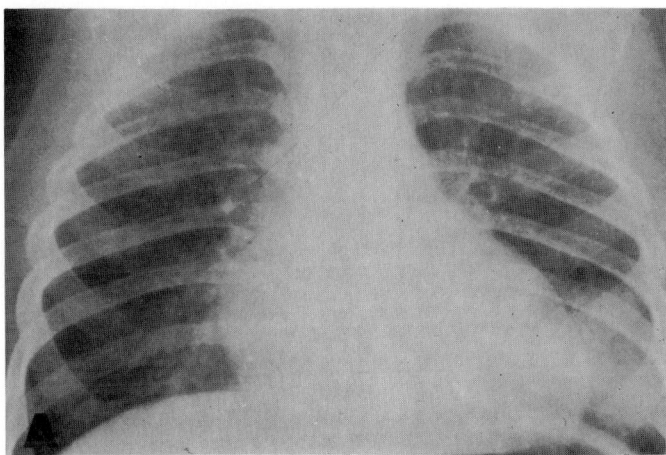

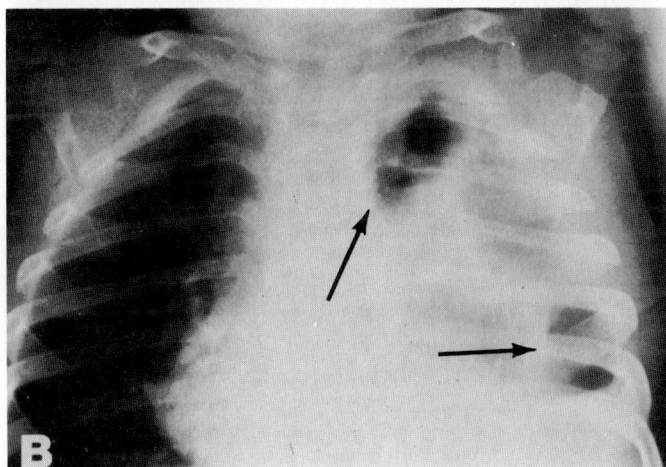

FIG. 1-2. *Chest x-rays of a patient with staphylococcal pneumonia.* (A) *The x-ray shows left lower pneumonia.* (B) *The second examination three days later shows many pneumatoceles* (**arrows**).

Laboratory Approach

Smear. Gram-positive cocci in clumps or clusters should be presumed to be staphylococci until culture results are available (Fig. 1-3). Gram-positive cocci in singles or pairs might be staphylococci, but could also be streptococci or pneumococci. Tetrads or packets are likely to be planococci or sarcina, but might be staphylococci. The main value of the Gram stain is to detect chains, which indicate pneumococcus or streptococcus (Fig. 1-3).

Culture. Staphylococci are readily recovered on simple media, such as a sheep blood agar plate, where they are often beta-hemolytic (Fig. 1-7). The colony color of *Staph. aureus* is usually golden, cream, or buff, but can be porcelain white. Characteristic golden-pigmented colonies are almost always coagulase-positive, and may be reported by the laboratory as *Staph. aureus* without

FIG. 1-3. *Gram-positive cocci: possible arrangements.* (A) *In direct smear, could be staphylococci, micrococci, streptococci or pneumococci.* (B) *In broth culture, probably streptococci, but could be pneumococci.* (C) *Could be streptococci (including enterococci) or pneumococci.* (D) *Typical of pneumococci, but could be streptococci.* (E) *Typical of staphylococci or micrococci.* (F) *Typical of planococci or aerococci, but in clinical specimens likely to be reported as micrococcus group.*

further testing. However, white colonies may not be *Staph. epidermidis*, and should be tested for coagulase production.

Staph. aureus is usually defined by coagulase testing, but mannitol fermentation is also used in some laboratories (Table 1-1).[3,13] The mannitol fermentation must be done under anaerobic conditions.[13] Occasionally an organism ferments mannitol, but is coagulase-negative, and may be called a coagulase-negative *Staph. aureus*. Only about two percent of *Staph. aureus* are coagulase-negative and mannitol-positive.[13,17] Therefore, for routine diagnostic laboratory work, separation of mannitol and coagulase variants is not justified.[17] Coagulase testing may be done by two methods and tests for two forms of coagulase:[5,20]

1. *Slide Coagulase Test.* This takes only a few minutes. When a drop of plasma is added to a saline suspension of staphylococci on a slide, clumps of fibrinogen appear around the bacteria in less than a minute. This tests for bound coagulase, which is also called clumping factor.

2. *Tube Coagulase Test.* The organism is added to rabbit plasma and allowed to incubate at 37° C. This requires about 4 hours incubation for a preliminary reading for clotting, tested for by tipping the tube, but must be incubated overnight for a final reading. This tests for free coagulase, and is generally regarded as the definitive test for coagulase, although adequate comparison of the two types of coagulase for correlation with virulence has not been done.

Serology. Detection of serum antibodies against teichoic acid has been shown to be of some clinical value in the detection of *Staph. aureus* in serious human infections.[24] Teichoic acid is a prominent component of the cell wall of many gram-positive cocci, including *Staph. aureus*. It can be extracted from *Staph. aureus* using an enzyme, lysostaphin. Serum antibodies against teichoic acid can be detected in patients with staphylococcal bacteremia, osteomyelitis, and especially endocarditis, as often found in heroin addicts. Counterimmunoelectrophoresis (Fig. 4-2), using a patient's serum as antibody and extracted teichoic acid as antigen, is the method usually used to detect staphylococcal antibody in patients.

Serum antibodies produced against several other antigens of *Staph. aureus* can be measured, but are apparently not important in protecting humans from staphylococcal infections.[14]

Biologic Characteristics of Clinical Interest

Normal flora. Both *Staph. aureus* and *Staph. epidermidis* should be regarded as normal human flora. *Staph. epidermidis* is found on the skin of most normal humans. *Staph. aureus* is found in the nose of about 50 percent of normal individuals, and on the skin of about 20 percent of normal individuals.[27] Individuals with *Staph. aureus* in the nose are called nasal carriers. Antibiotic ointment or lysostaphin nasal spray applied to the nose reduces the carrier rate, but only for a period of about 10 days.[19]

Transmission and Contagion. Hospital employees and patients tend to become colonized by the staphylococcal strains, which are usually penicillin-resistant, in their environment. The nasal carrier does not seem to be much risk to others, unless the carrier transfers the staphylococci by hand contact. Most transmission of staphylococci appears to be via the hands, with limited transmission by aerosol in the immediate vicinity. Infected open wounds, with *Staph. aureus* in the pus, are important sources for transmission by hands, as are contaminated dressings and clothing.

Newborn nurseries also appear to be important reservoirs of *Staph. aureus*.

Penicillinase. This is an enzyme in the general group of beta-lactamases, which inactivates penicillin by disrupting its beta-lactam ring. Penicillinase production is the mechanism of penicillin-resistance of most strains of *Staph. aureus*. The appearance of staphylococci that produce penicillinase can result from a variety of mechanisms, including transfer of genetic material by bacteriophage (transduction), as well as by mutation.[8]

Beta-lactamases. Many different beta-lactamases are produced by bacteria, and can be classified in several ways. Those enzymes that hydrolyze the beta-lactam ring of penicillin derivatives can be called penicillinases, while those that destroy the beta-lactam ring of cephalosporin derivatives can be called cephalosporinases. However, beta-lactamases are often partially active against both of these two broad groups.

The most commonly used classification, often referred to as Richmond's classification, is based on the activity of the enzyme against five or more substrates (such as penicillin, ampicillin, carbenicillin, cloxacillin, cephaloridine, and cephalexin). At least 15 types of beta-lactamases from gram-negative bacteria are classified into Richmond's Roman numeral-letter groups. Class I enzymes are predominantly active against penicillins; Class III enzymes have approximately equal activity against penicillins and cephalosporins, but are inhibited by cloxacillin.[21] For example, Type IIIa enzyme is a common beta-lactamase found in gram-negative bacteria.

Beta-lactamases are also sometimes named for the plasmid (R factor) that carries them. (See R factor, in Shigella, Chap. 3.) The beta-lactamase specified by the R factor called TEM appears to be a Type IIIa enzyme, and is very frequent in gram-negative bacteria. Other R factors, such as those called RI 1, RP 1, and RGN 823, also can carry the genetic coding for Type IIIa beta-lactamase.

Another fundamental difference between beta-lactamases is whether they are produced by gram-positive or gram-negative bacteria. The various beta-lactamases produced by gram-positive bacteria appear to be closely related to each other, just as those produced by gram-negatives appear to be closely related to each other.[21] Beta-lactamases produced by gram-positive bacteria are typically released in large quantities into the environment of the bacteria, whereas a beta-lactamase produced by a gram-negative bacterium is typically strategically located in the bacterial cell wall, so it can inactivate the antibiotic as it approaches the cell wall.

Beta-lactamases of gram-positive bacteria are all inducible, but most of those produced by gram-negative bacteria are not inducible (i.e., they are constitutive). The genetic material coding for beta-lactamases in gram-positives can be transmitted by a bacteriophage (transduction), whereas the gram-negatives often acquire a beta-lactamase via a plasmid (conjugation).

Beta-lactamase can be inhibited by clavulante (clavulanic acid). This is useful in the study of these enzymes, but as yet has no clinical application.

Methicillin Resistance. *Staph. aureus* strains are now rarely resistant to methicillin or other penicillinase-resistant penicillins, such as nafcillin. These strains are also usually resistant to cephalosporins but susceptible to vancomycin. The susceptibility to gentamicin is variable. Sometimes these strains are susceptible to the synergistic effect of methicillin and gentamicin.

Methicillin-resistance is not mediated by a beta lactamase. It is sometimes called *intrinsic resistance*. It has been gradually increasing in frequency since the early 1960s and is more likely to appear in hospitals.

The mechanism of this intrinsic resistance is not clear, but does not involve inactivation of the methicillin. Antibiotic tolerance, as described below, is not involved, as the methicillin does not even inhibit the microorganism.

Gentamicin Resistance. Although gentamicin is seldom used to treat *Staph. aureus* infections, resistance to gentamicin is increasing. It is clearly due to plasmid-transmitted enzymes that inactivate the drug.

Antibiotic Tolerance. In this situation, the antibiotic inhibits, but does not kill the microorganism. Some strains of *Staph. aureus* exhibit resistance to methicillin on the basis of tolerance. Tolerance in *Staph. aureus* can occur independently in the three classes of antibiotics usually used: penicillinase-resistant penicillin, cephalosporins, or vancomycin. Tolerance usually occurs in only part of the *Staph. aureus* population, the remainder being killed by the antibiotic. The mechanism appears to be a deficiency of autolytic enzyme activity in the *Staph. aureus*. Tolerance appears to be a mechanism of antibiotic failure in some patients.

Staphylococcal Epidermolytic Toxin. This toxin is spread by the blood from a primary focus of infection to the skin where it produces redness followed by exfoliation (Fig. 1-4), called the scalded-skin syndrome.[12]

Staphylococcal Toxic-Shock Syndrome. This recently described syndrome is characterized by fever, headache, confusion, conjunctivitis, scarlatinaform rash, renal insufficiency, and severe prolonged shock.[23] It resembles scalded-skin syndrome. It appears to be associated with an exotoxin produced by the phage Group I of staphylococci, as distinct from scalded-skin syndrome, caused by an exotoxin usually produced by phage Group II staphylococci.

Enterotoxins. These are produced by a number of types of *Staph. aureus*.[4] Several serologic types of enterotoxin have been identified, and can be

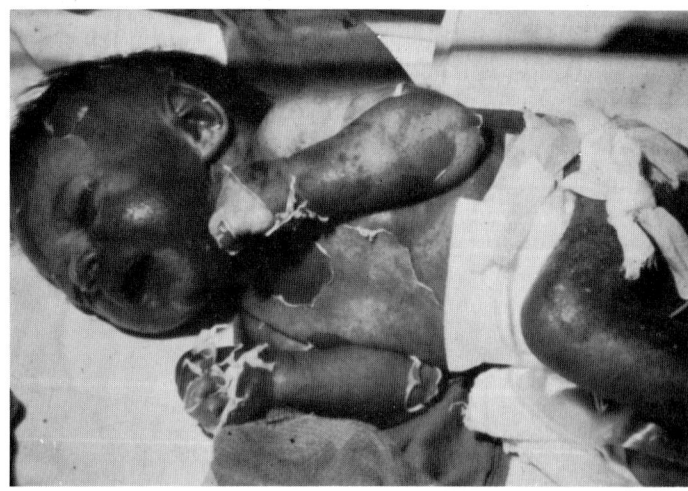

FIG. 1-4 *Scalded-skin syndrome, produced by staphylococcal epidermolytic toxin. (Photo from Dr. Lowell Glasgow. Reproduced by permission of New England Journal of Medicine)*

detected and assayed by several serologic techniques. Enterotoxins can produce vomiting in kittens, monkeys, or human volunteers.

Other Virulence Factors. *Staph. aureus* produces many virulence factors. Other factors of lesser clinical importance in human infections include hemolysins, hyaluronidase, leucocidin, and staphylokinase.

Localization of Transient Bacteremia. Often there is a history of previous mild trauma to the area of a deep skin or bone staphylococcal abscess. Probably a transient bacteremia is more likely to localize and infect such a place of lesser resistance, called a *locus minoris resistentiae*.

Animal Disease. Many animals, particularly cattle, may be sources or reservoirs of human staphylococcal disease.

Treatment

Drainage. If enclosed pus is present, as in a boil or abscess, operative drainage is usually essential.

Antibiotics. Penicillin is the drug of choice for penicillin-sensitive *Staph. aureus*. However, in recent years about 80 percent of community-acquired staphylococcal infections are penicillin-resistant.[7] Penicillin is useless in patients with *Staph. aureus* bacteremia, if the organism is penicillin-resistant.

A penicillinase-resistant penicillin, such as nafcillin, methicillin, or oxacillin, is best for penicillin-resistant *Staph. aureus*. Strains of *Staph. aureus* in the United States[9] are rarely resistant to penicillinase-resistant penicillins.[9] Vancomycin is useful for treatment of methicillin-resistant *Staphylococci*. Aminoglycosides such as gentamicin are generally not recommended for therapy of *Staph. aureus*, but are synergistic with penicillins. Rifampin is being used more frequently for difficult staphylococcal infection such as endocarditis because it penetrates the heart valve vegetations and the cell walls of macrophages, unlike other antistaphylococcal antibiotics. It should be used with a second antistaphylococcal antibiotic to delay emergence of resistance.

In adults with positive blood cultures for *Staph. aureus*, prolonged antibiotic therapy is usually recommended, in order to provide effective therapy for possible endocarditis. However, in children, staphylococcal bacteremia virtually never results in endocarditis and often can be adequately treated with appropriate intravenous antibiotics for an average of 2 to 3 weeks.[10]

Prevention

Technique. The most important methods of prevention of staphylococcal infections are careful handwashing in hospitals and nurseries, careful skin preparation, and sterile technique during operative procedures.

Interference. Eradication of the carrier state and prevention of infection by an original virulent carrier strain has been achieved by antibiotic therapy to suppress the original strain, followed by deliberate colonization of the nose by a low virulence strain; particularly phage type 502A. This is called interference. This procedure has been successful in interrupting outbreaks of staphylococcal disease in institutions or families.[2] Occasionally, however, the low virulence strain causes serious disease,[6] and this form of therapy has been abandoned.

Vaccines. Inoculation of patients with killed preparations of their own infecting orgranism has been used for patients with recurrent staphylococcal infections. However, vaccine has not been clearly proved to be of value, and is seldom used.[14]

PNEUMOCOCCI (STREPTOCOCCUS PNEUMONIAE)

Objectives

1. Explain the significance of pneumococcal types and explain how the typing is done. Describe the role of the capsule in the virulence of the organism.
2. Describe five illnesses in addition to pneumonia that can be caused by the pneumococcus.
3. Describe methods for recognition of pneumococcal infection other than recovery of the organism on culture. Include a description of how *Strep. pneumoniae* is usually distinguished from alpha-hemolytic streptococci.
4. Describe the principles and problems in the development and use of pneumococcal vaccines.

Names and Types

The pneumococcus was formerly called *Diplococcus pneumoniae*. This name was derived from its typical appearance as a pair of gram-positive cocci, and from the early recognition that it is an important cause of pneumonia. The pneumococcus is now called *Streptococcus pneumoniae*, because it shares many characteristics with streptococci, including the formation of short chains. However, the clinical and laboratory features are sufficiently different so that the traditional name pneumococcus will be retained here.

There are about 80 types of pneumococci, but most infections with bacteremia are due to fewer than ten types.[44] Pneumococcal types are defined by a Quellung (swelling) reaction, in which the capsule of the organism swells when in contact with type-specific antiserum prepared in rabbits. Typing was important before antibiotics were available, because patients were treated with type-specific rabbit antiserum. In order to identify the infecting type, lung puncture was done to obtain a small amount of lung fluid with organisms, which were then typed using the Quellung reaction. These lung puncture studies done before antibiotic therapy was available have indicated that the pneumococcus could be cultured from the lung of the majority of patients with lobar pneumonia.

Immunity is type-specific, so that infection with one type does not protect the patient from any other type.

Frequency and Importance

Pneumococci are the most frequent cause of bacterial pneumonia, and the most frequent cause of purulent meningitis in adults. Most pneumococcal infections respond dramatically to penicillin therapy, so accurate early diagnosis is important.

Clinical Patterns of Illness

Pneumococci are a *very frequent* cause of:

Pneumonia. Pneumococcal pneumonia is typically characterized by cough, high fever, and pain on inspiration. Adults typically spit up blood-tinged sputum, which reveals many segmented neutrophils and gram-positive diplococci on Gram stain. The white blood count is often higher than 15,000, with a shift to neutrophilic forms. Chest films typically reveal lobar or segmental consolidation (Fig. 1-5).

Pneumococci are also a frequent cause of pneumonia that is not lobar and does not have all of the typical clinical and laboratory findings described above.

Purulent Meningitis. *Strep. pneumoniae* is the most frequent cause of purulent meningitis in adults over 25 years of age. It is also the most frequent cause in any age group of meningitis complicating skull fracture with spinal fluid leaking from the nose or ear canal.

Purulent Otitis Media and Sinusitis.

Pneumococci are an *occasional* cause of:

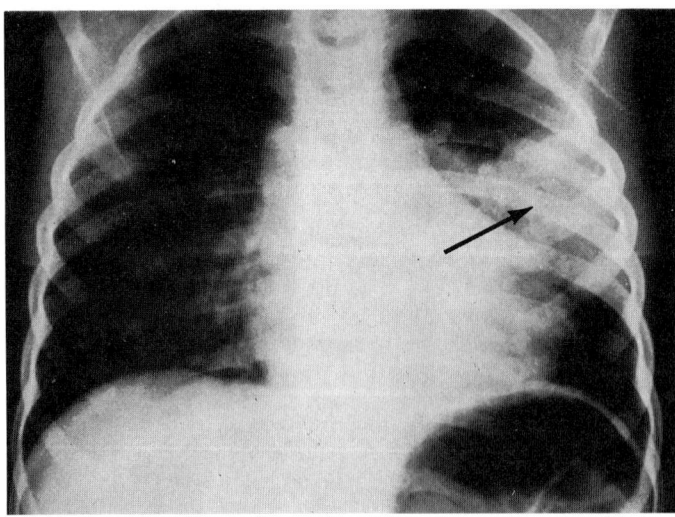

FIG. 1-5. Segmental pneumonia, involving the superior segment of the left lower lobe (**arrow**). Segmental pneumonia, as well as lobar pneumonia, is often pneumococcal.

Occult pneumococcal bacteremia is a diagnosis used to describe a syndrome manifested by high fever with only minimal upper respiratory findings.[46] Typically, this syndrome is observed in children less than 3 years old, with a white blood count over 25,000. Some patients recover without antibiotic treatment, but treatment should be given.[32]

Fulminating pneumococcal septicemia may occur in individuals who have no spleen, or whose splenic function is inadequate, such as patients with sickle cell anemia.[30] Pneumococcal antigen can be detected in the blood in fulminant cases, and can be associated with purpura and an increased bleeding tendency similar to that seen in meningococcemia.[33]

Primary peritonitis is especially common in children with nephrotic syndrome.

Pneumococci are a *rare* cause of:

Osteomyelitis, Septic Arthritis, and Acute Bacterial Endocarditis.

Laboratory Diagnosis

Gram Stain. Pneumococci are often described as lancet-shaped diplococci. A lancet is a surgical knife with a short, double-edged blade. The shape of the diplococcus is that of two pointed arches, base to base. Occasionally pneumococci appear ovoid, club-shaped, short-chained, or gram-negative in old cultures or in spinal fluid. Alpha-hemolytic streptococci, enterococci, and staphylococci are gram-positive cocci that may resemble pneumococci; but the lancet-shaped, diploid form is usually fairly typical of the pneumococcus. Gram stain is most useful for recognition of pneumococci in spinal fluid or sputum. The organism can sometimes be seen in a Gram stain of the peripheral blood in fulminating disease.[33]

Quellung Reaction With Omniserum. Omniserum, developed in 1963, is a pool of antisera of about 80 pneumococcal types. The Quellung reaction with omniserum has a higher correlation with recovery of pneumococci on culture than does the Gram stain.[43]

Culture. Carbon dioxide is essential for the primary isolation of about eight percent of pneumococci, so a candle jar should be used.[29] A few strains are obligate anaerobes. On sheep blood agar plates, pneumococci produce alpha-hemolytic colonies; that is, green hemolysis. Occasionally, the mucoid appearance and depressed center of the colony will lead the technician to suspect pneumococci. Pneumococci can be distinguished from alpha-hemolytic streptococci in that they are liquified by bile, or are sensitive to optochin paper discs.[31] Optochin is a chemical that can be impregnated into a filter paper disc which can be dropped on a freshly streaked plate and interpreted in the same way as an antibiotic susceptibility test.[31] Alpha-hemolytic streptococci grow right up to the edge of the disc, but the growth of pneumococci is inhibited around the disc (Fig. 1-6).

Detection of Antigens in Body Fluids. Polysaccharide capsular antigen detected in spinal fluid or urine can be regarded as evidence of infection. Counterimmunoelectrophoresis (CIE) uses type-specific pneumococcal antibody made in rabbits to detect type-specific capsular antigen in a patient's sputum, spinal fluid, or urine.[45] The diffusion of antigen and antibody in wells cut in agar is speeded by passing an electrical current through the agar (Fig. 4-2).

Latex particles can also be coated with pneumococcal antibody in order to detect pneumococcal antigen in a patient's specimen.[34] The antigen-antibody reaction is detected by observing the clumping together of the latex particles (Fig. 8-8).

Serum Antibodies. Demonstration of a type-specific antibody rise is possible, but not practical, because of the many types. Pneumococcal antibody can be detected in the blood of some infected patients by use of the enzyme-linked immunosorbent assay technique (see Fig. 13-7). Pneumococcal antigens are attached to polystyrene tubes and then incubated with the patient's serum. Antihuman IgG and IgM conjugated with alkaline phosphatase are then added, and adhere to any specific antibody attached to the antigen on the tube. After washing the tubes, alkaline phosphatase (attached to antihuman IgG or IgM attached to the patient's antibody attached to the antigen) can be detected by adding substrate for the enzyme alkaline phosphatase.[41] The clinical application of antibody detection by this method is being investigated.

Biologic Characteristics of Clinical Interest

Autolysis. After continued incubation in the laboratory, and sometimes in treated or untreated infections, the pneumococcus may become gram-negative and lyse, a phenomenon called autolysis.

Capsule. The polysaccharide capsule of the pneumococcus is clearly correlated with virulence, and protects the organism from phagocytosis. Anticapsular antibodies in the serum

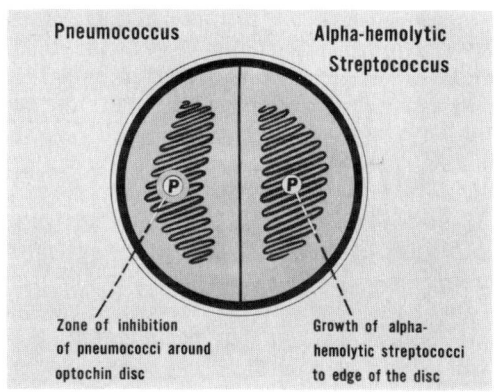

FIG. 1-6. *Pneumococci are inhibited by an optochin disc, whereas alpha-hemolytic streptococci are not.*

have been demonstrated to be protective in both experimental and naturally occurring infections.

Normal Flora. As pneumococci are part of the normal flora of the nasopharynx, other infectious agents or host factors are probably primarily responsible for invasion and infection. For the same reason, recovery of pneumococci from a sputum culture does not necessarily indicate that they are the cause of a patient's disease. When sensitive methods (such as sheep agar plates with gentamicin, or mouse inoculation) are used to test normal individuals, about 20 percent of adults and about 30 to 40 percent of children were found to be colonized by pneumococci in the nose or throat.[39]

Infections in Mice and Rats. Pneumococci are pathogenic for mice. Inoculation of a mouse with sputum aids in the recovery of pneumococci, as the many other organisms which may be present in sputum may interfere with growth of pneumococci on artificial media. However, mouse inoculation is rarely used in clinical laboratories because the presence of pneumococci in sputum does not necessarily indicate a causal relationship.

Experimental infections in rats have been useful in the study of the pathogenesis and treatment of pneumococcal pneumonia. Such studies have demonstrated that penicillin effectively kills the organism in the outer edema zone of the pneumonia where the organisms are multiplying rapidly.[47] However, phagocytosis is the major mechanism of killing the pneumococci in the central portion of the pneumonia where the organisms are not multiplying.

Transformation. This method of DNA transfer was discovered in pneumococci. Rough colonies can be transformed into smooth colonies when the rough colonies take up fragments of naked chromosomal DNA from smooth colonies. The soluble DNA is then integrated into the recipient cell's chromosome.

The use of the word transformation to describe DNA transfer should be distinguished from transformation which can occur in cell culture. In that case, cells gain the ability to proliferate much longer than normal. (see Tumor Viruses, Chap. 14)

Transformation of DNA as studied in the pneumococcus is one of several mechanisms of DNA transfer. A second mechanism, conjugation, the transfer of DNA between two bacterial cells across a bridge, is clinically important in plasmid-mediated antibiotic resistance (see Shigella, Chap. 3). A third mechanism, transduction, the transfer of DNA via a bacteriophage, is clinically important in the ability of Group A streptococci to produce erythrogenic toxin, and the ability of the diphtheria bacillus to produce diphtheria toxin.

Treatment

Penicillin is highly effective against pneumococci and is the most important part of treatment of pneumococcal infections. For patients allergic to penicillin, erythromycin is usually recommended for minor infections, and vancomycin for serious infections. About five percent of pneumococcal isolates are resistant to tetracycline, so this drug should not be used for suspected pneumococcal disease. Gentamicin is included in selective bacteriologic media used to grow pneumococci, and development of pneumococcal pneumonia can occur in patients receiving gentamicin. Therefore, gentamicin is contraindicated as single drug therapy of possible pneumococcal pneumonia. Trimethoprim is also effective against pneumococcus, and trimethoprim–sulfamethoxazole is useful against pneumococcal otitis media.

Penicillin-resistant pneumococci can be produced in the laboratory by serial transfer in the presence of increasing concentrations of penicillin.[37] Naturally occurring strains of pneumococci, which required about 25 times as much penicillin as usual to inhibit growth, were observed in aborigines in New Guinea in 1969, where use of prophylactic penicillin was widespread.[38] However, these strains were still susceptible to easily achievable serum concentrations of penicillin.

In 1977, highly resistant strains of pneumococci were observed in South Africa, Canada, and the United States.[40] The 1969 New Guinea strains required penicillin concentrations of about 0.5 μg per ml to inhibit the pneumococcus. However, the 1977 United States strains required about 4 μg per ml, a much higher level of resistance and one which is clinically significant. Interestingly, most resistant strains were originally observed in patients receiving penicillin prophylaxis for various reasons.

Prevention

Vaccines. Multivalent pneumococcal vaccine has been available since 1978 for 14 pneumococcal types. Studies done in the 1940s indicated that protective antibody can be produced in humans by injection of the purified capsular antigen.[42] Immunization with capsular vaccine produces a decrease in both the carrier state and clinical disease. However, the duration of protection has not been studied for periods longer than 6 months.[42] Since recurrences occur due to the same type even after natural infection,[36] protection may not last long enough to make the vaccine practical.

Since immunity is type-specific, immunization or infection with one type does not protect against infection by the many other types. Although most serious disease appears to be caused by only eight or so types,[44] disease due to other types might theoretically become more frequent if the usually common types were eliminated.

A review of types recovered from bacteremia patients at Boston City Hospital between 1935 and 1974 indicates most types recovered were the low numbered types (types 1–8 and 14).[35] The ten most frequent types accounted for about 75 percent of the 1543 typed isolates, but 25 other types accounted for the remaining 25 percent of cases.

The efficacy of the new pneumococcal vaccine containing capsular antigen for 14 types has not yet been adequately evaluated. It is strongly recommended for patients without splenic function—either because of splenectomy because of traumatically ruptured spleen, for patients born without spleens, or for patients whose spleens are non-functional because of multiple infarctions, as in sickle cell disease. A second injection of vaccine may produce a strong local reaction.

GROUP A STREPTOCOCCI

Objectives

1. Explain why sheep blood agar plates, rather than human blood agar plates, should be used for the primary isolation of beta-hemolytic streptococci from throat cultures.
2. Explain when it is important to determine if beta-hemolytic streptococci are Group A, and describe several ways the laboratory does this.
3. Explain the clinical importance of streptolysin O, erythrogenic toxin, and M protein.
4. Explain how the clinical manifestations of Group A streptococcal infections in infants and young children may differ from those of older children and adults.
5. Describe the mechanism whereby Group A streptococci become capable of elaborating erythrogenic toxin.
6. Describe the features of a streptococcal illness associated with an increased risk of acute rheumatic fever.
7. List the antibiotic regimens recommended by the American Heart Association for the treatment of streptococcal pharyngitis.

Definitions

Streptococci are gram-positive cocci that readily form chains in broth cultures; this can be used as a presumptive method to distinguish them from staphylococci, which form clumps (Fig. 1-3). Streptococci can be distinguished from staphylococci with greater certainty by the catalase test, in which a single colony of the organism is added to a drop of ordinary 3 percent hydrogen peroxide solution placed on a glass slide. Staphylococci contain peroxidases that decompose the H_2O_2, producing vigorous bubbling (catalase-positive). Streptococci are catalase-negative. This test is a practical office procedure.

Streptococci can be classified on the basis of the type of hemolysis they produce on sheep blood agar plates. Groups and types of streptococci are ultimately defined by precipitin tests (Lancefield), which use a streptococcal cell wall extract as the antigen, and specific antisera produced in animals as the antibody (see Fig. 1-9). It should be noted that "grouping" and "typing" are technical terms which should not be used in the general sense of "classifying".

Frequency and Importance

The most important kind of streptococcus is the Group A beta-hemolytic streptococcus, as defined later. Group A beta-hemolytic streptococci are important because they cause severe, life-threatening illnesses, such as pneumonia, septicemia, or meningitis. They are also important because a small percentage of patients with untreated pharyngitis due to Group A streptococci will develop acute rheumatic fever. Group A streptococcal pharyngitis is very common in children, and prevention of acute rheumatic fever by detection and treatment of streptococcal pharyngitis is an important public health goal. It is interesting to note that Group A streptococcal infection of skin does not result in acute rheumatic fever unless Group A streptococcal pharyngitis is also present.

Group A streptococcal pharyngitis or skin infection can cause acute glomerulonephritis. Even early antibiotic therapy of the streptococcal infection is often not early enough to prevent glomerulonephritis.

Classification of Hemolysis

Alpha hemolysis is partial hemolysis. It is often green in appearance. From this green appearance comes the name viridans (green or verdant) group of streptococci. These streptococci are discussed in a later section entitled Other Streptococci.

Beta hemolysis is complete hemolysis. The area around the colony appears clear, and the zone of beta-hemolysis is large compared to the colony size (Fig. 1-7). On microscopic examination of the zone of hemolysis, using low power magnification, no intact red cells can be seen. However microscopic examination is rarely needed to confirm complete hemolysis because of the clear appearance.

If hemolyzed blood is a red color, why does the area around a beta-hemolytic colony appear clear, rather than red? The hemoglobin has been released from the erythrocyte and diffuses throughout the agar. An isolated beta-hemolytic colony is

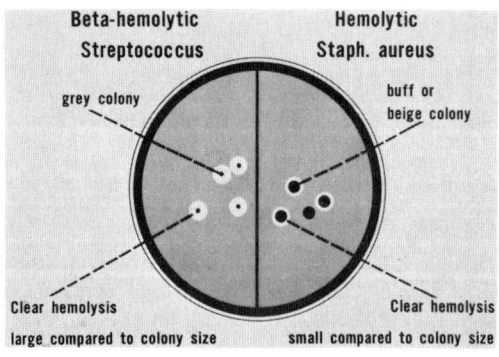

FIG. 1-7. *Comparison of hemolysis size produced by beta-hemolytic streptococci with that produced by hemolytic* Staph. aureus.

surrounded by clearer hemolysis than a group of many hemolytic colonies.

Alpha-prime hemolysis is a form of alpha hemolysis which is nearly complete and which may resemble beta hemolysis if human blood agar plates are used.[78] It is not green. It is often responsible for a false-positive diagnosis of beta streptococci on human blood agar plates.

Gamma hemolysis is no hemolysis. It is characteristically produced by enterococci, which are discussed later in this chapter (see Other Streptococci).

Streptococcal Groups

Beta-hemolytic streptococci can be grouped into Group A, B, C, D, G, and other lettered groups. Group A has special importance because it is the only group that causes rheumatic fever.

Group A streptococci are defined as streptococci that contain Group A antigen in their cell wall, as determined by a precipitin reaction. The antiserum for each group is usually prepared by infecting a rabbit with streptococci of that group. The group antigen is extracted from a pure culture of the organisms. Some of the streptococcal extract is used to fill part of a capillary tube, and antiserum is then added to the tube. If the antigen extracted from a particular organism is Group A, a specific precipitin reaction occurs at the interface of the Group A antiserum and the extract, causing the appearance of a cloudy band.

There are several other methods for defining streptococcal groups (Table 1-2). Many clinical laboratories report a presumptive diagnosis of Group A streptococci by testing the organism for bacitracin sensitivity (Fig. 1-8). More than 95 percent of Group A streptococci are bacitracin

TABLE 1-2. METHODS OF GROUPING STREPTOCOCCI*

	Group A	Group B	Group D	Other Groups
Precipitin method of Lancefield	Standard	Standard	Standard	Standard
Fluorescent antibody (Fig. 1-9)		May be available	NA	NA
Counterimmuno— electrophoresis (Fig. 4-2)	Possible	Possible	Possible	Possible
Co-agglutination (Fig. 1-12)	Practical	Practical	NA	Practical for C & G
Chemical and metabolic methods	Bacitracin susceptibility (95% accurate) (Fig. 1-8)	Bacitracin & trimethoprim– sulfamethoxazole resistance (95% accurate) CAMP tests[87] Pigment production[93]	Bile-esculin test (95% accurate)	Bacitracin resistant & trimethoprim– sulfamethoxazole sensitive (95% accurate)
Latex agglutination (Fig. 8-8)	Practical	Practical	NA	Practical

NA = Not available

*Note that grouping is a technical, not a general, word in this context.

sensitive.[78] Beta-hemolytic streptococci that are bacitracin resistant are very rarely Group A. Therefore, bacitracin sensitivity correlates very closely with the precipitin reaction; in many clinical laboratories, susceptibility to bacitracin discs is still used to define Group A streptococci. The bacitracin disc can be placed on the primary isolation plate to give an early reading, and is useful provided sparse growth of streptococci or many other bacteria in the area do not interfere.[71]

Many clinical laboratories are replacing the bacitracin susceptibility method with other, more rapid and more accurate methods. The fluorescent antibody (FA) method is rapid and reliable, and can be done by picking colonies from the primary isolation plate and testing them immediately for Group A antigen (Fig. 1-9).[58]

Coagglutination, shown in diagrammed in Figure 1-12, is another rapid method for detecting Group A antigen. The test can be done using about five colonies from a sheep blood agar plate.[77] It appears promising for use in office laboratories, but may require reincubation of selected beta-hemolytic colonies for complete accuracy.

A latex fixation test, diagrammed in Figure 8-8, has recently been developed to group beta-hemolytic streptococci obtained from a sheep blood agar plate. The principles and problems are similar to coagglutination. Groups B, C, G and F antisera are available. Knowledge of the clinical significance of Groups C, G, and F will undoubtedly be expanded by the newly available simple grouping methods for these groups (Table 1-2).

Group A streptococci are almost always beta-hemolytic. A non-hemolytic Group A streptococcus has been observed, but this must be regarded as an extraordinarily rare exception.[62]

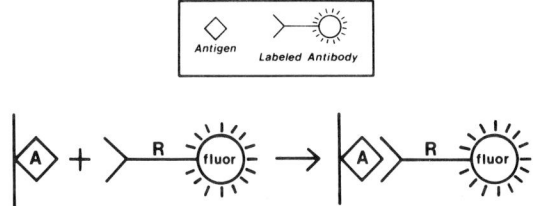

FIG. 1-9. *Direct fluorescent antibody method of grouping beta-hemolytic streptococci. A = patient's Group A streptococci on a slide. Fluor = fluorescein-conjugated rabbit antiserum against Group A antigen. R = rabbit antiserum.*

In most clinical situations in which a throat culture is done because a patient has pharyngitis, grouping is unnecessary. Grouping is especially useful for patients with few colonies or minimal symptoms, when the coincidental recovery of non-group A streptococci is suspected. Recently there has been a greater need to do grouping of beta-hemolytic isolates from throat cultures, because of the increased recovery of unimportant non-Group A streptococci from commercial blood agar plates, which have more nutrients than in the past.

Streptococcal Types

The M types of Group A streptococci are defined by a precipitin reaction similar to that used to define groups. The antigen is the M protein found in the cell wall. Only Group A streptococci produce M protein, which is closely associated with the virulence of the organism.

Immunity to streptococcal infection is generally regarded as being type-specific; that is, a person can presumably be infected with one type of Group A streptococci only once. However, reinfection with the same type can occur in patients who have received penicillin treatment for the first infection, indicating type-specific immunity sometimes does not develop.[53] Since there are about 60 types of Group A streptococci, one individual can become infected by Group A streptococci many times, each time presumably with a different type.

Some types are known to be nephritogenic types; that is, especially likely to produce glomerulonephritis. Type 12 is the best example of a nephritogenic type. The reason for some types being nephritogenic is unknown. However, a reasonable hypothesis is that the bacteria have antigens which closely resemble human glomerular antigens.

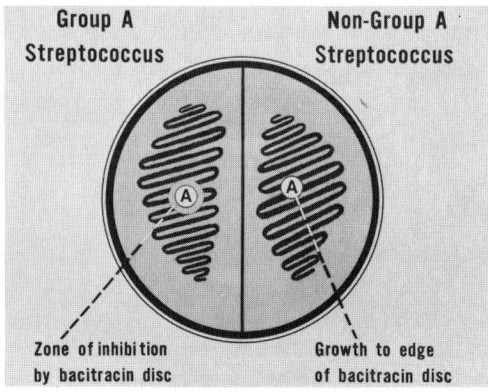

FIG. 1-8. *Group A streptococci are inhibited by bacitracin discs, but other groups of streptococci are not.*

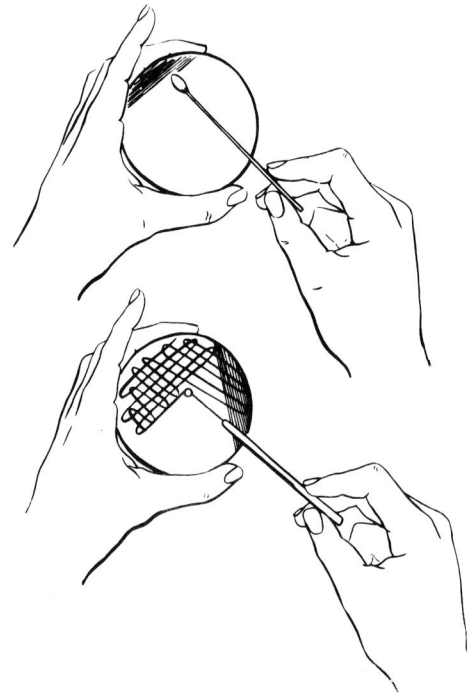

FIG. 1-10. How to streak a throat swab for culture for beta-hemolytic streptococci.

T-typing is another form of typing based on T antigens, which are not closely associated with virulence, as are the M proteins. There are fewer T types, and the typing procedure is easier than M-typing, because it can be done by simple slide agglutination.

Laboratory Diagnosis

Culture. Laboratory evidence suggesting a Group A streptococcal infection is almost always developed by culture of the organism. Methods of culture vary in sensitivity for recovering the organism. The usual method is to swab the throat and spread this swab directly on the surface of the sheep blood agar plate (Fig. 1-10). Then the plate is streaked, using a flamed wire loop and stabbing the agar to increase the chance of observing beta hemolysis. Sheep blood agar is used for several reasons:[78]

1. *Avoidance of Haemophilus hemolyticus.* Sheep erythrocytes contain an enzyme which destroys DPN and TPN (V factor), which are growth factors for *Haemophilus* species. Sheep blood agar is useful to inhibit the growth of *Haemophilus hemolyticus*, which requires V factor, and which is beta-hemolytic and therefore may be confused with beta-hemolytic streptococcus colonies.

2. *Avoidance of antibiotics or antistreptococcal antibodies.* Human blood also may contain antibodies or antibiotics from the blood donor which may inhibit the growth of the streptococci.

3. *Avoidance of alpha-prime hemolysis.* Some alpha-hemolytic streptococci can produce a form of hemolysis called alpha-prime hemolysis on human blood agar plates. This closely resembles beta hemolysis, and makes recognition of true beta hemolysis more difficult.

Serum Antibodies. It is possible to make the diagnosis of streptococcal infection by demonstrating an antibody rise in the host using one serum specimen obtained early in the disease and another serum specimen obtained several weeks later. However, antibiotic therapy tends to interfere with a rise in antibody titer,[55,68] and therefore antibiotic therapy would have to be withheld in order to detect an antibody titer rise (Fig. 1-11). By the time antistreptococcal antibodies have appeared or increased, it may be too late to prevent rheumatic fever. Therefore, serologic diagnosis of streptococcal infection is not practical, except retrospectively. Antistreptococcal antibody titers have also proved useful as a test of the accuracy of negative cultures.

Antistreptolysin O (ASO) is an antibody produced by humans in response to a beta-hemolytic streptococcal infection. Streptolysin O is an anti-

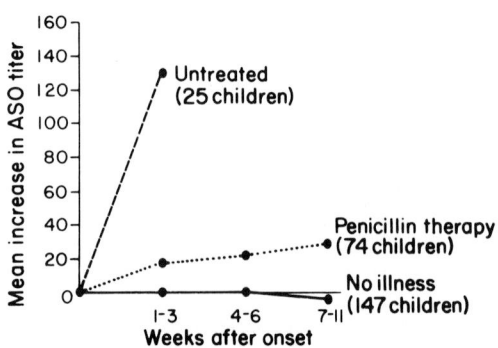

FIG. 1-11. Mean ASO titer rise: 147 children with no illness; 74 children with pharyngitis, fever, and Group A streptococci in their throat culture treated with penicillin; and 25 consecutive children with fever, pharyngitis not treated with an antibiotic because inadequate culture methods (on human blood agar) did not detect any beta-hemolytic colonies.[68] Early adequate penicillin therapy suppresses the ASO titer rise.

gen found in the streptococcus and is a hemolysin. The O refers to the fact that this hemolysin is oxygen-labile. Streptolysin S (S refers to serum) is not oxygen-labile and is the hemolysin that causes the beta-hemolysis on the surface of a sheep blood agar plate. Streptolysin S is not sufficiently antigenic in humans to produce a reliably detectable antibody.

The ASO titer is a test frequently used in clinical medicine. There are several other antibodies that humans make in response to streptococcal infection, but the ASO titer is the best standardized and the most commonly used.

The titer is often expressed as Todd units. A titer of 1:250 is called 250 Todd units. There must be a rise in titer, using two sera, to prove an infection. As is usually the case in diagnosis by serologic methods, a single measurement of the ASO titer is very difficult to interpret. However, a very high titer, such as 1200 Todd units, suggests recent streptococcal infection. A very low titer, such as 50 units, is evidence against a recent streptococcal infection. Intermediate titers in the range of 125 to 500 Todd units are really not helpful, as they may reflect an infection many months previously.

It is generally assumed, but not proved, that the streptococcal antibody response parallels the antigen–antibody reaction of acute rheumatic fever. Acute rheumatic fever is sometimes observed in the absence of an ASO titer rise, and in this situation, measurement of other streptococcal antibodies is indicated. If the ASO titer has risen, indicating that the patient has responded to the streptococci infection with antibodies, there is an increased risk of acute rheumatic fever.

Other Streptococcal Antibodies. Other streptococcal antigens that stimulate the production of antibodies in the human include hyaluronidase, streptokinase, NADase (DPNase), and streptodornase (DNase B). These antigens are discussed later as virulence factors.

Recently, a commercial test called Streptozyme has become available for simultaneous detection of antibodies to five streptococcal antibodies (anti-streptolysin O, anti-hyaluronidase, anti-streptokinase, anti-NADase, and anti-DNase B). The commercial kit provides sheep erythrocytes which are sensitized to all five of these antigens, and clumping of the erythrocytes by the patient's serum indicates a moderate titer of antibody against one or more of these five antigens. The test is simple as a screen for recent infection, but a rise in titer is still necessary to confirm a streptococcal infection with certainty, so that this test is really of little practical value.

Fluorescent Antibody Techniques. Group A streptococci can be detected and identified using fluorescent antibody techniques. One of the earliest uses of the fluorescent antibody (FA) method was for the detection of Group A streptococci.[51] Specific Group A antiserum is labeled (conjugated) with isothiocynate, which fluoresces when exposed to a particular wave length of light (Fig. 1-9).

A suspension of suspected Group A streptococci is put on a glass slide, dried, fixed by rinsing with ethanol, and incubated a few minutes with fluorescein-labeled Group A antiserum. The antiserum is then rinsed off. The antiserum (antibody) can be observed through a microscope, which has a light source filtered to give the proper wave length.

The FA technique would be more useful for rapid detection of Group A streptococci if it were accurate when throat swabs are smeared directly on the slide. Unfortunately, for maximum accuracy, the throat swab must be incubated in broth to increase growth of the organism, so the test cannot be read in less than about 4 hours, which is nearly as inconvenient a delay for the physician as the 18 hours required for a conventional culture. However, the FA technique is useful for rapid grouping of beta-hemolytic streptococci recovered on sheep blood agar plates.[58] Other streptococcal group antigens, such as G and C, occasionally produce weakly fluorescent cross reactions, so controls are needed.

Excessively Sensitive Methods. The surface streaking of a sheep blood agar plate has been used in practice for about 20 years and is adequate for detection of streptococcal pharyngitis so that antibiotic treatment can be given and acute rheumatic fever prevented.[54] More sensitive methods may detect more carriers, but have not been adequately studied in practice. In fact, more sensitive methods may be excessively sensitive and lead to unnecessary treatment of a low-grade carrier state without preventing any additional cases of acute rheumatic fever.

Throat Smears. Gram or other stains of throat swabbings have not been demonstrated to be more reliable than sheep blood agar cultures to detect streptococcal pharyngitis in practice. Smears have not been shown to be a useful guide to the prevention of acute rheumatic fever.

Clinical Patterns of Illness

The type of illness produced by beta-hemolytic streptococci is related to age. In some ways, this relation to age resembles a similar age relationship in tuberculosis. The term streptococcosis was coined to emphasize this analogy to tuberculosis.[50] In infants and very young children, both tuberculosis and streptococcal infection tend to have a gradual onset, a chronic course, and a tendency to spread, with poor localization and occasional dissemination. In older children, both diseases are more acute, with faster localization and less likelihood of dissemination. However, in recent usage, the term streptococcosis has also been used to mean any of the various patterns of beta-hemolytic streptococcal infection.

Group A streptococci are a *frequent* cause of:

Exudative Pharyngitis in School-Age Children. Typically, there is fever, headache, sore throat, and vomiting at the onset, with red throat, exudate on the tonsils or pharynx, and tender cervical lymph nodes. Rarely, an outbreak occurs in which a large number of individuals develop pharyngitis, with the onset at almost exactly the same time. This should make the physician suspect ingestion of food or milk contaminated with Group A streptococci.[60]

Skin Infections. Cellulitis, impetigo, and surgical or traumatic wound infections are commonly caused by beta-hemolytic streptococci. Severe gangrene is a rare complication of streptococcal cellulitis.

Group A streptococci are an *occasional* cause of otitis media, orbital cellulitis, and sinusitis. Postpartum endometritis, also called puerperal sepsis, is still occasionally seen.

Streptococcal pneumonia is uncommon, is usually severe, and typically responds unexpectedly slowly to penicillin therapy.[70] Purulent pleural effusion (empyema) or pneumatoceles (Fig. 1-2) are occasionally observed. Extrarespiratory streptococcal diseases include septic arthritis, osteomyelitis, and vaginitis.

Diseases caused by Groups B and D are discussed later in this chapter. Groups C, F, G are rare causes of wound infection or bacteremia.[56] Group G and Group C streptococci have been reported to cause outbreaks of pharyngitis,[48,57] but carry no risk of resulting in acute rheumatic fever, according to current belief.

Asymptomatic Infection and the Carrier State

Asymptomatic, subclinical, or mild streptococcal infections are probably relatively frequent.[67] When an individual has Group A streptococci recovered from the respiratory tract in the absence of any manifestations of disease, that person is called a carrier. Some carriers are convalescent carriers; that is, clinically recovered from a recent symptomatic infection, without treatment. Other carriers are transient carriers, with small numbers of Group A streptococci present for only a few days. Asymptomatic carriers are considerably less contagious than sick patients with beta-hemolytic streptococci.

Some individuals have small numbers of organisms present in the pharynx, but do not develop serologic evidence of infection. Usually the Group A streptococci are present in small numbers when the individual is asymptomatic. The organisms recovered from carriers are less likely to be M-typable than those recovered from sick individuals. Eradication of the carrier state by antibiotic therapy is sometimes indicated in closed populations, such as military bases, in order to reduce the frequency of rheumatic fever.

Nasal carriers have been called dangerous carriers, since they are highly contagious compared to pharyngeal carriers, who are relatively noncontagious.[59] Some streptococcal outbreaks have been traced to anal carriers.[74]

Surveys of asymptomatic school age children show a 5 to 15 percent carrier rate of Group A streptococci.[61] The frequency reported depends primarily on the ability of the culture method to detect small numbers of Group A streptococci, the amount of streptococcal disease in the community, how well the children are screened for being truly asymptomatic, and how susceptible the individual is to colonization.[79] Carrier rates over 30 percent probably reflect both poor screening for symptoms and the use of very sensitive culture methods. Studies conducted in an office setting, using surface streaking of a sheep blood agar plate show carrier rates of two to 10 percent.[52,69]

Some sick children will coincidentally have Group A streptococci recovered on throat culture. In order to avoid "treating positive cultures", the physician should not take a throat culture unless the plan is to treat the child if a positive culture is found. Furthermore, excessively sensitive culture methods should not be used. The surface streaking of sheep blood agar plates has been shown to be sufficiently sensitive to detect streptococcal pharyngitis to be effective in prevention of acute rheumatic fever in office practice. It is possible that FA methods are more sensitive than necessary. These methods have not been adequately studied in office practice, compared to the standard conventional cultures, and cannot yet be recommended as a standard for clinical practice.

Biologic Characteristics of Clinical Interest

Non-suppurative Sequelae. Acute rheumatic fever and acute glomerulonephritis are the two major non-suppurative sequelae of Group A streptococcal infection. Acute rheumatic fever can occur about 2 to 4 weeks after streptococcal pharyngitis, but apparently does not occur after streptococcal skin infections.[49] Acute glomerulonephritis occurs about 1 to 2 weeks after either skin or pharyngeal streptococcal infection. Acute rheumatic fever results in residual heart valve damage (rheumatic heart disease) in about half of all cases, whereas recovery from glomerulonephritis is rarely associated with functional renal impairment.[75]

Certain M types of Group A streptococci are particularly likely to produce acute glomerulonephritis and are referred to as nephritogenic. Types 12 and "Red Lake" (Type 49) are especially nephritogenic.

The reasons that some individuals develop acute rheumatic fever after untreated streptococcal pharyngitis are unknown. Known risk factors include previous rheumatic fever, presence of a typable strain, persistence of a positive throat culture, rise in titer of anti-streptococcal antibodies, and severity of the clinical illness (with fever, exudate, or scarlet fever rash).

The particular M serotype or strain also appears to vary in virulence, and many recent Group A streptococcal isolates have laboratory characteristics of decreased virulence, including lesser amounts of M protein and decreased capsule size.[65]

Heart-Reactive Antibodies. Group A streptococci contain an antigen or antigens that can be used to produce antibodies in rabbits and that can be shown to react with heart muscle.[64] These heart-reactive antibodies are usually detected by conjugating the antibody with fluorescein, and observing the fluorescent staining of the heart muscle. Heart-reactive antibodies are also found in humans after streptococcal infection, and are found in higher titer in patients with acute rheumatic fever than in patients with glomerulonephritis or uncomplicated streptococcal infection.

Heart-reactive antibodies are also found in patients undergoing heart surgery. These antibodies can be absorbed by heart tissue but not by Group A streptococcal membranes. Either heart tissue or Group A streptococcal membranes will absorb the heart-reactive antibodies found in patients with acute rheumatic fever.

Streptococcal antigens that thus cross-react with heart tissue appear to be distinct from the streptococcal M protein.

Normal Flora. Beta-hemolytic streptococci of Groups B, C, F, and G are sometimes found in the pharynx of normal individuals, and can be considered normal flora. Group A streptococci are also found in about five to ten percent of normal children, but are not usually regarded as normal flora. Instead, the children are regarded as carriers, because of the importance of Group A streptococci as a cause of respiratory disease.

Some individuals are anal carriers of Group A streptococci.[74]

Erythrogenic Toxin. This is the cause of the rash in scarlet fever (scarlatina) and can be produced by any type of Group A streptococci. A bacteriophage can transform any Group A streptococcus into an erythrogenic toxin-producing strain by lysogeny. Repeated attacks of scarlet fever can occur, since a number of types can be transformed and produce erythrogenic toxin.

Virulence Factors. Streptokinase activates plasminogen in normal serum to form a plasmin which lyses fibrin. This may be one reason the pus in streptococcal infections is thin, rather than thick, as in staphylococcal infections. Streptodornase depolymerizes DNA and formerly was used to liquify thick viscous exudates, as in empyema. It is also probably another reason for the thin type of exudate associated with streptococcal infection. Hyaluronidase, originally called spreading factor, was sometimes given along with subcutaneous fluids before intravenous administration of replacement fluids became standard practice. It enhances the subcutaneous spread of streptococci in cellulitis. The *capsule* may protect the organism from phagocytosis, although this is much less important than with the pneumococcus.

M Protein. This is usually regarded as the major virulence factor of Group A streptococci. Isolates that are not typable and lack M protein usually will regain typability and M protein after passaging in mice. However, some nontypable Group A streptococcal isolates have been associated with typical clinical streptococcal illness, acute rheumatic fever, and nephritis, and have not become typable even after passaging in mice.[73] This is evidence for the belief that M protein (typability) is not essential for virulence in humans. However, typability or the presence of certain types, such as Type 12, are generally associated with more frequent and more severe disease.

Plasmid-Mediated Resistance. Beta-hemolytic streptococci which are resistant to tetracycline or chloramphenicol often lose the resistance after treatment with acriflavine.[72] This is evidence that this resistance is mediated by plasmids (see Shigella in Chap. 3).

Adherence. Group A streptococci have increased avidity for adherence to pharyngeal cells obtained from persons with previous acute rheumatic fever compared to normal controls.[76] The clinical or pathogenetic significance of this observation is as yet unknown.

Decreasing Frequency of Acute Rheumatic Fever. Recently it has been suggested that there may have been a decrease in the virulence of Group A streptococci since the 1940s. Recent isolates of Group A streptococci have less M protein and less of a tendency to form long chains in broth—both regarded as virulence factors associated with an increased risk of acute rheumatic fever. It has also been suggested that the frequency of acute rheumatic fever has decreased since the 1940s because the virulence of Group A streptococci has decreased.

Virulence as Related to Passaging. Interruption of passaging of Group A streptococci by the widespread use of antibiotics may be a possible cause of decreased virulence. Before antibiotics were available, rapid human passaging of the Group A streptococcus in epidemics may have resulted in an increase in the virulence of the organism.

The concept of increased virulence because of rapid passaging in the natural host is supported by several observations. Serious complications are most frequent near the end of an outbreak in a closed population; for example, in mumps outbreaks in boys' schools. In outbreaks in herds of animals, the fatality rate increases with passaging. Passaging of polio-vaccine virus in vaccine recipients may increase its virulence (p. 221).

The opposite result (attenuation) is observed by passaging a virus in an unnatural host, such as the production of vaccines in cell cultures. The interruption of rapid passaging can also be achieved by isolation or segregation of contagious patients. For example, poliomyelitis appeared to be decreasing in frequency before the introduction of vaccine, and the frequency and severity of scarlet fever was decreasing before the introduction of chemotherapy—both possibly a result of the recognition of the contagiousness of these diseases and increased use of segregation, isolation, crowd avoidance, and other public health measures to interrupt the rapid spread or passaging.

Treatment

Penicillin is the antibiotic of choice for Group A streptococcal infections. For patients allergic to penicillin, erythromycin is an acceptable substitute. Tetracycline should not be used for therapy because about five percent of Group A streptococci are resistant. Sulfonamides are not effective in the treatment of streptococcal pharyngitis, but are effective in prophylaxis.

The American Heart Association recommends any of the following regimens as acceptable for the treatment of streptococcal pharyngitis.[63]

1. Benzathine penicillin, 600,000 to 1.2 million units, intramuscularly, depending on the patient's weight. This provides very low serum levels for about a month. The dose of benzathine penicillin should not be reduced by substituting procaine penicillin for part of the total dose.

2. Oral penicillin, 125 mgm three or four times a day for 10 days. This is effective if the patient takes medication reliably.

3. Erythromycin, in appropriate dosages, orally for 10 days. This can be used if the patient is allergic to penicillin.

Higher doses of intravenous penicillin are needed for serious infections, such as pneumonia or meningitis.

Prevention

Chemoprophylaxis. Daily oral penicillin or monthly injections of long-acting benzathine penicillin are indicated for most patients with past rheumatic fever. These are usually continued for the lifetime of the patient, so the initial diagnosis of rheumatic fever should be made with as much certainty as possible, with hospitalization, consultation, and thorough laboratory evaluation.

Vaccines. Streptococcal M protein vaccines given in order to try to prevent acute rheumatic fever are still investigational. The protection is type-specific and effective. However, in one study, normal siblings of patients with rheumatic heart disease were protected from Type 3 streptococcal infection by Type 3 vaccine, but had a higher attack rate of rheumatic fever per Group A streptococcal infection than did unvaccinated siblings.[66]

GROUP B STREPTOCOCCI

Objectives

1. Describe the clinical diseases produced by Group B streptococci.
2. List the areas of the body that may be colonized by Group B streptococci and indicate the approximate range of frequency of colonization of each area.
3. Discuss the problems involved in prevention of Group B streptococcal disease in the newborn infant.
4. Discuss the difference between early and late-onset neonatal septicemia.

Definitions and Groups

Group B streptococci are also called *Strep. agalactiae*, and can cause mastitis in cows. Group B streptococci are almost always beta-hemolytic, like Group A streptococci. Group B streptococci can be grouped by several methods (Table 1-2). They can be identified by the standard Lancefield method of extracting the cell wall antigen and showing it reacts with group B specific antisera prepared in rabbits. Group B streptococci can be rapidly identified by using the fluorescent antibody method shown in Figure 1-9.

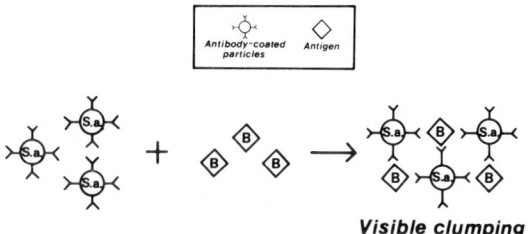

FIG. 1-12. *Coagglutination method as used to group beta-hemolytic streptococci. s.a.* = Staph. aureus *coated with antibody to Group B streptococci. B* = *Group B streptococci in a saline suspension. Compare with Fig. 8-8.*

Group B streptococci can be grouped by a technique called coagglutination (Fig. 1-12).[95] Protein A containing staphylococci are coated with antibodies specific for Group B streptococci. When Group B streptococci are added to the antibody-coated staphylococci reagent on a glass slide, agglutination can be observed. Sometimes colonies taken from the plate can be used directly, but for complete accuracy in detecting Group A streptococci from throat cultures, the beta-hemolytic colonies must be treated with an enzyme, or grown overnight before testing. This method is commercially available for the grouping of Groups A, B, C and G.

Beta-hemolytic streptococci can also be presumptively classified into Group A, Group B, or neither Group A or B on the basis of susceptibility testing using bacitracin and trimethroprim-sulfamethoxazole disks (Table 1-2).[89] Counterimmunoelectrophoresis can be used for rapid detection of Group B antigen in the spinal fluid or urine in patients with meningitis (Fig. 4-2).[88] Group B streptococci also can be recognized by an arrow-shaped zone of hemolysis which occurs when they are streaked perpendicular to a hemolytic *Staph. aureus*, a test often abbreviated as CAMP (for the initials of the authors who described the method).[87] Almost 100 percent of Group B streptococci produce CAMP factor, which produces the synergistic lysis with *Staph. aureus* beta-hemolysin. Other methods include hippurate hydrolysis, and production of a yellow to orange pigment in stabs of agar incubated anaerobically.[93]

Frequency and Importance

Group B streptococci emerged in the 1970s as an extremely frequent cause of septicemia, meningitis, and death in otherwise normal newborn infants.[83] Apparently the newborn infant can be colonized from the mother's cervix in utero, or at the time of birth.

About five to twenty percent of women are asymptomatic cervical carriers, the frequency depending on the sensitivity of the culture method and the type of women cultured.

Types

Group B streptococci can be typed into Types Ia, Ib, Ic, II and III, using antiserum prepared in rabbits. Some Group B isolates are nontypable. Most severe disease is caused by Type III.

Clinical Patterns of Illness

Group B streptococci are a *frequent* cause of:

Neonatal Septicemia and Meningitis. In the past ten years, this organism has become the first or second most frequent cause of these newborn syndromes. Early onset disease (in the first 5 days of life) is secondary to colonization in utero or during delivery, and has a mortality rate over 50 percent. Late onset disease can be ascribed to disease complicating colonization after birth, but still has a high mortality rate.

Group B streptococcal sepsis often resembles RDS (Respiratory Distress Syndrome), a noninfectious disease usually secondary to lung immaturity.[81] RDS is typically manifested by a rapid respiratory rate and a diffusely hazy chest roentgenogram.

Recent publications present evidence that a high percentage of band forms in the differential leukocyte count is useful for the recognition of Group B disease,[92] which resembles RDS.

The frequency of neonatal infection with this organism is related to the high frequency with which it is found in the maternal cervix or vagina. The severity of neonatal Group B streptococcal infections has led to exploration of the use of treatment of maternal carriers of the organism.

Group B streptococci are an *occasional* cause of osteomyelitis, septic arthritis, otitis media, orbital cellulitis, or pneumonia in young infants.[90]

Group B streptococci are a *rare* cause, in adults, of wound infections, urinary infections, pneumonia, endometritis, meningitis, or peritonitis.[86]

Biologic Characteristics of Clinical Interest

Colonization and Carrier State. The frequency of colonization depends on the number of swabs taken from a single source and also on the sensitivity of the laboratory method used. Thus, any particular percentage or range should be considered only as an approximation.

About ten percent of vaginal cultures and about 15 percent of rectal cultures of pregnant women are positive for Group B streptococci.[82] About 75 percent of children born to colonized women will themselves become colonized.[89] In college women, colonization with Group B streptococci was significantly related to sexual experience, use of an intrauterine device, the first half of the menstrual cycle, and age less than 20 years.[84]

Adherence. There appears to be variable adherence to Group B streptococci to vaginal epithelial cells in different times of pregnancy. Adherence (and hence colonization) has also been postulated as an important variable in the pathogenesis of Group A streptococcal pharyngitis, gonococcal urethritis (Chap. 5), and toxigenic *Escherichia coli* diarrhea (Chap. 3).

& Bordetella pertussis phase variations

Contagion. Newborn infants who are not colonized at birth may become colonized in the newborn nursery. This presumably occurs because of transmission from one infant to another by the hands of attendants.[80,94] Breast milk has been demonstrated to be a source of infection in nursing infants.[91]

Transplacental (Maternal) Antibody. Infants with invasive Group B streptococcal disease almost always lack protective type-specific antibody, which is normally transferred transplacentally, suggesting that this maternally transferred antibody is important in protecting the infant from Group B disease.[85]

Treatment

Group B streptococci are uniformly susceptible to penicillin or ampicillin. The addition of gentamicin (as is usually done in a neonatal meningitis of unknown cause) appears to be synergistic with ampicillin, both in a few human cases and in studies using rabbits as an experimental model.

Prevention

Penicillin therapy does not eradicate Group B streptococci from the mucous surfaces of newborn infants, although it eradicates the organism from blood, spinal fluid, or joint fluid of a newborn with clinical illness. Penicillin therapy is also unreliable for permanent eradication of the organism from colonized pregnant women.[83]

A vaccine made from the capsular antigen might theoretically be of value, and might be given to colonized pregnant women, but has not been investigated.

Prevention may be enhanced by daily treatment of the newborn infant's umbilical cord stump with triple dye (an antiseptic) or silver sulfadiazine. This reduces colonization of the cord stump by Group B streptococci, but has not been shown to reduce the frequency of Group B streptococcal disease.

Administration of intravenous ampicillin during labor appears to prevent transmission of Group B streptococci to the newborn infant.[96] Women can be screened at 34 to 36 weeks gestation to detect those with cervical colonization.

OTHER STREPTOCOCCI

Objectives

1. Describe the viridans group of streptococci and indicate their clinical importance.
2. Distinguish between enterococci and Group D streptococci that are not enterococci. Indicate the clinical importance of this distinction.

Classification of Other Streptococci

The term other streptococci refers to any streptococci other than beta-hemolytic streptococci.

Viridans Group of Streptococci. These organisms produce incomplete hemolysis (alpha-hemolysis). Viridans refers to the green hemolysis this organism produces on blood agar plates. The green discoloration is due to the formation of a by-product of hemoglobin.

Streptococcus Mutans. This is a species of alpha-hemolytic streptococci which is normal flora of the throat, and is a rare cause of pneumonia with purulent pleural fluid.[102] It occasionally is mistaken for an enterococcus. *Strep. mutans* probably is a contributing factor in dental caries and periodontal disease. *Strep. sanguis* and *Strep. salivarus* are other species in the viridans group of streptococci.

Group D Streptococci.[101,104] These streptococci are ultimately defined by a precipitin reaction between an extract of the streptococcal cell wall and antisera prepared in rabbits, using the same methods described earlier for Group A streptococci. However, since extraction of cell wall antigen is cumbersome, metabolic and growth characteristics are usually used. Thus, blackening of bile-eschulin agar is often used by clinical laboratories to define Group D in much the same way that bacitracin sensitivity is used to define Group A.[101]

Group D streptococci are usually alpha-hemolytic or nonhemolytic, although a few are beta-hemolytic.

There are two major subgroups of Group D streptococci: enterococci and non-enterococcal Group D streptococci. This distinction is clinically important because it correlates with antibiotic susceptibility.

Enterococci. This group of streptococci was originally named for its usual habitat in the enteric tract. The actual definition of this group has varied somewhat through the past 40 years.[97] In the 1960s, enterococcus was sometimes used synonymously with Group D streptococcus, but this terminology is no longer accepted. Enterococci are usually defined by various metabolic and growth tests, especially salt tolerance (ability to grow in six percent sodium chloride solution). The most common species is *Strep. faecalis*, but most clinical laboratories do not determine or report species within the enterococcus group. Most enterococci are resistant to penicillin or penicillinase-resistant penicillins.

Non-Enterococcal Group D Streptococci. These organisms are also normal flora of the human bowel, but are much less numerous than enterococci. The clinical laboratory usually distinguishes these bacteria from enterococci on the basis of growth in salt broth (salt tolerance), but other metabolic characteristics can be used.[104] The differentiation between non-enterococcal Group D streptococci and enterococci is of clinical importance only because enterococci are more likely to be resistant to penicillin and other antibiotics, and this is particularly important in the treatment of infective endocarditis.[101,103,104]

Streptococcus bovis and *Streptococcus equinis* are examples of non-enterococcal Group D streptococci. However, most clinical microbiology laboratories do not take the time and expense to report these species. The clinically important fact is whether or not a Group D streptococcus isolated from a patient with endocarditis is susceptible to penicillin. A preliminary disk susceptibility test gives a rapid probable answer, but the isolate should also be tested by tube dilution susceptibility against penicillin if possible. *Strep. bovis* infections can occur as a complication of carcinoma of the colon.

Anaerobic Streptococci. These streptococci are defined by adequate growth only under anaerobic conditions. They may be nonhemolytic or alpha-hemolytic. This group also usually includes the microaerophilic streptococci, which grow to tiny colonies under aerobic conditions, but grow better anaerobically.

Clinical Patterns of Illness

Other streptococci are *frequently* a cause of:

Subacute bacterial endocarditis is most frequently due to enterococci or the viridans group of streptococci. The source of infection with viridans streptococci or non-enterococcal Group D streptococci is usually the mouth, as in transient bacteremia during dental work.

The source of endocarditis due to enterococci is usually presumed to be the bowel or urinary tract, presumably caused by transient bacteremia due to mild injury, diarrhea, straining to defecate, or urinary tract instrumentation. Anaerobic streptococci *rarely* are a cause of endocarditis.

Lung abscess or empyema is often due to anaerobic streptococci.

Other streptococci are *occasionally* a cause of:

Urinary infection is occasionally caused by enterococci. Anaerobic streptococcal urinary infections are exceedingly rare. Wound infections are occasionally caused by anaerobic streptococcai or enterococci. Viridans streptococci are rarely a cause of wound infections.

These streptococci are rarely a cause of meningitis, cellulitis, septic arthritis, puerperal sepsis, septic abortion, and abscesses of the brain or soft tissues.

Biologic Features of Clinical Interest

Normal Flora. Enterococci are normal flora of the bowel. Of the aerobic bacteria of the bowel, only *E. coli* is found more frequently and in higher concentrations. Most of the fecal streptococci are enterococci, as defined by growth characteristics, but some are not Group D. Since streptococci are found in large numbers in the normal human stool, it is not surprising that they are frequently a cause of urinary tract infections or other infections secondary to fecal contamination.

The viridans group of streptococci are normal flora of the throat.[98] When alpha-streptococci are not found in the throat culture, recent antibiotic therapy is the most likely explanation.

Anaerobic streptococci are normal flora of the bowel, and also of the mouth.

Synergistic Infection. Postoperative progressive bacterial synergistic gangrene is an example of a synergistic infection with anaerobic streptococci and another organism, usually *Staph. aureus.* Typically, the infection occurs in the incision site of an abdominal or chest abscess. The wound edges become purple and tender. The anaerobic streptococcus and *Staph. aureus* together produce the typical lesion in experimental infections in guinea pigs, but either organism alone does not.[99]

Treatment

Treatment of these infections depends on the sensitivity of the organism to penicillin. Most strains of viridans streptococci are highly sensitive to penicillin. Most strains of anaerobic streptococci are sensitive to penicillin, but resistant to kanamycin and gentamicin.[100] However, most enterococci are penicillin-resistant, but are usually susceptible to ampicillin. A combination of ampicillin and gentamicin appears to be the most effective therapy for enterococcal endocarditis. Non-enterococcal Group D streptococci are typically susceptible to penicillin.

REFERENCES

Staphylococci

1. Barrett, F.F., Casey, J.I., Wilcox, C., and Finland, M.: Bacteriophage types and antibiotic susceptibility of *Staphylococcus aureus*. Arch. Intern. Med., *125:*867–873, 1970.

2. Boris, M., et al.: Bacterial interference. Protection against recurrent intrafamilial staphylococcal disease. Am. J. Dis. Child., 115:521–529, 1968.

3. Branson, D.: Identification of Micrococcaceae in clinical bacteriology. Appl. Microbiol., 16:906–911, 1968.

4. Casman, E.P.: Staphylococcal enterotoxin. Ann. N.Y. Acad. Sci., 128:124–131, 1971.

5. Duthie, E.S.: Evidence for two forms of staphylococcal coagulase. J. Gen. Microbiol., 10:427–436, 1962.

6. Houck, P.W., Nelson, J.D., and Kay, J.L.: Fatal septicemia due to Staphylococcus aureus 502A. Am. J. Dis. Child., 123:45–48, 1972.

7. Hughes, G.B., Chidi, C.C., and Macon, W.L., IV: Staphylococci in community-acquired infections. Increased resistance to penicillin. Ann. Surg., 183:355–357, 1976.

8. Jessen, O., et al.: Changing staphylococci and staphylococcal infections. A ten-year study of bacteria and cases of bacteremia. New Engl. J. Med., 281:627–635, 1969.

9. Klimek, J.J., et al.: Clinical, epidemiologic, and bacteriologic observations of methicillin-resistant Staphylococcus aureus at a large community hospital. Am. J. Med., 61:340–345, 1976.

10. Ladisch, S., and Pizzo, P.A.: Staphylococcus aureus sepsis in children with cancer. Pediatrics, 61:231–234, 1978.

11. Sabath, L.D., et al.: A new type of penicillin resistance of Staphylococcus aureus. Lancet, 1:443–447, 1977.

12. Melish, M.E., and Glasgow, L.A.: The staphylococcal scalded-skin syndrome. Development of an experimental model. New Engl. J. Med., 282:1114–1119, 1970.

13. Mossell, D.A.A.: Attempt in classification of catalase-positive staphylococci and micrococci. J. Bacteriol., 84: 1140–1147, 1962.

14. Mudd, S.: Resistance against Staphylococcus aureus. J.A.M.A., 218:1671–1673, 1971.

15. Murray, H.W., Tuazon, C.U., and Sheagren, J.N.: Staphylococcal septicemia and disseminated intravascular coagulation. Staphylococcus aureus endocarditis mimicking meningococcemia. Arch. Int. Med., 137:844–847, 1977.

16. Musher, D.M., and McKenzie, S.O.: Infections due to Staphylococcus aureus. Medicine, 56:383–409, 1977.

17. Person, D.A., Yu, P.K.W., and Washington, J.A., II: Characterization of Micrococcaceae isolated from clinical sources. Appl. Microbiol., 18:95–97, 1969.

18. Perkins, R.E., and Kundsin, R.B.: Comparison of heat shocking and acridine orange treatment in phage-typing of non-typable strains of Staphylococcus aureus. J. Clin. Microbiol., 4:334–337, 1976.

19. Quickel, K.E., Jr., et al.: Efficacy and safety of topical lysostatin treatment of persistent nasal carriers of Staphylococcus aureus. Appl. Microbiol., 22:446–450, 1971.

20. Raymond, E.A., and Traub, W.H.: Identification of staphylococci isolated from clinical material. Appl. Microbiol., 19:919–922, 1970.

21. Richmond, M.H., and Sykes, R.B.: The beta-lactamases of gram-negative bacteria and their possible physiologic role. Adv. Microb. Physiol., 9:31–88, 1973.

22. Smith, I.M., Beals, P.D., Kingsbury, K.R., and Hansenclever, H.F.: Observations on Staphylococcus albus septicemia in mice and men. Arch. Intern. Med., 102:375–388, 1958.

23. Todd, J., Fishaut, M., Kaprall, F., and Welch, T.: Toxic-shock syndrome associated with phage-group-I staphylococci. Lancet, 2:1116–1118, 1978.

24. Tuazon, C.U., and Sheagren, J.N.: Teichoic acid antibodies in the diagnosis of serious infections with Staphylococcus aureus. Ann. Intern. Med., 84:543–546, 1976.

25. Wentworth, B.B.: Bacteriophage typing of the staphylococci. Bacteriol. Rev., 27:253–272, 1963.

26. Williams, D.N., Lund, M.E., and Blazevic, D.J.: Significance of urinary isolates of coagulase-negative Micrococcaceae. J. Clin. Microbiol., 3:556–559, 1976.

27. Williams, E.O.: Healthy carriage of Staphylococcus aureus: its prevalence and importance. Bacteriol. Rev., 27:56–71, 1963.

28. Wilson, T.S., and Stuart, R.D.: Staphylococcus albus in wound infection and in septicemia. Canad. Med. Assoc. J., 93:8–16, 1965.

Pneumococci

29. Austrian, R., and Collins, P.: Importance of carbon dioxide in the isolation of pneumococci. J. Bacteriol., 92:1281–1284, 1966.

30. Bisno, A.L., and Freeman, J.C.: The syndrome of asplenia, pneumococcal sepsis and disseminated intravascular coagulation. Ann. Intern. Med., 72:389–393, 1970.

31. Bowers, E.F., and Jeffries, L.R.: Optochin in the identification of Streptococcus pneumoniae. J. Clin. Pathol., 8:58–60, 1955.

32. Bratton, L., Teale, D.W., and Klein, J.O.: Outcome of unsuspected pneumococcemia in children not initially admitted to the hospital. J. Pediatr., 90:703–706, 1977.

33. Coonrod, J.D., and Leach, R.P.: Antigenemia in fulminant pneumococcemia. Ann. Intern. Med., 84:561–563, 1976.

34. Coonrod, J.D., and Rylko-Bauer: Latex agglutination in the diagnosis of pneumococcal infection. J. Clin. Microbiol., 4:168–174, 1976.

35. Finland, M., and Barnes, M.W.: Changes in occurrence of capsular serotypes of Streptococcus

pneumoniae at Boston City Hospital during selected years between 1935 and 1974. J. Clin. Microbiol., 5:154–166, 1977.

36. Finland, M., and Winkler, A.W.: Recurrences in pneumococcus pneumonia. Am. J. Med. Sci., 188:309–320, 1934. Reprinted in J. Infect. Dis., 125 (Suppl.) 15–21, 1972.

37. Gunnison, J.B., Fraher, M.A., Pelcher, E.A., and Jawetz, E.: Penicillin-resistant variants of pneumococci. Appl. Microbiol., 16:311–314, 1968.

38. Hansman, D., et al.: Increased resistance to penicillin of pneumococci isolated from man. New Engl. J. Med., 284:175–177, 1971.

39. Hendley, J.O., Sande, M.A., Stewart, P.M., and Gwaltney, J.M., Jr.: Spread of *Streptococcus pneumoniae* in families. I. Carriage rates and distribution of types. J. Infect. Dis., 132:55–61, 1975.

40. Jacobs, M.R., et al.: Emergence of multiple resistant pneumococci. New Engl. J. Med., 299:735–740, 1978.

41. Kaijser, B., Berntsson, E., and Broholm, K-A.: The use of enzyme-linked immunosorbent assay (ELISA) for diagnosis of pneumococcal pneumonia. Abstract 178. Interscience Conference on Antibiotics and Antimicrobial Chemotherapy, 1974.

42. MacLeod, C.M., Hodges, R.G., Heidelberger, M., and Bernhard, W.G.: Prevention of pneumococcal pneumonia by immunization with specific capsular polysaccharides. J. Exp. Med., 82:445–465, 1945.

43. Merrill, C.W., Gwaltney, J.M., Jr., Hendley, J.O., and Sande, M.A.: Rapid identification of pneumococci. Gram stain vs the Quellung reaction. New Engl. J. Med., 288:510–512, 1973.

44. Mufson, M.A., Kruss, D.M., Wasil, R.E. and Metzger, W.I.: Capsular types and outcome of bacteremic pneumococcal disease in the antibiotic era. Arch. Int. Med., 134:505–510, 1974.

45. Sottile, M.I., and Rytel, M.W.: Application of counterimmunoelectrophoresis in the identification of *Streptococcus pneumoniae* in clinical isolates. J. Clin. Microbiol., 2:173–177, 1975.

46. Torphy, D.E., and Ray, C.G.: Occult pneumococcal bacteremia. Am. J. Dis. Child., 119:336–338, 1970.

47. Wood, W.B., Jr., and Smith, M.R: An experimental analysis of the curative action of penicillin in acute bacterial infections. I. The relationship of bacterial growth rates to the antimicrobial effects of penicillin. J. Exp. Med., 103:487–498, 1956.

Group A Streptococci

48. Benjamin, J.T., and Perriello, V.A., Jr.: Pharyngitis due to Group C hemolytic streptococci in children. J. Pediatr., 89:254–256, 1976.

49. Bisno, A.L., et al.: Contrasting epidemiology of acute rheumatic fever and acute glomerulonephritis. New Engl. J. Med., 283:561–565, 1970.

50. Boisvert, P.L., Darrow, D.C., Powers, G.F., and Trask, J.D.: Streptococcosis in children. A nosographic and statistical study. Am. J. Dis. Child., 64:516–534, 1942.

51. Breese, B.B.: The use of the fluorescent antibody technic for identification of Group A streptococci in pediatric practice. Am. J. Public Health, 58:2295–2305, 1968.

52. Breese, B.B., and Disney, F.A.: The accuracy of diagnosis of beta-streptococcal infections on clinical grounds. J. Pediatr., 44:670–673, 1954.

53. Breese, B.B., Disney, F.A., and Talpey, W.B.: The prevention of type specific immunity to streptococcal infections due to the therapeutic use of penicillin. Occurrence of second attacks due to the same type of Group A hemolytic streptococci. Am. J. Dis. Child., 100:353–359, 1960.

54. Breese, B.B., and Hall, C.B.: Beta Hemolytic Streptococcal Diseases. Boston, Houghton Mifflin, 1978. p. 9–19.

55. Brock, L.L., and Siegel, A.C.: Studies on the prevention of rheumatic fever: the effect of time of initiation of treatment of streptococcal infections on the immune response of the host. J. Clin. Invest., 32:630–632, 1953.

56. Broome, C.V., Moellering, R.C., Jr., and Watson, B.K.: Clinical significance of Lancefield Groups L–T streptococci isolated from blood and cerebrospinal fluid. J. Infect. Dis., 133:382–392, 1976.

57. Center for Disease Control: Foodborne epidemic of Group G streptococcal pharyngitis—Vermont. Morb. Mort. Week. Rep., 17:406–407, 1968.

58. Ederer, G.M., and Chapman, S.S.: Simplified fluorescent-antibody staining method for primary plate isolates of Group A streptococci. Appl. Microbiol., 24:160–161, 1972.

59. Hamburger, M., Green, M.J., and Hamburger, V.G.: The problem of the "dangerous carrier" of hemolytic streptococci. I. Number of hemolytic streptococci expelled by carriers with positive and negative nose cultures. J. Infect. Dis., 77:68–81, 1945.

60. Hill, H.R., et al.: Foodborne epidemic of streptococcal pharyngitis at the United States Air Force Academy. New Engl. J. Med., 280:917–921, 1969.

61. Jackson, H.: Streptococcal control in grade schools. Am. J. Dis. Child., 130:273–279, 1976.

62. James, L., and McFarland, R.B.: An epidemic of pharyngitis due to a nonhemolytic Group A streptococcus at Lowry Air Force Base. New Engl. J. Med., 284:750–752, 1971.

63. Kaplan, E.L., et al.: A.H.A. Committee Report. Prevention of rheumatic fever. Circulation, 55:1–4, 1977.

64. Kaplan, M.H.: Immunologic relation of streptococcal and tissue antigens. I. Properties of an antigen in certain strains of Group A streptococci exhibiting an immunologic cross-reaction with human heart tissue. J. Exp. Med., 146:579–599, 1977.

65. Krause, R.M.: Prevention of streptococcal sequelae by penicillin prophylaxis: a reassessment. J. Infect. Dis., 131:592–601, 1975.

66. Massell, B.F., Honikman, L.H., and Amezcua, J.: Rheumatic fever following streptococcal vaccination. Report of three cases. J.A.M.A., 207:1115–1119, 1969.

67. Miller, J.M., Stranger, S.L., and Massell, B.F.: A controlled study of beta-hemolytic streptococcal infection in rheumatic families. I. Streptococcal infection among healthy siblings. Am. J. Med., 25:825–844, 1958.

68. Moffet, H.L., et al.: Erythromycin estolate and phenoxymethyl penicillin in the treatment of streptococcal pharyngitis. Antimicrob. Agents Chemother., 1963:759–764, 1964.

69. Moffet, H.L., Cramblett, H.G., and Smith, A.: Group A streptococcal infections in a children's home. II. Clinical and epidemiologic patterns of illness. Pediatrics, 33:11–17, 1964.

70. Molteni, R.A.: Group A beta-hemolytic streptococcal pneumonia. Am. J. Dis. Child., 131:1366–1371, 1977.

71. Murray, P.R., Wold, A.D., Hall, M.M., and Washington, J.A., II: Bacitracin differentiation for presumptive identification of group A β-hemolytic streptococci: comparison or primary and purified plate testing. J. Pediatr., 89:576–579, 1976.

72. Nakae, M., Inoue, M., and Mitsuhashi, S.: Artificial elimination of drug resistance from Group A beta-hemolytic streptococci. Antimicrob. Agents Chemother., 7:719–720, 1975.

73. Quinn, R.W., and Lowry, P.N.: Some characteristics of nontypable Group A streptococci. Yale J. Biol. Med., 45:572–583, 1972.

74. Richman, D.D., Breton, S.J., and Goldmann, D.A.: Scarlet fever and group A streptococcal surgical wound infection traced to an anal carrier. J. Pediatr., 90:387–390, 1977.

75. Schacht, R.C., Gluck, M.C., Gallo, G.R., and Baldwin, D.S.: Progression to uremia after remission of acute poststreptococcal glomerulonephritis. New Engl. J. Med., 295:977–981, 1976.

76. Selinger, D.S., Julie, N., Reed, W.P., and Williams, R.C., Jr.: Adherence of Group A streptococci to pharyngeal cells: a role in the pathogenesis of rheumatic fever. Science, 201:455–457, 1978.

77. Slifkin, M., Engwall, C., and Pouchet, G.R.: Direct-plate serological grouping of beta-hemolytic streptococci from primary isolation plates with the Phadebact streptococcus test. J. Clin. Microbiol., 7:356–360, 1978.

78. Taranta, A., and Moody, M.D.: Diagnosis of streptococcal pharyngitis and rheumatic fever. Pediatr. Clin. North Am., 18:125–143, 1971.

79. Zimmerman, R.A., Horn, K.A., Meyer, W.T., and Klesins, P.H.: Applying an old concept to a new method of streptococcal surveillance. Pediatrics, 53:275–279, 1974.

Group B Streptococci

80. Aber, R.C., et al.: Nosocomial transmission of Group B streptococci. Pediatrics, 58:346–353, 1976.

81. Ablow, R.C., et al.: A comparison of early-onset Group B streptococcal neonatal infection and the respiratory-distress syndrome of the newborn. New Engl. J. Med., 294:65–70, 1976.

82. Badri, M.S., et al.: Rectal colonization with Group B streptococcus: relation to vaginal colonization of pregnant women. J. Infect. Dis., 135:308–312, 1977.

83. Baker, C.J.: Summary of the workshop on perinatal infections due to Group B streptococcus. J. Infect. Dis., 136:137–152, 1977.

84. Baker, C.J., et al.: Vaginal colonization with Group B streptococcus: a study in college women. J. Infect. Dis., 135:392–397, 1977.

85. Baker, C.J., and Kasper, D.L.: Correlation of maternal antibody deficiency with susceptibility to neonatal Group B streptococcal infection. New Engl. J. Med., 294:753–756, 1976.

86. Bayer, A.S., Chow, A.W., Anthony, B.F., and Guze, L.B.: Serious infections in adults due to Group B streptococci. Clinical and serotypic classification. Am. J. Med., 61:498–503, 1976.

87. Darling, C.L.: Standardization and evaluation of the CAMP reaction for the prompt, presumptive identification of *Streptococcus agalactiae* (Lancefield Group B) in clinical material. J. Clin. Microbiol., 1:171–174, 1975.

88. Fenton, L.J., and Harper, M.H.: Direct use of counterimmunoelectrophoresis in detection of Group B streptococci in specimens containing mixed flora. J. Clin. Microbiol., 8:500–502, 1978.

89. Gunn, B.A.: SXT and Taxo A disks for presumptive identification of Group A and B streptococci in throat cultures. J. Clin. Microbiol., 4:192–193, 1976.

90. Howard, J.B. and McCracken, G.H., Jr.: The spectrum of Group B streptococcal infections in infancy. Am. J. Dis. Child., 128:815–818, 1974.

91. Kenny, J.F., and Zedd, A.J.: Recurrent group B streptococcal disease in an infant associated with the ingestion of infected mother's milk. J. Pediatr., 91:158–159, 1977.

92. Manroe, B.L., Rosenfeld, C.R., Weinberg, A.G., and Browne, R.: The differential leukocyte count in the assessment and outcome of early-onset neonatal Group B streptococcal disease. J. Pediatr., 91:632–637, 1977.

93. Merritt, K., and Jacobs, N.J.: Characterization and incidence of pigment production by human clinical Group B streptococci. J. Clin. Microbiol., 8:105–107, 1978.

94. Paredes, A., et al.: Nosocomial transmission of Group B streptococci in a newborn nursery. Pediatrics, 59:679–682, 1977.

95. Szilagyi, G., Mayer, E., and Eidelman, A.I.: Rapid isolation and identification of Group B streptococci from selective broth media by slide coagglutination test. J. Clin. Microbiol., 8:410–412, 1978.

96. Yow, M.D., et al.: Ampicillin prevents intrapartum transmission of Group B streptococcus. J.A.M.A., 241:1245–1247, 1979.

Other Streptococci

97. Deibel, R.H.: The Group D streptococci. Bacteriol. Rev., 28:330–366, 1964.

98. Facklam, R.R.: Physiological differentiation of viridans streptococci. J. Clin. Microbiol., 5:184–201, 1977.

99. Meleney, F.L.: A differential diagnosis between certain types of infectious gangrene of the skin. With particular reference to haemolytic streptococcus gangrene and bacterial synergistic gangrene. Surg. Gynecol. Obstet., 56:847–867, 1933.

100. Pien, F.D., Thompson, R.L., and Martin, W.J.: Clinical and bacteriologic studies of anaerobic gram-positive cocci. Proc. Mayo Clinic, 47:251–257, 1972.

101. Ravreby, W.D., Bottone, E.J., and Keusch, G.T.: Group D streptococcal bacteremia, with emphasis on the incidence and presentation of infections due to *Streptococcus bovis*. New Engl. J. Med., 289:1400–1403, 1973.

102. Sattler, F.R., and Ruskin, J.: Empyema due to *Streptococcus mutans*. Chest, 71:229–231, 1977.

103. Toala, P., McDonald, A., Wilcox, C., and Finland, M.: Susceptibility of Group D streptococcus (enterococcus) to 21 antibiotics in vitro, with special reference to species differences. Am. J. Med. Sci., 258:416–430, 1969.

104. Watanakunakorn, C.: *Streptococcus bovis* endocarditis. Am. J. Med., 56:256–260, 1974.

2

Gram-Positive Rods

CORYNEBACTERIA

Objectives

1. Describe the typical appearance of three clinical patterns of illness produced by *Corynebacterium diphtheriae*.
2. Describe how infectious mononucleosis can be distinguished from diphtheria and indicate what other microrganisms may be found in the throat which may resemble the diphtheria bacillus.
3. Explain why laboratory confirmation of diphtheria is exceedingly important, but why treatment should not be delayed while awaiting laboratory confirmation.
4. Describe what material is used in the Schick test and explain what a positive Schick test means.
5. Describe the mechanism of action of diphtheria toxin.
6. List two species of *Corynebacterium* other than *C. diphtheriae*, and indicate their clinical importance.

Name, Species and Types

Corynebacteria are club-shaped, gram-positive rods (See Fig. 2-5). *Corynebacterium diphtheriae* is also called the diphtheria bacillus. The word diphtheria is derived from a Greek word for membrane.

There are three types of *C. diphtheriae:* mitus, intermedius (minimus) and gravis, differentiated by the degree of black or brown color of the colonies on special media. However, the colony type is not well correlated with severity of the disease. Any of these types may be toxigenic, producing the toxin responsible for most of the clinical manifestations of diphtheria. Even non-toxigenic strains have, on rare occasions, produced disease.[6] *C. diphtheriae* also can be typed further by the use of bacteriophage, which allows more exact and detailed epidemiologic studies.

Corynebacterium diphtheriae should not be considered normal flora of the respiratory tract or skin. However, many other species of corynebacteria are found in the nose and throat of normal people. These organisms are often called diphtheroids. The anaerobic organism *Corynebacterium acnes*, now called *Propionibacterium acnes*, is common and normal flora of the skin. It occurs frequently as a contaminant of blood cultures, but also rarely causes disease (see Chap. 6).

Diphtheroids, also called coryneform bacteria or *Corynebacterium* species, are a rare cause of endocarditis of prosthetic heart valves,[16,18] or septicemia in compromised hosts such as leukemics.[8] Of clinical importance is the observation that these species are typically resistant to penicillin, but susceptible to vancomycin.[8,18]

Corynebacterium vaginale (at times called *Haemophilus vaginalis*) appears to be the usual cause of "non-specific" vaginitis (called non-specific because recognized causes of vaginitis, such as gonococcus, *Candida albicans*, or *Trichomonas vaginalis* have been excluded).[17] *C. vaginale*, however, could not be associated with vaginitis in another study.[11] *C. vaginale* is also a rare cause of bacteremia, especially in postpartum women or newborn infants.[13]

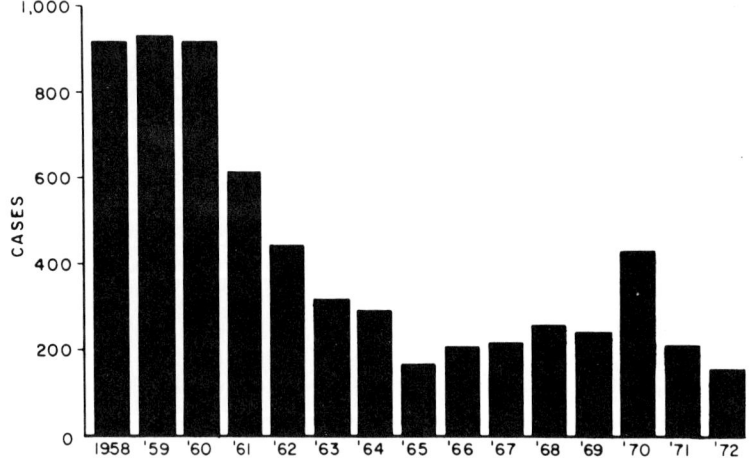

FIG. 2-1. *Diphtheria: annual reported cases in the United States, 1958–1972.*[4]

Frequency and Importance

About 200 cases of diphtheria, with about ten deaths, are reported each year in the United States. There were 304 cases reported in 1975, of which 203 were cutaneous. The annual frequency from 1958 through 1972 is shown in Figure 2-1. The increased frequency of cutaneous diphtheria is shown in Figure 2-2. A single suspected case should be regarded as an urgent public health situation. Cultural confirmation is exceedingly important to justify the large efforts needed for control.

In the Chicago outbreak of 1970, most patients were from the near northwest side of the city. Ten of the 21 confirmed cases were hospitalized at Children's Memorial Hospital, where both the clinical course and cultural characteristics of the organism were observed by the author. In most outbreaks in the United States, most patients are of a single ethnic group (especially native Americans), cultural subgroups which have had no immunizations, especially lower socioeconomic groups. Typically, the outbreaks are quite focal, involving only families, extended families, or small areas within a city.

Clinical Patterns of Illness

C. diphtheriae should be the presumptive diagnosis in the following syndromes:

Membranous pharyngitis in an individual without past diphtheria immunization, and in absence of evidence for infectious mononucleosis by slide test or peripheral blood smear. A membrane extending onto the soft palate is particularly typical. The patient does not usually have high fever, but appears very sick ("toxic"). Cervical adenitis is common. Slight bleeding occurs when the membrane is peeled back with a tongue blade. Proteinuria is typically present early in the disease.

Myocarditis, with arrhythmias and congestive heart failure, usually does not occur until the second week of the illness, and is the usual mechanism of death. Paralysis of peripheral nerves, especially paralysis of palate, occasionally occurs.

Membranous laryngitis is usually not recognized until the time of tracheal intubation or tracheotomy, unless the membrane extends up to the posterior pharynx.

C. diphtheriae is a *very rare* cause of the following syndromes:

Bleeding Rhinitis. In nasal diphtheria, there is typically a thin, slightly bloody nasal discharge. This type of rhinitis is usually not due to diphtheria, but a membrane should be looked for in the nose. If a membrane is present, diphtheria

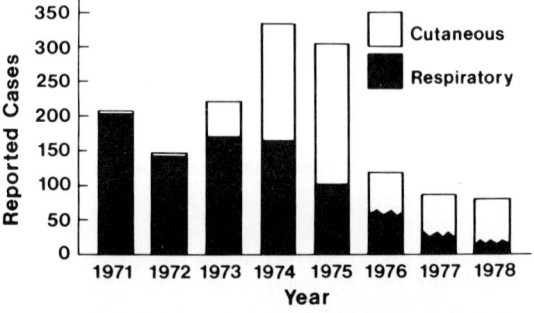

FIG. 2-2. *Diphtheria in the United States, 1971–1975, showing increase in cutaneous diphtheria, most of which occurred in the Seattle, Washington area.*[4] *1976–1978 data are provisional.*

is likely. The membrane may be overlooked because the diagnosis of diphtheria is not suspected, and the membrane thought to be exudate.

Skin Infections. Cutaneous diphtheria occurs in the tropics, but has been generally regarded as uncommon in the United States. Recently, cutaneous diphtheria has been recognized more frequently (Fig. 2-2). Toxigenic *C. diphtheriae* can sometimes be recovered from infected insect bites or impetigo-like skin lesions, which are usually shallow, oozing or crusted ulcers.[2] These lesions may serve as a reservoir for the organism, and may be a source of contagion for many months.[1]

Other Syndromes. Otitis media, vaginitis, myocarditis, or peripheral nerve paralysis are usually observed as complications of typical upper respiratory diptheria, but very rarely may occur as an isolated manifestation of *C. diphtheriae* infection.

Asymptomatic or Subclinical Infection

Many adults have serologic and skin test evidence of past infection with *C. diphtheriae* without any history of diphtherial disease or immunization. This is the basis for the belief that asymptomatic infection is rather common.[5]

Laboratory Diagnosis

Smear. Because of the presence of diphtheroids in the throat of normal individuals, the diagnosis of *Corynebacterium diphtheriae* infection based on a stain of a throat smear is unlikely to be accurate. When given smears made from a variety of species of corynebacteria, most laboratory technicians cannot distinguish *C. diphtheriae* from the other species of corynebacteria.

Fluorescent antibody methods to detect *C. diphtheriae* on throat smear may be useful if there is no cross reaction with diphtheroids. A "weakly positive" report is useless, and creates unnecessary alarm.

Culture. The preliminary recognition of corynebacteria in the laboratory is based on the appearance of typical colonies on tellurite or Tinsdale media.[14] The organism also grows satisfactorily on sheep blood or chocolate blood agar plates, so these media can be used if special media is not immediately available.

Toxigenicity of a *Corynebacterium diphtheriae* isolate can be determined by injecting the organism intradermally into a rabbit, by agar diffusion tests or counterimmunoelectrophoresis, which demonstrate precipitin bands, or by observing for destruction of a tissue monolayer.

The classification of *Corynebacterium diphtheriae* into gravis, intermedius, and mitus does not have as close a relation to virulence as was originally thought. The final test of virulence is the production of toxin.

Serology. Antitoxin levels can be measured by animal protection tests. However, diphtheria toxin is a poor antigen, and serologic diagnosis of diphtheria infection is not practical.

Biologic Characteristics of Clinical Interest

Exotoxin. This is the major factor in virulence of *C. diphtheriae*. The exotoxin is produced by the bacillus in the necrotic tissue where antibiotics are not able to reach the organism. The toxin is responsible for myocarditis, proteinuria, and peripheral nerve paralysis.

Diphtheria toxin is a protein that acts to inhibit protein synthesis in hosts cells by inactivating the elongation factor, EF-2, and thus inhibiting polypeptide chain elongation.[15]

Special Media. The recovery of the organism is aided by the use of special media. *C. diphtheriae* does grow on sheep blood agar plates, where the colonies are alpha-hemolytic, grey or white, and hard. However, a laboratory technician is unlikely to identify these colonies unless the clinical diagnosis is suggested. Because of the public health importance of a laboratory-confirmed diagnosis, special selective media should be used.

Potassium tellurite (K_2TeO_3) is incorporated into culture media to detect the diphtheria bacillus by its ability to reduce the tellurite (colorless) to tellurium (black or brown).[14] Staphylococci, enteric bacteria, and diphtheroids also can produce this reaction, so Gram stain and testing for toxin are necessary to confirm the identification.

Treatment

Airway. Maintaining an airway is the most important early therapy, if the membrane is producing tracheal or laryngeal obstruction. Tracheotomy may be necessary.

Antitoxin. This is extremely important therapy, and should never be delayed while waiting for laboratory reports. The frequency of death due to diphtheria is directly related to the duration of the disease before receiving antitoxin. Antitoxin is a horse serum product and requires testing for allergy before use.

Antibiotic Therapy. This is of minimal value for treatment, and should never be used instead of antitoxin. Erythromycin is used to eradicate the organism in carriers.[10]

Prevention

Diphtheria Toxoid. This is made by adding small amounts of formalin to toxin, which makes the toxin inactive but does not destroy the antigenicity of the material. Toxin is a poor antigen, and having the disease may not produce immunity. If immunization is "full" (complete), fatal diphtheria almost never occurs.

Herd immunity does not appear to be involved in diphtheria, as even a very high percentage of immune individuals (more than 85 percent) does not prevent spread of the disease.[12]

Schick Test. This test is named for Bela Schick, a distinguished pioneer in the study of diphtheria. It is sometimes used to determine susceptibility to diphtheria, to determine the need for a diphtheria toxoid booster, or to determine the capability to form serum antibodies as part of the diagnostic evaluation for immunologic defects. Unfortunately, the Schick test is not completely accurate for any of these uses, and better alternatives are available.

An understanding of the principles involved in the Schick test is useful to appreciate its limitations. A small amount of diphtheria toxin is injected intradermally, and the site examined in 5 days. If no local reaction occurs (Schick negative), the individual has sufficient circulating antitoxin to neutralize the toxin injected. Unfortunately, the level of antitoxin necessary to produce a negative Schick test may not be sufficient to prevent clinical diphtheria, and diphtheria can occur in Schick-negative individuals.[9]

If an area of redness occurs where the toxin was injected (Schick-positive), the individual probably has insufficient antitoxin, unless an allergic reaction to the toxin or diluent has occurred. False-positive Schick tests due to allergy can be detected in some cases by use of a control consisting of heat-inactivated toxin. In addition, allergic false-positive reactions are usually gone by 5 days, but exclusion of a false-positive reaction cannot be done with certainty without testing for the actual antitoxin level. When antitoxin levels were measured in a study of medical students, about 10 percent had a false-positive Schick test not detected by the use of a heat-inactivated control, and apparently due to allergy to the intact toxin.[3]

In the past, the Schick test was used to avoid giving diptheria toxoid to individuals who did not need it and might have an allergic reaction to the toxoid. In recent years, lower doses of diphtheria toxoid are used for adults, and the Schick test is not necessary for this purpose, since the Schick test itself is more likely to give a severe local reaction than the toxoid.

More specific and more accurate tests than the Schick test are available for testing an individual's ability to make circulating antibodies.

LISTERIA MONOCYTOGENES

Objectives

1. Describe in what circumstances *Listeria monocytogenes* is likely to be mistaken for a contaminant.
2. Describe the typical clinical illness of listeria infection in newborns, and the circumstances under which it is more likely to infect older children and adults.
3. Discuss the methods used for laboratory confirmation of the diagnosis.

Definitions and Types

Listeria monocytogenes is a gram-positive rod sometimes mistakenly identified as "diphtheroids". The organism is named for the bacteriologist Lister, and for the monocytosis produced in infections in animals. Many serologic types and subtypes can be identified by agglutination reactions, and such serotyping may provide some clues as to the source of the infection.[22]

Frequency and Importance

There are about 100 cases of listerosis with about 25 fatalities reported each year in the United States. The disease is observed in the first month of life more frequently than in any ten-year period of life. In older age groups, patients with Hodgkin's disease and hematologic cancers appear to be particularly susceptible,[24] but normal adults also become infected.

Clinical Patterns of Illness

Listeria monocytogenes is an *occasional* cause of:

Neonatal Septicemia. An early onset pattern (granulomatosis infantiseptica), occurring in the first 4 days of life, has an extremely high mortality, probably because the infection began before birth.[26] Late onset (after 4 days) neonatal listeriosis has a slightly better prognosis.

Retrospective analysis of perinatal infections may elicit a history of fever during pregnancy, accompanied by mild respiratory symptoms, and myalgia or urinary symptoms, but the clinical illness in mother is usually not specific, and not clinically distinguishable.[22]

Septicemia also can occur in pregnant women, or adults with hematologic malignancies.

Purulent Meningitis.[28] Most cases of purulent meningitis occur in the neonatal period, but may occur in older individuals, particularly if debilitated.[23,28] The spinal fluid typically shows a predominancy of neutrophils, with low glucose and a high protein, but occasionally lymphocytes predominate. The prognosis is relatively good for survival and complete recovery, if treated.

Endocarditis has been reported, especially in normal persons, and has a high mortality rate.[20]

Laboratory Diagnosis

Smear. Sometimes the smear appears to show gram-positive cocci, or even gram-negative rods or diplococci.[23]

Culture. The organism grows readily on sheep blood agar plates, usually producing a beta-hemolysis. It can be distinguished from *Corynebacterium* species (diphtheroids) by its motility at room temperature. Serologic typing can be done by a reference laboratory for epidemiologic information.

tumbling motility at 25°C

Serology. Diagnosis by antibody titer rise is possible in reference laboratories, but is not practical.

White Blood Count. Although the organism produces monocytosis in animals, monocytosis is *not found in the* peripheral blood in septicemia. Monocytosis or lymphocytosis has been reported in the spinal fluid in meningitis, in a small proportion of cases.[28]

Biologic Characteristics of Clinical Interest

Sources of Infection. Listeriosis is primarily a disease of animals, fish, and birds. Listeria have also been recovered from soil specimens. Clustering of cases among newborns may indicate spread from one newborn infant to another, but investigation of outbreaks has not clearly demonstrated nosocomial transmissions.[19,21]

Perinatal Infections. Pregnant animals are more susceptible to infection, and abortion or infection of the fetus is frequent. This has rarely been reported in humans.[19,27]

Refrigeration. The organism is best grown from infected tissues after refrigeration of the tissue. The importance of this with respect to pathogenesis is unknown. The laboratory should be informed when this organism is suspected in tissues.

Cell-Mediated Immunity. Resistance to infection with *L. monocytogenes* depends on cell-mediated immunity, according to animal experiments. This is also consistent with the observation that infection is more frequent in patients who are immunosuppressed or who have malignancies.

Treatment

Ampicillin appears to be the best antibiotic.[23] Good results have also been obtained with penicillin or tetracycline. Penicillin or ampicillin and gentamicin are synergistic against *L. monocytogenes* in experimental infections in animals, and in vitro.[25,29]

CLOSTRIDIUM SPECIES

Objectives

1. Describe the typical clinical picture of tetanus and give the differential diagnosis.

32 GRAM-POSITIVE RODS

2. Describe the usual mechanisms of death in tetanus before and after respirator therapy, and describe emergency management.

3. Describe the effects of tetanus toxin on the central nervous system.

4. Describe the effects of antitoxin given before and after the onset of tetanus.

5. Describe the typical clinical pictures of gas gangrene, botulism, and clostridial food poisoning.

6. Discuss laboratory confirmation of botulism.

7. Discuss treatment of gas gangrene and botulism.

8. Discuss the evidence that pseudomembraneous enterocolitis following antibiotics treatment is often caused by clostridial toxins.

Species and Types

Clostridium is a genus of anaerobic, gram-positive spore-forming bacilli. *Clostridium tetani* is the cause of tetanus. Other important clostridia include *Cl. perfringens,* the most frequent species of clostridia found in gas gangrene, and *Cl. botulinum,* which produces botulism, a form of food poisoning.

Numerous species of clostridia can produce gas gangrene;[46] but only *Cl. tetani* produces tetanus, and only *Cl. botulinum* produces botulism. *Cl. tetani* has only one type. *Cl. botulinum* has six types, A through F, of which A, B, and E are the usual types found in human botulism.

Various species of *Clostridium* can produce bacteremia, especially in patients with malignancy. *Clostridium perfringens* accounted for over half of the isolates in one study which reported 65 isolates (three percent) from 2168 positive blood cultures, representing 12 clostridial species.[4]

Frequency and Importance

Tetanus. Approximately 100 cases are reported annually in the United States. In 1977 there were 87 reported cases. Tetanus is important because it is completely preventable by immunization. Treatment of tetanus is difficult and requires specialized care. The mortality rate is high, but if the patient survives, recovery is usually complete.

Botulism. About 20 to 30 cases were reported annually, until infantile botulism was recognized

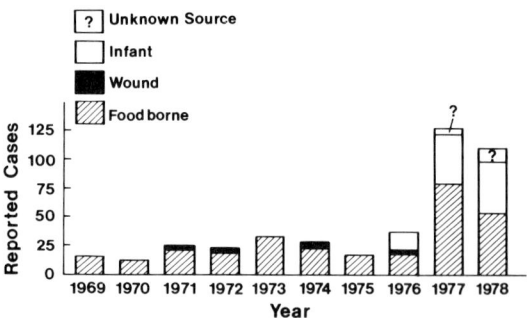

FIG. 2-3. *Botulism in the United States,*[36] *with provisional data for 1978.*

in 1976 (Fig. 2-3). The disease is important because early diagnosis and specific treatment can be lifesaving.

Clostridial Food Poisoning. Food poisoning is usually suspected only when several individuals have the simultaneous onset of an illness, after having eaten a common meal. *Cl. perfringens* is one of the most frequent confirmed causes of food poisoning outbreaks in the United States. Approximately 50 outbreaks of *Cl. perfringens* food poisoning, involving about 3000 patients, are reported each year. Outbreaks of food poisoning of all kinds are grossly underreported, so the total number of outbreaks is probably much larger.

Pseudomembraneous Enterocolitis. *Clostridium* species, particularly *Cl. difficile* (rhymes with seal), have been recently recognized to cause severe diarrhea and colitis.

Clinical Syndromes Produced by Clostridium Species

Tetanus. In its usual clinical pattern, tetanus is characterized by episodes of muscular rigidity which become more frequent and more prolonged over a period of a few days. If untreated by sedation, death results from inadequate respiration because of spasm of respiratory muscles. After the patient is begun on mechanical ventilation, death is usually due to aspiration pneumonia.

The diagnosis is suspected on the basis of episodic muscle spasms, especially the unopposed muscles of the abdomen and jaw (trismus). Convulsions and stiff neck are common, and often lead the physician to do a lumbar puncture, but the spinal fluid is normal. A grossly infected wound is often, but not always, found.

Gangrene. Several patterns of clostridial infections of soft tissue are observed: anaerobic cellulitis, myonecrosis, and gas gangrene. *Cl. perfringens*, formerly called *Cl. welchii*, is the most frequent species found. Damaged muscle provides the focus for anaerobic growth; but the involvement may spread to healthy muscle, particularly when gas production, toxins, and enzymes such as collagenase attack the normal muscle. The odor of the diseased muscle is foul and a crackling sensation called crepitus is felt when the gas-infiltrated areas are palpated.

Botulism. This word is derived from *botulus*, the Latin word for sausage. It usually results from ingestion of spoiled home-canned foods.[49] Since the toxin is heat-labile, the disease will not occur if the food is heated (boiling for 10 minutes). The early manifestations are typically associated with cranial nerve paralysis, particularly paralysis of eye muscles, (Fig. 2-4) pupillary muscles, and the muscles of swallowing. Generalized flaccid paralysis occurs later. Wound infection is a rare cause of botulism.[47]

Infant Botulism. The syndrome of infant botulism was described in 1976 in at least 20 infants about 1 to 8 months of age.[32] The clinical findings are similar to botulism in older individuals, with cranial nerve paralysis, but generalized muscle weakness was also prominent, resulting in a "floppy infant". Presumably the organisms are ingested as part of the infant's food, and the toxin is formed in the infant's gastrointestinal tract. Botulinum toxin is typically found in the feces, but rarely in the serum of these infants. Electromyography shows a characteristic pattern useful for rapid diagnosis.

Bacteremia. This typically complicates postoperative infections, particularly involving the biliary tract.[50] The most frequent species involved is *Cl. perfringens*.[41] Septic abortion, with hemolysis and hemoglobinuria, is another possible source of clostridial bacteremia.[37] Patients with necrotic or ulcerated lesions, particularly decubitus ulcers or necrotic tumors, also may develop clostridial bacteremia.[52] The detection of gram-positive rods suspected to be clostridia in a blood culture should raise the question of an abdominal abscess with *Bacteroides* species and *E. coli* (see Chap. 3) also present, and antibiotic therapy also effective against these latter two organisms should be considered.

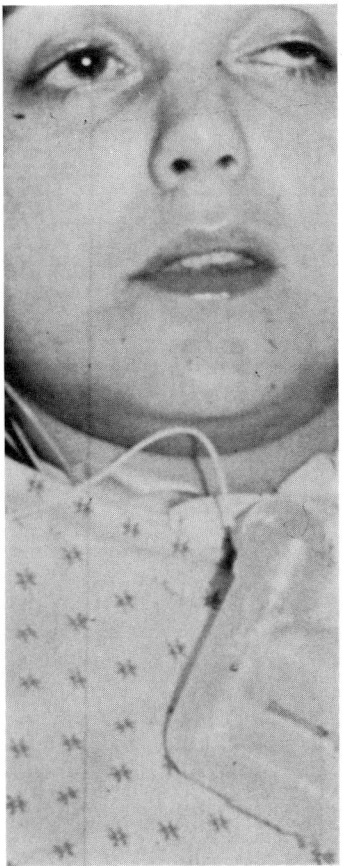

FIG. 2-4. *Cranial nerve paralysis in a patient with botulism. Note the deviation of the left eye, droopy eyelid (ptosis), and facial asymmetry. Tubing from a tracheotomy to a ventilator can also be seen. (Photo from Drs. William Terranova and Joel Breman)*

Diarrheal Food Poisoning. *Clostridium perfringens* is a major cause of diarrheal food poisoning in the United States.[44] It is usually associated with meat dishes, although burritos containing only beans have also been implicated. Typically, the meat has been cooked sufficiently to destroy the vegetative organisms, but not the spores, which germinate and produce toxin during a period of inadequate refrigeration. The meat is then served without heating sufficient to inactivate the toxin, and diarrhea occurs about 12 hours after ingestion.[38,44] Vomiting and abdominal pain are sometimes prominent.[55]

Uterine Infections.[51] Postpartum or postabortal localized uterine infections may occur without bacteremia. Localized or generalized peritonitis sometimes occurs, but gas formation is rare. In some cases, leucocyte counts of 25,000 to 50,000 have been observed.[51]

Enteritis Necroticans. This is a fulminating clostridial infection of the bowel, with toxicity, bloody diarrhea, and a high mortality rate.[35]

Antibiotic-Induced Colitis. *Clostridium difficile* and *Cl. sordelli* each appear to be associated with antibiotic-induced colitis, although *Cl. difficile* appears to be more important.[34,53] The toxin of *Cl. difficile* is cytotoxic in tissue culture. The enterocolitis it produces in hamsters is neutralized by gas gangrene antitoxin (which contains *Cl. sordelli* antitoxin), which is often used because *Cl. difficile* antitoxin is not available.[34] In humans with severe diarrhea after antibiotics (especially clindamycin), there is often a pseudomembrane that resembles yellow-white plaques seen on sigmoidoscopy. The diarrhea will often persist after stopping the antibiotic. Probably some, and possibly many, of the patients with pseudomembraneous enterocolitis attributed to *Staph. aureus* in the 1950s really had disease caused by *Cl. difficile*. Both *Staph. aureus* and *Cl. difficile* enterocolitis in humans respond well to oral vancomycin therapy.[53]

Sub-lethal antibiotic concentrations which *inhibit* but do not *kill* this spore-forming organism may enhance toxin production. Therefore, some *Cl. difficile* isolates may be reported as susceptible to the antibiotic the patient was receiving when the colitis developed, but were only inhibited, not killed.

Laboratory Diagnosis

Culture. Strict anaerobic conditions are needed to grow *Clostridium* species. *Cl. perfringens* often produces a double zone of beta hemolysis on blood agar, and often can be identified promptly. *Cl. tetani* is unusually difficult to grow because it is so oxygen-sensitive; but the clinical diagnosis of tetanus can be made with a high degree of certainty without bacteriologic proof.

The bacteriology laboratory may report anaerobic gram-positive rods growing in a blood culture, which raises the question of a clostridium. Usually, however, it is an anaerobic diphtheroid, *Corynebacterium (Propionibacterium) acnes*, a common skin organism, the most common contaminant of blood cultures (Chap. 6). Usually, the bacteriologist is able to distinguish the square-ended large bacillus of clostridia, which often has spores on Gram stain, from the club-shaped bacillus of propionibacteria

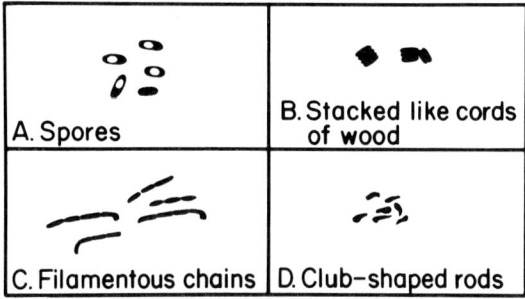

FIG. 2-5. *Gram-positive rods: possible arrangements. (A.) If aerobic, usually bacillus; if anaerobic, often clostridia. (B.) Could be listeria or corynebacteria. (C.) If anaerobic, or from cervical or vaginal area, probably lactobacillus. (D.) Typical of corynebacteria or propionibacterium (anaerobic).*

(Figure 2–5). Clostridia often produce gas in the broth. Some isolations of clostridia in a clinical bacteriology laboratory may be contaminants, but antibiotic therapy with penicillin is usually indicated pending clarification.[35]

Smear. Gram-positive bacilli can be found in histologic specimens in fatal septicemia, but are rarely found in direct smears of pus from living patients. Blood cultures often show spores on Gram stain.

Animal Inoculation. This is useful in the diagnosis of botulism. The patient's serum or a suspension of the suspected food can be injected into mice, and if botulinum toxin is present, the mice die. The specific toxin type is identified by mouse protection tests, using type-specific antitoxins. *Cl. botulinum* can be detected rapidly in food, using gas chromatographic techniques. Recently, it has been shown that it is useful to examine the patient's stool for the organism and for the toxin.[40]

Biologic Properties of Clinical Interest

Exotoxins. Several clostridia species produce potent exotoxins. These include the neurotoxins of *Cl. tetani* and *Cl. botulinum*; and the hemolysin, lecithinase, and enterotoxins of *Cl. perfringens*. Botulinum toxin acts to prevent the release of acetylcholine at the synapse.[42] Tetanus toxin acts on the anterior horn cells of the spinal cord, and once fixed to these cells, it cannot be released or neutralized by antitoxin, which acts only to neutralize any toxin produced at the site of the wound.⁴³ *Cl. perfringens* enterotoxin produces secretion of fluid in isolated segments of rabbit bowel.⁴⁵

Necrotic Tissue. This usually has a low pH, and can allow many *Clostridium* species to survive at higher oxygen concentrations than usual.

Normal Flora. *Clostridium perfringens* is often found in the colon of normal individuals. *Cl. perfringens* is also found on the skin of about 20 percent of normal individuals, and the genital tract of about five percent of normal individuals.³⁵

Gas Production. Necrotizing clostridial infections often produce gas, apparently due to muscle digestion rather than fermentation of carbohydrates.

Electrophysiological Studies. Botulism is associated with characteristic electrophysiological studies.⁴⁸

Antigen Detection. Counterimmunoelectrophoresis can be used to detect *Clostridium perfringens* enterotoxin in the patient's feces, and antibodies to the enterotoxin can be detected in the patient's serum in diarrheal food poisoning.

Treatment

Sedation. The most important early treatment of tetanus is sedation to prevent respiratory muscle spasm and consequent respiratory failure. Physicians should use a sedative with which they are familiar, such as phenobarbital, in doses usually used for severe convulsions. Other sedatives or muscle relaxants are equally effective.

Patients with tetanus eventually require mechanical ventilation while the physician keeps the patient paralyzed by muscle relaxants, so all patients should be transferred promptly to a medical center where continuous supervision by physicians is possible.

Antitoxin. Tetanus antitoxin is of secondary importance to sedation. Gas gangrene polyvalent antitoxin is used in treatment of gas gangrene. Trivalent botulism antitoxin for types ABE is useful for the treatment of botulism.³⁹ These antitoxins are horse serum products, and the patient should be tested for allergy to the antitoxin before its use.

Antibiotics. Penicillin is usually recommended for therapy of *Cl. perfringens* bacteremia, primarily because of the greater amount of clinical and experimental evidence for its efficacy. Ampicillin, chloramphenicol, or clindamycin are also effective according to sensitivity tests.⁵⁴

Hyperbaric Oxygen. Dramatic results have been reported in patients with gas gangrene who have been treated with hyperbaric oxygen after failing to respond to other measures. However, hyperbaric oxygen therapy has not been studied in humans in a controlled fashion.³⁰

Prevention

Tetanus toxoid is effective and safe, and universal immunization against tetanus should be mandatory. Tetanus antitoxin is effective in prevention of tetanus, but is not a satisfactory substitute for active immunization with tetanus toxoid since wounds are often unrecognized or minor. Gas gangrene antitoxin is of no value in the prevention of gas gangrene.³⁰

Prophylactic antibiotics are recommended for most patients before cholecystectomy, because of the risk of postoperative clostridial bacteremia,⁵⁰ and other infection. Prophylactic antibiotics are sometimes used for tetanus-prone injuries, defined as extensive crushing, contaminated, or necrotic wounds. Oxytetracycline appeared to be more effective than penicillin in protecting mice from tetanus when vegetative organisms were injected.³¹

Botulism can often be prevented by avoiding unnecessary exposure. Patients with botulism often report that the home-canned food looked or smelled bad, but they ate it anyway. *Cl. perfringens* diarrheal food poisoning can sometimes be avoided by adequate refrigeration of meat after cooking.

Infantile botulism may be associated with the ingestion of honey containing *Cl. botulinim* spores, and honey should not be fed to infants.³³

ACTINOMYCES AND NOCARDIA SPECIES

Objectives

1. List three diseases that can be caused by *Actinomyces* species, and three that can be caused by *Nocardia* species.

2. Describe similarities and differences between actinomyces and nocardia which are important in their recognition in tissue sections or by culture.

3. Discuss differences in treatment for these two genera.

Actinomyces

Actinomyces species are anaerobic gram-positive rods. The name means "ray-fungi". However, these organisms are classified as bacteria because they contain cell wall constituents (such as muramic acid), and like bacteria, are susceptible to antibiotics.[61] *A. bovis* produces "lumpy jaw" in cattle. *A. israelii*, named for the bacteriologist Israel, is the usual cause of actinomycosis in humans. Many other species such as *A. viscosus*, which is often found in dental plaque, occasionally produce disease in humans.[61]

Actinomycosis is rare in humans, although 24 culturally confirmed cases were observed in one referral center in a period of 24 years.[59] *Cervical adenitis* is the most characteristic form of actinomycosis. Usually there is a firm tender mass in the submandibular area, which often results in draining sinuses. Rarely, the disease extends to the mandible, and produces osteomyelitis. Typically, the center of the mass is black and necrotic.

Focal chronic pneumonia is rarely caused by *Actinomyces* species. The disease can extend to produce empyema, lung abscess, or draining sinuses of the chest wall. Chronic draining sinuses of the abdomen can occur, especially complicating operative or traumatic abdominal wounds.

Brain abscess is rarely due to actinomyces. Definitive cure is possible, using systemic antibiotics and excision or local antibiotics.[67]

The diagnosis can often be suspected by finding "sulfur granules" in pus. These granules do not contain sulfur, but are composed of masses of bacterial filaments, protein, and polysaccharides. A confirmed laboratory diagnosis is based on culture of the organisms, using strict anaerobic conditions.

Poor oral hygiene is generally regarded as a possible predisposing factor in human actinomycosis.

Treatment consists of excision of the diseased tissue, if possible, and penicillin.

Nocardia

Nocardia are bacteria named after Nocard, who recovered the organism from chronic skin and lymph node abscesses in cattle. *Nocardia* is the aerobic genus of the *Actinomycetaceae* family, in which *actinomyces* is the anaerobic genus. *Nocardia asteroides* is the species that causes most systemic nocardial infections, although other species can cause human disease.

Nocardia resemble the tubercle bacillus and other mycobacteria in several ways. Nocardia are weakly acid-fast, chromogenic (often red or orange), and are antigenically related to the mycobacteria.[64] Nocardiosis is uncommon in the United States, although it has been estimated that between 500 and 1000 cases occur each year.[57] It is most frequently found in immunosuppressed patients.

Nocardia species are a rare cause of several diseases.[60,63] Nocardial *pneumonia* tends to be chronic, producing nodules, cavities, or purulent pleural effusion, especially in compromised hosts treated with corticosteroids or immunosuppressive agents.[56] *Pulmonary alveolar proteinosis* may occur as an associated disease, and presumably the nocardiosis is a complication of the proteinosis. *Fulminating disseminated infection* can occur in patients receiving corticosteroids or immunosuppressive agents, but may occur in apparently normal individuals. *Brain abscess* is rarely due to a *Nocardia* species, and can complicate penetrating head injuries.[65]

Gram stain of pus reveals gram-positive coccal, rod, or filamentous forms which are often confused with diphtheroids. Acid-fast stain may reveal bacilli difficult to distinguish from tubercle bacilli.

Nocardia grow well on ordinary antibiotic-free media for fungi. Colonies may appear in a few days, but sometimes do not appear until 1 to 3 weeks. When the organism is recovered from sputum, a fungal blood culture should be obtained to look for hematogenous dissemination.[66]

The diagnosis is often made by biopsy of the infected area. Granulomatous lesions resembling those of tuberculosis are produced. Gram stain of the tissue section is essential to be sure the filamentous forms are not overlooked. Sulfur granules are occasionally found in tissue, but not frequently as in actinomycosis.

The reservoir of nocardia appears to be soil and plants, rather than humans or animals, although naturally occurring infections of animals do oc-

13. Monif, G.R.G., and Baer, H.: *Haemophilus (Corynebacterium) vaginalis*, septicemia. Am. J. Obstet. Gynecol., *120:*1041–1045, 1974.
14. Moore, M.S., and Parson, E.I.: A study of a modified Tinsdale's medium for the primary isolation of *Corynebacterium diphtheriae*. J. Infect. Dis., *102:*88–93, 1958.
15. Pappenheimer, A.M., Jr., and Gill, D.M.: Diphtheria. Science, *182:*353–358, 1973.
16. Reid, J.D., and Greenwood, L.: Corynebacterial endocarditis. A report of two cases with review. Arch. Intern. Med., *119:*106–110, 1967.
17. Smith, R.F., Rodgers, H.A., Hines, P.A., and Roy, R.M.: Comparisons between direct microscopic and cultural methods for recognition of *Corynebacterium vaginale* in women with vaginitis. J. Clin. Microbiol., *5:*268–272, 1977.
18. Van Scoy, R.E., Cohen, S.N., Geraci, J.E., and Washington, J.A., II: Coryneform bacterial endocarditis. Difficulties in diagnosis and treatment, presentation of three cases, and review of literature. Mayo Clin. Proc., *52:*216–219, 1977.

Listeria Monocytogenes

19. Azimi, P.H., and Cramblett, H.G.: *Listeria monocytogenes* infection in newborn siblings. Am. J. Dis. Child., *131:*398–399, 1977.
20. Bayer, A.S., Chow, A.W., and Guze, L.B.: *Listeria monocytogenes* endocarditis: report of a case and review of the literature. Am. J. Med. Sci., *273:*319–323, 1977.
21. Filice, G.A., et al.: *Listeria monocytogenes* infection in neonates: investigation of an epidemic. J. Infect. Dis., *138:*17–23, 1978.
22. Gray, M.L., and Killinger, A.H.: *Listeria monocytogenes* and listeric infections. Bacteriol. Rev., *30:*309–382, 1966.
23. Lavetter, A., et al.: Meningitis due to *Listeria monocytogenes*. New Engl. J. Med., *285:*598–603, 1971.
24. Louria, D.B., et al.: Listeriosis complicating malignant disease: a new association. Ann. Intern. Med., *67:*261–281, 1967.
25. Mohan, K., et al.: Synergism of penicillin and gentamicin against *Listeria monocytogenes* in ex vivo hemodialysis culture. J. Infect. Dis., *135:*51–53, 1977.
26. Ray, C.G., and Wedgwood, R.J.: Neonatal listeriosis. Pediatrics, *34:*378–392, 1964.
27. Shackelford, P.G., and Feigin, R.D.: *Listeria* revisited. Am. J. Dis. Child., *131:*391–392, 1977.
28. Visintine, A.M., Oleske, J.M., and Nahmias, A.J.: *Listeria monocytogenes* infection in infants and children. Am. J. Dis. Child., *131:*393–397, 1977.
29. Wiggins, G.L., Albritton, W.L., and Feeley, J.C.: Antibiotic susceptibility of clinical isolates of *Listeria monocytogenes*. Antimicrob. Agents Chemother., *13:*854–860, 1978.

Clostridium Species

30. Altemeier, W.A. and Fullen, W.D.: Prevention and treatment of gas gangrene. J.A.M.A., *217:*806–813, 1971.
31. Anwar, A.A. and Turner, T.B.: Antibiotics in experimental tetanus: in vitro and in vivo studies. Johns Hopkins Med. J. *98:*85–101, 1956.
32. Arnon, S.S., et al: Infant botulism. Epidemiological, clinical, and laboratory aspects. J.A.M.A., *237:*1946–1951, 1977.
33. Arnon, S.S., et al.: Honey and other environmental risk factors for infant botulism. J. Pediatr., *94:*331–336, 1979.
34. Bartlett, J.G., et al.: Role of *Clostridium difficile* in antibiotic-associated pseudomembraneous colitis. Gastroenterology, *75:*778–782, 1978.
35. Bornstein, D.L., Weinberg, A.N., Swartz, M.N. and Kunz, L.J.: Anaerobic infections—a review of current experience. Medicine, *43:*207–232, 1964.
36. Center for Disease Control: Morbidity Mortality Weekly Rep., 25–27: Annual supplements. Summary, 1976–1978.
37. Davenport, O.W. and Oken, D.E.: Pathologic physiology and treatment of postabortal *Clostridium welchii* sepsis: report of survival of patient with massive hemolysis and anuria. Am. J. Obstet. Gyn., *80:*512–515, 1960.
38. Dische, F.E. and Elek, S.D.: Experimental food poisoning by *Clostridium welchii*. Lancet, *2:*71–74, 1957.
39. Donadio, J.A., Gangarosa, E.J. and Faich, G.A.: Diagnosis and treatment of botulism. J. Infect. Dis., *124:*108–112, 1971.
40. Dowell, V.R., et al.: Coproexamination for botulinal toxin and *Clostridium botulinum*. A new procedure for laboratory diagnosis of botulism. J.A.M.A., *238:*1829–1832, 1977.
41. Gorbach, S.L. and Thadepalli, H.: Isolation of *Clostridium* in human infections: evaluation of 114 cases. J. Infect. Dis., *131* (Suppl.):s81–s85, 1975.
42. Kao, I., Drachman, D.B. and Price, D.L.: Botulinum toxin: mechanism of presynaptic blockade. Science, *193:*1256–1258, 1976.
43. Laurence, D.R. and Webster, R.A.: Pathologic physiology, pharmacology, therapeutics of tetanus. Clin. Pharmacol. Ther., *4:*36–72, 1963.
44. Lowenstein, M.S.: Epidemiology of *Clostridium perfringens* food poisoning. New Engl. J. Med., *286:*1026–1028, 1972.
45. McDonel, J.L. and Duncan, C.L.: Regional localization of activity of *Clostridium perfringens* Type A enterotoxin in the rabbit ileum,

jejunum, and duodenum. J. Infect. Dis., 136:661–666, 1977.

46. MacLennan, J.D.: The histotoxic clostridial infections of man. Bacteriol. Rev., 26:177–276, 1962.

47. Merson, M.H. and Dowell, V.R., Jr.: Epidemiologic, clinical, and laboratory aspects of wound botulism. New Engl. J. Med., 289:1005–1010, 1973.

48. Oh, S.J.: Botulism: electrophysiological studies. Ann. Neurol., 1:481–485, 1976.

49. Petty, C.S.: Botulism: the disease and the toxin. Am. J. Med. Sci., 249:345–359, 1965.

50. Pyrtek, L.J. and Bartus, S.H.: *Clostridium welchii* infection complicating biliary tract surgery. New Engl. J. Med., 266:689–693, 1962.

51. Ramsay, A.M.: The significance of *Clostridium welchii* in the cervical swab and bloodstream in postpartum and postabortum sepsis. J. Obstet. Gy., 56:247–258, 1949.

52. Ratbun, H.K.: Clostridial bacteremia without hemolysis. Arch. Intern. Med., 86:496–501, 1968.

53. Rifkin, G.D., Fekety, F.R., Silva, J. and Sack, R.B.: Antibiotic-induced colitis. Implication of a toxin neutralised by *Clostridium sordelli* antitoxin. Lancet, 2:1103–1106, 1977.

54. Schwartzman, J.D., Reller, L.D. and Wang, W-L.L.: Susceptibility of *Clostridium perfringens* isolated from human infections to twenty antibiotics. Antimicrob. Agents Chemother., 11:695–697, 1977.

55. Thomas, M., et al.: Hospital outbreak of *Clostridium perfringens* food poisoning. Lancet, 1:1046–1048, 1977.

Actinomyces and Nocardia Species

56. Balikian, J.P., Herman, P.G., and Kopit, S.: Pulmonary nocardiosis. Radiology, 126:569–573, 1978.

57. Beaman, B.L., Burside, J., Edwards, B., and Causey, W.: Nocardial infections in the United States, 1972–1974. J. Infect. Dis., 134:286–289, 1976.

58. Cook, F.V., and Farrar, W.E., Jr.: Treatment of *Nocardia asteroides* infection with trimethoprim-sulfamethoxazole. South Med. J., 71:512–515, 1978.

58. Cox, F., and Hughes, W.T.: Contagiousness and other aspects of nocardiosis in the compromised host. Pediatrics, 55:135–138, 1975.

59. Eastridge, C.E., et al.: Actinomycosis: a 24-year experience. South. Med. J., 65:839–843, 1972.

60. Idriss, Z.H., Cunningham, R.J., and Wilfert, C.M.: Nocardiosis: report of three cases and review of the literature. Pediatrics, 55:479–484, 1975.

61. Larsen, J., Bottone, E.J., Dikman, S., and Saphir, R.: Cervicofacial *Actinomyces viscosus* infection. J. Pediatr., 93:797–801, 1978.

62. Maderazo, E.G., and Quintiliani, R.: Treatment of nocardial infection with trimethoprim and sulfamethoxazole. Am. J. Med., 57:671–675, 1974.

63. Murray, J.F., Finegold, S.M., Froman, S., and Will, D.W.: The changing spectrum of nocardiosis. A review and presentation of nine cases. Am. Rev. Respir. Dis., 83:315–330, 1961.

64. Palmer, D.L., Harvey, R.L., and Wheeler, J.K.: Diagnostic and therapeutic considerations in *Nocardia asteroides* infection. Medicine, 53:391–401, 1974.

65. Portetz, D.M., Smith, M.N., and Park, C.H.: Intracranial suppuration secondary to trauma. Infection with *Nocardia asteroides*. J.A.M.A., 232:730–731, 1975.

66. Roberts, G.D., Brewer, N.S., and Hermans, P.E.: Diagnosis of nocardiosis by blood culture. Mayo Clin. Proc., 49:293–296, 1974.

67. Teng, P.: Actinomycotic cerebral abscess. A report of two cases with recovery. J.A.M.A., 175:807–810, 1961.

Bacillus Species

68. Brachman, P.S.: Anthrax. Ann. N.Y. Acad. Sci., 174:577–582, 1970.

69. Farrar, W.E., Jr.: Serious infections due to "non-pathogenic" organisms of the genus *Bacillus*. Review of their status as pathogens. Am. J. Med., 34:134–141, 1963.

70. Gold, H.: Anthrax. A report of one hundred seventy cases. Arch. Intern. Med., 96:387–396, 1958.

71. Ihde, D.C. and Armstrong, D.: Clinical spectrum of infection due to *Bacillus* species. Am. J. Med., 55:839–845, 1973.

72. LaForce, F.M., et al.: Epidemiologic study of a fatal case of inhalation anthrax. Arch. Environ. Health, 18:798–805, 1969.

73. Melling, J., Capel, B.J., Turnbull, P.C.B. and Gilbert, R.J.: Identification of a novel enterotoxogenic activity associated with *Bacillus cereus*. J. Clin. Pathol., 29:938–940, 1976.

74. Portnoy, B.L., Goepfert, J.M. and Harmon, S.M.: An outbreak of *Bacillus cereus* food poisoning resulting from contaminated vegetable sprouts. Am. J. Epidemiol., 103:589–594, 1976.

75. Terranova, W., and Blake, P.A.: *Bacillus cereus* food poisoning. New Engl. J. Med., 298:143–144, 1978.

76. Turnbull, P.C.B., French, T.A. and Dowsett, E.G.: Severe systemic and pyogenic infections with *Bacillus cereus*. Br. Med. J., 1:1628–1629, 1977.

Enteric Gram-Negative Rods

GENERAL CLASSIFICATION

Definitions

Enteric bacteria are defined as any bacteria often recoverable from the intestinal tract. Enteric gram-negative rods are defined here as including all nonfastidious gram-negative rods, which are sometimes recovered from the bowel. *Enterobacteriaceae* is a large family of enteric bacteria. The term enteric pathogen usually refers to salmonellae or shigellae, although many other bacteria (especially diarrheagenic *E. coli*) can cause diarrhea.

Non-lactose fermenter is used to describe enteric rods that do not ferment lactose, particularly on the primary isolation plate. *Shigella* and *Salmonella* species are the important non-lactose fermenters, but *Proteus* species and some other normal flora of the bowel also do not ferment lactose.

Coliform refers to lactose fermenters that resemble *Escherichia coli*, including *Klebsiella* and *Enterobacter* species. Paracolon is an obsolete term formerly used to refer to late lactose fermenters. These organisms ferment lactose slowly over several days and include some species of *E. coli*, and genera usually called *Enterobacter, Hafniae, Serratia, Providencia, Arizona, Citrobacter,* and *Edwardsiella*.

Nonfermenter is a term that should not be confused with non-lactose fermenter. Nonfermenter refers to a group of organisms that do not *ferment* carbohydrates, but either *oxidize* or *fail to utilize* carbohydrates (see p. 60).

Classification

Classifications of enteric bacteria change periodically, but the genus names have remained fairly constant. The fundamental properties used to classify microorganisms, such as polar flagellation in pseudomonads, are of little practical value to the clinician. Table 3-1 shows the simplified clinical classification used in this section, which follows most microbiological classifications in simplified terms.

Frequency and Importance

The space allotted here for the various enteric gram-negative rods is roughly proportional to the clinical importance of the organisms. Organisms discussed in a short paragraph, rather than having their own section are of minimal clinical importance and are included only for reference. Some uncommon enteric bacteria are discussed in Chapter 6, Miscellaneous Bacteria (see Index).

Diseases

Enteric gram-negative rods can produce a variety of diseases, discussed here in general terms, with more detail given under the headings for individual organisms.

Diarrhea. *Shigella* and *Salmonella* species produce diarrhea by invasion of the bowel mucosa. *Vibrio cholerae* produces diarrhea by the

42 ENTERIC GRAM-NEGATIVE RODS

TABLE 3–1. CLASSIFICATION AND SYNONYMS FOR ENTERIC GRAM-NEGATIVE RODS

ENTEROBACTERIACEAE FAMILY
    Escherichia—Citrobacter—Shigella
    Salmonella—Arizona—Edwardsiella
    Klebsiella—Enterobacter (formerly Aerobacter)—Serratia
    Proteus—Providencia

PSEUDOMONADS
    Pseudomonas aeruginosa
    Pseudomonas species

VIBRIO GROUP
    Vibrio cholerae—Vibrio parahemolyticus—other vibrios
    Campylobacter fetus (Vibrio fetus) subspecies *jejunii* or *intestinalis*

NONFERMENTER GROUP
    Achromobacter
    Acinetobacter calcoaceticus var. anitratus (Herellea vaginicola)
    Acinetobacter calcoaceticus var. lwoffi (Mima polymorpha)
    Alcaligenes
    Bordetella bronchoseptica
    Flavobacterium
    Moraxella

BACTEROIDES FAMILY
    Bacteroides
    Fusobacterium
    Leptotricha

mechanism of an enterotoxin. Some serotypes of *E. coli* produce diarrhea by invasion; others, by an enterotoxin, as discussed in the section on *E. coli*.

Urinary Tract Infections. Any enteric bacteria can cause urinary infections. The most abundant aerobic enteric bacteria are also the most frequent cause of urinary infections. Most urinary infections occur in females, and are caused by aerobic enteric bacteria colonizing the perineum and periurethral area.

Intra-Abdominal Infections. These infections are usually the result of perforation of the bowel—as from a ruptured appendix—and usually begin as mixed infections containing *E. coli*, bacteroides, and other enteric bacteria.

Septicemia and Meningitis. These serious infections from enteric bacteria typically occur in a compromised host, such as a newborn infant, or an older person who is immunosuppressed or who has a malignancy.

Laboratory Identification of Enteric Bacteria

Cultures of feces or a rectal swabbing are first plated on a selective medium—such as EMB (eosin-methylene blue), S–S (Shigella–Salmonella) or MacConkey's agar—which inhibit bacteria other than enteric gram-negative rods. Typically, these media contain substrates and indicators that indicate H_2S production or lactose fermentation.

Lactose Fermentation. *Shigella* and *Salmonella* species are non-lactose fermenters (NLF). Thus, colonies that do not ferment lactose and change the indicator are easily recognized and picked for further study. (*Proteus* species also are non-lactose fermenters.)

Hydrogen Sulfide Production. Any bacteria that produce H_2S appear as black colonies as on media containing iron sulfide as a substrate. *Salmonella* species typically appear as black colonies, but some *Proteus* and *Citrobacter* species also produce H_2S.

Triple Sugar Iron (TSI) Agar. This agar contains lactose, sucrose, glucose, and iron sulfide and is classically used as the first test of a colony that produces H_2S or is a non-lactose fermenter on a primary isolation plate.

Other Tests. Approximately 10 to 30 other tests might be done in a clinical laboratory to identify enteric bacteria. These include testing the ability of the isolate to ferment or oxidize a variety of sugars, production of gas (CO_2), production of enzymes or utilization of substrates such as nitrate or citrate. Motility, pigment production, and growth at various temperatures may also be considered. Production of various products, which indicate a particular metabolic pathways, may also be used. For example, the Voges–Proskauer test detects bacteria that produce acetyl methyl carbinol, such as klebsiella and enterobacter, and distinguishes them from *E. coli*.

SALMONELLA

Objectives

1. List some of the most common sources of salmonella gastroenteritis in humans.
2. Discuss the evidence that antibiotic therapy may not be indicated in mild salmonella illnesses.
3. Explain why it takes several days to identify salmonellae in the bacteriology laboratory.
4. Describe the typical clinical pattern of typhoid fever.

Species and Types

The genus *Salmonella* was named after Daniel Salmon (Săm ŭn), an American veterinarian who first isolated *Salmonella choleraesuis* from hogs with diarrhea. At one time, salmonellae were classified into over 600 species, usually named after a geographic area, such as *S. heidelberg*, *S. montevideo*, and *S. newport*. Recently, many of these serologically distinct strains have been classified as serotypes of the species *S. enteritidis*. For example, *Salmonella enteritidis*, serotype *typhimurium*, is the most common serotype in the United States.

In addition to *S. enteritidis*, the other major species are *S. typhi*, the cause of typhoid fever; *S. paratyphi*, a species with three serotypes that cause paratyphoid fever; and *S. choleraesuis*, an important cause of salmonella bacteremia.

The Arizona group of bacteria are also sometimes classified as *Salmonella* species. These organisms usually cause disease in reptiles, but occasionally infect humans and produce and same clinical manifestations as salmonellae.[15]

Salmonellosis is defined as any disease caused by a salmonella. *S. typhi* also has been called *Eberthella typhosa*, after Eberth, who described the organism.

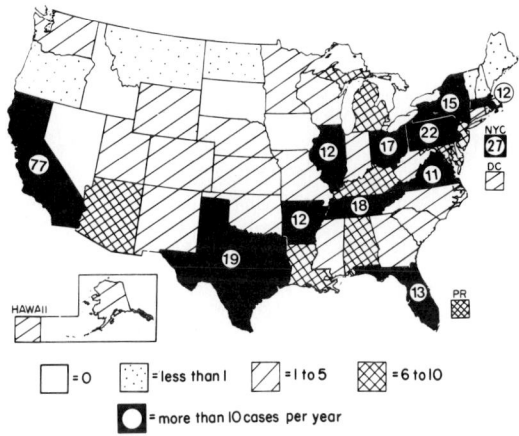

FIG. 3-1. *Typhoid fever: average annual reported cases, 1970–1972.*[4]

Frequency and Importance

Salmonella species are important because they are the most frequent cause of food-borne diarrhea in the United States. About 20,000 cases of salmonellosis are reported each year, but probably the real frequency is 100 times as great.[2]

About 200 new cases of typhoid fever are reported each year in the United States, although the exact number is unknown since reports often do not specify if the patient is a new case. Each new case is important because it usually stimulates a considerable public health effort to identify the source. If typhoid carriers with positive cultures are included, about 400 cases are reported annually in the United States (Fig. 3-1).

About 65 percent of salmonella isolations are from individuals less than 20 years of age (Fig. 3-2). About half of the fatal cases in salmonella epidemics occur in individuals less than 1 year of age, or more than 60 years of age.[2]

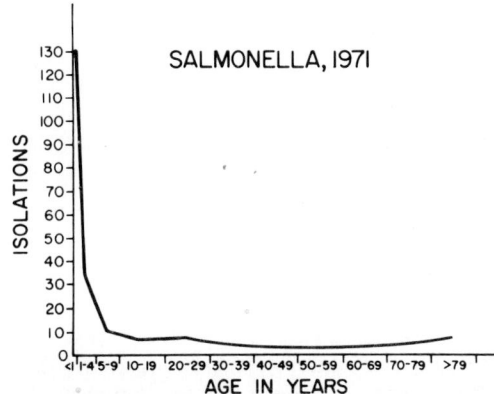

FIG. 3-2. *Salmonella: age frequency.*[5]

Clinical Patterns of Illness

Salmonella species are an *occasional* cause of:

Acute Diarrhea. Fever is usually present in these cases, and the stool is sometimes grossly bloody. This is the most frequent illness due to salmonella, and is most severe in infants, young children, and debilitated elderly patients.

Fever Without Localizing Signs. This pattern is also called the enteric fever or typhoidal pattern. However, the diagnosis of fever without localizing signs is a more likely preliminary problem-oriented diagnosis than is enteric fever or septicemic salmonellosis. Headache is prominent in salmonella enteric fever, just as it is in typhoid fever.

The prototype example of the enteric fever pattern of salmonella infection is typhoid fever. In typhoid fever, diarrhea is often absent or not prominent, but abdominal tenderness and distension may be present.[7,8] Headache and delirium are prominent. Typhoid means like typhus, which means cloud, referring to clouded sensorium. The white blood count is usually 5000 to 10,000; but about 20 to 40 percent band forms are often found in the differential.[7] Massive gastrointestinal hemorrhage is a rare complication of typhoid fever.[31]

The same pattern of fever without other signs can also be caused by *S. paratyphi*, and is called paratyphoid fever.[20] About 20 to 60 percent of patients with paratyphoid fever have diarrhea.[20]

Salmonella species are a rare cause of meningitis, especially in the newborn;[3,22] osteomyelitis, especially complicating sickle cell anemia;[3] mycotic aneurysm or arteritis, especially in atherosclerotic arteries;[6] and abscesses, especially in the rectal area.

Asymptomatic Infections

Unsuspected carriers are sometimes discovered on routine cultures, as for employment as food handlers. Presumably these individuals have had minor episodes of diarrhea due to salmonella in the past.

Experimental Infections

Experimental Typhoid Fever. Human volunteers have provided useful information, particularly with respect to pathogenesis, treatment, and the effects of vaccines. The median incubation period was about 7 days, but ranged from 3 to 56 days. Stools were often positive for the organism during the incubation period, but were usually not positive during the first week of the febrile illness.

Diagnostic Approach

Culture. Recovery of the organism is the most reliable method of laboratory diagnosis. About 24 hours incubation are required to detect suspicious colonies on primary isolation. These suspicious colonies, which show hydrogen sulfide production or failure to ferment lactose, are subcultured for further tests. Thus, another 24 hours are necessary to demonstrate the definitive metabolic characteristics of salmonellae.

After the isolate has been shown to have metabolic characteristics compatible with salmonella, a saline suspension of the organism is put on a glass slide and tested for agglutination with several salmonella group antisera.

Once the isolate has been grouped in the local hospital laboratory, it is usually sent to a reference laboratory, where a large battery of reference antisera can be used to identify the species and serotype. Further typing can be done using bacteriophage.[13] This may be useful in epidemiologic studies of *S. typhimurium*, a serotype so common it is often difficult to trace its source.

Lactose-positive salmonellae are rare, but have been observed.

Serum Antibodies. Commercial salmonella antigens are available and can be tested against the patient's serum. However, many individuals have moderate titers of antibodies to these antigens, and often no increase in antibodies is detected after diarrhea with a rectal culture positive for salmonellae. Therefore, this laboratory test is not reliable and probably should be abandoned.

Coagglutination. Suspected colonies of salmonellae can be presumptively identified on the original isolation plate using the technique of coagglutination. Specific antisalmonella antiserum is added to formaldehyde and heat-treated protein A-containing staphylococci and this suspension is added to the suspect colony. Coagglutination of the staphylococci–antiserum–salmonella mixture is recognized under low power of the microscope (Fig. 1-12).[10]

Biologic Characteristics of Clinical Interest

Antigenic Structure. The complex antigenic structure of salmonellae allows elaborate serotyping of isolates, and may aid epidemiologists in tracing outbreaks to sources. However, there is little need for the clinician to memorize the location or significance of the H, O, and Vi antigens.

Transferable Drug Resistance. Resistance to antibiotics due to a transferable R factor (a plasmid) can occur in salmonellae. Antibiotic therapy of a patient clearly increases the chances that plasmid-mediated antibiotic resistance will occur in the salmonella recovered from that patient.[28] This is discussed further in this chapter in the section on shigella.

Carrier State. The chronic carrier state is usually defined for salmonella as excretion of the organism for more than a year after the illness. Most individuals have a convalescent carrier state for several weeks to a few months after the illness.
 1. *Typhoid carriers.* It is useful to attempt to eradicate the chronic carrier state for S. typhi because this organism is found only in humans, who are the direct or indirect source of other human infections. Intensive parenteral ampicillin therapy is sometimes effective, but typhoid carriers often have a focus of chronic infection in the gallbladder, so that cholecystectomy is sometimes necessary.[29]
 2. *Other salmonella carriers.* Family members or contacts of a patient with salmonellosis may be found to be excreting the same serotype, but it is likely that they were infected by the same food source as the patient, rather than being the source of the patient's infection. Such carrier contacts should be regarded as convalescent carriers, and usually should not be treated.
 Chronic salmonella carriers are probably of little risk to others, except in situations of poor hygiene, so attempts to eradicate the carrier state for *Salmonella* species other than typhoid are rarely indicated. In one study, severely debilitated individuals who were carriers of *Salmonella derby* for 2 to 11 months apparently had no complications related to the carrier state.[19] However, for food handlers or individuals who may expose infants or debilitated individuals, a course of ampicillin may be indicated.[12]

Animal Reservoirs. Poultry are especially important sources of salmonella. Some species are named after their principle animal host. For example, S. typhimurium (typhus of mice), which produces severe natural disease in mice, is the most frequently recovered serotype in humans. S. choleraesuis (cholera of swine), once thought to be the cause of hog cholera, is a frequent cause of the septicemic form of salmonellosis in humans.

Other Sources. Salmonella infections in man have been traced to:
 1. *Commercially-solid foods,* particularly egg and meat products, such as eggnog, nonfat dry milk, imitation ice cream, dried eggs, cake mixes, candy bars, beef jerky, barbecued chicken or turkey, canned corn beef, and other prepared meat products. The majority of epidemics have been due to poultry and poultry-related products.[2] Other sources have included apple cider and rare roast beef.
 The economic impact of food-borne salmonellosis is very great and justifies vigorous preventive measures.[18]
 2. *Pet animals,* particularly turtles (even turtles certified as salmonella-free), parakeets, horses, dogs, and cats.
 3. *Animal products* not used for human foods, particularly plant foods (containing bone meal) and enzymatic drain cleaners.
 4. *Medical products,* such as dried yeast, vitamin preparations, and carmine red.
 5. *Water* was the apparent source of the 1973 outbreak of typhoid fever in Florida.
 6. *Human carriers* are an occasional source of salmonella other than typhoid, but are the only source of typhoid fever. Human milk was shown to be the source of an outbreak of salmonellosis in newborn infants.[26]
 7. *Foreign travel* is an important source of typhoid fever. From 1970 through 1976, about 30 to 50 percent of cases of typhoid fever reported in the United States were apparently acquired during foreign travel, especially Mexico.[23]

Treatment

Antibiotic Therapy.
 1. *Typhoid fever* should be treated with either ampicillin or chloramphenicol.[24] An outbreak of typhoid fever in Mexico in 1972 was due to chloramphenicol-resistant S. typhi, and ampicil-

lin resistance was also occasionally noted, so different therapy may be needed. The combination of trimethoprim and sulfamethoxazole (co-trimoxazole) appeared to be as effective as chloramphenicol in one study of typhoid fever,[27] as effective as amoxicillin in another study,[14] and may be useful for chloramphenicol-resistant, ampicillin-resistant strains.

In another study, oral chloramphenicol was more effective in reducing the duration of fever than parenteral ampicillin or oral co-trimoxazole.

2. *Acute salmonella diarrhea* in young infants or debilitated patients should probably be treated with ampicillin. This area is somewhat controversial since no large prospective study has been done comparing specific antibiotic therapy with no antibiotic therapy. However, one study of shigellosis included a few patients with salmonellosis and showed that ampicillin was more effective than a placebo in the few patients with salmonellosis.[17]

Several retrospective studies have indicated that antibiotic therapy results in somewhat longer duration of excretion of salmonellae. In such studies it is not clear to what extent the clinical severity of the illness influenced the decision to seek medical aid or receive antibiotic therapy. An outbreak of S. typhimurium in school children in Suffolk, England in 1954 treated with antibiotics was compared with an outbreak in South Wales in 1964 in which no antibiotics were given.[9] A larger proportion of the Suffolk children excreted the organism for a longer time, but neither the treated nor the untreated group excreted the organism after 18 weeks. Sporadic cases of salmonella diarrhea in Rhode Island in 1964–1965 were reviewed and the antibiotic-treated cases were compared to the non-antibiotic treated cases.[25] The patients treated with ampicillin excreted the organism for about 4 weeks longer than the untreated patients.

An outbreak of S. typhimurium occurred at a town festival and turkey barbecue in Nebraska in 1967 due to inadequately cooked, salmonella-contaminated turkey rolls.[21] This outbreak involved about 1900 persons, many of whom were treated with ampicillin or chloramphenicol.[1] Follow-up cultures of treated patients were compared with untreated individuals, and the duration of excretion of salmonella was somewhat longer in antibiotic treated patients. Antibiotic-resistant strains with resistance mediated by transfer factor developed in the antibiotic treated patients, but not in the untreated patients.[1]

Corticosteroids. Cortisone has been used along with antibiotics to treat typhoid fever for brief periods (less than three days) with dramatic improvement, and without any increased frequency of intestinal perforation.[32]

Complications

Perforation of the bowel is a rare complication. Encephalopathy has been reported in children, but also appears to be rare.[33] Febrile convulsions occasionally occur.

Prevention

Avoid Exposure.
1. *Public caution* in avoiding likely sources of salmonella, such as turtles, dried eggs, dried milk, and undercooked poultry is useful.
2. *Food production standards* can probably be improved, especially with foods of known high risk, such as dried eggs or dried milk. Addition of a single contaminated unit of meat, milk, or eggs to a large pool of otherwise uncontaminated food results in contamination of the entire pool, a situation which can probably be avoided.
3. *Isolation of hospital patients with diarrhea* should be done, especially in special risk areas such as newborn nurseries, and wards with infants with other chronic diseases. If an obstetrical patient has diarrhea, her newborn infant should be isolated and segregated from other newborn infants.

Immunization. Typhoid vaccine is rarely indicated in the United States. Paratyphoid vaccines are not effective. Other salmonella vaccines are not available.

SHIGELLA

Objectives

1. Define dysentery and describe the typical illness of shigella dysentery.
2. Explain why identification of shigellae in the laboratory may take 2 or 3 days.
3. Discuss the therapy of shigellosis, including resistant strains.

Definitions, Species and Types

Shigella is a genus named for the Japanese bacteriologist Shiga, who demonstrated specific agglutinating antibodies against the organism in the blood of patients who recently had dysentery. Diarrhea caused by a *Shigella* species is called shigellosis or bacillary dysentery, to distinguish it from amebic dysentery.

The species Shiga studied produces the most severe disease, and is now known as *Shigella dysenteriae* type 1, or the Shiga bacillus. The other three species of Shigella are named after the individuals who studied these organisms: Flexner, Sonne, and Boyd.

The four species of *Shigella* can be further subdivided into various types. *S. flexneri* is typed on the basis of serologic agglutinations, for example, *S. flexneri* type 2a. *S. sonnei* can be subdivided on the basis of colicin typing (see bacteriocins, under pseudomonas), for example, *S. sonnei* type 4.[48] Such typing of Shigella isolates is done in reference laboratories, and may be useful in tracing epidemics, recognizing common sources, and identifying resistant strains.[48]

Some variants of *E. coli* have many of the metabolic characteristics of shigella and often cross-react with shigella antisera. These organisms do not cause diarrhea, and are occasionally found in asymptomatic normal individuals. They were once called *Alkalescens dispar* or *Shigella dispar*.

Frequency and Importance

About 15,000 cases of shigellosis are reported annually in the United States.[61] In severe cases, especially in small children, the patient may die if fluid losses are not replaced.

Most shigellosis in the United States is due to *S. flexneri* or *S sonnei*.[61] *S. dysenteriae* type 1, the Shiga bacillus, produces more severe disease, and was the cause of an epidemic in Central America in 1969.[56] The number of cases of Shiga dysentery occurring in United States travelers to Mexico and Central America has recently increased to about 70 per year by 1972, with evidence of rare spread of *S. dysenteriae* within the United States.[35]

The age distribution of shigellosis indicates it is most frequent in children and young adults, with about 50 percent of cases occurring in children 4 years of age or younger (Figure 3-3).

Shigellosis typically occurs in individuals ex-

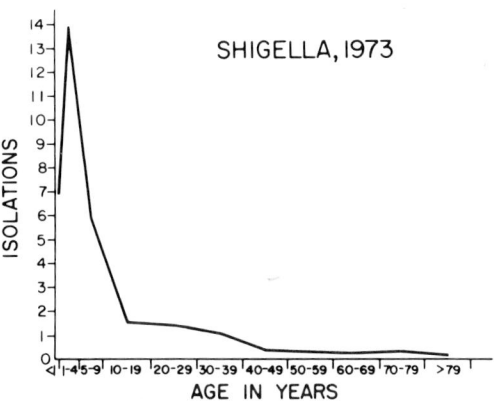

FIG. 3-3. *Shigellosis: age frequency.*[35]

posed to poor sanitary conditions. Outbreaks of shigellosis occur frequently in institutions for the mentally retarded, Native American reservations, day-care centers and in other communities or groups with crowding or poor sanitation. Shigellosis also occurs in individuals in upper income groups who travel to foreign areas with poor sanitation, such as Mexico. Other sources include swimming in contaminated rivers,[60] and sexual transmission between male homosexuals. Staff members have been implicated in the transmission of shigellosis in a custodial institution.[68]

Clinical Patterns of Illness

Shigella species are a *frequent* cause of:

Dysentery-like Diarrhea. Classical shigellosis is manifested by fever and dysentery (diarrhea with blood, pus, and mucous in the stools). In the United States, dysentery-like diarrhea is usually due to shigellae.

Shigella species are an *occasional* cause of:

Acute Gastroenteritis. Shigellosis is not often associated with a typical dysentery. Other clinical features that suggest shigellosis include diarrhea in the household, high fever, and lax anal sphincter tone. The patient is often seen by physician soon after the onset, and typically has little or no improvement if untreated.[57] Vomiting may be present early. Abdominal pain and fever may occur one or two days before the onset of the diarrhea. Appendicitis is occasionally suspected because of the abdominal pain. Lumbar puncture is sometimes done in patients with shigellosis when the onset of diarrhea does not occur until after the fever and convulsions. The spinal fluid is usually normal, as discussed under complications.

Shigella species are a *rare* cause of:

Fulminating fatal diarrhea rarely may occur, particularly in children 1 to 4 years old. This pattern is called Ekiri in the Far East. The cause of death in this pattern is not clear, even with an autopsy examination.

Colitis without high fever consists of severe diarrhea with bright red blood and mucus, but without high fever. It is frequently confused with amebic dysentery or the acute onset of idiopathic ulcerative colitis. It is typical of *S. dysenteriae* type 1 (Shiga's bacillus).[56]

Asymptomatic Infection

A brief convalescent carrier state of less than one month may occur, but a more chronic carrier state is extremely unusual.[55] When the laboratory reports the recovery of suspected shigellae from an asymptomatic individual (a foodhandler, for instance) or from a urine culture, the clinician should suspect the organism is really a variant of *E. coli*.

Experimental Infections

In one study with adult prisoner volunteers in 1946, it was necessary to give 10 billion organisms[64] to produce clinical disease. This suggests that the infection may sometimes require a rather high dose of organisms in adults, who may have had previous infection. After 12 hours, headache, abdominal cramps, and vomiting occurred. Within 24 hours of ingestion, fever, toxicity, prostration, and diarrheal stools streaked with mucus were noted. With lower dosages of organisms, the incubation period was about 2 days, and the illness was milder. Nausea and vomiting were prominent early findings.

In more recent studies of adult prisoner volunteers in 1969, as few as 200 *Shigella flexneri* organisms were sufficient to produce dysentery.[39] The mean incubation period to fever was 2 days; to diarrhea, 4 days. Thus, fever preceded diarrhea by about 2 days, on the average. The duration of excretion of shigellae was as long as 78 days, with a mean of 27 days.

In a river-associated outbreak of *S. sonnei*, coliform counts in the area of swimming about 8 kilometers below a sewage treatment plant outfall were about 400,000 organisms per 100 ml.[60]

Laboratory Approach

Culture. A colony on a culture plate may be suspected as a possible shigella because it is a non-lactose fermenter, as recognized by the color of the colony. The shigella colony does not ferment the lactose in the medium, so acid is not produced, as indicated by the pH indicator color. However, many fecal organisms, such as *Proteus* species, also are non-lactose fermenters; so recognition of a non-lactose fermenter is only a preliminary test. Identification of the metabolic characteristics of shigella usually requires another 24 hours for subculture on additional test media. If these metabolic tests are compatible with shigella, a saline suspension of the isolate is placed on a glass slide and tested for agglutination with four groups of shigella and with *Shigella dispar* antisera. If the organism is agglutinated by one of the shigella group antisera, it can be sent to a reference laboratory for further serotyping.

The clinician should not hurry the laboratory into doing agglutination testing of non-lactose fermenters before other metabolic tests are done, since non-lactose fermenters of other species may cross react with the shigella antisera and produce a false positive preliminary report.

Serum Antibodies. The diagnosis of shigella infection can be confirmed in some research laboratories by demonstration of a rise in antibodies, but this procedure is ordinarily not available.

Smear. Fluorescent antibody techniques have been developed which can identify shigellae in a smear of feces, but this test is generally not available at the present time.

Nonspecific Tests. If the peripheral blood differential has more band forms than segmented neutrophils, and the patient has severe diarrhea, shigella is much more likely than other etiologies.[59] Studying a smear of the stool for leukocytes is useful to distinguish bacillary dysentery from amebic dysentery, but has not been adequately tested in acute diarrhea that is not dysentery-like diarrhea. Some reports have suggested that methylene blue smears of feces in a patient with diarrhea will help to identify shigellosis.[49] However, the clinical features of the illness, especially the gross findings of dysentery, are a better guide to whether or not antibiotics should be used before culture results are available.

Characteristics of Clinical Interest

Cell Penetration. Using hybrid shigella strains that have either cell-penetrating ability or toxin production, cell penetration can be shown to be the major factor in virulence in the toxin-producing S. dysenteriae,[54] as well as in other species.[53] This may explain the ineffectiveness of nonabsorbed oral antibiotics in therapy.

Toxins. All species produce endotoxin. S. dysenteriae produces a potent exotoxin. However, exotoxins do not appear to be important in the pathogenesis of human disease. Cell penetration (invasion) appears to be more important.

Some strains of S. flexneri and S. sonnei produce a cytotoxin which resembles that produced by S. dysenteriae type 1, so that toxin production may play a role in pathogenesis in some strains.[51]

Host Defenses. In the guinea pig, gastrointestinal hypermotility is a major factor in host defense, and the use of opiates to stop the diarrhea produces more severe disease.[43] This provides a theoretical argument against the use of opiates to relieve symptoms in human shigellosis. Furthermore, in human volunteer experiments, an opiate–atropine remedy (Lomotil) had an adverse effect on the clinical course of shigellosis.[41]

Gastric acidity is also a host defense, and administration of bicarbonate with the organism increases the severity of disease.[40]

Treatment

Ampicillin, 100 mg per kg per day by the intravenous route, is the treatment of choice for hospitalized patients with shigellosis. It is clearly more effective than sulfadiazine or a placebo.[45] Non-absorbable oral antibiotics are not effective, compared to a placebo.[46] For outpatients, a dose of 50 ampicillin mg per kg per day for 5 days is sufficient.

Ampicillin-resistant shigellae have recently become much more frequent,[63] possibly because of widespread use of ampicillin, with production of resistant strains by the mechanism of transferable resistance factor. These resistant strains respond to trimethoprim–sulfamethoxazole.[36,58]

Complications of Shigellosis

Convulsions. In one series, convulsions were observed in 11 percent of children with diarrhea due to shigellae, compared to five percent with diarrhea due to salmonellae, and four percent with diarrhea due to a nonspecific etiology.[62] The spinal fluid is almost always normal,[34,52,62] but a few patients have been reported with 10 to 400 leukocytes per cubic millimeter.

Bacteremia With a Different Enteric Organism. Enterobacter bacteremia has been reported to occur 5 or 6 days after the onset of the illness, and was associated with the new appearance of high fever, leukocytosis, and toxicity.[47]

Extra-gastrointestinal Infection.[34] Bacteremia due to shigellae is extremely rare.[66] Conjunctivitis, vaginitis, and urinary infection have been documented by culture, and probably are secondary to fecal contamination. Thrombopenia, pneumonia, arthritis, or nephritis are rare, but have been reported in some cases due to fulminating disease with bacteremia, although some of these complications may be coincidental.

Other Complications of Diarrhea. Shock, hyponatremia with encephalopathy, hemolytic–uremic syndrome, and disseminated intravascular coagulation are rare complications of shigellosis.[66]

Prevention

Avoid Exposure. Most shigellosis can be related to poor hygienic conditions, or contact with other individuals who have recently had shigellosis. Some major outbreaks have been traced to water. Travelers to areas where shigellosis is common should avoid fresh unwashed food. Isolation of hospitalized patients with shigellosis is important, and attendants should wear gloves when in contact with the patient or feces.

Prophylactic Chemotherapy. Antibiotic therapy of exposed children in the family has been advocated,[50] but is not the usual practice.

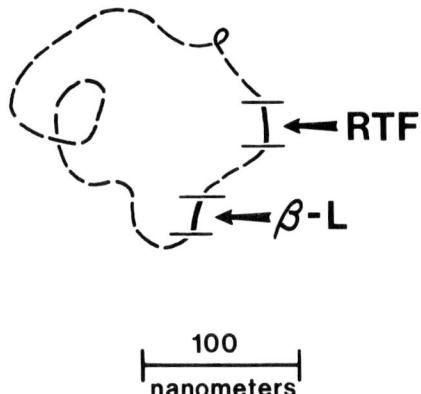

FIG. 3-4. *A plasmid as seen by electron microscopy. It contains genetic material which codes for a beta-lactamase (β-L). This plasmid also contains a resistance transfer factor (RTF), the genetic material which codes for the transfer of the plasmid.*

Immunization. Attenuated vaccines are under investigation.[65] Attenuation involves selecting mutants or creating hybrids with *E. coli*. Unfortunately, the vaccine strain may fail to adhere and colonize the recipient's intestinal tract, may fail to provide protection when the recipient is challenged with the wild strain, or may even revert to a virulent form.[65]

Transferable Antibiotic Resistance

Antibiotic resistance is said to be transferable in instances when mixing of a sensitive organism (such as *E. coli* sensitive to ampicillin) with a resistant organism (such as a shigella strain resistant to ampicillin), results in the sensitive organism becoming resistant after only a few hours in contact with the resistant strain.[67] Most transferable drug resistance is transferred by cell to cell contact (conjugation) with the transfer of nonchromosomal DNA. This DNA package is usually circular and is called by various names: episome, plasmid, or R factor. Plasmid is the most common name used (Fig. 3-4).

A plasmid, which may be relatively large, contains many different genes, each of which may mediate resistance for a different antibiotic. Large plasmids usually contain genes that code for transfer of resistance, while small plasmids do not. Thus, a multiply resistant organism can transfer resistance to several drugs to a sensitive organism via a single plasmid. "Multiply resistant" has come to mean that an organism is resistant to two or more drugs to which the species is usually susceptible.

History. Multiple drug resistance was first recognized in Japan in 1955 in shigella outbreaks, and first recognized in the United States in 1967, also in an outbreak of shigellosis.[42] Subsequent studies have clarified the genetics and bacterial physiology of transferable resistance.[37,38,61] Transferable resistance may occur in staphylococci and most enteric bacteria. Resistance to sulfonamides and most of the commonly used antibiotics has been shown to be transferrable.

Characteristics of R Factor (Resistance Plasmids). A plasmid is DNA that is found within the bacterial cell but outside the chromosomes. Large plasmids control their own replications, but replication errors often occur. Plasmids tend to be unstable in that they often disappear if there are no external factors which allow them to give the cell an advantage for survival.

The transfer of a plasmid is initiated by conjugation, which is more likely to occur between similar species. Two bacterial cells conjugate by making contact via a pilus, the synthesis of which is often controlled by the plasmid. The contact triggers replication of the plasmid, a copy of which is transferred to the recipient cell. Plasmids can infect sensitive cells rapidly. Newly resistant cells appear at a faster rate than the cells divide, indicating that the plasmid must replicate faster than the chromosomes, which divide synchronously with the cell.[67]

Relation to Antibiotic Therapy. Use of antibiotics appears to increase the probability of the appearance of multiply resistant strains. When therapy is stopped, the transferred resistance is often lost. The routine use of antibiotics in animal feeds also appears to increase the probability of multiply resistant bacteria. Transfer of antibiotic resistance from resistant to sensitive strains in the gastrointestinal tract is probably not as frequent a source of antibiotic-resistant pathogens as is environmental selection in a hospital.[44]

DIARRHEAGENIC E. COLI

Objectives

1. Describe the evidence that some strains of *E. coli* can cause diarrhea.

2. Describe current methods used to determine whether a strain of *E. coli* is a likely cause of diarrhea.
3. Describe the clinical patterns of diarrhea that *E. coli* can produce in experiments with adult volunteers.
4. Give the evidence that some serotypes of *E. coli* produce an enterotoxin.
5. List the kinds of enterotoxin produced, and indicate how they are measured.
6. List two other virulence factors which appear to be important in relation to diarrhea.
7. Describe the antibiotic regimens shown to be effective in therapy of *E. coli* diarrhea in newborn infants.
8. Give the evidence that some strains of *E. coli* are sometimes the cause of traveler's diarrhea.
9. Explain why serotyping of *E. coli* to determine enteropathogenic serotypes is rarely indicated.

Definitions

Escherichia coli, named for the German bacteriologist Escherich, is usually abbreviated *E. coli*. It is an important cause of diarrhea and other infections. Non-diarrheal diseases produced by *E. coli* are discussed later in this chapter.

Enteropathogenic *E. coli* is a term formerly used to refer to certain serotypes which were discovered to be a cause of outbreaks of diarrhea in newborn nurseries.[94] In this section, the expression "traditional enteropathogenic serotypes" will be used to indicate those historically important serotypes. However, recent observations have shown that these traditional enteropathogenic serotypes are not closely correlated with production of enterotoxin.

Toxigenic *E. coli* are strains that have been shown to produce an enterotoxin as discussed further on p. 52. Heat-liable and heat-stabile enterotoxins have been identified.

Invasive *E. coli* are strains which can produce diarrhea in humans by their ability to invade intestinal mucosa.

"Diarrheagenic *E. coli* strains" is a broad expression used in this section to include any strains which produce diarrhea, whether by enterotoxin, invasion, or unknown mechanisms. The term *diarrheagenic* is the most appropriate term to use until the mechanism of the diarrhea is known.

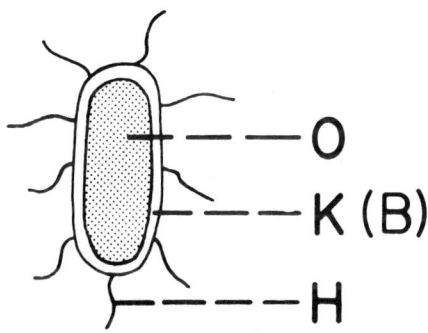

FIG. 3-5. *E. coli* antigens. O = somatic. K or B = capsular. H = flagellar. Example serotypes using this terminology include O111/B4 and O142/K86/H6.

Serotypes

E. coli serotypes are named for the O (somatic) polysaccharide antigen and the K or B (capsular) antigen; for example, O111:B4. Occasionally, flagellar antigens are also used in typing, and are labeled with H and a number, as O78:K80:H12. If the *E. coli* serotype has no flagellae, it is nonmotile, and is labeled NM (Fig. 3-5). The B antigen is also called the K antigen (from the German word Kapsul). The K-1 antigen is related to virulence, as discussed on p. 56.

Traditional Enteropathogenic Serotypes

There were approximately 12 serotypes traditionally regarded as enteropathogenic. Because traditional enteropathogenic serotypes could be found in normal individuals, their etiologic role in infantile diarrhea was difficult to prove. The first evidence that some serotypes of *E. coli* could produce diarrhea was noted in outbreaks of diarrhea in newborn nurseries.[92] A single serotype of *E. coli*, such as O127:B8, was recovered in about ten percent of the infants with diarrhea compared to one percent of normal control infants.[83]

However, data obtained in the 1970s suggested that serotype was not the best marker to indicate the ability of a strain of *E. coli* to cause diarrhea. Therefore, serotyping of *E. coli* to see if they are a traditional enteropathogenic serotype is no longer routinely done on organisms recovered from patients with diarrhea.

Even though enteropathogenic serotypes do not correlate with toxin production, they do correlate with diarrhea.[79,80] Several studies suggest that traditional enteropathogenic serotypes do produce diarrhea, not by toxin production, but presumably by other mechanisms.

Diarrheal Patterns of Illness

E. coli is probably a *frequent* cause of:

Traveler's Diarrhea. Some outbreaks of diarrhea in newly arrived travelers to various areas have been associated with toxin-producing serotypes of *E. coli*.[87] The exact frequency and importance of toxigenic serotypes of *E. coli* in traveler's diarrhea has not been accurately determined. However, the discovery that traveler's diarrhea was often due to toxigenic *E. coli* stimulated important studies of these toxins. A key observation occurred when British soldiers flew to Aden, in the Persian Gulf, got diarrhea there, and then a technician in England got diarrhea while working with the *E. coli* strain (0148:K?:H28) that was isolated from the soldiers.

E. coli is an *occasional* cause of:

Epidemic diarrhea of the newborn typically occurs in newborn nurseries and may cause the death of previously normal newborn infants. Historically, such outbreaks have been associated with the traditionally recognized enteropathogenic serotypes of *E. coli*. However, some recent outbreaks appear to have been caused by toxigenic strains of other enteric bacteria, such as klebsiella, which may have had plasmid-mediated toxin production (see p. 54).[78]

Sporadic Diarrhea in Infants. Approximately ten percent of infants less than 2 years of age with diarrhea will have one of the 12 or so enteropathogenic serotypes, compared to about one to two percent of normal infants.[83]

Food and Water-Borne Outbreaks. Outbreaks of diarrhea due to *E. coli* in food[93] or water[85] have been documented, but the frequency is undetermined.

Experimental Infections in Humans

Several patterns of illness can be produced by feeding enteropathogenic *E. coli* to prisoner volunteers.[72] One serotype, which can be demonstrated to produce an enterotoxin, can produce cholera-like illness with mild watery diarrhea. Another serotype which has the property of invasiveness, and can be demonstrated to have the ability to destroy cell cultures, can produce a dysentery-like illness, with bloody diarrhea, fever and toxicity. Lower doses of this invasive organism can produce mild diarrhea.[72] In experimental studies, the disease is made more severe by prior ingestion of sodium bicarbonate, which decreases gastric acidity and increases the viability of the ingested bacteria.[72]

These experimental studies showed the value of classifying diarrhea as cholera-like (watery, and mediated by an enterotoxin) or dysentery-like (bloody, and secondary to invasive properties of the microorganism).

Laboratory Approach

Culture. *E. coli* are readily recovered using simple media. Enteric culture methods are described in the section of this chapter on shigella. Colonies ferment lactose are picked if selective media such as EMB (eosin-methylene blue) are used. The organism can be identified as *E. coli* on the basis of a series of metabolic tests.

Escherichia Coli Serotyping. It is not necessary for the clinician to understand details of methods used by the laboratory to type *E. coli*, which are described in more detail elsewhere.[75] The time-consuming, complex procedure of serotyping has generally been abandoned by clinical diagnostic laboratories.

Enterotoxin Production. *E. coli* can produce at least two kinds of enterotoxin; heat-labile toxin (LT) and heat-stable toxin (ST). Tests for enterotoxin production usually involve animal or cell culture systems, and such tests are not routinely available in clinical microbiology laboratories. However, recovery of an enterotoxin-producing strain of *E. coli* from a patient with diarrhea usually indicates a causal relationship. Therefore, simpler tests for toxin production are being sought. Heat-labile enterotoxins from different strains of *E. coli* are very similar antigenically. LT strains appear to be a frequent cause of traveller's diarrhea, but rarely cause sporadic diarrhea in the United States.

Heat-Labile Toxin (LT) is antigenic, filterable and resembles the heat-labile toxin of cholera (choleragen). It can be detected by several animal or cell culture models. Chinese hamster ovary cell cultures or mouse adrenal tumor cell cultures show a cytopathic effect when exposed to (LT). The classic model is the rabbit isolated ileal loop, which dilates and fills with fluid when a

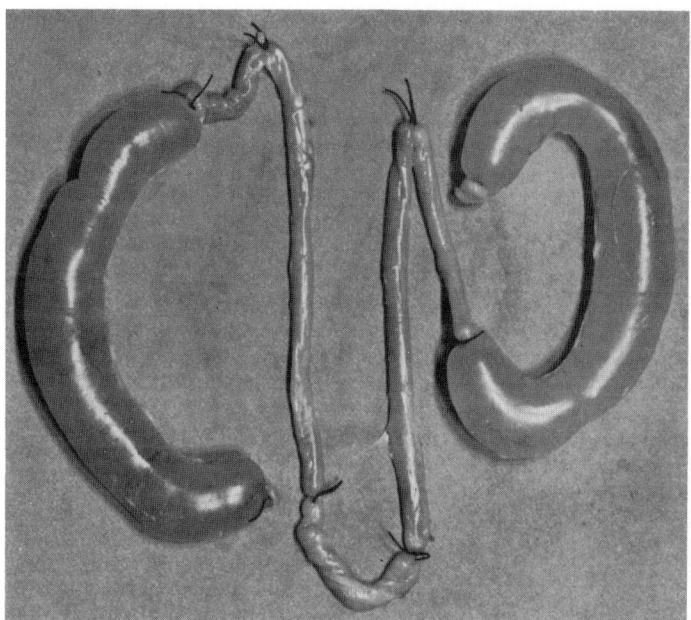

FIG. 3-6. *Ileal loop test for enterotoxin. Ligated sections were inoculated before removal from the animal. Center small loops are normal and lateral loops are full of fluid after injection of enterotoxin from toxigenic E. coli.*[82] *(Photo from Professor Harley W. Moon)*

bacteria-free, heat-labile enterotoxin containing cultural products are injected (Fig. 3-6).[76, 77] The mechanism of action appears to be stimulation of intestinal adenylate cyclase, resulting in a block of sodium and water absorption.

LT can also be detected in adrenal cell cultures.

Using dogs as an experimental model, heat-labile and heat-stable *E. coli* toxins could be clearly distinguished, and heat-labile *E. coli* enterotoxin challenge protected the dog from the effects of a later challenge with *V. cholerae* enterotoxin, but the converse was not true.[88]

Recently the ELISA method has been used to detect heat-labile enterotoxin in *E. coli* isolates (see Fig. 3-9). The *Vibrio cholerae* heat-labile enterotoxin is so antigenically similar to the *E. coli* toxin, that antisera to the *V. cholerae* toxin can be used to detect the *E. coli* toxin.[95] Staphylococcal coagglutination (Fig. 1-12) is also useful for detecting the heat-labile enterotoxin of *E. coli*.

Heat-Stable toxin (ST) resists boiling for 30 minutes, is nonantigenic, short acting, and has a low molecular weight. It has been clearly shown to be a cause of diarrhea in newborn infants.[87] Heat-stable enterotoxin is also best detected by biologic methods, at the present time. Supernatants of broth cultures produce fluid accumulation in the stomach and intestine when injected into the stomachs of infant mice.[71] ST appears to act by stimulation of guanylate cyclase.

Heat-labile toxin-related diarrhea appears to be less frequent than heat-stable toxin-related diarrhea in the United States, but may be a more frequent cause of infantile diarrhea in other countries.[74] Some bacteria produce both forms of toxin. In countries with high levels of sanitation, such as the United States, rotaviruses (Chapter 14) appear to be a much more frequent cause of diarrhea than diarrheagenic *E. coli*.

Invasiveness. The ability of particular strains of *E. coli* to invade intestinal mucosa can be detected by several biological methods.[72] The ability to penetrate and produce corneal inflammation in guinea pig eyes is one measure of the virulence factor of invasiveness (sometimes called the Sereny test). Another is the cytopathic effect produced on HeLa cell cultures. Disease produced by invasive *E. coli* is a dysentery-like diarrhea resembling shigellosis.

An outbreak of *E. coli* dysentery in adults produced by an invasive strain was related to mushroom caps stuffed with imported contaminated Camembert cheese.[93] Invasive *E. coli* diarrhea appears to be uncommon in the United States.

Rabbits can be used as an animal model to detect invasiveness in *E. coli* strains, as manifested by mucosal ulcers and bacteria and neutrophils in the lamina propria.[72]

Invasiveness should be distinguished from adherence and colonization of *E. coli*, as discussed below. Invasiveness appears to be mediated by a multiplicity of genes rather than by plasmids.

Biologic Characteristics of Clinical Interest

Plasmids as Transmitters of Enterotoxin.[90] Genes controlling the production of both heat-stable and heat-labile enterotoxin are transmitted by plasmids. Often the same plasmid carries genes controlling other functions such as antibiotic resistance. It appears that transmissibility of enterotoxin does not occur readily, as it was observed in only one of 27 strains studied.[90] Possibly certain serotypes of *E. coli* are more efficient recipients or carriers of plasmids than other serotypes.

Plasmid-Controlled Adhesiveness or Colonization. The ability of *E. coli* strains to cause diarrhea may also be related to the ability of the bacteria to adhere to small intestine (adhesiveness) and colonize it.[73] This property of adherence is plasmid-mediated and is related to a pilus-like surface antigen which has been found in a strain of *E. coli* from a human with cholera-like diarrhea. This strain adhered to an infant rabbit small intestine, grew to a high concentration and produced diarrhea.[73] There is a corresponding virulence factor in piglets associated with the K88 antigen.

Human volunteers have been used to study the role of colonization factor antigen (CFA) in the production of diarrhea.[89] Using variants of a single strain which differed only in the presence of colonization factor antigen, it was demonstrated that diarrhea occurred only in those who received high doses of CFA-positive strains. This colonization factor was distinct from type 1 pili (the so-called common pili), K88, or K99 antigens—which also have been related to colonization.

Colonization-factor antigens stimulate antibodies in humans who acquire traveller's diarrhea due to toxigenic *E. coli*.

K-1 Antigen. This antigen is associated with bacteremia and meningitis, but not with diarrhea. This is discussed in more detail in the next section, on enteric bacteria.

Incubation Period. In volunteers pretreated with bicarbonate, severe diarrhea occurred 5 to 12 hours after the invasive *E. coli* strain was fed.[72] In the Camembert cheese-related outbreak of invasive *E. coli* diarrhea, the incubation period was 24 to 72 hours, with the shorter incubation period associated with a more severe illness, and possibly a larger infecting dose.

Sources of Diarrhea-Producing Escherichia Coli. Animals can be a source of *E. coli* diarrhea in humans. Although humans are presumed to be the usual source of new cases, cats and dogs can be healthy carriers of *E. coli* diarrhea strains and may possibly be a source of *E. coli* diarrhea in humans.[81] Naturally occurring diarrhea in calves and swine may be caused by some serotypes of *E. coli*.[82]

Respiratory transmission has been postulated to be a source of *E. coli* diarrhea on the basis of recovery of the organism from the pharynx.[70]

Treatment

Fluids and Electrolytes. Supportive therapy to correct dehydration and any electrolyte disturbances is more important than antibiotic therapy (Fig. 3-7). Since laboratory identification of enterotoxogenic *E. coli* is not readily available, attempts to eradicate *E. coli* in the stool of a patient with diarrhea are not warranted. Adults and older children do well without antibiotics if supportive care is given with fluids and electrolytes.

Antibiotic Therapy. About the only circumstance where antibiotic therapy may be warranted is in an outbreak of serious diarrheal illness in a newborn nursery, where traditional enteropathogenic serotypes are being recovered.

Oral neomycin suspension, which is not absorbed, 100 mg per kg per day for 3 to 5 days, is standard therapy in infants.[84] When neomycin is used frequently in a newborn nursery, neomycin-resistant strains may be selected and become prevalent. In this situation, oral colistin, 5 mg per kg per day is usually used.

Ampicillin therapy aborted the illness in experimental infections of volunteers with a strain of diarrheagenic *E. coli* which produced dysentery-like diarrhea.[72] However, in the outbreak of invasive *E. coli* dysentery in adults traced to imported Camembert cheese, all 28 adults recovered within one week without hospitalization or antibiotics.

Prevention

Avoid Exposure. Careful handwashing in newborn nurseries may decrease exposure, as indirect evidence has shown that hands are an important means of colonization by diarrhea-producing *E.*

OTHER ENTERIC BACTERIA 55

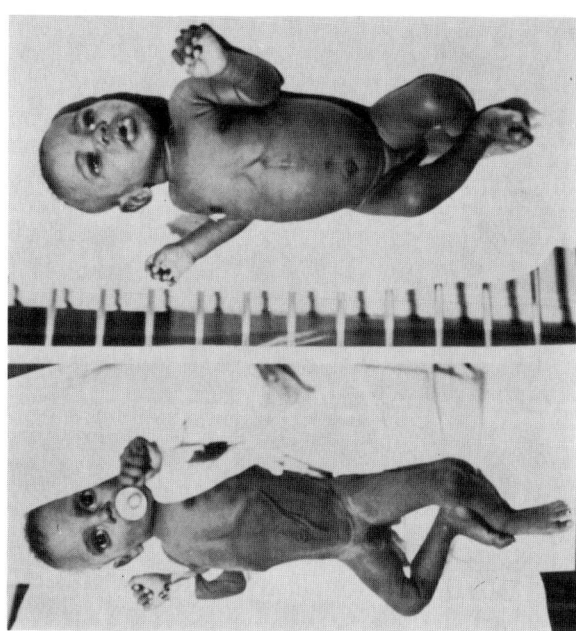

FIG. 3-7. *Dehydration in infants is often associated with poor tissue turgor, recognized by the failure of pinched skin to snap back to normal. Top: dehydrated baby with poor turgor (demonstrated by pinching the skin on the abdomen), and sunken eyes. Bottom: same infant 48 hours later after intravenous fluids.*

coli in the nursery.[69] However, the newborn baby is probably colonized by the mother, although the baby may also receive some protective antibodies from her.[86]

Nursery Outbreaks.[94] Immediate neomycin therapy of all exposed infants in a newborn nursery is usually indicated as soon as an outbreak of *E. coli* diarrhea is recognized. Adherence to strict isolation precautions and segregation of infants is also indicated.[94]

Breast Feeding. Colostrum, which is secreted by the breast for a few days after delivery, and breast milk contain some IgA antibody which may coat the bowel wall and may be effective in protecting against *E. coli* infection.[91]

OTHER ENTERIC BACTERIA

Most of the aerobic gram-negative rods found as normal flora of the bowel are opportunistic pathogens, which means disease occurs primarily because of a lapse in host defenses, rather than because of the virulence of the organism. Enteric gram-negative bacilli are a frequent cause of wound infections or bacteremia. Other enteric bacteria which rarely cause human disease are discussed in Chapter 6.

Escherichia Coli

E. coli are the most numerous aerobic bacteria found in the bowel. The majority of *E. coli* involved in disease are not enteropathogenic, and are not recovered from patients with diarrhea. The mechanisms of production of diarrhea by *E. coli* are discussed in the preceding section.

E. coli is a *frequent* cause of:

Urinary Tract Infections. As *E. coli* is the most numerous aerobic bacterial species of the bowel, it is not surprising that they are the most common cause of those urinary infections which are secondary to transmission of bowel flora to the bladder via the uretha, especially in females.

The definitive diagnosis of a urinary tract infection is usually based on the recovery of more than 100,000 organisms per milliliter of a clean voided urine. Quantitative cultures can be done by inoculation of a measured volume of urine (usually 0.001 ml), onto a culture plate (Fig. 3-8).

Peritonitis and Abdominal Abscesses. After bowel perforations, *E. coli* and *Bacteroides* species are the most common cause of peritonitis and abdominal abscesses.

Neonatal Septicemia. *E. Coli* and Group B streptococci (see p. 20) are the most common causes of septicemia and meningitis in the infant in the first few weeks of life.

E. coli is an *occasional* cause of:

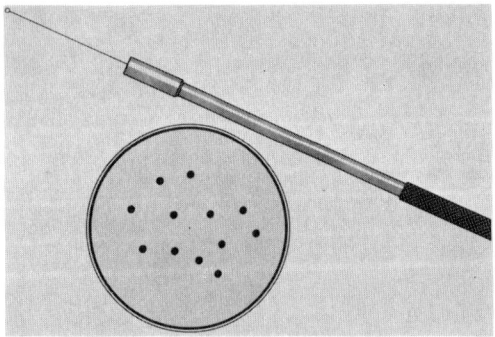

FIG. 3-8. *A quantitative loop that delivers 0.001 ml of urine, and a culture plate that has been inoculated with the loop. Twelve colonies can be counted, indicating the urine contained 12,000 organisms per ml when inoculated, a number usually indicative of contamination. Note the small size of the loop.*

Wound infections (especially postoperatively) and **septicemia in compromised hosts** (especially with liver disease).[99]

The recovery of *E. coli* from a culture usually indicates an opportunistic infection caused by the organism present in largest numbers in an area of obstruction or contamination.

K-1 Antigen. The presence of this mucopolysaccharide capsular antigen is closely related to virulence in *E. coli*, particularly for septicemia and meningitis in newborn infants.[103] Antibodies raised against this antigen in mice protect the mouse from experimental infection with *E. coli* containing K-1 antigen. The K-1 mucopolysaccharide antigen cross reacts antigenically with the capsular antigen of Group B streptococci (another important cause of neonatal disease), and serotype B meningococci. The virulence effect of K-1 capsular antigen appears to be related to increased resistance to phagocytosis, and increased resistance to the bactericidal activity of normal adult serum.[113]

The K-1 antigen does not appear to be an important virulence factor for non-neonatal *E. coli* infections, such as urinary tract infections.

Klebsiella

The genus *Klebsiella* was named after the bacteriologist Klebs. The term klebsiella is also widely used to mean the species *K. pneumoniae*, which also has been called Friedlander's bacillus after the pathologist who associated the organism with fatal necrotizing pneumonia.

Klebsiella species are normal flora of the human bowel, and are presumptively distinguished in the bacteriology laboratory from *Enterobacter* species by lack of motility due to a mucoid capsule. However, other biochemical tests are also necessary to confirm this distinction. There are about 70 serotypes of *K. pneumoniae*, defined by antiserum made against the capsular specific soluble substance.[106] Other species of *Klebsiella* include *K. ozaenae* and *K. rhinoschleromatis*. *K. ozaenae* has been recognized more frequently as a cause of serious infections.

In the past, *Klebsiella* and *Aerobacter* (now *Enterobacter*) species were not distinguished, and were reported as klebsiella–aerobacter. Now most laboratories report such organisms as either *Klebsiella* or *Enterobacter* species.

Klebsiella species have several special features:

Mucoid Capsule. This makes the organism usually appear typically mucoid on culture plates, and prevents the organism from being motile, which is a function of flagellae.

Normal Flora of the Throat. Studies done before 1970 reported a moderately high frequency of "klebsiella" as normal flora of the throat, but failed to exclude enterobacter from this designation and did not define the species of klebsiella. *Klebsiella pneumoniae* had a prevalence of one percent or less in recent studies of healthy children and adults, but was recovered from about 40 percent of alcoholic adults.[102]

Severe Pneumonia. *Klebsiella pneumoniae* is an occasional cause of severe necrotizing pneumonia, especially in alcoholic patients.[102]

Other Infections. *Klebsiella pneumoniae* is a much less frequent cause of other infections than *E. coli*, but can cause urinary infections and wound infections. It is a rare cause of arthritis and meningitis.[111] *Klebsiella* species are an important cause of hospital-acquired infections.[112]

Sensitivity to Cephalosporins. About 80 percent of *Klebsiella* species are sensitive to cephalosporins, whereas most *Enterobacter* species are resistant. Resistance is usually due to production of a beta-lactamase (see p. 4).

Bacteriocin Production. *Klebsiella pneumoniae* can be typed on the basis of sensitivity to their bacteriocins (klebocins)—see p. 60.

Antibody Detection. Using the ELISA method for detection of antibody, antibodies to various *Klebsiella pneumoniae* capsular polysaccharides can be demonstrated after human infections.

Enterobacter

Enterobacter, formerly called *Aerobacter* species, are an occasional cause of hospital-acquired infections, especially urinary infections.[107,110]

Enterobacter species include *E. hafniae* (also called the genus *Hafnia*), *E. cloacae*, *E. aerogenes*, and *E. liquefaciens*. All of these species can be considered normal flora of the human bowel and have no special clinical significance. *Enterobacter agglomerans* (also called an *Erwinia* species) was recovered in a nationwide outbreak of commercially contaminated intravenous fluids. This yellow-pigmented organism is now classified as *Chromobacterium typhiflavum* and is discussed in Chapter 6.

Serratia

The genus *Serratia* is named for Serafino Serrati, an Italian physicist and inventor. Therefore the genus should be pronounced "Se-rot-ia," not "Se-race-ia."

Serratia marcescens has been used in the past in human experiments involving colonization of the urinary tract, because the colonies of the strains used were red and easy to trace, and the organism was regarded as nonpathogenic and safe for such experiments. However, in recent years, many nonpigmented strains have been recognized, and the organism has been recognized as an occasional cause of hospital-acquired infections.[97,114] *Serratia marcescens* isolates can sometimes be typed in reference laboratories, using bacteriophage or bacteriocin (marcescin) typing, for epidemiologic investigation of outbreaks.

Diseases caused by serratiae include endocarditis and septic arthritis (especially in drug addicts).[98]

Amikacin or gentamicin appears to be effective therapy for this organism, which is resistant to most antibiotics.[98] The combination of polymyxin B and rifampin also appears to be effective.[108]

Proteus

Proteus species are an occasional cause of urinary infections, and *rarely* cause wound infections or bacteremia.[96] *Proteus* species include *P. mirabilis*, *P. vulgaris*, and *P. morganii* (also called *Morganella morganii*).

Proteus species have several features:

Urea Utilization. *Proteus* species produce a urease and so can utilize urea as an energy source, producing ammonia and an alkaline pH. When a patient with a urinary infection has an alkaline urine, the infecting organism is almost always a *Proteus* species. The urease appears to provide proteus with a great ability to invade renal epithelium. This invasiveness and alkaline pH predisposes the patient to urinary stones.

Swarming. *Proteus* species typically have numerous flagellae, and sometimes spread over the surface of agar plates as they grow, producing a thin film, a phenomenon called swarming. Many enteric media incorporate various chemicals to inhibit swarming by proteus, which otherwise spread over the colonies of other organisms, which are then hard to pick as a pure colony.

Antigens Shared With Rickettsiae. *Proteus* species share some antigens with rickettsiae. This is the basis for the serologic diagnosis of some rickettsial diseases, especially Rocky Mountain spotted fever, by detection of proteus antibodies, (called the Weil–Felix reaction).

Susceptibility to Penicillin. *Proteus mirabilis*, the most frequent *Proteus* species, is often very susceptible to penicillin, which is relatively unusual for enteric bacteria. However, ampicillin is usually more effective than penicillin, and is the treatment of choice if the isolate is susceptible. *Proteus mirabilis* is sometimes referred to as indole-negative *Proteus*.

Guthrie Test. *Proteus* species utilize phenylalanine, and this reaction can be used as a biologic test (called the Guthrie test) for the detection of small amounts of phenylalanine. This test is used most frequently to screen the blood of newborn infants for phenylalanine, which is found in phenylketonuria, a disease causing mental retardation. Phenylketonuria can be treated by dietary restriction of phenylalanine.

Providencia

This genus resembles *Proteus* species in many metabolic tests, and is a rare cause of serious infections.[101]

Citrobacter

This organism was once classified as *Escherichia freundii*.[104] It has also been regarded as a variant of *E. coli* which produces H_2S. It can cause urinary tract infections, cellulitis, severe neonatal meningitis,[109] and various hospital-acquired infections.[100]

Citrobacter species include *C. intermedium* and *C. diversus*.

Edwardsiella

This is an organism which resembles *E. coli*, except for the production of H_2S. It is named for the American bacteriologist Edwards, who studied and classified enteric bacteria. The only species is *Edwardsiella tarda*. It occasionally produces serious disease, such as meningitis.[105]

Hafnia

These bacteria also have been classified as *Enterobacter hafniae*, and are similar to other *Enterobacter* species (see Table 3-1).

PSEUDOMONAS SPECIES

Objectives

1. Describe how the laboratory distinguishes between *Pseudomonas aeruginosa* and other *Pseudomonas* species.
2. Describe two methods of typing *Ps. aeruginosa*, and list the indications for typing.
3. List three clinical syndromes often caused by *Ps. aeruginosa*.
4. List the most frequent sources of infection due to nonpigmented *pseudomonas* species.
5. List the antibiotics that are usually effective against *Ps. aeruginosa*.
6. Describe investigational modes of treatment of *Ps. aeruginosa*.

Definitions, Species, and Types

The *Pseudomonas* genus of bacteria are gram-negative bacilli with a polar flagellum. Based on their clinical patterns, pseudomonads can be classified into two groups: *Pseudomonas aeruginosa* and other *Pseudomonas* species. *Ps. aeruginosa* is primarily characterized by the production of water soluble pigments; particularly pyocyanine (blue) and fluorescein (green). The other *Pseudomonas* species usually do not have water soluble pigments.

Almost all human disease due to pseudomonads is due to *Ps. aeruginosa*. However, other medically important species of *Pseudomonas* are *Ps. maltophilia*, *Ps. cepacia*, and *Ps. pseudomallei*.

Typing of *Ps. aeruginosa* is occasionally useful to trace spread of disease, similar to the use of typing of Group A streptococci or *Staph. aureus*. This can be done by phage typing, or by pyocin typing, using the capacity of pyocin of one stain to inhibit the growth of another.[121,122] Commercial antisera are available for typing isolates by their O antigens, using slide agglutination.

Frequency and Importance

Pseudomonas aeruginosa is one of the most frequent causes of hospital-acquired infections. It is also important because only a few antibiotics are effective against *Pseudomonas aeruginosa*, and most of these antibiotics are rather toxic and must be given parenterally. The nonpigmented *Pseudomonas* species are a less common cause of infections, and are characteristically associated with an exposure to water or a moist environment.

Clinical Patterns of Illness

Pseudomonas aeruginosa is a *frequent* cause of:

Burn Infections. Severe burns are often colonized by pseudomonas soon after the patient is hospitalized. Bacteremia occurs most frequently about 5 days after the burn, when fluid in the tissues is resorbed into the circulation.[121, 132]

Otitis externa, which is ear discharge, with a painful oozing ear canal, is often due to *Ps. aeruginosa*. Water in the ear canal is a predisposing factor for otitis externa, which is often called "swimmer's ear", because of its relation to swimming.

Pneumonia in Compromised Hosts. Pneumonia often occurs in patients with cystic fibrosis of the pancreas, and is often caused by *Ps. aeruginosa*, particularly with unusually mucoid strains.[134] Pneumonia in immunosuppressed patients with malignancy also is frequently due to *Ps.*

aeruginosa. Bacteremia in such patients is associated with a mortality rate of about 50 percent.[117]

Colonization Without Illness. Surveillance data indicate that colonization may occur without disease, but that the risk of subsequent disease is great in colonized patients with serious underlying disease.[115] *Pseudomonas aeruginosa* is normal flora of the bowel. There is some evidence that vegetables can be a dietary source of intestinal pseudomonas.[125]

Pseudomonas aeruginosa is an *occasional* cause of:

Septicemia. Typically, this occurs as a complication of malignancy, but may also occur in the newborn infant. Characteristic skin lesions are often noted, which is first a papular rash, then vesicles, then black or purple ulcerations on the skin, called ecthyma gangrenosum.[120] Occasionally, deep subcutaneous abscesses occur, which need incision and drainage. Histologically, invasion of blood vessels (vasculitis) is characteristic. Occasionally, *Ps. aeruginosa* sepsis can be related to the use of poor sterilization technique by drug addicts.

Meningitis. Typically, this occurs in patients with a damaged central nervous system, particularly meningomyeloceles, or brain tumors.[119]

Urinary Infections. *Ps. aeruginosa* is rare as a primary cause of urinary infections, which can occur after urinary tract instrumentation, or during or after antibiotic therapy.[129]

Secondary Pneumonias. Typically, these occur in patients with other lung disease, especially those with tracheostomies.[128]

Osteomyelitis. Typically, this occurs as a complication of puncture wounds of the foot, or as hematogenous infection of a vertebra or clavicle, especially complicating heroin addiction.[126]

Swimmer's Rash. *Pseudomonas aeruginosa* has been implicated as a cause of an itching, sometimes pustular rash, occurring in swimmers in heated swimming pools.[137] The organism was recovered from the skin rash, and from the inadequately chlorinated whirlpool and swimming pool water.

Other *Pseudomonas* species are an *uncommon* cause of:

Pneumonia. *Pseudomonas pseudomallei* can produce chronic pneumonia with cavitation, or acute fulminating pneumonia.[138] It is seen in the United States in individuals recently returned from Asia. Infection with this organism is called melioidosis. The term melioid means "like melis", the Greek name for what we now call glanders, a disease of horses caused by *Pseudomonas mallei*, which is characterized by pneumonia or skin abscesses.

Several unclassified, nonpigmented *Pseudomonas* species can cause pneumonia. Typically, the patient has been exposed to contaminated liquids, inhalation therapy equipment, or mist.[123]

Other Infections. Wound infections, meningitis, and urinary infections are occasionally produced. The most frequently encountered species are *Ps. maltophilia, stutzeri, putida,* and *cepacia*.[123]

Laboratory Approach

Culture. This is the most useful method to detect pseudomonas infection. *Pseudomonas* species grow readily on all kinds of media, and *Ps. aeruginosa* colonies can often be recognized immediately by their odor and its pigment.

Serum Antibodies. Antibodies to *Pseudomonas aeruginosa* can be found in low concentration in the sera of normal individuals, and in higher concentration in patients with pseudomonas infections.[118] These serologic techniques are not yet practical for serologic diagnosis of infection. Antibodies to pseudomonas exotoxin can also be detected.

Detection of Pigment. Early diagnosis can often be made by seeing a green or blue-green color to the pus, as from a chest tube, or by detecting a green fluorescence with an ultraviolet (Wood's) light on the surface of a burn or in the urine.[131]

Biologic Characteristics of Clinical Interest

Virulence factors. Hemolysin, lecithinase, and protease are responsible for the tissue effects of pseudomonas.[127] Endotoxin is produced and can be responsible for septic shock. Exotoxin is also important in the virulence of *Ps. aeruginosa*.[135]

Resistance to Disinfectants. Quarternary ammonium disinfectants compounds, such as Zephiran, are relatively ineffective against *Ps. aeruginosa*.

Mucoid Strains. Extremely mucoid colonies are sometimes observed, especially in strains recovered from patients with cystic fibrosis of the pancreas, but the clinical significance is unknown.[119,134]

Bacteriocin Production. Bacteriocins are substances produced by bacteria which inhibit the growth of other strains of the same species. The bacteriocins produced by *Ps. aeruginosa* are called pyocins, and are useful to type different strains of this species.[121,122] Pyocins should not be confused with pyocyanine, the water-soluble blue-green pigment which is typically produced by almost all strains of *Ps. aeruginosa*. Bacteriocins are named for the bacteria which produce them; for example, colicins from *E. coli* and klebocins from klebsiella.

Strict Aerobe. *Pseudomonas aeruginosa* is a strict aerobe, and often will not produce visible growth in vacuum exhausted blood culture bottles, unless the bottles have air introduced.[124]

Relation to Water. *Ps. aeruginosa* and other pseudomonas species thrive well in a moist environment, and are frequently recovered from water reservoirs in inhalation therapy equipment. These organisms can even grow in heated pools with some chlorination, which can be a source of swimmer's rash and swimmer's ear.

Relation to Neutrophils. *Ps. aeruginosa* is a very frequent cause of infection in neutropenic patients, especially those with leukemia. Neutrophils appear to be particularly important to host defenses against *Ps. aeruginosa*.

Treatment

Chemotherapy. *Pseudomonas aeruginosa* is usually susceptible to tobramycin, gentamicin, amikacin, polymyxin B, or carbenicillin. Nonpigmented *Pseudomonas* species are more likely to be susceptible to gentamicin than to other antibiotics.[136]

Vaccines. Killed pseudomonas vaccines may be of value in burned patients,[116] and possibly in patients with leukemia or cystic fibrosis.[130]

Immunotherapy. Convalescent human plasma or hyperimmune human globulin from immunized volunteers may be useful treatments, if available.[116]

NONFERMENTER GROUP

Definitions

The nonfermenter group of gram-negative rods is characterized by its mode of utilization of carbohydrates.[150] Unlike the *Enterobacteriaceae*, which *ferment* hexoses to form two 3-carbon acids, the nonfermenters either *oxidize* hexoses to form a 6-carbon acid, or fail to utilize hexoses. The definitive test is performed by inoculating the organism into two tubes of the hexose, usually glucose. One tube is incubated anaerobically and the other aerobically. If the organism utilizes the hexose anaerobically, producing acidification of the medium as shown by a color change in the indicator, it is a fermenter. If it fails to utilize the hexose anaerobically, but does so aerobically, it is an oxidizer. Some organisms do not ferment or oxidize hexoses, and are sometimes called *nonoxidizers*.

Pseudomonoas aeruginosa and other *Pseudomonas* species are nonfermenters. This section on nonfermenters deals with organisms that are not members of the *Pseudomonas* family, which is defined by polar flagella.

Many nonfermenters have been difficult to classify, and have been known by several different names. Table 3-1 shows the older and newer and current names of non-fermenting bacteria.

Frequency and Importance

Many of the nonfermenter group of bacteria are found in standing water. *Acinetobacter lwoffi* (formerly called *Mima polymorpha*) is often found as normal flora of the axilla.[140] *Acinetobacter anitratus* (formerly called *Herellea vaginicola*) is also found on the skin of normal individuals. However, acute infections due to nonfermenters are relatively uncommon in individuals who are not hospitalized.

Nonfermenters are important because they are an occasional cause of serious infections, especially hospital-acquired infections. Recovery of an organism of this group should raise the question of a water source, such as mist or inhalation therapy equipment.[152]

The relationship of these bacteria to water was first recognized in infants treated with humidity or mist. In 1961, an editorial entitled "Water Bugs in the Bassinet" was influential in calling attention to these risks and stimulating further study.[139]

Classification

Clinical isolates are often not assignable to any defined species. These have been classified at a reference laboratory at the Center for Disease Control in Atlanta, Georgia.[146,152,153]

Genera of Nonfermenters

Alcaligenes. This group is named for the alkaline reaction they often produce in test media. *Alcaligenes* species do not ferment hexoses, but utilize protein components in the media to form ammonia, so the culture medium becomes alkaline. The usual species recovered from clinical specimens is *Alcaligenes faecalis*, which can be considered normal flora of the human bowel.

A variety of serious infections with alcaligenes have been reported, including meningitis and septicemia.[158] An organism classified as Group IVe, which resembles alcaligenes, is a rare cause of human disease.[156]

Flavobacterium. This genus is named for the yellow color (flavo = yellow) of the colonies of many of the species in the genus. *Flavobacterium meningosepticum* is a species which has produced meningitis in young infants exposed to mist contaminated with this organism, and continues to be a problem in neonatal intensive care units.[148,155] Therapy with intraventricular erythromycin or vancomycin were used in addition to systemic antibiotics.[148,155]

Achromobacter. This group of organisms is not generally accepted as a genus. The name refers to lack of pigmentation of the colonies. Outbreaks have been reported using the name achromobacter; for example, septicemia in newborn infants, apparently from water used to wash the eyes.[145] *Achromobacter xylosoxidans* has been recently reported as a cause of bacteremic pneumonia.[141]

Acinetobacter. This genus name refers to the observation that these organisms are non-motile. There are two clinically important species of *Acinetobacter: A. calcoaceticus* var. *anitratus* and *A. calcoaceticus* var. *lwoffi.*

Acinetobacter anitratus was formerly called *Herellea vaginicola.* Anitratus refers to the failure of the organism to utilize nitrates. *Acinetobacter anitratus* is a frequent cause of burn infections, and an occasional cause of urinary and respiratory infections, especially those acquired in the hospital.[142,146,157,159] It is found on the skin of about 25 percent of normal persons.[146] It has also been called *Moraxella vaginicola* and *Achromobacter anitratus.*[154] It is resistant to many antibiotics but carbenicillin and an aminoglycoside such as tobramycin are synergistic against this organism.[147]

Acinetobacter lwoffi was formerly called *Mima polymorpha.* The word Mima was used because the organism mimics the gonococcus, since it is a small gram-negative coccobacillus, resembling the diplococcus of *N. gonorrheae*, and is sometimes found in a urethral discharge. It also has been called *Achromobacter lwoffii, Acinetobacter lwoffii,* or *Moraxella lwoffii.*[154] Lwoffii refers to the French bacteriologist Lwoff.

Acinetobacter lwoffi is a rare cause of meningitis, endocarditis, and other serious infections.[146] It is distinguished from *Acinetobacter anitratus* by testing for acidification of ten percent lactose and ten percent glucose.

Minocycline, a tetracycline, appears to be very active against *Acinetobacter* species in agar dilution antibiotic susceptibility tests.[151]

Moraxella. *Moraxella duplex* is the usual species in this group, and is also called *Mima polymorpha* var. *oxidans,* and *Moraxella osloensis.*[144,149] It is a rare cause of serious infections such as meningitis, endocarditis, abscess, vaginitis, or septic arthritis.[154] It also can cause purulent conjunctivitis which may progress to corneal ulceration and penetration, producing pus in the anterior chamber of the eye (hypopyon).[143]

Bordetella Bronchiseptica. This organism has laboratory characteristics which are very different from *Bordetella pertussis* (Chap. 4), because it is not fastidious and usually is classified as a nonfermenter. Most isolates are from the respiratory tract and the organism has been rarely associated with bronchitis and endocarditis, but appears to be much more important in respiratory disease of animals.[146]

VIBRIO SPECIES

Objectives

1. Describe the classical picture of cholera.
2. Discuss the biologic characteristics of *Vibrio cholerae* which account for its spread by water and its relationship to gastric acidity.
3. Discuss the mechanism of action of killed cholera bacteria vaccine and cholera toxoid.
4. Describe the clinical illnesses that can be produced by *Vibrio parahaemolyticus*, lactose fermenting (L+) halophilic vibrios, and *Campylobacter fetus*.

Species and Types

Vibrio is a genus of gram-negative rods that are curved and have a polar flagellum. Vibrios can be recognized under a darkfield microscope by a characteristic darting movement.

Vibrio cholerae is the classic cause of Asiatic cholera. The El Tor biotype (variant) of *V. cholerae* is named for the El Tor quarantine station on the Sinai peninsula, where it was first recovered. It differs from the classic *V. cholerae* by the production of a soluble hemolysin. There are two antigenic serotypes, called Ogawa and Inaba.[177]

Halophilic (salt-loving) vibrios are found in seawater along the coastal regions of the United States. *Vibrio parahaemolyticus* is a cause of diarrheal food poisoning in the United States along the seacoasts. Usually the source is seafood, such as crabs or shrimp.[168] It is also a rare cause of wound infections, especially wounds contaminated by sea water.[183] It is the most frequent cause of diarrhea in Japan. *Vibrio alginolyticus* is a rare cause of wound infections in the United States.[176]

Campylobacter fetus (formerly called *Vibrio fetus*) is a rare cause of abortion and neonatal septicemia in humans,[173] and produces similar illnesses in cattle and sheep.[175] This organism also can be a cause of diarrhea, salpingitis, and pneumonia.[162,178,180]

Lactose-fermenting halophilic vibrios (called L+ vibrios) have recently been distinguished from the other halophilic vibrios *V. parahemolyticus* and *V. alginolyticus*.[162] Two clinical presentations have been observed. Septicemia has occurred after eating raw oysters contaminated with the L+ vibrio, especially in patients with liver disease.[162] Wounds contaminated with seawater can become infected with the L+ vibrio. Like the other halophilic vibrios, it is probably part of normal marine flora.

Non-cholera vibrios are other vibrios that resemble *V. cholerae* in some ways, but are not agglutinated by *V. cholerae* antisera. They are called non-agglutinable (NAG) vibrios or non-cholera vibrios (NCV).[174] Disease produced by these vibrios closely resembles cholera and is associated with a toxin closely resembling cholera toxin.

In summary, vibrios are now classified into four general groups: *V. cholerae*, halophilic vibrios, *Campylobacter fetus* (with its subspecies *jejunii* and *intestinalis*) and non-cholera vibrios.

Frequency and Importance

The frequency of halophilic vibrios, *Campylobacter* species, and non-cholera vibrios in the United States is not clear, but these species have been recognized with gradually increasing frequency in the past 15 years.

The importance of cholera outside United States as a severe diarrheal disease is obvious. It could be imported to North America, although spread within the United States is presumed to be unlikely.

Cholera had not occurred in the United States since 1911, except for rare cases in laboratory workers,[177] until a case occurred in Texas in 1973.[181] Tourists recently arrived from cholera areas may bring the disease into the United States, but the patient in Texas in 1973 had no known exposure to cholera. Another case, manifested as cholecystitis without an identified source, occurred in Alabama in 1977 in a man who had traveled overseas often. In 1978, a cluster of about 11 cases (Inaba serotype) occurred in Louisiana. All appeared to be secondary to eating boiled crabs.

Clinical Patterns of Illness

Cholera-like Diarrhea. Classical cholera is characterized by "rice-water stools" (clear, brown-tinged watery stools with flecks of mucous).[161,166] An immense amount of water is lost in the stools in severe cholera, and this dehydration produces a clinical picture not found in any other disease. "Washer-woman's hand" describes the wrinkled skin due to severe dehydration.[165] Cramping abdominal pain is severe. The isotonic dehydration is severe in adults and can be fatal in children.

Diarrheal Food Poisoning. Outbreaks of *V. parahemolyticus* food poisoning are characterized by severe diarrhea, cramps, and vomiting, with little fever. The frequency in the United States is unknown, but it is probably more frequent than generally appreciated.

Dysentery-like Diarrhea. *Campylobacter fetus* subspecies *jejunii* and *V. parahemolyticus* are rare causes of dysentery in the United States. *C. fetus* subspecies *intestinalis* can produce disseminated disease.

Laboratory Diagnosis

Culture. *V. cholerae* grows adequately on blood agar plates, but does not grow well on most selective enteric media. *Campylobacter fetus ss. jejunii*, formerly called *C. jejunii* grows better at 42°C than at 37°C; and grow better in a candle jar with five to ten percent CO_2.[175] Selective media developed by Skirrow is used to detect campylobacter.[178]

Enzyme Linked Immunosorbent Antibody (ELISA) Method. Cholera toxin can be detected by an ELISA method (Fig. 3-9), which also can be used to detect the heat-labile enterotoxin of *E. coli*. This method (pronounced ē lī' sa)[164] can be used to detect either antigen or antibody.[164,170] Antibody is linked (conjugated) to an enzyme such as alkaline phosphatase or peroxidase. When the enzyme-labeled antibody is bound to its specific antigen, it is not removed by a rinse. The enzyme is detected by adding a substrate, which changes color when the enzyme acts on it. The color change is detected grossly or by using a colorimeter. Thus, the enzyme acts as a very sensitive label to detect small amounts of antibody. The method is called the immunoperoxidase assay if peroxidase is used as the enzyme.

Either antigen or antibody can be absorbed to the surface of a microtiter well or slide. Then the patient's specimen is added. Then the detecting system of antibody is added. Rinsing is done after

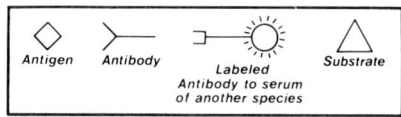

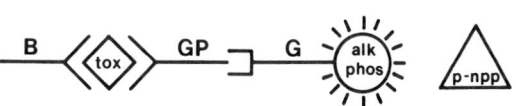

FIG. 3-9. *ELISA method to detect antigen (enterotoxin of* V. cholerae *or* E. coli*). B=burro antiserum to* V. cholerae *enterotoxin. tox =heat-labile toxin of* E. coli *or* V. cholerae. *GP=guinea pig antiserum to* V. cholerae *enterotoxin. G=goat antiserum to guinea pig antibody. alk phos=alkaline phosphatase conjugated to goat antibody. p-npp=p-nitrophenyl phosphate. The symbols are shown in the key, and are the same throughout this book for antigen, antibody, labeled antibody, and other kinds of antibody, as shown.*

each step to remove unbound reagent. When the antigen is trapped between two different antibodies (as in the detection of enterotoxin as in Fig. 3-9), this is sometimes called an "antibody sandwich".

The ELISA technique is similar to two other labeled-antibody methods: the fluorescent antibody (FA) and the radioimmunoassay (RIA). In the first method, antibody is be conjugated with fluorescein, which is detected by filtered-light microscopy (see Fig. 1-9). Antibody can also be conjugated with radioactive iodine (^{125}I), which is detected by measuring radioactive emissions.[182] The radioimmunoassay method has the advantage of great sensitivity, but is subject to the precautionary restrictions of radioactive materials. Detection of hepatitis B surface antigen (Fig. 14-2) is an example of a useful RIA method.

The major uses of the ELISA and other antibody-labeled methods, are detection of viruses which have not yet been cultivated in cell cultures (such as hepatitis A virus and rotaviruses, Chap. 14), detection of antigens, which previously required cumbersome biologic methods (such as enterotoxin), and determination of immunity by detection of low titers of antibody (as in chickenpox).

Smear. Dark field examination of stool water reveals the darting movements of *Vibrio cholerae* or *Campylobacter* species. Fluorescent antibody techniques may be available in epidemic areas.

Serology. Antibodies to cholera organisms or to cholera enterotoxin can be measured in research laboratories, but these measurements are usually not used for ordinary diagnostic purposes.

Biologic Characteristics of Clinical Interest

Enterotoxin. A heat-labile exotoxin is produced by cholera vibrios. This is called choleragen, and reproduces the disease in humans when introduced into the small intestine using an intestinal biopsy capsule.[161]

Cholera toxin acts by increasing the amount of chloride ion excreted by the small intestine and decreasing the sodium ion absorbed.[160] This effect is produced by activating or stimulating the activity of the enzyme adenylate cyclase, with a resultant increase in the concentration of cyclic AMP.

Glucose and Intestinal Transport. Glucose and amino acids such as glycine increase sodium and water transport across the bowel, so that oral therapy of cholera is relatively effective. The effect of cholera toxin can be reversed by the oral feeding of glucose. The recognition of this specific effect of glucose has led to effective treatment of formerly fatal diarrhea with oral glucose–electrolyte solutions in underdeveloped countries, and has been called one of the most important medical advances of the century.

Halophilic Fastidious Organisms. V. cholerae and V. parahaemolyticus are halophilic (salt-loving), and require a salty alkaline environment, such as the human bowel or tidal estuaries. Spread from person to person is very rare, but the sputum may contain the organism for several days after vomiting ceases. Spread of the disease usually occurs from drinking contaminated water, eating vegetables kept moist by contaminated water, seafoods harvested from contaminated water, or swimming in contaminated water.

Inhibition by Acidity. Gastric acidity protects against disease, and bicarbonate ingestion increases the susceptibility to infection in volunteers.[166]

Animal Reservoirs. Dogs and cats can get diarrhea when infected with *Campylobacter* species, and may be a source of infections in humans.

Treatment

Fluid Replacement. Immense water losses must be replaced, and this is the most important aspect of treatment. Intravenous fluids are essential for severe disease, but glucose-containing oral fluids are adequate in many cases, even in children.[165]

Antibiotics. Tetracycline eradicates *V. cholerae* and *Campylobacter fetus*. Chloramphenicol or tetracycline are effective against *V. parahaemolyticus* and *V. alginolyticus*.

Prevention

Avoid Exposure. Humans are the only reservoir for cholera. Cholera has spread from a focus in the South Pacific since 1958, covering most of Asia and Africa by 1971.[171] In 1977, about 58,000 cases were reported in Africa, Oceania, and Asia. In Europe, 24 imported cases and in North America three imported cases were reported. In countries where the disease is endemic, water and undercooked seafood are the usual sources of infection.

Raw meat and raw milk appear to be sources of *V. fetus*.[169,179] Steamed crabs, boiled shrimp, and clams have been sources of outbreaks of food poisoning due to this organism, presumably in association with inadequate cooking.

Immunization. Cholera vaccine is usually given as a killed whole cell vaccine, which has some protective effect, primarily due to an antibacterial mechanism.[171] Antibodies to the cholera toxin (choleragen) are not produced by these vaccines. However, a cholera toxoid is being investigated, and appears to be of greater value than killed bacterial vaccine.[160]

Management of Introduced Cholera.[172] If a cholera case is confirmed in the United States, it is unlikely to spread. It is not a highly communicable disease, and health service personnel are not at unusual risk provided ordinary procedures of enteric isolation, including handwashing, are observed. Mass immunization, chemoprophylaxis, and quarantine are unnecessary.

BACTEROIDES SPECIES

Objectives

1. Explain why bacteroides infections are now being recognized more frequently.
2. List three clinical situations in which bacteroides infection might be expected.

3. List two antibiotics usually effective against *Bacteroides* species.

Definitions and Species

Bacteroides is a genus of anaerobic gram-negative bacilli of the human large intestine. *Fusobacterium*, and *Leptotrichia* are other genera of anaerobic gram-negative enteric bacilli in the family *Bacteroidaceae*, but *Bacteroides* is a term often used by clinicians to refer to the entire *Bacteroideaceae* family of anaerobic gram-negative rods.[188]

B. fragilis is the species most frequently recovered from clinical specimens. There are numerous subspecies of *B. fragilis*. The most virulent subspecies is *B. fragilis* subsp. *fragilis*. Its increased virulence is probably related to its polysaccharide capsule, and its relative resistance to the bactericidal activity of normal human serum.[193,195]

Frequency and Importance

Bacteroides are the most numerous bacteria in the human bowel, where they are about 1000 times more numerous than *E. coli*. *Bacteroides* species are also normal flora of the mouth and female genital tract. The frequency of bacteroides infections is not yet fully defined, because adequate anaerobic methods have been widely available only recently. However, it is likely that many "sterile abscesses" and presumptive infections with negative cultures in the past were due to bacteroides, which are exceedingly difficult to grow without special anaerobic techniques.

Clinical Patterns of Illness

Bacteroides are an *occasional* cause of:

Abscesses. Abdominal abscesses, particularly after bowel perforation, or perforated appendicitis, are frequently caused by bacteroides.[196,197] Other abscesses occasionally associated with bacteroides include cervical abscess and brain abscess.

Uterine and Genital Infections and Pelvic Thrombophlebitis. *Bacteroides* are normal inhabitants of the female genital tract, and are an occasional cause of Bartholin's abscesses and adnexal abscesses, as well as postoperative uterine infections.[194] Septic abortion may be associated with bacteroides bacteremia.[189] Acute salpingitis, when it is not gonococcal, is usually associated with mixed aerobic and anaerobic flora, especially bacteroides.[187]

Bacteremia. Bacteroides bacteremia is typically associated with bowel disease, particularly postoperatively, but also can occur as a complication of mucosal ulcerations, adenocarcinoma, or diverticulitis.[190,196,199] In one series, many patients had a marked leukocytosis (greater than 25,000).[196] In a more recent series, bacteremia was noted to have a much higher mortality rate in patients over 40 years of age, and was more frequently associated with jaundice and liver abscess than bacteremia due to other organisms.[189]

Human Bites. Tooth injuries of the fist and human bites are often associated with crushed and necrotic skin which has been inoculated with mouth flora. *Fusobacterium* species are fastidious gram negative rods which are normal oral flora and which may infect human bites.[192]

Other Purulent Infections. Meningitis, septic arthritis, chronic otitis media, chronic sinusitis or mastoiditis, and empyema are rarely due to bacteroides infection.[189]

Laboratory Diagnosis

Culture. Bacteroides need very anaerobic conditions. A candle jar is not anaerobic, but is used to provide carbon dioxide. Vacuum bottles used to obtain blood cultures are not anaerobic enough to support the growth of most anaerobes. Thioglycollate broth, a standard anaerobic broth, is clearly inferior to other prereduced liquid media for recovering bacteroides. Although most clinically important anaerobes are relatively aerotolerant, and survive some delay in processing,[184] the clinician should constantly insist on prompt transport and processing of specimens which are likely to contain anaerobes.

Gram Stain. It is particularly useful to examine a Gram stain of pus from abscesses, as gram-negative rods may be seen, but will not be found on culture if aerobic techniques are used.

Fluorescent antibody staining can be done of smears of pus and body fluids for rapid detection of *B. fragilis* and some other bacteroides. This technique has practical value because of the slowness of anaerobic culture methods.

Serum Antibodies. Patients with severe bacteroides infections often develop antibodies, which are being studied for possible diagnostic value.

Biologic Features of Clinical Interest

Fluorescence. B. melaninogenicus is named for the black colonies it produces on blood agar. The colonies also produce a red fluorescence, under a Wood's light, which produces a portion of the ultraviolet spectrum. This fluorescence can also be observed in skin ulcers and abdominal abscesses where this organism is present, and can be useful in the clinical diagnosis of infection by this organism.[191]

Mixed Infections. Bacteroides infections are often associated with aerobic organisms; for example, in acute nongonococcal salpingitis and abdominal abscesses. When a mixture of fecal flora is inoculated intraperitoneally in experimental animals, antibiotic treatment of the aerobes only produces a chronic abdominal abscess.[200] Effective treatment of the anaerobes only allows the aerobes to produce an acute abscess.

Treatment

Antibiotics. Chloramphenicol or clindamycin appear to be the most effective antibiotics against bacteroides, by in vitro methods.[198] Metronidazole is synergistic with clindamycin against B. fragilis.[186]

Aspiration pneumonia or lung abscess typically responds very well to parenteral penicillin therapy, although the *Bacteroides* species in the oral flora are occasionally resistant to penicillin.[185]

REFERENCES

Salmonella

1. Aserkoff, B., and Bennett, J.V.: Effect of antibiotic therapy in acute salmonellosis on the fecal excretion of salmonellae. New Engl. J. Med., 281:636–640, 1969.
2. Aserkoff, B., Schreoder, S.A., and Brachman, P.S.: Salmonellosis in the United States—a five year review. Am. J. Epidemiol., 92:13–24, 1970.
3. Black, P.H., Kunz, L.J., and Swartz, M.N.: Salmonellosis—a review of some unusual aspects. New Engl. J. Med., 262:811–817, 864–870, 921–927, 1960.
4. Center for Disease Control: Morbidity Mortality Weekly Report Annual supplements. Summaries, 1970–1972.
5. ———— Salmonella surveillance. p. 5–6. Annual summary—1971, issued October, 1972.
6. Cohen, P.S., O'Brien, T.F., Schoenbaum, S.C., and Medeiros, A.A.: The risk of endothelial infection in adults with salmonella bacteremia. Ann. Intern. Med., 89:931–932, 1978.
7. Collins, R.N., Marine, W.M., and Nahmias, A.J.: The 1964 epidemic of typhoid fever in Atlanta. Clinical and epidemiologic observations. J.A.M.A., 197:179–184, 1966.
8. Colon, A.R., Gross, D.R., and Tamer, M.A.: Typhoid fever in children. Pediatrics, 56:606–609, 1975.
9. Dixon, J.M.S.: Effect of antibiotic treatment on duration of excretion of *Salmonella typhimurium* by children. Br. Med. J., 2:1343–1345, 1965.
10. Edwards, E.A., and Hilderbrand, R.L.: Method for identifying *Salmonella* and *Shigella* directly from the primary isolation plate by coagglutination of Protein A-containing staphylococci sensitized with specific antibody. J. Clin. Microbiol., 3:339–343, 1976.
11. Gangarosa, E.J., et al.: An epidemic-associated episome? J. Infect. Dis., 126:215–218, 1972.
12. Garcia de Olarte, D., et al.: Treatment of diarrhea in malnourished infants and children. Am. J. Dis. Child., 127:379–388, 1974.
13. Gershman, M.: Single phage-typing set for differentiating salmonellae. J. Clin. Microbiol., 5:302–314, 1977.
14. Gilman, R.H., et al.: Comparison of trimethoprim–sulfamethoxazole and amoxicillin in therapy of chloramphenicol-resistant and chloramphenicol-sensitive typhoid fever. J. Infect. Dis., 132:630–636, 1975.
15. Guckian, J.C., Byers, E.H., and Perry, J.E.: *Arizona* infection of man. Report of a case and review of the literature. Arch. Intern. Med., 119:170–175, 1967.
16. Hornick, R.B., et al.: Typhoid fever: pathogenesis and immunologic control. New Engl. J. Med., 283:686–691, 739–746, 1970.
17. Haltalin, K.C., Kusmiexz, H.T., Hinton, L.V., and Nelson, J.D.: Treatment of acute diarrhea in outpatients. Double-blind study comparing ampicillin and placebo. Am. J. Dis. Child., 124:554–561, 1972.
18. Levy, B.S.: The economic impact of a food-borne salmonellosis outbreak. J.A.M.A., 230:1281–1282, 1974.
19. McCall, C.E, et al.: Delineation of chronic carriers of *Salmonella derby* within an institution for incurables. Antimicrob. Agents Chemother., 4:717–721, 1964.

20. Meals, R.A.: Paratyphoid fever. A report of 62 cases with several unusual findings and a review of the literature. Arch. Intern. Med., 136:1422–1428, 1976.

21. National Communicable Disease Center: *Salmonella* surveillance. A large outbreak of salmonellosis following a turkey barbecue. Report No. 67:2–3, 1967.

22. Rabinowitz, S.G., and MacLeod, N.R.: *Salmonella* meningitis. A report of three cases and review of the literature. Am. J. Dis. Child., 123:259–262, 1972.

23. Rice, P.A., Baine, W.B., and Gangarosa, E.J.: *Salmonella typhi* infections in the United States, 1967-1972: increasing importance of international travellers. Am. J. Epidemiol., 106:160–166, 1977.

24. Robertson, R.P., Wahab, M.F.A., and Raasch, F.O.: Evaluation of chloramphenicol and ampicillin in salmonella enteric fever. New Engl. J. Med., 278:171–176, 1968.

25. Rosenstein, B.J.: Salmonellosis in infants and children. J. Pediatr., 70:1–7, 1967.

26. Ryder, R.W., Crosby–Ritchie, A., McDonough, B., and Hall, W.J., III: Human milk contaminated with *Salmonella kottbus*. A cause of nosocomial illness in infants. J.A.M.A., 238:1533–1534, 1977.

27. Sardesai, H.V., Karandikar, R.S., and Harshe, R.G.: Comparative trial of co-trimoxazole and chloramphenicol in typhoid fever. Br. Med. J., 1:82–83, 1973.

28. Schroeder, F.A., Terry, T.M., and Bennett, J.V.: Antibiotic resistance and transfer factor in salmonella, United States, 1967. J.A.M.A., 205:903–906, 1968.

29. Simon, H.J., and Miller, R.C.: Ampicillin in the treatment of chronic typhoid carriers. Report of fifteen treated cases and a review of the literature. New Engl. J. Med., 274:807–815, 1966.

30. Snyder, M.J., et al.: Comparative efficacy of chloramphenicol, ampicillin, and co-trimoxazole in the treatment of typhoid fever. Lancet, 2:1155–1157, 1976.

31. Strate, R.W., and Bannayan, G.A.: Typhoid fever causing massive lower gastrointestinal hemorrhage. A reminder of things past. J.A.M.A., 236:1979–1980, 1976.

32. Woodward, T.E., and Smadel, J.E.: Management of typhoid fever and its complications. Ann. Intern. Med., 60:144–157, 1964.

33. Zellweger, H., and Idriss, H.: Encephalopathy in salmonella infections. Am. J. Dis. Child., 99:770–777, 1960.

Shigella

34. Barrett–Connor, E., and Connor, J.D.: Extra-intestinal manifestations of shigellosis. Am. J. Gastroenterol., 53:234–245, 1970.

35. Center for Disease Control: *Shigella dysenteriae* 1–Colorado. Morbidity Mortality Weekly Rep., 22:101–102, 1973.

36. Chang, M.J., et al.: Trimethoprim–sulfamethoxazole compared to ampicillin in the treatment of shigellosis. Pediatrics, 59:726–729, 1977.

37. Davies, J.E., and Rownd, R.: Transmissible multiple drug resistance in *Enterobacteriaceae*. Science, 176:758–768, 1972.

38. Dulaney, E.L., and Laskin, A.I. (eds.): The problems of drug-resistant pathogenic bacteria. Ann. N.Y. Acad. Sci., 182:1–415, 1971.

39. Dupont, H.L., et al.: The response of man to virulent *Shigella flexneri* 2a. J. Infect. Dis., 119:296–299, 1969.

40. ———: Immunity in shigellosis. I. Response of man to attenuated strains of shigella. II. Protection induced by oral live vaccine or primary infection. J. Infect. Dis., 125:5–11, 12–16, 1972.

41. DuPont, H.L., and Hornick, R.B.: Adverse effect of Lomotil therapy in shigellosis. J.A.M.A., 226:1525–1528, 1973.

42. Farrar, W.E., and Dekle, L.C.: Transferable antibiotic resistance associated with an outbreak of shigellosis. Ann. Intern. Med., 67:1208–1215, 1967.

43. Formal, S.B., Abrams, G.D., Schneider, H., and Sprinz, H.: Experimental shigella infections. VI. Role of the small intestine in an experimental infection in guinea pigs. J. Bacteriol., 85:119–125, 1962.

44. Gardner, P., and Smith, D.H.: Studies on the epidemiology of Resistance (R) factors. Ann. Intern. Med., 71:1–9, 1969.

45. Haltalin, K.C., et al.: Double-blind treatment study of shigellosis comparing ampicillin, sulfadiazine, and placebo. J. Pediatr., 70:970–981, 1967.

46. ———: Comparison of orally absorbable and nonabsorbable antibiotics in shigellosis. J. Pediatr., 72:708–720, 1968.

47. Haltalin, K.C., and Nelson, J.D.: Coliform septicemia complicating shigellosis in children. J.A.M.A., 192:441–443, 1965.

48. Haltalin, K.C., Woodman, E., and Nelson, J.D.: Colicin types of *Shigella sonnei* in relation to antibiotic resistance. J. Infect. Dis., 132:307–310, 1976.

49. Harris, J.C., DuPont, H.L., and Hornick, R.B.: Fecal leukocytes in diarrheal illness. Ann. Intern. Med., 76:697–703, 1972.

50. Hoefnagel, D.: Fulminating, rapidly fatal shigellosis in children. New Engl. J. Med., 258:1256–1257, 1958.

51. Keusch, G.T., and Jacewicz, M.: The pathogenesis of Shigella diarrhea. VI. Toxin and antitoxin in *Shigella flexneri* and *Shigella sonnei* infections in humans. J. Infect. Dis., 135:552–556, 1977.

114. Wilfert, J.N., Barrett, F.F., and Kass, E.H.: Bacteremia due to *Serratia marcescens*. New Engl. J. Med., 279:286–289, 1968.

Pseudomonas

115. Abbe, J.S., and Moffet, H.L.: Surveillance of *Pseudomonas aeruginosa* infections in a children's hospital. Antimicrob. Agents Chemother., 1970:303–308, 1971.
116. Alexander, J.W., and Fisher, M.W.: Immunization against *Pseudomonas* in infections after thermal injury. J. Infect. Dis., 130 (Suppl.):S152–S162, 1974.
117. Baltch, A.L., and Griffin, P.E.: *Pseudomonas aeruginosa* bacteremia: a clinical study of 75 patients. Am. J. Med. Sci., 274:119–129, 1977.
118. Crowder, J.C., and White, A.: A serologic response in human pseudomonas infection. J. Lab. Clin. Med., 75:128–136, 1970.
119. Dalton, A.C., and Smith, G.M.: Mucoid *Pseudomonas aeruginosa* causing meningitis. Am. J. Clin. Pathol., 55:723–725, 1971.
120. Dorff, G.J., Geimer, N.F., Rosenthal, D.R., and Rytel, M.W.: Pseudomonas septicemia. Illustrated evolution of its skin lesion. Arch. Intern. Med., 128:591–595, 1971.
121. Edmonds, P., Saskind, R.R., MacMillan, B.G., and Holder, I.A.: Epidemiology of *Pseudomonas aeruginosa* in a burns hospital: evaluation of serological, bacteriophage, and pyocin typing methods. Appl. Microbiol., 24:213–218, 1972.
122. Farmer, J.J., and Herman, L.G.: Epidemiological fingerprintings of *Pseudomonas aeruginosa* by the production of and sensitivity to pyocin and bacteriophage. Appl. Microbiol., 18:760–765, 1969.
123. Gilardi, G.L.: Infrequently encountered *Pseudomonas* species causing infections in humans. Ann. Intern. Med., 77:211–215, 1972.
124. Knepper, J.G., and Anthony, B.F.: Diminished growth of *Pseudomonas aeruginosa* in unvented blood-culture bottles. Lancet, 2:285–287, 1973.
125. Kominos, S.D., Copeland, C.E., Grosiak, B., and Postic, B.: Introduction of *Pseudomonas aeruginosa* into a hospital via vegetables. Appl. Microbiol., 24:567–570, 1972.
126. Lewis, R., Gorbach, S., and Altner, P.: Spinal Pseudomonas chondroosteomyelitis in heroin users. New Engl. J. Med., 286:1303, 1972.
127. Liu, P.V.: Extracellular toxins of *Pseudomonas aeruginosa*. J. Infect. Dis., 130(Suppl.):S94–S99, 1974.
128. Lowbury, E.J.L., et al.: Sources of infection with *Pseudomonas aeruginosa* in patients with tracheostomy. J. Med. Microbiol., 3:39–56, 1970.
129. Moore, B., and Forman, A.: An outbreak of urinary *Pseudomonas aeruginosa* infection acquired during urological operations. Lancet, 2:929–931, 1966.
130. Pennington, J.E., et al.: Use of a *Pseudomonas aeruginosa* vaccine in patients with acute leukemia and cystic fibrosis. Am. J. Med., 58:629–636, 1975.
131. Polk, H.C., Jr., Ward, C.B., Clarkson, J.G., and Taplin, D.: Early detection of *Pseudomonas* burn infection. Clinical experience with Wood's light fluorescence. Arch. Surg., 98:292–295, 1969.
132. Rabin, E.R., et al.: Fatal *Pseudomonas* infection in burned patients. A clinical, bacteriologic and anatomic study. New Engl. J. Med., 265:1225–1231, 1961.
133. Reed, R.K., Larter, W.E., Sieber, O.F., Jr., and John, T.J.: Peripheral nodular lesions in pseudomonas sepsis: the importance of incision and drainage. J. Pediatr., 88:977–979, 1976.
134. Reynolds, H.Y., DiSant'Agnese, P.A., and Zierdt, C.H.: Mucoid *Pseudomonas aeruginosa*. A sign of cystic fibrosis in young adults with chronic pulmonary disease? J.A.M.A., 236:2190–2192, 1976.
135. Saelinger, C.B., Snell, K., and Holder, I.A.: Experimental studies on the pathogenesis of infections due to *Pseudomonas aeruginosa*: direct evidence for toxin production during pseudomonas infection of burned skin tissues. J. Infect. Dis., 136:555–561, 1977.
136. Tilton, R.C., Steingrimsson, O., and Ryan, R.W.: Susceptibilities of *Pseudomonas* species to tetracycline, minocycline, gentamicin, and tobramycin. Am. J. Clin. Pathol., 69:410–413, 1978.
137. Washburn, J., Jacobson, J.A., Marston, E., and Thorsen, B.: *Pseudomonas aeruginosa* rash associated with a whirlpool. J.A.M.A., 235:2205–2207, 1976.
138. Weber, D.R., Douglass, L.E., Brundage, W.G., and Stallkamp, T.C.: Acute varieties of melioidosis occurring in U.S. soldiers in Vietnam. Am. J. Med., 46:234–244, 1969.

Non-Fermenter Group

139. Anon.: Water bugs in the bassinet (editorial). Am. J. Dis. Child., 101:273–277, 1961.
140. Donald, W.D.: Studies on *Mimeae* organisms as related to infants and children. Pediatrics, 37:756–761, 1966.
141. Dworzack, D.L., Murray, C.M., Hodges, G.R., and Barnes, W.G.: Community-acquired bacteremic *Achromobacter xylosoxidans* Type IIIa pneumonia in a patient with idiopathic IgM deficiency. Am. J. Clin. Pathol., 70:712–717, 1978.

142. Elston, H.R., and Hoffman, K.C.: Identification and clinical significance of *Achromobacter (Herellea) anitratus* in urinary tract infections. Am. J. Med. Sci., *251*:75–80, 1966.

143. Fedukowicz, H., and Horwich, H.: The gram-negative diplobacillus in hypopyon deratitis. Arch. Ophthalmol., *49*:202–211, 1953.

144. Feigin, R.D., San Joaquin, V., and Middelkamp, J.N.: Septic arthritis due to *Moraxella osloensis*. J. Pediatr., *75*:116–117, 1969.

145. Foley, J.F., Gravelle, C.R., Englehard, W.E., and Chin, T.D.Y.: *Achromobacter* septicemia—fatalities in prematures. Am. J. Dis. Child., *101*:279–288, 1961.

146. Gardner, P., Griffin, W.B., Swartz, M.N., and Kunz, L.J.: Non-fermentative gram-negative bacilli of nosocomial interest. Am. J. Med., *48*:735–749, 1970.

147. Glew, R.H., Moellering, R.C., Jr., and Buettner, K.R.: In vitro synergism between carbenicillin and aminoglycosidic aminocyclitols against *Acinetobacter calcoaceticus* var. *anitratus*. Antimicrob. Agents Chemother., *11*:1036–1041, 1977.

148. Hazuka, B.T., Dajani, A.S., Talbot, K., and Keen, B.M.: Two outbreaks of *Flavobacterium meningosepticum* Type E in a neonatal intensive care unit. J. Clin. Microbiol., *6*:450–455, 1977.

149. Hendriksen, S.D.: Moraxella, Acinetobacter, and the Mimae. Bacteriol. Rev., *37*:522–561, 1973.

150. Hugh, R., and Leifson, E.: The toxonomic significance of fermentative versus oxidative metabolism of carbohydrates by various gram-negative bacteria. J. Bacteriol., *66*:24–26, 1953.

151. Kuck, N.A.: In vitro and in vivo activities of minocycline and other antibiotics against *Acinetobacter (Herellea-Mima)*. Antimicrob. Agents Chemother., *9*:493–497, 1976.

152. Moffet, H.L., and Williams, T.: Bacteria recovered from distilled water and inhalation therapy equipment. Am. J. Dis. Child., *114*:7–12, 1967.

153. Oberhofer, T.R., Rowen, J.W., and Cunningham, G.F.: Characterization and identification of gram-negative, nonfermentative bacteria. J. Clin. Microbiol., *5*:208–220, 1977.

154. Pickett, M.J., and Manclark, C.R.: Tribe Mimeae. An illegitimate epithet. Am. J. Clin. Pathol., *43*:161–165, 1965.

155. Rios, I., Klimek, J.J., Maderazo, E., and Quintiliani, R.: *Flavobacterium meningosepticum* meningitis: report of selected aspects. Antimicrob. Agents Chemother., *14*:444–447, 1978.

156. Rockhill, R.C., and Lutwick, L.I.: Group IVe-like gram-negative bacillemia in a patient with obstructive uropathy. J. Clin. Microbiol., *8*:108–109, 1978.

157. Robinson, R.G., Garrison, R.G., and Brown, R.W.: Evaluation of the clinical significance of the genus *Herellea*. Ann. Intern. Med., *60*:19–27, 1964.

158. Sherman, J.D., Ingall, D., Wiener, J., and Pryles, C.V.: *Alcaligenes faecalis* infection in the newborn. Am. J. Dis. Child., *100*:212–216, 1960.

159. Wallace, R.J., Jr., Awe, R.J., and Martin, R.R.: Bacteremic *Acinetobacter (Herellea)* pneumonia with survival. Am. Rev. Respir. Dis., *113*:695–699, 1976.

Vibrio Species

160. Adams, M.M.E.: Cholera: new aids in treatment and prevention. Science, *179*:552–555, 1973.

161. Benyajati, C.: Experimental cholera in humans. Br. Med. J., *1*:140–142, 1966.

162. Blake, P.A., et al.: Disease caused by a marine vibrio. Clinical characteristics and epidemiology. New Engl. J. Med., *300*:1–5, 1979.

163. Brown, W.J. and Sautter, R.: *Campylobacter fetus* septicemia with concurrent salpingitis. J. Clin. Microbiol., *6*:72–75, 1977.

164. Bullock, S.L., and Wallis, K.W.: Evaluation of some of the parameters of the enzyme-linked immunospecific assay. J. Infect. Dis., *136(suppl.)*:S279–S285, 1977.

165. Carpenter, C.C.J. and Hirschhorn, N.: Pediatric cholera: current concepts of therapy. J. Pediatr., *80*:874–878, 1972.

166. Cash, R.A., et al.: Response of man to infection with *Vibrio cholerae*. I. Clinical, serologic, and bacteriologic responses with a known inoculum. J. Infect. Dis., *129*:45–52, 1974.

167. Center for Disease Control: International notes. Cholera—worldwide. Morbidity Mortality Weekly Rep., *21*:170–171, 1972.

168. Dadisman, T.A., Jr., Nelson, R., Molenda, J.R. and Garber, H.J.: *Vibrio parahaemolyticus* gastroenteritis in Maryland. I. Clinical and epidemiologic aspects. Am. J. Epidemiol., *96*:414–426, 1973.

169. Eden, A.N.: Perinatal mortality caused by *Vibrio fetus*. J. Pediatr., *68*:297–304, 1966.

170. Engvall, E., and Perlmann, P.: Enzyme-linked immunosorbent assay, ELISA. III. Quantitatiion of specific antibodies by enzyme-labeled, anti-immunoglobulin in antigen-coated tubes. J. Immunol., *109*:129–135, 1972.

171. Gangarosa, E.J.: The epidemiology of cholera: past and present. Bull. N.Y. Acad. Med., *47*:1140–1151, 1971.

172. Gangarosa, E.J.: Cholera. Implications for the United States. J.A.M.A., *227*:170–171, 1974.

173. Guerrant, R.L., Lahita, R.G., and Washington, W.C., Jr.: Campylobacteriosis in man: pathogenic mechanisms and review of 91 bloodstream infections. Am. J. Med., *65*:584–592, 1978.

174. Hughes, J.M., Hollis, D.G., Gangarosa, E.J., and Weaver, R.E.: Non-cholera vibrio infections in the United States. Ann. Int. Med., *88*:602–606, 1978.

175. King, E.O.: The laboratory recognition of *Vibrio fetus* and a closely related Vibrio isolated from cases of human vibriosis. Ann. N.Y. Acad. Sci., *98*:700–711, 1962.

176. Rubin, S.J. and Tilton, R.C.: Isolation of *Vibrio alginolyticus* from wound infections. J. Clin. Microbiol., *2*:556–558, 1975.

177. Sheehy, T.W., Sprinz, H., Augerson, W.S. and Formal, S.B.: Laboratory *Vibrio cholerae* infection in the United States. J.A.M.A., *197*:321–326, 1966.

178. Skirrow, M.B.: Campylobacter enteritis: a new disease. Br. Med. J., *2*:9–10, 1977.

179. Soonattrakul, W., Andersen, B.R. and Bryner, J.H.: Raw liver as a possible source of *Vibrio fetus* septicemia in man. Am. J. Med. Sci., *261*:245–249, 1971.

180. Targan, S.R., Chow, A.W., and Guze, L.B.: *Campylobacter fetus* associated with pulmonary abscess and empyema. Chest, *71*:105–108, 1977.

181. Weissman, J.B., et al.: A case of cholera in Texas, 1973. Am. J. Epidemiol., *100*:487–498, 1975.

182. Yalow, R.S.: Radioimmunoassay: a probe for the fine structure of biologic systems. Science, *200*:1236–1245, 1978.

183. Zide, N., Davis, J. and Ehrenkranz, N.J.: Fulminating *Vibrio parahaemolyticus* septicemia. Arch. Int. Med., *133*:479–481, 1974.

Bacteriodes

184. Bartlett, J.G., Sullivan-Sigler, N., Louie, T.J., and Gorbach, S.L.: Anaerobes survive in clinical specimens despite delayed processing. J. Clin. Microbiol., *3*:133–136, 1976.

185. Bartlett, J.G., and Gorbach, S.L.: Treatment of aspiration pneumonia and primary lung abscess. Penicillin G vs. clindamycin. J.A.M.A., *234*:935–937, 1975.

186. Busch, D.F., Sutter, V.L., and Finegold, S.M.: Activity of combinations of antimicrobial agents against *Bacteroides fragilis*. J. Infect. Dis., *133*:321–328, 1976.

187. Eschenbach, D.A., et al.: Polymicrobial etiology of acute pelvic inflammatory disease. New Engl. J. Med., *293*:166–171, 1975.

188. Felner, J.M., and Dowell, V.R., Jr.: Bacteroides bacteremia. Am. J. Med., *50*:787–796 1971.

189. Gelb, A.F., and Seligman, S.J.: *Bacteriodaceae* bacteremia. Effect of age and focus of infection upon clinical course. J.A.M.A., *212*:1038–1041, 1970.

190. Mathias, R.G., et al.: Bacteremia due to Bacteroidaceae: a review of 92 cases. J. Infect. Dis., *135(Suppl.)*:S69–S73, 1977.

191. Meyers, M.B., Cherry, G., Bornside, B.B., and Bornside, G.H.: Ultraviolet red fluorescence of *Bacteroides melaninogenicus*. Appl. Microbiol., *17*:760–762, 1969.

192. Murphy, R., Katz, S., and Massaro, D.: *Fusobacterium* septicemia following a human bite. Arch. Intern. Med., *111*:51–53, 1963.

193. Onderdonk, A.B., Kasper, D.L., Cisneros, R.L., and Bartlett, J.G.: The capsular polysaccharide of *Bacteroides fragilis* as a virulence factor: comparison of the pathogenic potentials of encapsulated and unencapsulated strains. J. Infect. Dis., *136*:82–89, 1977.

194. Pearson, H.E., and Anderson, G.V.: Genital bacteroidal abscesses in women. Am. J. Obstet. Gynecol., *107*:1264–1265, 1970.

195. Polk, B.F., and Kasper, D.L.: *Bacteroides fragilis* subspecies in clinical isolates. Ann. Intern. Med., *86*:569–571, 1977.

196. Saksena, D.S., Block, M.A., McHenry, M.C., and Truant, J.P.: *Bacteroidaceae:* anaerobic organisms encountered in surgical infections. Surgery, *63*:261–267, 1968.

197. Sanders, D.Y., and Stevenson, J.: Bacteroides infections in children. J. Pediatr., *72*:673–677, 1968.

198. Sutter, V.L., Kwok, Y-Y., and Finegold, S.M.: Susceptibility of *Bacteroides fragilis* to six antibiotics determined by standardized antimicrobial disc susceptibility testing. Antimicrob. Agents Chemother., *3*:188–193, 1973.

199. Tynes, B.S., and Frommeyer, W.B., Jr.: Bacteroides septicemia. Cultural, clinical, and therapeutic features in a series of twenty-five patients. Ann. Intern. Med., *56*:12–65, 1962.

200. Weinstein, W.M., et al.: Antimicrobial therapy of experimental intra-abdominal sepsis. J. Infect. Dis., *132*:282–286, 1975.

4
Small Gram-Negative Rods

HAEMOPHILUS INFLUENZAE

Objectives

1. Describe how *Haemophilus influenzae* is distinguished from other species of *Haemophilus*.
2. Describe what special procedures are necessary in order to detect *Haemophilus influenzae* from clinical specimens.
3. Describe how *H. influenzae* is typed, and indicate which type is the major cause of disease.
4. Describe two life-threatening acute illnesses that can be caused by *H. influenzae*.
5. Describe the frequency of *H. influenzae* infection in different age groups.
6. Discuss some of the problems involved in testing *H. influenzae* for antibiotic susceptibilities.
7. Discuss the usual antibiotic therapy for *H. influenzae*, the antibiotics that are considered adequate or acceptable, and the antibiotics that are considered not effective. Describe the mechanism by which *H. influenzae* develops resistance to ampicillin.
8. Discuss protective antibodies to *H. influenzae*, their production by disease, immunization, and cross-reacting bacteria.
9. Discuss the problems of prevention of transmission of *H. influenzae* type b disease.

Definitions

Haemophilus influenzae is a small gram-negative rod, or coccobacillus, which may be mistaken for a diplococcus. Haemophilus (formerly spelled hemophilus) means blood-loving, and refers to the fact that members of the genus Haemophilus require growth factors found in blood. *H. influenzae* was recovered from the nasopharynx of many patients during the influenza epidemic of 1890. It was thought to be the cause of influenza, and so was called the influenza bacillus.

Species and Types

Six types (a through f) can be defined using rabbit antisera prepared against the soluble capsular substance. Type b is the type responsible for almost all cases of severe disease, although any of the other types may be pathogenic.[10] Type b differs from all other types in that it contains polyribose phosphate (PRP) in its capsule.[33] Antibodies to PRP protect against disease, and presumably the PRP capsule is the main reason type b is the major virulent type of *H. influenzae*. In this chapter, *H. influenzae* is often used to mean *H. influenzae* type b.

Other *Haemophilus* species, such as *H. parainfluenzae*, may occasionally be associated with severe disease such as epiglottitis, endocarditis, or meningitis, but this is very uncommon.[15,16] *H. aegyptius*, which closely resembles *H. influenzae*, is an occasional cause of purulent conjunctivitis. *Haemophilus ducreyi* is the cause of a veneral disease called chancroid, which produces genital lesions resembling the chancre of syphilis (see Fig. 7-3).[11]

Other *Haemophilus*-like organisms that are rare causes of disease include *H. aphrophilis*, *Actinobacillus actinomycetemcomitans*, and other small gram-negative rods resembling *Haemophilus* species are discussed in Chapter 6. *Haemophilus vaginalis* is also called *Corynebacterium vaginale* and is discussed briefly in Chapter 2. This organism, which causes vaginitis, does not fit neatly into either genus.

Frequency and Importance

H. influenzae type b is an extremely frequent and important pathogen of children. It is the most frequent cause of purulent meningitis in children, and thus may produce brain damage or death. It is the most frequent cause of fatal respiratory obstruction due to supraglottic laryngitis (also called epiglottitis).

Clinical Patterns of Illness

In children, *H. influenzae* type b is a *very frequent* cause of:

Purulent Meningitis. *H. influenzae* is the most frequent cause of purulent meningitis in the United States. It occurs most frequently in young children, but occasionally occurs in older children and adults.[25] In one hospital which admitted patients of all ages with meningitis, 97 percent of the *H. influenzae* meningitis occurred in children less than 15 years old, whereas about 65 percent of the meningococcal meningitis, and about 36 percent of the pneumococcal meningitis were in children less than 15 years of age.[22] Almost all *H. influenzae* meningitis is due to type b.

Laryngitis with epiglottitis is an important emergency in children, as the edema and inflammation may produce fatal obstruction of the larynx.[18] Typically, the child has a high fever, inspiratory stridor, and hoarseness or aphonia. The epiglottis typically appears cherry red and swollen. Tracheostomy or tracheal intubation is usually needed as an emergency procedure.

Ear and Eye Infections. Purulent otitis media,[32] conjunctivitis, and sinusitis are typically caused by non-typable *H. influenzae*, and orbital cellulitis and facial cellulitis[29] are almost always caused by type b.

Septic Arthritis. *H. Influenzae* ranks second after *Staph. aureus* as a cause of septic arthritis in preschool children.[12] Cellulitis without underlying arthritis is also often due to *H. influenzae*.

Bacteremia. *H. influenzae* is a frequent cause of bacteremia. It often recovered from the blood in patients with purulent meningitis, acute epiglottitis, septic arthritis, facial cellulitis, or orbital cellulitis. It also is recovered from the blood of children with no source of infection seen, so-called occult bacteremia.[40]

H. influenzae type b *occasionally* causes:

Pneumonia. *H. influenzae* was recovered in about 10 to 15 percent of pneumonia in young children who had lung punctures in the pre-antibiotic era. Acute pneumonia in adults with severe chronic lung disease often is due to non-typable *H. influenzae*.[17] Acute lobar pneumonia in young adults is occasionally caused by *H. influenzae* as proved by recovery of the organism from the blood.[41]

Bronchitis. *H. influenzae* recovered from the respiratory tract in bronchitis is frequently nontypable. However, in adults with chronic pulmonary disease, *H. influenzae* is a common cause of acute exacerbations.

Genitourinary Infections in Adults. Bacteremia with *H. influenzae* (usually untypable) has been observed in adults with postpregnancy or postabortion endometritis, acute salpingitis and urinary tract instrumentation.[1] It has also rarely been recovered from children with urinary tract infections, and infants with epididymitis.

Asymptomatic or Subclinical Infections

In the past, most individuals had protective antibodies by the time they entered primary school at about 6 years of age. A study in 1933 showed most adults had bactericidal antibodies to type b *H. influenzae*, but recent studies indicate that approximately 30 percent of adults have no protective antibody.[7] This is consistent with the observation that type b *H. influenzae* infections are increasing in adults, and also in newborn infants, who are less likely to be protected by transplacental antibody.[7]

Most individuals with type b *H. influenzae* bactericidal antibody have not had an illness, such as meningitis or epiglottitis, that can be clearly related to *H. influenzae*. A possible explanation for their antibody is that some immunity to type b *H. influenzae* is due to antibody stimulated by cross-reacting antigens from enteric bacteria or pyogenic bacteria.[2] This heterologous antibody may provide some protection, resulting in a milder illness when the patient is exposed to type b *H. influenzae*.

Since most individuals have a minor illness when infected with type b *H. influenzae*, the patient with a severe *H. influenzae* infection is usually not regarded as a contagion risk. However, a child carries *H. influenzae* type b after a severe illness such as meningitis or epiglottitis for several months.[23] Since meningitis due to type b *H. influenzae* has been observed in siblings of a child with such meningitis, antibiotic prophylaxis for siblings has been suggested. Unfortunately, antibiotic prophylaxis for siblings or close contacts may not be effective, because it may not eradicate the carrier state permanently, and therefore may not be effective for the long term, because it may result in permanent immunity. Close contacts in a day care center or nursery school also may be candidates for ampicillin prophylaxis,[42] although close observation and prompt treatment of symptomatic children is a reasonable alternative.

Laboratory Approach

Culture. *H. influenzae* has special growth requirements. The medium usually used is chocolate blood agar, in which blood has been heated and chocolatized (turned brown). Additional amounts of the growth factors already present in blood are sometimes added to the chocolate agar: X factor, which is hemin, and V factor, which is DPN or TPN.

After the specimen has been inoculated, the chocolate plate should be incubated in 10 percent CO_2 in a candle jar. *H. influenzae* may not be isolated unless the laboratory routinely utilizes cultural methods sufficient for growth of this organism, or unless the physician notifies the laboratory that the organism is suspected. Fortunately, almost all laboratories routinely set up spinal fluid specimens with conditions adequate to grow *H. influenzae*. However, most laboratories do not routinely set up respiratory specimens on chocolate agar with CO_2; in such situations the physician should inform the laboratory if the patient has a clinical illness known to be associated with *H. influenzae*.

Early identification of the organism after overnight incubation can sometimes be made on the basis of the characteristic mousey smell of the colonies, a stain showing gram-negative coccobacilli, and agglutination by type-specific antiserum of a colony suspended in saline. Laboratory identification of the various *Haemophilus* species is based on the factors required for growth of the organism. The X and V factors can be provided by commercially available disks, which are useful in distinguishing *H. influenzae* from *H. parainfluenzae*[27] (Fig. 4-1).

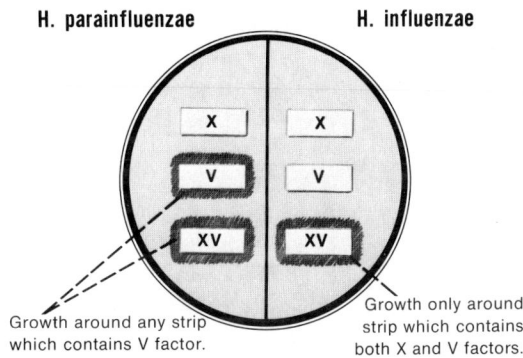

FIG. 4-1. *Identification of* Haemophilus *species by use of X, V, and XV strips. The isolate is streaked over a plate of a medium which lacks X and V factors. H. influenzae grows only around the strip containing both X and V factors.*

Gram Stain. In *H. influenzae* meningitis, Gram stain of the spinal fluid typically reveals gram-negative coccobacilli. These short, small rods are sometimes mistaken for the diplococci of the meningococci.

Quellung Reaction. When type b antiserum is added to spinal fluid containing *H. influenzae* type b, capsular swelling (Quellung reaction) occurs. This is specific, and can also be done using polyvalent antiserum for all six types. Individual antisera are available for types other than b, but are usually not used, since almost all isolates from serious illnesses are type b.

Serum Antibodies. Serologic study is not practical for diagnosis of *H. influenzae* infection. However, antibodies against *H. influenzae* in patient's sera can be quantitated, and their use as a diagnostic technique is theoretically possible, and may be available in some research laboratories for special situations.

Antigen Detection. The capsular polysaccharide antigen can be detected by its reaction with antisera prepared in animals. This may be useful in testing spinal fluid of patients with meningitis, especially if the patient has been receiving antibiotics, and no organism is found on smear or culture.[35,38] Type b antigen or polyribosephos-

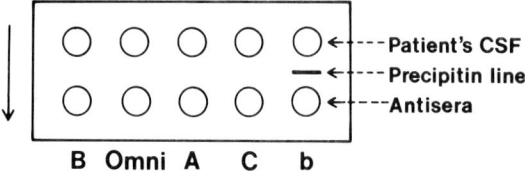

FIG. 4-2. *Countercurrentimmunoelectrophoresis (CIE) of a patient's CSF, which shows a precipitin line with type b* H. influenzae *(b). Antisera: B = Group B streptococci; Omni = pneumococcal omniserum; A and C = meningococcal serogroups A and C. The arrow shows the direction of the electric current.*

phate (PRP) can be detected by several methods, including coagglutination[35] (Fig. 1-12), latex fixation (Fig. 8-8) the ELISA method for antigen detection (Fig. 3-9), or counterimmunoelectrophoresis (Fig. 4-2).[38] Counterimmunoelectrophoresis (CIE) is usually less sensitive than latex particle agglutination.[38]

Biologic Features of Clinical Interest

Normal Flora. Non-typable *Haemophilus influenzae* is found in the nose or throat of about 20 to 35 percent of well babies.[27] However, type b *H. influenzae* is found in only about three to five percent of normal children.[24] Children recovered from severe *H. influenzae* type b disease such as meningitis or epiglottitis have a colonization rate as high as 80 percent.[23]

H. influenzae types other than type b need not be treated when found in the nose or throat, unless there is a clinical disease present, such as otitis media or bronchitis. Ordinarily, routine typing of *H. influenzae* recovered from respiratory isolates is not done. We do not know if antibiotic therapy should be given, when type b *H. influenzae* is found in asymptomatic individuals or patients with mild disease except for young children in closed populations like day care centers. Antibiotic therapy or close observation and antibiotic therapy for symptomatic patients has been advocated, but the choice should depend on the clinical details of the situation. Rifampin therapy to eradicate the carrier state appears promising.[9]

Inhibitory Effect of Sheep Blood. Sheep blood agar plates should be used for the diagnosis of streptococcal pharyngitis. TPN-ase and DPN-ase in the sheep blood destroy the V factor, thus inhibiting the growth of most *Haemophilus* species.[14] This is usually regarded as an advantage of sheep blood in blood agar plates, because it inhibits *Haemophilus hemolyticus (Haemophilus hemoglobinophilus)*. This species does not cause disease, but is a nuisance because it is beta-hemolytic, and the colony must be Gram stained to distinguish it from beta-hemolytic streptococci. Similarily, sheep blood agar plates, which should be used for throat cultures for beta-hemolytic streptococci, will usually not allow the detection of *Haemophilus influenzae*.

Satellite Phenomenon. *Haemophilus influenzae* may sometimes grow on a sheep blood agar plate near a colony of a hemolytic staphylococci, which release growth factors from the sheep erythrocytes by hemolysis. The growth of *Haemophilus* species near another colony of another species, usually a staphylococcus, is called the satellite phenomenon, and is sometimes used for preliminary laboratory identification of the organism.

Capsule. Young encapsulated coccobacillary forms are relatively virulent compared to older noncapsulated, filamentous forms. Since typability is usually related to the presence of a capsule, nontypable *H. influenzae* are usually regarded as relatively less virulent than typable isolates. Nevertheless, nontypable strains may have clinical importance.

Typability. The capsule and typability can easily be lost during passage in the laboratory. The type can be changed by transformation experiments with DNA, and type b antigen can be demonstrated by immunofluorescence even when the presence of capsules and typability by agglutination with antisera cannot be demonstrated.[3]

Experimental and natural viral infections in chimpanzees apparently produce a change in *H. influenzae* from the noncapsulated form to the capsulated and typable form.[4] These observations provide theoretical support for the clinical suspicion that some virus infections may increase the virulence of *H. influenzae*.

H. influenzae isolates that are not typable by slide agglutination, which involves the capsule, can often by typed by a microtiter agglutinin method, which apparently involves noncapsular somatic antigens, and indicates that commercial antibody contains somatic as well as capsular antibodies.[19] Thus, somatic antigens may provide a method of typing isolates previously regarded as nontypable. This method has not yet been adequately applied to clinical situations.

Cross Reactions with Other Species. The capsular polysaccharide of type b *H. influenzae* cross reacts with antigens of other species, including strains of *E. coli*, streptococci, and staphylococci.[2,31] This cross reaction is detected by observing a precipitin halo around colonies of other species, such as *E. coli*, grown in a medium containing type b antiserum, and is confirmed by demonstrating precipitin bands between type b antiserum and the *E. coli* antigen. These shared antigens can be used to stimulate protective antibodies, called heteroimmunization, in humans, as described below.

Antibiotic Susceptibility. These tests are sometimes difficult to perform accurately with *H. influenzae*, and results using different methods are conflicting. Typable strains appear to have antibiotic susceptibilities differing from nontypable strains. Capsulated type b strains recovered from spinal fluid are more susceptible to penicillin than are nontypable strains recovered from the respiratory tract.[8] Nontypable strains appear to be more susceptible to sulfonamides than are typable strains.

Other factors involved in conflicting susceptibility reports include the inoculum size and the problem of neutralization of sulfa inhibitors.[21]

Ampicillin Resistance. In 1974, an ampicillin-resistant type b *H. influenzae* was first detected in the United States. In general, the frequency of ampicillin-resistant *H. influenzae* appeared to be increasing in the 1970s, and ranges from 5 to 30 percent of CSF isolates.[28]

The ampicillin resistance has been demonstrated to be due to a beta-lactamase which disrupts the beta-lactam ring of ampicillin.[6] This beta-lactamase resembles the enzymes found in gram-negative enteric bacilli such as *Enterobacter* species (in which the enzyme is often a cephalosporinase), rather than the enzyme produced by *Staph. aureus*, (which is commonly called penicillinase) (Chap. 1). Thus, type b *H. influenzae* which are resistant to ampicillin are also resistant to the penicillinase-resistant penicillins such as methicillin.[39]

Beta-Lactamase Test. Ampicillin resistance can be rapidly detected in *H. influenzae* isolates by a test using the bacteria as a source of the enzyme, a cephalosporin (or penicillin) as the substrate, and an indicator to show if acidity is produced as a result of the disruption of the beta-lactam ring. This test is reliable when adequate controls are used, and correlates very well with susceptibility testing using the reliable standard tube dilution techniques. In general, a strain of *H. influenzae* which produces beta-lactamase will require, by the tube dilution susceptibility test, concentration of more than 4 µg per ml of ampicillin to inhibit the organism. Such high ampicillin concentrations are generally not attainable in the spinal fluid using very high intravenous dosages of ampicillin.

Chloramphenicol-Resistant Haemophilus Influenzae. In 1976, a chloramphenicol-resistant type b was detected in the United States.[20] At the present time, such resistant organisms are extremely rare. Resistance to chloramphenicol appears to be plasmid-mediated, and due to the production of an enzyme which inactivates chloramphenicol, apparently an acetyl transferase. Chloramphenicol resistance is detected by standard plate dilution or tube dilution susceptibility methods. In general, chloramphenicol resistance is defined by the requirement for a concentration of 16 micrograms per ml or more for inhibition of the organism. Chloramphenicol resistance appears to be independent of ampicillin resistance. In theory, since chloramphenicol is rarely used in outpatients, there should be little chloramphenicol "pressure", and chloramphenicol resistance should be much less frequent in the future than ampicillin resistance.

Plasmid Mechanism. Ampicillin resistance in *H. influenzae* is plasmid-mediated. The beta-lactamase enzyme is coded for by part of a plasmid (a piece of extrachromosomal genetic material) that can be transferred by conjugation, a mating between bacteria where genetic material is transferred across a bridge between them[30] (Fig. 3-4).

The frequency of plasmids that mediate resistance to a particular antibiotic appears to be increased by widespread use of that antibiotic. In the absence of such antibiotic pressure, many organisms lose their plasmids (which are often inherently unstable) and become susceptible. In the opposite situation, when ampicillin is used to treat a patient, the frequency of ampicillin-resistant normal flora (such as *H. parainfluenzae*) may increase. The plasmid containing beta-lactamase has been demonstrated to be transferable from *H. parainfluenzae* to *H. influenzae*.[5]

Treatment

Until 1974, ampicillin was considered the drug of choice for *H. influenzae* infections in children. Tetracycline has usually been effective, and can be used in adults when there is allergy to penicillin.

In 1978, combined therapy with both ampicillin and chloramphenicol was recommended for treatment of serious type b *H. influenzae* infections, such as meningitis, until accurate susceptibility tests are available on the isolates.

Cephalosporins are not clinically effective in *H. influenzae* meningitis infections, and should not be used. This includes cefamandole, which may be effective in soft tissue *H. influenzae* infections, but not for meningitis.

Studies comparing ampicillin and chloramphenicol have not shown any difference in the efficacy of these two antibiotics, if the organism is sensitive to each.

If an *H. influenzae* isolate should be resistant to both ampicillin and chloramphenicol, several antibiotics can be considered. Trimethoprim-sulfamethoxazole is virtually always effective against *H. influenzae*, but must be given by mouth. Some cephalosporins, such as cefamandole, also look promising, but penetrate the spinal fluid poorly. Some ampicillin-resistant strains are slightly more susceptible to carbenicillin, as the organism's beat-lactamase may be less effective in disrupting carbenicillin.[34]

Haemophilus ducreyi, like *H. influenzae*, is usually susceptible to ampicillin, but a few beta-lactamase strains have been observed, which are usually susceptible to sulfonamides.

Prevention

Antibiotics. Prophylaxis has been advocated for children exposed to siblings or day care contacts with *H. influenzae* meningitis. *H. influenzae* disease is not as fulminating as meningococcal disease, where prophylaxis is recommended because of the rapidity of the course of the illness, so sometimes close observation of the sibling contacts is a reasonable alternative.

Vaccine. A vaccine prepared from the capsular antigen is currently being investigated.[28] It stimulates antibiody production poorly in young children, and the duration of immunity and protective effects are unknown. Even infants who have previously had severe type b disease respond poorly to the vaccine.[26]

Heteroimmunization. Human volunteers have been fed a strain of *E. coli* (the "Easter" strain), which stimulates the production of antibodies that cross-react the type b capsular antigen of *H. influenzae*.[31] The reliability of and degree of protection provided by this procedure has not yet been adequately investigated.

Risk Factors. Some children are at higher risk for severe *H. influenzae* type b disease. These include young siblings,[23,37] young playmates in close contact as at a day care center,[42] and young infants with transplantation of the great arteries.[13]

BORDETELLA PERTUSSIS

Objectives

1. Describe the classical illness of whooping cough in children, and describe how the illness might differ in very young infants.
2. Discuss the methods available for the laboratory confirmation of *B. pertussis* infection.
3. Discuss the prevention and treatment of whooping cough.

Definitions

Bordetella pertussis is a gram-negative rod which is difficult to grow in the laboratory. The organism is named for Bordet, a Belgian bacteriologist, who with Gengou first recovered the organism, using a new culture medium, which was named after them. Other organisms in the genus *Bordetella* are *B. parapertussis* and *B. bronchiseptica*, which are rare causes of acute bronchitis.

Bordetella pertussis infection is the usual cause of whooping cough.[44,53,55] However, a whooping cough-like illness can be produced by other microorganisms, such as *Bordetella parapertussis*.[50] In addition, *B. pertussis* can produce bronchitis or pneumonia without a whoop.

Frequency and Importance

Approximately 2000 cases of whooping cough were reported in the United States in 1977. It is particularly severe, and may be fatal, in young infants who have not yet been immunized. Since fluorescent antibody techniques have become available, B. pertussis has been recognized as a cause of acute bronchitis without a typical whoop in older children or adults more frequently than was previously suspected, but the frequency of B. pertussis as a cause of acute bronchitis is not yet fully defined.

Clinical Patterns of Illness

B. pertussis is the *usual* cause of:

Whooping Cough-like Illness. In the classical illness, there is a history of exposure to a sibling with whooping cough. The patient has a series of staccato coughs, all in one exhalation (the paroxysmal cough), followed by an inspiratory whoop. This pattern can often be elicited by gagging the patient or by stimulating the child to cry. Vomiting, cyanosis, and apneic spells are common. Lymphocytosis is common, often in the area of 20,000 with 80 percent lymphocytes. Occasionally, lymphocytic leukemia is suspected because of a leukemoid reaction (white blood count greater than 50,000). Eosinophilia is often present.[55] The cough usually persists for many weeks, and may recur with subsequent minor respiratory infections for several months. Asymptomatic infection almost never occurs.[53,55]

B. pertussis is an *occasional* cause of:

Acute Bronchitis. In young infants, or in older children or adults with past immunizations, the cough is often not associated with a whoop. The illness is characterized by cough, without fever or evidence of pneumonia, and is usually diagnosed as acute bronchitis.

Perihilar pneumonia or atelectasis is occasionally due to B. pertussis as a complication of the bronchitis (Fig. 4-3).

Encephalopathy is a rare complication of severe whooping cough, and may be secondary to hypoxia or hypoglycemia, or possibly a direct effect of the bacterium.[55]

Laboratory Diagnosis

Smear. Fluorescent antibody methods are more sensitive than a culture for B. pertussis.[58] Nasal secretions are smeared on a slide, which can be sent to a laboratory for staining and examination. False positives are uncommon, but can occur.

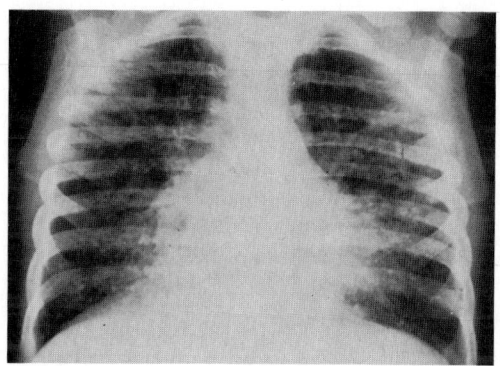

FIG. 4-3. *Chest film showing perihilar pneumonia obscuring the heart borders, sometimes called a "shaggy heart". This is not specific for pertussis, which is only one of the many causes of perihilar pneumonia. This patient had H. influenzae in his blood culture.*

Culture. Specimens for culture can be obtained by stimulating the patient to cough on a culture plate (cough plate). As an alternative, a flexible swab should be inserted into the nose to the posterior nasopharynx and then streaked on the plate. The standard method for culture is the use of Bordet-Gengou media, which usually must be made up fresh, when needed. Unfortunately, recovery of B. pertussis using conventional bacterial culture methods is often unsuccessful in typical clinical cases. In most hospitals, culture for B. pertussis can be regarded as a test of the laboratory's techniques, rather than as a useful way to confirm the clinical diagnosis.

Serum Antibodies. This method is not practical because only a few research laboratories do the studies, and because infants with B. pertussis recovered on culture often do not show a rise in antibody titer.[57]

Biologic Characteristics of Clinical Interest

Special Growth Requirements. The organism requires special media and often requires 3 to 5 days to grow to a recognizable colony.

Variation in Antigen Components. This is important in the production of vaccine. Freshly isolated bordetella have capsules. During repeated passages on artificial media, the capsule is lost as the organism undergoes a smooth to rough variation (Phase I to Phase IV). The rough variants are practically avirulent. Vaccines produced from recently isolated organisms are much more effective than vaccines produced from the subcultures of the rough form.

Serotypes have been defined on the basis of agglutination reactions. Poor protection by some British vaccines before 1968 appears to have been related to failure to use all serotypes in the manufacture of the vaccine.[56] However, the mechanisms of vaccine failure in the U.S. has not been adequately studied.

Toxic Factors. Lymphocytosis factor is a heat-labile toxin that can produce lymphocytosis in experimental animals, even when the organisms injected are dead.

Association with Adenovirus. Adenoviruses are sometimes recovered from patients with typical whooping cough, without concurrent recovery of *B. pertussis*. It is not yet clear whether adenovirus alone can cause the whooping cough syndrome, or whether *B. pertussis* was the cause but was not recovered because of inadequate laboratory techniques.[44,54,55] The marked lymphocytosis in some cases suggest that the *B. pertussis* was really the cause.

Animal Infections. Experimental whooping cough has has been produced in monkeys by nasal or intratracheal inoculation of the organism.[48] In all cases of experimental whooping cough in monkeys, marked lymphocytosis was observed during the course of the coughing. Antibody studies of experimentally infected animals indicate that permanent immunity after an attack of whooping cough may be related to local tissue immunity in the respiratory tract rather than to humoral immunity, because long before the appearance of antibodies, the organism could no longer be cultured. On the basis of these experimental animal infections, it has been suggested that the use of attenuated organism as a vaccine may be more practical than the use of killed organisms.[48]

Contagion. Use of FA techniques have demonstrated *B. pertussis* in many individuals with mild bronchitis which might not have been suspected to be due to *B. pertussis* except for known exposure. Hospital personnel with minor symptoms can transmit the disease, so hospitalized infants with the disease should be isolated and kept far from debilitated infants, and attendants should wear masks.[49]

Adherence. Fresh, virulent Phase I pertussis bacilli adhere to ciliated epithelial cell cultures much better than avirulent Phase III bacilli, suggesting adherence is an important factor in human infection.[52]

Treatment

Supportive Care. Hospitalization, with oxygen and resuscitation during apneic spells, is important in infants with severe illness. Steroids increase the mortality in mice.[55]

Antibiotics. Erythromycin is usually recommended. It does not modify the course of the disease, but eradicates the organism in a few days and makes the patient non-contagious.[44,45]

Pertussis Hyperimmune Globulin. This is prepared from blood obtained from adults who have recently been given pertussis vaccine. It appears to be of no value in changing the course of the illness.[43,45]

Prevention

Pertussis Vaccine. This is a killed bacterial vaccine, and is not completely effective. However, the severity of the disease in infants and young children, and the lack of effective therapy, makes it an important vaccine for the preschool child. Its use in partial doses has been recommended for adults during hospital outbreaks.[51] However, local reactions and the small possibility of encephalopathy from the vaccine make its risk greater than its benefits for adults or children over 6 years of age.[47]

Preventing Exposure. The highest risk and most severe disease is in young infants, who do not receive any protection from transplacental antibodies.[46] The young infant was once regarded as best protected by preventing exposure to the disease through immunization of older siblings, but in recent years adults seem to be a more important source of infection.[53]

Pertussis Hyperimmune Globulin. There are no controlled data on the efficacy in pertussis, as an attempt to prevent disease in the infant.

ZOONOSES

Diseases of animals which can be transmitted to humans are called zoonoses. Many of the small gram-negative rods have animals as their primary host, and are transmitted to humans by direct contact or through insect vectors. Examples include brucellosis (swine, cattle, goats, dogs), tularemia (rabbits, deer), plague (rats, rodents), and pasteurellosis (dogs, cats). Occuaptional or recreational contact with animals is an important clue that should help the physician think of these unusual bacteria.

YERSINIA SPECIES

Objectives

1. Describe three clinical patterns in which *Yersinia enterocolitica* should be considered as a possible cause.
2. Describe the usual clinical pattern of illness of *Yersinia pestis* in the United States.
3. Describe the source and frequency of plague in the United States.

Definitions and Species

Yersinia is a genus of small gram-negative rods that exhibit bipolar staining, like safety pins, when treated with special stains. The genus was named for Alexander Yersin, a nineteenth-century bacteriologist. *Pasteurella* is a genus named for Louis Pasteur. *Yersinia* species were formerly included in the genus *Pasteurella*; for example, *Yersinia pestis* was formerly called *Pasteurella pestis*.

Yersinia pseudotuberculosis and *Yersinia enterocolitica* are causes of mesenteric adenitis, a disease with severe abdominal pain, which resembles acute appendicitis. The name pseudotuberculosis is derived from the gross nodules resembling tubercles, although microscopically and clinically it does not resemble tuberculosis. *Y. enterocolitica* is also a cause of diarrhea, acute polyarthritis, and septicemia in compromised hosts, especially those with thallesemia or cirrhosis.

Yersinia pestis is the cause of plague.

Frequency and Importance

Yersinia enterocolitica is a frequent cause of diarrheal disease and severe abdominal pain with mesenteric adenitis in the United States. *Yersinia pseudotuberculosis* is a less common cause of mesenteric adenitis.

There are about 10 to 20 human cases per year of plague in the southwestern United States, primarily in New Mexico, where wild rodents are a reservoir (Fig. 4-4). Pneumonic plague is contagious and potentially is an epidemic disease.

Clinical Patterns of Illness

Yersinia species can cause the following syndromes:

Mesenteric Adenitis. *Yersinia pseudotuberculosis* can cause mesenteric adenitis. It has been recovered from mesenteric lymph nodes at the time of operation for suspected appendicitis.[75] *Y. enterocolitica*, closely related to *Y. pseudotuberculosis*, has been associated with fever, diarrhea, and abdominal pain, and some patients with this infection have been operated on for suspected appendicitis.[60,64]

Acute Diarrhea. *Y. enterocolitica* can produce acute diarrhea resembling shigellosis.[59,60,66] A rash, especially erythema nodosum or erythema multiforme, is often present.[66]

Suppurative Lymphadenitis. In *Yersinia pestis* infections there may be a local ulcerative lesion, but often the site of the flea bite is small or absent.[69] In plague, there is typically an exposure to rats, other rodents, or their arthropod vectors. Regional adenopathy usually is noted in the

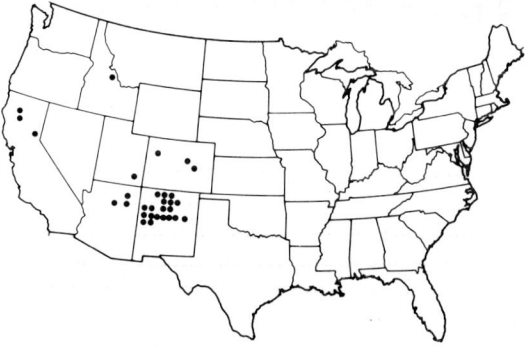

FIG. 4-4. *Plague: total reported cases in the United States, 1965–1970.*[62]

axilla or groin. When these nodes suppurate, they are called buboes (bubonic plague).

Pneumonia. *Y. pestis* occasionally results in pneumonia as part of a generalized septicemia; it then can be spread by the aerosol route, as has occurred in large epidemics.

Septicemia or Meningitis. *Y. pestis* was notorious as a cause of septicemia in epidemics in the past. Plague can be associated with disseminated intravascular coagulation,[63] with skin hemorrhages the probable source of the term "black" in epidemics of the Black Death.[67] Meningitis is a serious, but rare, complication of plague.

Sepsis can occur with *Y. pseudotuberculosis* or *Y. enterocolitica*.[76]

Abdominal Abscesses. Hepatic or splenic abscess is occasionally caused by *Y. enterocolitica*, and often resembles amebic hepatitis.[70]

Other clinical patterns caused by *Y. enterocolitica* include arthritis,[68,73] abscesses, febrile pharyngitis,[60] and myocarditis.[68] The arthritis can involve a single joint with a positive culture[73] or may involve multiple joints, which are sterile on culture, with the joint manifestations often lasting 1 to 3 months.

Laboratory Diagnosis

Smear. Gram stain of pus may reveal small, bipolar-stained gram-negative rods. Fluorescent antibody techniques are useful for the rapid diagnosis of *Y. pestis*. When plague is suspected in the United States, slides of pus can be flown to the Center for Disease Control in Atlanta, Georgia for examination using a fluorescent antibody stain.

Culture. *Yersinia* species can be readily recovered in the laboratory, but *Y. pestis* is hazardous to personnel. *Y. enterocolitica* requires incubation at room temperature for primary isolation. *Y. pseudotuberculosis* is rarely recovered from feces, but can be recovered from tissues removed surgically.

Serology. *Y. enterocolitica* infection can be detected by antibody studies available in some reference laboratories.

Biologic Characteristics of Clinical Interest

Animal Reservoirs. The source of human infections with these organisms is usually an animal host. Plague has its largest reservoir in rodents. In the United States, plague has occurred only in relation to exposure to wild rodents in the Southwest.[65] This is called sylvatic plague (sylvatic means forest). Most historical epidemics in humans were preceded by epidemics in rats.[61] Recent observations indicate that infected rabbits and even domestic dogs and cats can be a source of human plague in endemic areas.[74] *Y. enterocolitica* is found in animals, and in some cases family pets (especially dogs) appear to have been the source of human infections.

Water Reservoirs. *Y. enterocolitica* is an occasional contaminant of untreated water in wells or streams.

Preference for Cold. *Y. enterocolitica* grows better at colder temperatures, and produces a toxin at 25°C, but not at 37°C. It also is more frequently found in diarrhea in winter months.

Treatment

Y. enterocolitica is susceptible to tetracycline, trimethoprim–sulfamethoxazole, kanamycin, chloramphenicol, or aminoglycosides,[71] but few studies have been done treating clinical illnesses.[73]

For *Y. pestis*, streptomycin is the antibiotic which has had the most clinical demonstration of efficacy; penicillin is of no value.[72] Incision and drainage of a fluctuant, infected lymph node is not essential to therapy of plague, although needle aspiration may be useful for diagnosis. Great caution should be used to avoid spread by aerosol or direct contact during the incision.

Prevention

Control of Rodents. Rat control is the most important method of prevention of plague in urban areas, but the rodent reservoir of sylvatic plague is difficult to control. DDT can be used to kill the rat fleas, if an outbreak of plague is occurring in urban rats. Killing the rats without killing their fleas may result in the fleas moving to human hosts.

Vaccines. A killed vaccine is available for plague, but is of relatively low efficacy, and a local reaction at the injection site is common.

PASTEURELLA SPECIES

Objectives

1. Describe the common clinical patterns of *Pasteurella multocida* infection, the antibiotic treatment, and prevention.

Definitions and Species

Pasteurella is a genus named for Louis Pasteur. Some bacteria, such as the plague bacillus, formerly classified as pasteurellae, are now classified as yersiniae (see the preceding section).

Pasteurella multocida includes a number of different strains which infect different animal hosts, and is sometimes classified into several other species. Other *Pasteurella* species, which can apparently cause pneumonia, include *P. ureae, P. pneumotropica,* and *P. haemolytica*.[81]

Frequency and Importance

The frequency of human infections with *P. multocida* is not completely defined because the disease is not reportable, but it has been estimated that several thousand infections of animal bites with this organism occur each year.[78]

Clinical Patterns of Illness

Pasteurella multocida is a *frequent* cause of:

Infected Animal Bits. If dog or cat bites become infected, the organism is often *P. multocida,* which is normal mouth flora in many animals.[78]

Pasteurella species are a *rare* cause of:

Pneumonia with Cavitation. Hemoptysis may occur.[80]

Other Clinical Patterns. Osteomyelitis, urinary infection, meningitis, brain abscess, conjunctivitis, vaginitis, cervicitis, or endocarditis[77,79] are rarely caused by pasteurellae.

Biologic Characteristics of Clinical Interest

Pasteurella multocida has an important reservoir in dogs and cats. Outbreaks of *P. multocida* may occur in flocks or herds of animals when adverse environmental conditions increase host susceptibility.

Treatment and Prevention

Penicillin is the drug of choice for *P. multocida* infections. Penicillin has also been recommended for prevention of *P. multocida* infections in some severe bites, such as dog bites of the face. It also may be indicated in deep puncture bites, which are difficult to irrigate and debride adequately.

Animal bites can often be avoided. Small children should be taught not to put their face near a dog's mouth, especially when the dog is eating.

FRANCISELLA TULARENSIS

Objectives

1. Describe some of the typical exposures which might be elicited from a patient with tularemia.
2. Describe several of the clinical syndromes that may be caused by *Francisella tularensis.*
3. Describe how the clinical diagnosis of tularemia can be confirmed in the laboratory.

Names

These organisms are small gram-negative rods. The genus *Francisella* is named after Francis, who demonstrated transmission of tularemia by deerfly bites. The organism was formerly included in the genus pasteurella. Tularemia is named for Tulare County in California.

Frequency and Importance

About 150 cases of tularemia are reported annually in the United States (Fig. 4-5). The disease is rarely fatal, but may present considerable diagnostic difficulty unless an animal exposure is recognized.

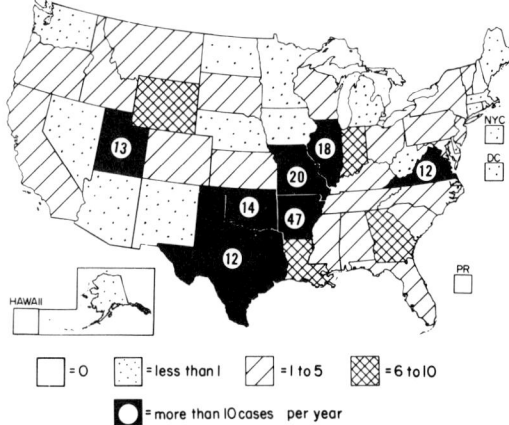

FIG. 4-5. *Tularemia: average annual reported cases, 1960–1972.* [84,85]

Clinical Patterns of Illness

F. tularensis is an *occasional* cause of:

Ulceroglandular syndrome is a shallow skin ulcer with regional lymphadenopathy, typically manifested by an ulcer on the hand and axillary adenopathy. Typically there is exposure to rabbits, muskrats, or other wild animals, or arthropod vectors, such as ticks or deerflies. High fever is usually noted for about a week. The course is usually self-limited, and was rarely fatal before antibiotics were available.

Suppurative lymphadenitis is the ulceroglandular syndrome without the local ulcer. Occasionally, tularemia is the clinical diagnosis when in reality the patient has plague.[91] These two diseases may be difficult to distinguish on clinical grounds, but plague is limited to the southwestern United States.

F. tularensis is a rare cause of several syndromes. In pneumonia due to tularemia, there is typically bronchopneumonia with hilar adenopathy.[89,90] Pleural effusion may also be present. Oculoglandular syndrome, which is purulent conjunctivitis with preauricular adenopathy, is occasionally due to tularemia.[82,86,87] Exudative pharyngitis or fever of unknown origin is rarely due to tularemia.[87,90]

Laboratory Diagnosis

Culture. *F. tularensis* is difficult to culture, because the organism is relatively fastidious, requires special media, and may be hazardous to laboratory personnel.

Serum Antibodies. If a serum obtained early in the illness is negative, a later serum should be tested, since the antibodies may not be detectable early in the illness.

Skin Test. This is valuable in the diagnosis of tularemia, as it is positive in about half of the patients at the time they seek medical care.[83]

Biologic Characteristics of Clinical Interest

Animal Reservoirs. Most human infections can be related to exposure to rabbits. Other animals, such as muskrats, can be a reservoir.[92] Ticks and rarely deerflies can serve as a vector of the disease to man.[88]

Treatment

Streptomycin is generally regarded as the drug of choice. Although kanamycin or gentamicin would probably be equally effective, insufficient clinical experience is available with these drugs for treatment of tularmemia.

Prevention

Avoid Unnecessary Exposure. Gloves should be used when skinning rabbits or other wild animals. Rabbits easily shot or captured by dogs or cats may be sick, and should be handled with caution.

BRUCELLA SPECIES

Objectives

1. Describe the usual sources and the frequency of brucellosis in the United States.
2. Describe the clinical patterns of brucellosis, and indicate how the diagnosis can be confirmed in the laboratory.

Name and Species

Brucella species are small gram negative coccobacilli. The genus was named for Bruce, who first recovered the organism from patients with "Malta fever", which was named for the island of Malta, where raw goat's milk was the usual source of infection.

The most common species recovered in the United States is *B. suis*, which is usually recovered from meat packers with exposure to swine. *B. abortus* typically produces abortions in cattle, but is now uncommon in the United States, since raw milk is seldom drunk and infected cattle are slaughtered. *B. melitensis* was named for the ancient term for Malta (Melitene). Ingestion of raw goat's milk or cheese is the typical mechanism for human infection with *B. melitensis*, but this organism is rarely found in the United States. *B. canis* is a recently recognized species in dogs, and characteristically produces abortions, especially in beagles. Human infections with *B. canis* have recently been recognized and may be more frequent than originally supposed.[99,100] In one study, about 73 percent of veterinarians and about 57 percent of male blood donors had an antibody titer high enough to indicate past infection with *B. canis*.

Frequency and Importance

Brucellosis is primarily a disease of animals, and is now uncommon in humans in the United States, with about 200 cases reported per year (Fig. 4-6). Most of these occur in individuals with occupational exposure to animals, especially meat packing (Table 4-1).

Clinical Patterns of Illness

Brucella species are a rare cause of several syndromes. In prolonged fever (of unknown origin) due to burcellosis, the patient typically has ingested raw milk or imported cheese, or has an occupational exposure to animals.[93,102] Fever is usually intermittent and chronic. Chills, sweat-

TABLE 4–1. OCCUPATIONS ASSOCIATED WITH LABORATORY-CONFIRMED BRUCELLOSIS IN THE UNITED STATES, 1976.

Occupation	Total Patients
Packing-house employee	144
Livestock producer	29
Veterinarian	10
Laboratory worker	4
Rendering-plant employee	3
Government meat-processing inspector	13
	203
Other	68
	271

(*Source:* Center for Disease Control, Oct. 1977.)[95]

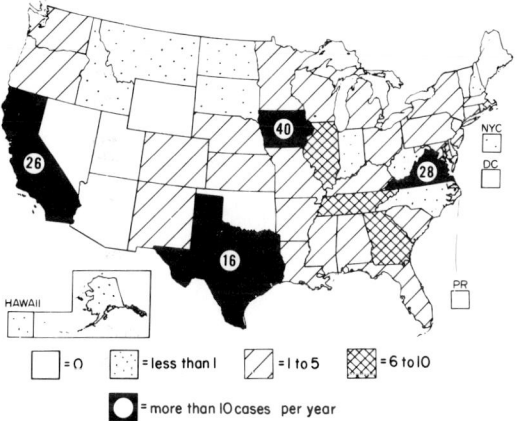

FIG. 4-6. *Brucellosis: average annual reported cases.*[94]

ing, weakness, weight loss, arthralgia, lymphadenopathy, and splenomegaly may be present. Leukopenia is not unusual. Testicular pain or swelling and difficult or frequent urination occasionally occur.

Chronic abscesses of liver or spleen may develop and may persist as long as 20 years after the initial exposure and febrile illness.[104] Calcification of an abdominal organ may be seen on radiograph.

Bone or joint infections due to brucellosis are often chronic, and include vertebral osteomyelitis, infection of the intervertebral disc (spondylitis), and septic arthritis.

Chronic skin ulcers can also be produced by contamination of abraded human skin with infected animal blood.[96]

Laboratory Diagnosis

Culture. Recovery of the organism from the blood is the usual method for making the diagnosis, and this is only likely during the acute febrile illness. When brucellosis is suspected, the laboratory should be informed, and may be advised to hold the blood culture for several weeks longer than usual.

Serology. Demonstration of a rise in agglutinating antibody titer is a possible method for demonstrating infection. A single high titer is of less value, but is evidence of infection in the undefined past. A titer of 1:250 or higher is often regarded as evidence for recent infection. Blocking antibodies may appear after the acute phase of the infection and interfere with serologic diagnosis, unless special techniques are used.

Commercially available brucella antigen may not detect *B. canis* antibody. If *B. canis* is suspected, the physician should be sure that this antigen is used.

Biologic Characteristics of Clinical Interest

Species Preference. Although there is some cross infection between species, the infection by a particular *Brucella* species is generally limited to a few preferred animal hosts. There appears to be low virulence of *Brucella* species for humans, and some patients with bacteremia recover without therapy.[98,104]

Intracellular Infection.[103] The chronicity of the infection is probably related to the intracellular location of the infection, particularly in reticuloendothelial cells. A brief course of corticosteroids alone may be effective therapy, and may also be used to release the organisms from the cells and allow the antibiotic to reach the organism.[98,104] However, corticosteroids should be used with caution and for only a few days, to avoid disseminated disease.

Dogs infected with *B. canis* can remain infected for many months in spite of intensive and prolonged antibiotic therapy.

Treatment

The combination of tetracycline and streptomycin is effective therapy, and is better than tetracycline alone. In vitro studies indicate ampicillin, gentamicin, kanamycin, or erythromycin should be effective,[97] although little clinical experience is available with these drugs in the treatment of brucellosis. Rifampin, which is able to penetrate cells very well, is very effective in mice and guinea pigs, but has not yet been studied in humans.[101]

A Jarisch–Herxheimer reaction can occur in the first 24 hours of antibiotic treatment, with increased fever and toxicity[105] (see the section on borrellia in Chap. 9).

REFERENCES

Haemophilus Influenzae

1. Albritton, W.L., Hammond, G.W., and Ronald, A.R.: Bacteremic *Haemophilus influenzae* genitourinary infection in adults. Arch. Intern. Med., *138*:1819–1821, 1978.
2. Bradshaw, M.W., et al.: Bacterial antigens cross-reactive with the capsular polysaccharide of *Haemophilus influenzae* type b. Lancet, *1*:1095–1097, 1971.
3. Catlin, B.W.: *Haemophilus influenzae* in cultures of cerebrospinal fluid. Noncapsulated variants typable by immunofluorescence. Am. J. Dis. Child., *120*:203–210, 1970.
4. Dochez, A.R., Mills, K.C., and Kneeland, Y., Jr.: Variation of *H. influenzae* during acute respiratory infection in the chimpanzee. Proc. Soc. Exp. Biol. Med., *30*:314–316, 1932.
5. Eickhoff, T.C., Ehret, J.M., and Baines, R.D.: Characterization of an ampicillin-resistant *Haemophilus influenzae* type b. Antimicrob. Agents Chemother., *9*:889–892, 1976.
6. Farrar, W.E., Jr., and O'Dell, N.M.: Beta-lactamase activity in ampicillin-resistant *Haemophilus influenzae*. Antimicrob. Agents Chemother., *6*:625–629, 1974.
7. Feigin, R.D., Richmond, D., Hosler, M.W., and Shackelford, P.G.: Reassessment of the role of bactericidal antibody in *Haemophilus influenzae* infection. Am. J. Med. Sci., *262*:338–346, 1971.
8. Gordon, M., and Zinnemann, K.: The in vitro sensitivity of *H. influenzae* to penicillin with special reference to meningeal strains of Pittman's type b. Br. Med. J., *2*:795–796, 1945.
9. Granoff, D.M., Gilsdorf, J., Gessert, C., and Basden, M.: *Haemophilus influenzae* type b disease in a day care center: eradication of carrier state by rifampin. Pediatrics, *63*:397–401, 1979.
10. Greene, G.R.: Meningitis due to *Haemophilus influenzae* other than type b: case report and review. Pediatrics, *62*:1021–1025, 1978.
11. Hammond, G.W., Lian, C.J., Wilt, J.C., and Ronald, A.R.: Comparison of specimen collection and laboratory techniques for isolation of *Haemophilus ducreyi*. J. Clin. Microbiol., *7*:39–43, 1978.
12. Harlow, M.S., Chung, M.K., and Plotkin, S.A.: *Haemophilus influenzae* septic arthritis in infants and children. Clin. Pediatr., *14*:1146–1149, 1975.
13. Hassink, S.G., et al.: Transposition of the great arteries: a new risk factor for *Haemophilus influenzae* type b meningitis. J. Pediatr., *94*:755–757, 1979.
14. Holt, L.B.: The growth factor requirements of *Haemophilus influenzae*. J. Gen. Microbiol., *27*:317–322, 1962.

15. Holt, R.N., Taylor, C.D., Schneider, H.J., and Hallock, J.A.: Three cases of *Haemophilus parainfluenzae* meningitis. Clin. Pediatri., *13*:666–668, 1974.

16. Jemsek, J.G., et al.: *Haemophilus parainfluenzae* endocarditis. Two cases and review of the literature in the past decade. Am. J. Med., *66*:51–57, 1979.

17. Jenne, J.W., et al.: The course of chronic hemophilus bronchitis treated with massive doses of penicillin and penicillin combined with streptomycin. Am. Rev. Resp. Dis., *101*:907–922, 1970.

18. Johnson, G.K., Sullivan, J.L., and Bishop, L.A.: Acute epiglottitis. Review of 55 cases and suggested protocol. Arch. Otolaryngol., *100*:300–337, 1974.

19. Kirkman, J.B., Jr., and Crawford, J.J.: Serotyping of noncapsular *Haemophilus influenzae*. Appl. Microbiol., *22*:133–134, 1971.

20. Long, S.S., and Phillips, S.E.: Chloramphenicol-resistant *Haemophilus influenzae*. J. Pediatr., *90*:1030–1031, 1977.

21. McLinn, S.E., Nelson, J.D., and Haltalin, K.C.: Antimicrobial susceptibility of *Haemophilus influenzae*. Pediatrics, *45*:827–838, 1970.

22. Mathies, A.W., Jr., et al.: Experience with ampicillin in bacterial meningitis. Antimicrob. Agents Chemother., *1965*:610–626, 1966.

23. Michaels, R.H., and Norden, C.W.: Pharyngeal colonization with *Haemophilus influenzae* type b: a longitudinal study of families with a child with meningitis or epiglottitis due to *H. influenzae* type b. J. Infect. Dis., *136*:222–228, 1977.

24. Michaels, R.H., Poziviak, C.S., Stonebraker, F.E., and Norden, C.W.: Factors affecting pharyngeal *Haemophilus influenzae* type b colonization rates in children. J. Clin. Microbiol., *4*:413–417, 1976.

25. Norden, C.W., Callerame, M.L., and Baum, J.: *Haemophilus influenzae* meningitis in an adult. A study of bactericidal antibodies and immunoglobulins. New Engl. J. Med., *282*:190–194, 1970.

26. Norden, C.W., Michaels, R.H., and Melish, M.: Effect of previous infection on antibody response of children to vaccination with capsular polysaccharide of *Haemophilus influenzae* type b. J. Infect. Dis., *132*:69–74, 1975.

27. Parker, R.H., and Hoeprich, P.D.: Disk method for rapid identification of *Haemophilus* species. Am. J. Clin. Pathol., *37*:319–327, 1962.

28. Peltola, H., Käyhty, H., Sivonen, A., and Mäkelä, P. H.: *Haemophilus influenzae* type b capsular polysaccharide vaccine in children: a double-blind field study of 100,000 vaccinees 3 months to 5 years of age in Finland. Pediatrics, *60*:730–737, 1977.

29. Rapkin, R.H., and Bautista, G.: *Haemophilus influenzae* cellulitis. Am. J. Dis. Child., *124*:540–542, 1972.

30. Saunders, J.R., and Sykes, R.B.: Transfer of a plasmid-specified beta-lactamase gene from *Haemophilus influenzae*. Antimicrob. Agents Chemother., *11*:339–344, 1977.

31. Schneerson, R., and Robbins, J.B.: Induction of serum *Haemophilus influenzae* type b capsular antibodies in adult volunteers fed cross-reacting *Escherichia coli* 075:K100:H5. New Engl. J. Med., *292*:1093–1096, 1975.

32. Schwartz, R., Rodriguez, W., Khan, W., and Ross, S.: The increasing incidence of ampicillin-resistant *Haemophilus influenzae*. A cause of otitis media. J.A.M.A. *239*:320–323, 1978.

33. Sell, S.H.W.: The clinical importance of *Haemophilus influenzae* infections in children. Pediatr. Clin. North Am., *17*:415–426, 1970.

34. Sinai, R., Hammerberg, S., Marks, M.I., and Pai, C.H.: *In vitro* susceptibility of *Haemophilus influenzae* to sulfamethoxazole–trimethoprim and cefaclor, cephalexin, and cephradine. Antimicrob. Agents Chemother., *13*:861–864, 1978.

35. Suksanong, M., and Dajani, A.S.: Detection of *Haemophilus influenzae* type b antigen in body fluids, using specific antibody-coated staphylococci. J. Clin. Microbiol., *5*:81–85, 1977.

36. Syriopoulou, V., et al.: Increasing incidence of ampicillin resistance in *Haemophilus influenzae*. J. Pediatr., *92*:889–992, 1978.

37. Tejani, A., Dobias, B., Nangia, B.S., and Velkura, H.: Intrafamilial spread of *Haemophilus* type b infections. Am. J. Dis. Child., *131*:778–781, 1977.

38. Thirumoorthi, M.C., and Dajani, A.S.: Comparison of staphylococcal coagglutination, latex agglutination, and counterimmunoelectrophoresis for bacterial antigen detection. J. Clin. Microbiol., *9*:28–32, 1979.

39. Thornsberry, C., Baker, C.N., Kirven, L.A., and Swenson, J.M.: Susceptibility of ampicillin-resistant *Haemophilus influenzae* to seven penicillins. Antimicrob. Agents Chemother., *9*:70–73, 1976.

40. Todd, J.K., and Bruhn, F.W.: Severe *Haemophilus influenzae* infections. Am. J. Dis. Child., *129*:607–611, 1975.

41. Wallace, R.J., Jr., Musker, D.M., and Martin, R.R.: *Haemophilis influenzae* pneumonia in adults. Am. J. Med., *64*:87–93, 1978.

42. Ward, J.I., Gorman, G., Phillips, C., and Fraser, D.W.: *Haemophilus influenzae* type b disease in a day care center. J. Pediatr., *92*:713–717, 1978.

Bordetella Pertussis

43. Balagtas, R.C., Nelson, K.E., Levin, S., and Gotoff, S.P.: Treatment of pertussis with pertus-

sis immune globulin. J. Pediatr., 79:203–208, 1971.

44. Baraff, L.J., Wilkins, J., and Wehrle, P.F.: The role of antibiotics, immunizations, and adenoviruses in pertussis. Pediatrics, 61:224–230, 1978.

45. Bass, J.W., et al.: Antimicrobial treatment of pertussis. J. Pediatr., 75:768–781, 1969.

46. Congeni, B.L., Orenstein, D.M., and Nankervis, G.A.: Three infants with neonatal pertussis. Clin. Pediatr., 17:113–114, 117–118, 1978.

47. Grady, G.H., and Wetterlow, L.H.: Pertussis vaccine: reasonable doubt? New Engl. J. Med., 298:966–967, 1978.

48. Huang, C.C., et al.: Experimental whooping cough. New Engl. J. Med., 266:105–110, 1962.

49. Kurt, T.L., Yeager, A.S., Guenette, S., and Dunlop, S.: Spread of pertussis by hospital staff. J.A.M.A., 221:264–267, 1972.

50. Linnemann, C.C., Jr., and Perry, E.B.: *Bordetella parapertussis*. Am. J. Dis. Child., 131:560–561, 1977.

51. Linnemann, C.C., Jr., et al.: Use of pertussis vaccine in an epidemic involving hospital staff. Lancet, 2:540–544, 1975.

52. Matsuyama, T.: Resistance of *Bordetella pertussis* Phase I to mucociliary clearance by rabbit tracheal mucous membrane. J. Infect. Dis., 136:609–616, 1977.

53. Nelson, J.D.: The changing epidemiology of pertussis in young infants. The role of adults as reservoirs of infection. Am. J. Dis. Child., 132:371–373, 1978.

54. Nelson, K.E., et al.: The role of adenoviruses in the pertussis syndrome. J. Pediatr., 86:335–341, 1975.

55. Olson, L.C.: Pertussis. Medicine, 54:427–469, 1975.

56. Preston, N.W.: Effectiveness of pertussis vaccines. Br. Med. J., 2:11–13, 1965.

57. Scottish group: Diagnosis of whooping cough. Comparison of serological tests with isolation of *Bordetella pertussis*. A combined Scottish study. Br. Med. J., 4:637–639, 1970.

58. Whitaker, J., Donaldson, P., and Nelson, J.D.: Diagnosis of pertussis by the fluorescent-antibody method. New Engl. J. Med., 263:850–851, 1960.

Yersinia Species

59. Asakawa, Y., Akahane, S., and Kagata, N.: Two community outbreaks of human infection with *Yersinia enterocolitica*. J. Hygiene, 71:715–723, 1973.

60. Black, R.E., et al.: Epidemic *Yersinia enterocolitica* infection due to contaminated chocolate milk. New Engl. J. Med., 298:76–79, 1978.

61. Camus, A.: The Plague. Modern Library, New York, 1947.

62. Center for Disease Control: Current trends. Plague—United States. Morbidity Mortality Weekly Rep., 19:298–299, 1970.

63. Finegold, M.J.: Pathogenesis of plague. A review of plague deaths in the United States during the last decade. Am. J. Med., 45:549–554, 1968.

64. Gutman, L.T., et al.: A inter-familial outbreak of *Yersinia enterocolitica* enteritis. New Engl. J. Med., 288:1372–1377, 1973.

65. Kartman, L., Goldenberg, M.I., and Hubbert, W.T.: Recent observations on the epidemiology of plague in the United States. Am. J. Pub. Health, 56:1554–1569, 1966.

66. Kohl, S., Jacobson, J.A., and Nahmias, A.: *Yersinia enterocolitica* in children. J. Pediatr., 89:77–79, 1976.

67. Langer, W.L.: The black death. Sci. Am., 210:114–121, 1964.

68. Leino, R., and Kalliomaki, J.L.: Yersiniosis as an integral disease. Ann. Intern. Med., 81:458–461, 1974.

69. Palmer, D.L., Kisch, A.L., Williams, R.C., Jr., and Reed, W.P.: Clinical features of plague in the United States: the 1969–1970 epidemic. J. Infect. Dis., 124:367–371, 1971.

70. Rabson, A.R., Hallett, A.F., and Koornhof, H.J.: Generalized *Yersinia enterocolitica* infection. J. Infect. Dis., 131:447–451, 1975.

71. Raevuori, M., Harvey, S.M., Pickett, M.J., and Martin, W.J.: *Yersinia enterocolitica*: in vitro antimicrobial susceptibility. Antimicrob. Agents and Chemother., 13:888–890, 1978.

72. Reed, W.P., Palmer, D.L., Williams, R.C., Jr., and Kisch, A.L.: Bubonic plague in the Southwestern United States—a review of recent experience. Medicine, 49:465–486, 1970.

73. Spira, T.J., and Kabins, S.A.: *Yersinia enterocolitica* septicemia with septic arthritis. Arch. Intern. Med., 136:1305–1308, 1976.

74. Von Reyn, C.F., et al.: Epidemiologic and clinical features of an outbreak of bubonic plague in New Mexico. J. Infect. Dis., 136:489–494, 1977.

75. Weber, J., Finlayson, N.E., and Mark, J.B.D.: Mesenteric lymphadenitis and terminal ileitis due to *Yersinia pseudotuberculosis*. New Engl. J. Med., 283:172–174, 1970.

76. Yamashiro, K.M., Goldman, R.H., Harris, D., and Uyeda, C.T.: *Pasteurella pseudotuberculosis*. Acute sepsis with survival. Arch. Intern. Med., 128:605–608, 1971.

Pasteurella Species

77. Doty, G.L., Loomus, G.N., and Wolf, P.L.: *Pasteurella* endocarditis. New Engl. J. Med., 268:830–832, 1963.

78. Francis, D.P., Holmes, M.A., and Brandon, G.: *Pasteurella multocida.* Infections after domestic bites and scratches. J.A.M.A., *233:*42–45, 1975.

79. Hubbert, W.T., and Rosen, M.N.: I. *Pasteurella multocida* infection due to animal bite. II. *Pasteurella multocida* infection in man unrelated to animal bite. Am. J. Pub. Health, *60:*1103–1117, 1970.

80. Maneche, H.C., and Toll, H.W., Jr.: Pulmonary cavitation and massive hemorrhage caused by *Pasteurella multocida.* New Engl. J. Med., *271:*491–494, 1964.

81. Starkebaum, G.A., and Plorde, J.J.: Pasteurella pneumonia: report of a case and review of the literature. J. Clin. Microbiol., *5:*332–335, 1977.

Francisella Tularensis

82. Bloom, M.E., Shearer, W.T., and Barton, L.L.: Oculoglandular tularemia in an inner city child. Pediatrics, *51:*564–566, 1973.

83. Buchanan, T.M., Brooks, G.F., and Brachman, P.S.: The tularemia skin test. Ann. Intern. Med., *74:*336–343, 1971.

84. Center for Disease Control: Tularemia—United States 1960–1972. Morbidity Mortality Weekly Rep., *19:*95, 1970.

85. Center for Disease Control: Morbidity Mortality Weekly Rep. vol. 18–21: Annual supplements. Summaries, 1969–1972.

86. Guerrant, R.L., Humpheries, M.K., Jr., Butler, J.E., and Jackson, R.S.: Tickborne oculoglandular tularemia. Case report and review of seasonal and vectorial associations in 106 cases. Arch. Intern. Med., *136:*811–813, 1976.

87. Hughes, W.T.: Tularemia in children. J. Pediatr., *62:*495–502, 1963.

88. Klock, L.E., Olsen, P.F., and Fukushine, T.: Tularemia epidemic associated with the deerfly. J.A.M.A., *226:*149–152, 1973.

89. Miller, R.P., and Bates, J.H.: Pleuropulmonary tularemia. A review of 29 patients. Am. Rev. Respir. Dis., *99:*31–41, 1969.

90. Overholt, E.L., Tigertt, W.D., Kadull, P.J., and Ward, M.K.: An analysis of forty-two cases of laboratory acquired tularemia. Am. J. Med., *30:*785–806, 1961.

91. Sites, V.R., Poland, J.D., and Hudson, B.W.: Bubonic plague misdiagnosed as tularemia. J.A.M.A., *222:*1642–1643, 1972.

92. Young, L.S., et al.: Tularemia epidemic: Vermont, 1968. Forty-seven cases linked to contact with muskrats. New Engl. J. Med., *280:*1253–1260, 1969.

Brucella Species

93. Buchanan, T.M., Faber, L.C., and Feldman, R.A.: Brucellosis in the United States, 1960–1972. Part I. Clinical features and therapy. Medicine, *53:*403–413, 1974.

94. Center for Disease Control: Morbidity Mortality Weekly Rep. Annual supplements. Summaries 1970–1972.

95. Center for Disease Control: Brucellosis surveillance. p. 5. Annual summary, Brucellosis—1976, Issued October, 1977.

96. Christianson, H.B., Pankey, G.A., and Applewhite, M.L.: Ulcers of the skin due to *Brucella suis.* Report of a case. Arch. Derm., *98:*175–176, 1968.

97. Hall, W.H., and Manion, R.E.: In vitro susceptibility of Brucella to various antibiotics. Appl. Microbiol., *20:*600–614, 1970.

98. Halpern, S.E., and Wolf, S.G.: An unusual case of brucellosis. J.A.M.A., *204:*679–681, 1968.

99. Monroe, P.W., Silberg, S.L., Morgan, P.M., and Adess, M.: Seroepidemiological investigation of *Brucella canis* antibodies in different human population groups. J. Clin. Microbiol., *2:*382–386, 1955.

100. Munford, R.S., et al.: Human disease caused by *Brucella canis.* A clinical and epidemiologic study of two cases. J.A.M.A., *231:*1267–1269, 1975.

101. Phillippon, A.M., et al.: Rifampin in the treatment of experimental brucellosis in mice and guinea pigs. J. Infect. Dis., *136:*482–488, 1977.

102. Schirger, A., et al.: Brucellosis: experiences with 224 patients. Ann. Intern. Med., *52:*827–837, 1960.

103. Spink, W.W.: Some biologic and clinical problems related to intracellular parasitism in brucellosis. New Engl. J. Med., *247:*603–610, 1952.

104. Spink, W.W.: Host-parasite relationship in human brucellosis with prolonged illness due to suppuration of the liver and spleen. Am. J. Med. Sci., *247:*129–136, 1964.

105. Young, E.J., and Suvannoparrat, U.: Brucellosis outbreak attributed to ingestion of unpasteurized goat cheese. Arch. Intern. Med., *135:*240–243, 1975.

5
Gram-Negative Cocci

MENINGOCOCCUS

Objectives

1. Discuss the frequency of meningococcal infection at various ages.
2. Describe the typical clinical manifestations of severe meningococcemia.
3. Describe the chemotherapy of meningococcemia.
4. Discuss chemprophylaxis for individuals exposed to patients with meningococcal disease.
5. Discuss the significance of serum antibodies to the meningococcus.

Definitions

The meningococcus *(Neisseria meningitidis)* is a gram-negative D-shaped diplococcus. The genus is named for Neisser, who first described the other important disease-producing member of the genus, *Neisseria gonorrheae*. The species name *meningitidis* reflects its importance as the usual cause of epidemics of meningitis.

Other *Neisseria* species are normal flora of the nose and throat, and have species names such as *N. flava* (yellow), *N. catarrhalis*, and *N. sicca* (dry). *N. catarrahalis* is now classified in the genus *Branhamella* (see Chapter 6).

Serogroups

The term serogroup is usually used instead of group or type. Meningococcal serogroups are usually defined by agglutination of a saline suspension of the bacteria with commercially available type-specific antisera against the capsular polysaccharide, prepared in animals. Coagglutination can also be used to group meningococci (see p. 19). The major serogroups are A through D, which have been recognized for many years, and X through Z, which have been recently recognized. Serogroup A is the usual cause of large outbreaks of meningitis and is usually susceptible to sulfa drugs.

Serogroup A was not prevalent in the United States during the 1960s, but reappeared at a low frequency in the 1970s. Serogroup B is of special interest because these types are often resistant to sulfa drugs. Serogroup A has been associated with severe epidemics in past decades.

Frequency and Importance

Meningococcal disease occurs predominantly in children and young adults (Fig. 5-1). Of the 2644 cases of meningococcal disease reported in the United States in 1969, 42 percent occurred in children less than 5 years of age, 21 percent occurred in children 5 to 15 years of age, and 21 percent occurred in young adults 16 to 24 years of age.[5] Infants under a year of age had the highest frequency of any year of life.[5] This age pattern continued throughout the 1970s.

Meningococcemia is important because it is one of the most frequent infectious causes of death in normal individuals. Outbreaks can occur, with a high mortality rate. Individuals exposed to meningococcal disease may be very anxious, and the physician must be able to deal with this situation.

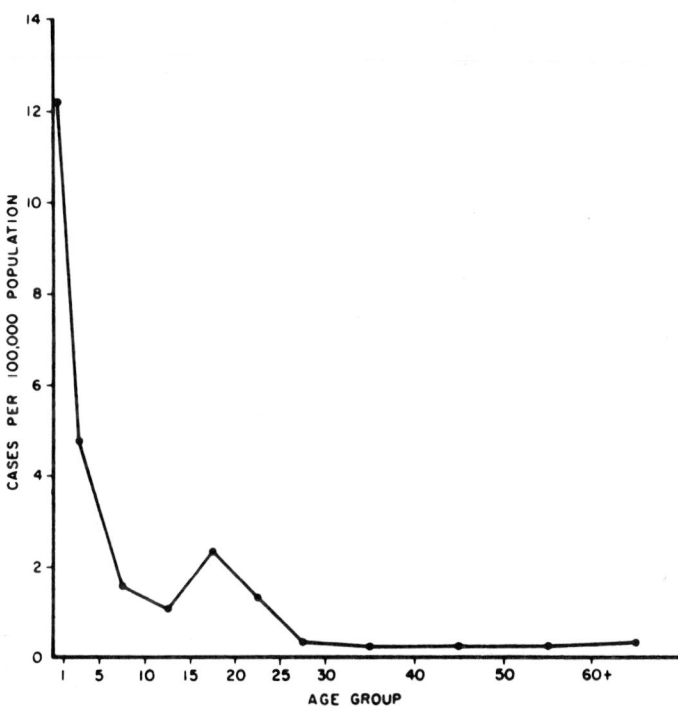

FIG. 5-1. *Meningococcal disease attack rate by age group, United States, 1969.*[5]

Clinical Patterns of Illness

N. meningitidis is a *frequent* cause of:

Primary Septicemia (Meningococcemia). This is an extremely serious, often fulminating, illness which is one of the most important of all medical emergencies. The patient usually appears anxious and seriously ill. High fever and a petechial or purpuric rash are early findings (Figs. 5-2 and 5-3). It is very important to follow the blood pressure closely and to treat shock aggressively with fluids. Arthritis or arthralgia may be present. Vomiting and diarrhea are occasionally prominent.[6] Prompt treatment is often life-saving, but the mortality rate remains high. Endotoxin shock, disseminated intravascular coagulation, and myocarditis are common serious complications. Meningitis is a frequent, but not invariable, part of the pattern.

Purulent Meningitis is usually sporadic in civilian populations, but outbreaks in the very young infant (less than 6 months old) may be atypical, with absence of petechial lesions, and a gradual course, probably because of the presence of some transplacentally transferred maternal antibodies.

N. meningitidis is an *uncommon* cause of:

Chronic Petechial Rash. Chronic meningococcemia is extremely unusual, and probably reflects a host response modified by antibiotics or partial immunity from antibodies.[10] Low grade fever, arthritis, and a petechial rash occur in this syndrome, which can be diagnosed with certainty only by a positive blood culture.

Pneumonia, Arthritis, Purulent Conjunctivitis, Exudative Pharyngitis, or Otitis Media. Often one or more of these syndromes occurs in addition to septicemia or meningitis, but each is occasionally the only clinical manifestation of meningococcal infection. Pneumonia without meningitis appears to be a frequent manifestation of serogroup Y meningococcal infection.[21]

Asymptomatic or Subclinical Infection

The Meningococcal Carrier State. This is defined as a nose or throat culture positive for meningococci in a pateint without clinical disease. The carrier state is found in about five to ten percent of civilians in general, but is higher (about 20 percent) in young adults, or in family contacts of

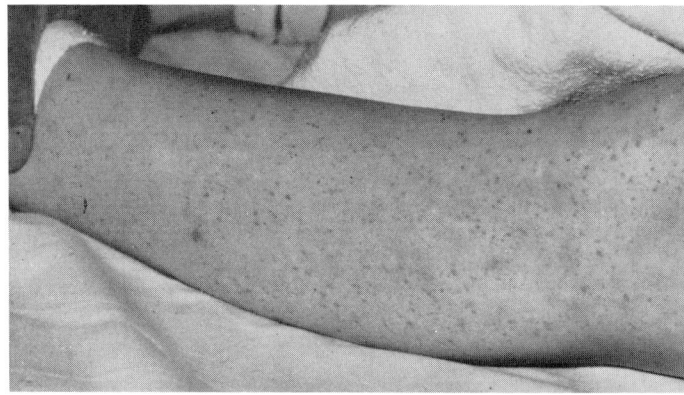

FIG. 5-2. Patient with meningococcemia and typical petechial rash. (Photo from Dr. Norman Fost)

patients with meningitis.[15] Only about five percent of young children are carriers.[13] The carrier state usually increases to above 50 percent in closed populations before an outbreak of meningococcal disease.

Neisseria meningitidis can also colonize the urethra, cervix, and anal canal, especially in relation to sexual contact.

Mild Respiratory Illnesses. Few civilian laboratories routinely inoculate nose or throat cultures on the special media needed to detect meningococci, and no studies have been done to determine how frequently meningococci are associated with minor respiratory illnesses. However, most adults have protective antibodies against the meningococci,[14] and are probably immune on the basis of past infection, perhaps with N. lactamica,[13] which may have been a mild illness rather than a totally asymptomatic infection. Infection with non-groupable meningococci also stimulate antibodies that protect against pathogenic strains, and this is probably important in developing and maintaining natural immunity to meningococcal disease.[13,14,27]

Relatively mild febrile illnesses have been reported in young infants with meningococcemia, so not all patients have fulminating illnesses.[2]

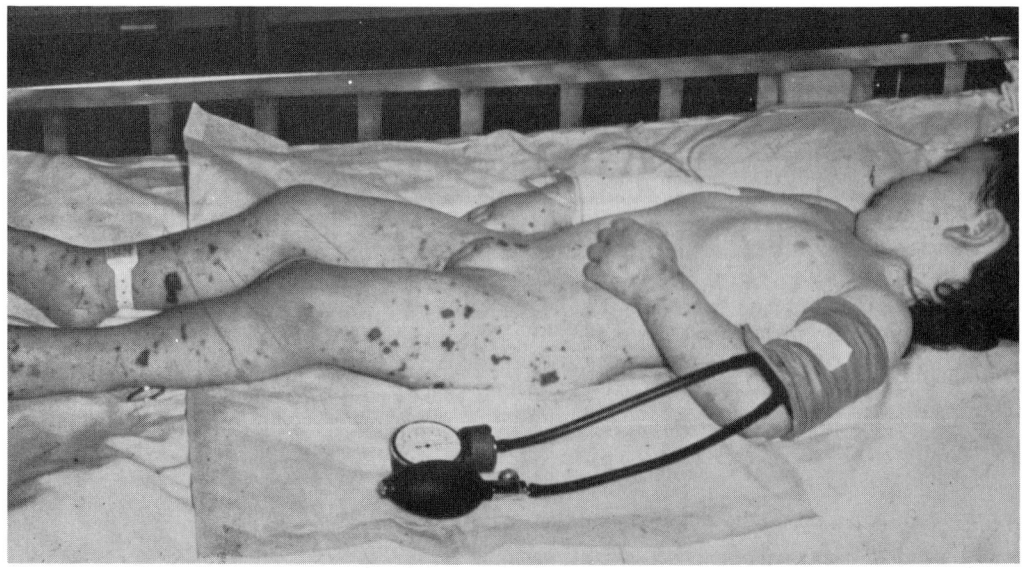

FIG. 5-3. Purpuric rash of meningococcemia. A blood pressure cuff should be left around the arm, and careful observations should be made to detect and treat shock. (Photo from Dr. Norman Fost)

Laboratory Diagnosis

Smear. Gram stain is useful in examining cerebrospinal fluid or a smear of petechiae. The organism appears as diplococci, with a typical shape, with flattened edges in contact. Fluorescent antibody methods for detecting the organism on smear may be available in some laboratories.

Culture. Primary isolation of the meningococcus is enhanced if the media contains blood or serum and is incubated in CO_2 (in a candle jar). When nose or throat cultures are taken to detect carriers, special media, such as Thayer–Martin medium, are usually used, which contain antibiotics to decrease the recovery of the normal flora of the nasopharynx.

Rapid identification of the meningococcus is often possible as soon as colonies are present (often within 24 hours after inoculation) by demonstration of agglutination of a colony suspended in saline, using type-specific antisera. Coagglutination can be used for rapid identification and grouping of isolate (p. 19). *N. meningitidis* is distinguished from other *Neisseria* species because it produces acid from maltose and glucose, but not from sucrose, fructose, or lactose. However, these sugar fermentations require another 24 hours. *N. lactamica* is a recently recognized species that acidifies the same sugars as *N. meningitidis*, but also produces acid from lactose.[13] It is found in the nasopharynx of about 10 to 20 percent of young children and may sometimes be mistaken for *N. meningitidis*.[13]

Occasionally, the fastidious meningococcus fails to grow on subculture and is lost. Although *Neisseria* species other than *N. meningitidis*, such as *N. subflava* and *N. catarrhalis*, can occasionally cause purpura, septicemia, and meningitis,[11] any gram-negative diplococcus recovered from cerebrospinal fluid (CSF) is almost always a meningococcus, especially if the organism is lost on subculture.

Oxidase Test. All *Neisseria* species are positive in a simple color test for the presence of cytochrome oxidase (the oxidase test). This test is of some value for rapid recognition that a colony may be a *Neisseria* species.

Serum Antibodies. Serologic diagnosis is possible, but is usually available only in research laboratories. Recent studies with the ELISA method indicate that this may be a highly sensitive test for detection of meningococcal antibodies.

Detection of Antigen. Detection of meningococcal antigens is possible in spinal fluid or synovial fluid, using meningococcal antisera in counterimmunoelectrophoresis (Fig. 4-2) for a precipitin reaction. It is not yet widely used.

Limulus Lysate Test. This is used to detect and quantitate endotoxin in blood or other body fluids. *Limulus polyphemus*, the horseshoe crab, has "blood" cells called amebocytes. A fluid lysate of these amebocytes is solidified by endotoxin, providing a sensitive and rapid assay for endotoxin. Endotoxin can often be rapidly detected by this method in the blood with bacteremia due to endotoxin-producing gram-negative bacteria. Endotoxin may be detected in the CSF in meningococcal or *H. influenzae* meningitis and in septic arthritis using this method. False positives (for example, with gram-positive bacteremias) occur frequently enough that the test is of limited value. Penicillins can inhibit the test and produce false-negative results. Thus, this test is still investigational and somewhat controversial.

Biologic Characteristics of Clinical Interest

Endotoxin. Meningococcemia can be considered the prototype disease producing endotoxin shock. Endotoxin is a complex lipopolysaccharide which is a component of the cell wall of most gram-negative bacteria. It can produce fever, leukopenia, and thrombocytopenia. It produces coagulation disturbances which result in focal hemorrhages (as in the skin and adrenals), or clotting within the blood vessels (disseminated intravascular coagulation), especially in the small blood vessels of the lungs.[7]

Predisposing Factors. About 40 percent of individuals without antibodies develop systemic disease when infected. The factors responsible for the occurrence of meningitis in exposed individuals without serum antibodies are unknown. Fatigue and alcohol ingestion may predispose the host to serious disease, but viral respiratory disease probably does not.

Resistance to Sulfa Drugs. Because of the possibility of resistance, and the seriousness of the disease, sulfonamides should not be used alone for therapy of meningococcal disease.

IgA Proteinase. In neisseria, this enzyme appears to be important in colonization and probably in disease.[24]

Difficult Susceptibility Testing. Fastidious growth requirements make the use of standard disc susceptibility tests difficult. Some laboratories prefer to use the more cumbersome plate dilution technique, because of its increased reliability.[16]

Treatment

Supportive Measures. Treatment of shock and disseminated intravascular coagulation are as important and as urgent as antibiotic therapy. A large intravenous needle or plastic catheter should be inserted immediately, both for blood volume expansion and for antibiotic therapy. Plasma or isotonic fluids should be given rapidly to replenish plasma volume loss due to splanchnic pooling or arteriovenous shunting.[28] Corticosteroids are usually given in high doses to produce vasodilation and to protect cells from the effect of endotoxin.[28] In rabbits, heparin does not appear to be useful to prevent the pulmonary microthrombi, which are more important in producing death than adrenal hemorrhage or renal cortical necrosis.[7]

Antibiotic Therapy. Penicillin or ampicillin should be given without delay as soon as the clinical diagnosis of meningococcemia is suspected. The intravenous route should be used. The organism is always quite susceptible to either penicillin or ampicillin. Sulfonamides should not be used alone, as the organism may be sulfa resistant.

In penicillin-allergic individuals, chloramphenicol is the best alternative. Chloramphenicol is effective in vitro, enters the spinal fluid well, and is clinically effective.[9] Tetracyclines or erythromycin are effective in vitro, but have had less clinical usage.

Prevention

Vaccines. Vaccine for serogroup A and serogroup C meningococci appears to be effective.[3] Immunization appears to be practical for groups with a high risk of meningococcal disease, such as military recruits. The efficacy and duration of protection for children is under investigation and appears to be effective.[26]

There is a possibility that immunization of recruit populations with serogroups A and C may result in an increase of disease due to other serogroups, such as Y.[25]

Avoid Exposure. Separation of recruits to avoid overcrowded conditions appears to have been successful in decreasing both carrier rates and disease. Recruits newly arrived at a training center are at greater risk of acquiring serious disease than are seasoned veterans, who presumably have had past exposures and are likely to be immune.

Management of Exposed Individuals

Intimate Exposure Increases Risk. Exposed family or intimate associates have a higher risk of meningococcemia than the general population.[14] The secondary attack rate in the United States was about 3 per 1000 household members in one study.[23] However, management of exposed individuals is unfortunately often delayed.[23]

Hospital attendants who have uncontrolled contact with a patient's oral secretions are also considered candidates for chemoprophylaxis.

Schoolmates in nursery schools and preschools are also at an increased risk, and should receive prophylaxis if the contact is as close as that with a family member.[19]

Throat Cultures. Culturing family or medical contacts of the patient with meningococcemia is not recommended.[1] Many laboratories are not capable of doing it accurately and several days are needed for complete results.

The use of chemotherapeutic agents to prevent secondary cases in the family is a moderately controversial issue. There is no prospective evidence that such use of drugs prevents secondary cases, although several drugs are effective in eradicating the carrier state. This also reduces the risk of spreading the meningococcus beyond the family.

Eradication of the Carrier State. Most authorities recommend a course of an antimicrobial drug, such as rifampin, for the family of a patient with meningococcemia.[20] This is usually referred to as chemprophylaxis, although there is no evidence that such drugs prevent development of meningococcal disease. They are, however, effective in eliminating the carrier state. Rifampin, for example, is effective in eliminating the carrier state, but is ineffective for therapy of meningococcal disease. At the opposite extreme, penicillin is excellent therapy of meningococcal disease, but is relatively ineffective in eliminating the carrier state. This apparent paradox appears to be explained by the fact that drugs effective in eradicating the carrier state are secreted into pharyngeal secretions, saliva, and tears.[8,17]

Studies of Prophylactic Drugs. Rifampin appears to be the most effective drug for eradication of the carrier state, and is the one currently recommended.[12]

Minocycline is equally effective, but has been associated with vestibular toxicity (vertigo, dizziness) frequently enough so that it is not recommended, because it is thought that many individuals will not continue to take the drug when symptoms develop.

Penicillin has been advocated by a few experts because of the belief that most secondary cases are already incubating the disease, and such drug therapy is really early treatment. However, penicillin has a minimal effect on reducing the frequency of the carrier state and is sometimes ineffective in preventing incubating meningococcemia.[4,22] For example, in an outbreak in Hawaii in 1968, four individuals were given intramuscular penicillin because of exposure to an individual with meningococcal meningitis, but they developed purulent meningitis 24 to 48 hours after the injection. All four recovered.[4]

In a study of suppression of the carrier state in naval recruits, ampicillin, tetracycline, and erythromycin were each more effective than penicillin, which had the same effect as a placebo, but none of the antibiotics led to permanent eradication of the organisms.[22] Thus, penicillin, tetracycline, or erythromycin can be considered unreliable for eradication of the carrier state.

Sulfonamide prophylaxis is of value only if the organism is susceptible, and many meningococci are resistant to sulfonamides. Therefore, sulfonamides should *not* be used for prophylaxis unless the meningococcus isolate has been shown to be sulfa senstivie by a reliable susceptibility testing method, such as plate dilution. Sulfonamide therapy is very effective in eliminating the carrier state if the meningococcus in sufa-sensitive.[8]

Observation of Exposed Family Members.[12] This has been recommended by some authorities. It is not practical with persons of doubtful reliability. Furthermore, fear and anxiety often lead the family members to search for a doctor who will give them the reassurance of a medication. Prophylaxis is often given because of the psychological needs of the patient or the physician rather than on any evidence of its value.

Vaccine. If the meningococcus isolated from the index case is identified as type A or type C, then immunization of household contacts with the appropriate vaccine is also recommended.[12]

NEISSERIA GONORRHEAE (GONOCOCCUS)

Objectives

1. Describe the typical clinical picture of gonococcal infection in the male and contrast it with that in the female.
2. Describe the clinical manifestations of gonorrhea in the male and female. Include discussion of asymptomatic and disseminated infections.
3. Discuss the collection and transportation of cultures for gonococcus. Describe the value and limitations of the Gram stain for this organism.
4. Discuss the clinical importance of pili and plasmids in the gonococcus.
5. Describe the relationship of disseminated neisserial infections to deficiency of a serum complement component.
6. Describe a regimen of antibiotic therapy for gonorrhea which is currently recommended and indicate an alternate therapy which can be used if the patient is allergic to standard therapy.

Definitions

Neisser was a bacteriologist who described the agent of gonorrhea, a word derived from the Greek word for "flow of seed." *Neisseria gonorrheae* is a gram-negative diplococcus.

Frequency and Importance

In the 1960s, the frequency of gonococcal infections increased so dramatically that it became an important health problem (Fig. 5-4). In the 1970s, at least one million cases, and possibly five million cases, occurred each year in the United States, but the rapid increase in reported cases appeared to be leveling off by 1980. Many serious complications of gonococcal infections have be-

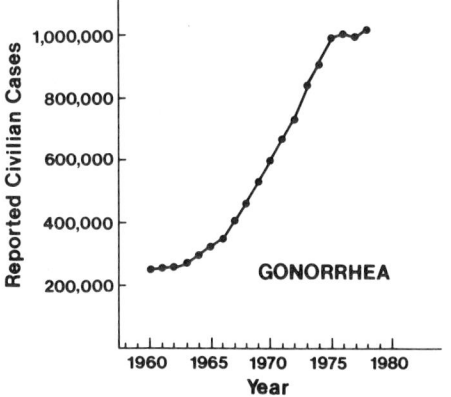

FIG. 5-4. *Gonorrhea: reported civilian cases in the United States, 1960–1978.*

come much more frequent, including septic arthritis, infective endocarditis, and neonatal conjunctivitis.

Clinical Patterns of Illness

N. gonorrheae is a *frequent* cause of:

Purulent Urethritis. In the male, this is manifested by a burning sensation on urination, and a purulent urethral discharge. Frequency and urgency of urination become prominent, and the patient usually seeks treatment. Gonococcal urethritis is less common in the female.

Purulent vaginitis and purulent urethritis occur in some women and can occur in children. In women, however, the cornified epithelium and pH provide protection from gonococcal infection not present in children.[60]

Salpingitis, also called pelvic inflammatory disease (PID), may be acute or chronic.[37] Lower quadrant abdominal pain is the usual presenting system. The infection can usually be localized to the fallopian tubes by bimanual palpation of the area.

Purulent arthritis or tenosynovitis is frequently caused by N. gororrheae.[45,67] The tendon sheaths of the hands, feet, or knees are most frequently involved. Arthritis may be serous or grossly purulent.

N. gonorrheae is an *occasional* cause of several syndromes:

Purulent conjunctivitis typically occurs in newborn infants in whom conjunctival prophylaxis has been omitted or inadequately done. It also can occur in young children or adults.[62] Typically, the discharge is extraordinarily purulent. (Fig. 5-5).

Gonococcal proctitis occurs in women as well as in homosexual men. It may be manifested by bloody diarrhea with copious amounts of mucopus, or may be mild or asymptomatic.[48]

Infection of the liver (perihepatitis) can complicate rectal or genital infection (called the Fitz-Hugh Curtis syndrome).[57]

The gonococcus can be recovered from the pharynx in some patients, and appears to be associated with sore throat.[68]

Disseminated gonorrhea bacteremia can be manifested in several forms:

Septicemia with pustular skin lesions is a gonococcal syndrome in which the patient appears acutely ill, and the pustular skin lesions represent dissemination of the bacteria.[29,44] Infective endocarditis may occur as a complication of the bacteremia.

Fever and arthralgia is a gonococcal syndrome that may be mistaken for acute rheumatic fever.[44] Bacteremia is usually present, but the patient usually does not look as ill as in the septicemic pattern.

Chronic prostatitis, epididymitis, or urethral stricture are sequalae of untreated gonococcal urethritis in the male.

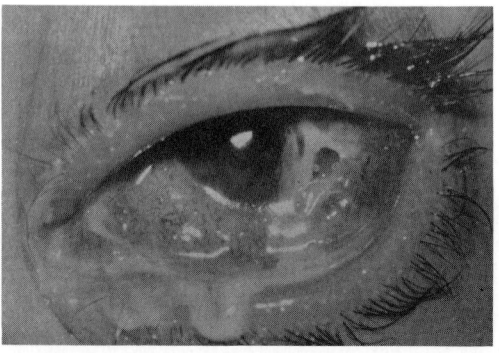

FIG. 5-5. *Purulent conjunctivitis due to gonorrhea. (Photo from Dr. Fred Brightbill)*

Asymptomatic or Subclinical Infections

Females. Routine cultures of the cervix of young pregnant women reveal gonococci in about two to five percent, depending on the group studied.[35] The majority of infected females are asymptomatic.

Males. In one study, about 10 percent of asymptomatic male contacts of infected women had positive urethral cultures, and about 50 percent of suspected homosexuals without a urethral discharge had a positive rectal culture.[54]

Experimental Infections

Experiments in transmission of gonorrhea in humans are complicated by variable host response, possibly on the basis of past infection.[53] The usual incubation period (between inoculation and first symptoms) is 3 to 5 days, but may be as long as 30 days, and as short as 1 day. About 80 percent of experimentally exposed volunteers developed mild urethritis, which receded without any treatment.

In more recent volunteer studies, the relationship of colony type to virulence was studied.[46] Some colony types of *N. gonorrheae* did not produce disease. After intraurethral inoculation of virulent colonies, the human volunteers developed burning and urethral discharge within 48 hours. Tender inguinal lymph nodes were noted for a few days in most subjects, with a tender testis or epididymis in a few subjects. A rise in antibody titer in most subjects could be demonstrated using fluorescent antibody methods. Further development of human and animal experimental models is needed.

Laboratory Diagnosis

Smear. This is a valuable way to make a presumptive laboratory diagnosis, but the organism is fastidious in its growth requirements and is often lost in subculture. Other organisms, such as *Mimae* or other *Neisseria* species, may mimic the appearance of gonococcus on smear, but are more easily cultured and identified (that is, less likely to die on subculture) than the gonococcus. The smear is very reliable if positive from the male urethra or female cervix, but can be negative in the female when the culture is positive.[65]

Culture. For primary isolation, special media are required, usually a chocolate agar, with antibiotics added (variations of Thayer–Martin medium), and incubation must be done in a candle jar.[63] Colonies are gram-stained and tested with the oxidase test, which turns purple with *Neisseria* (and *Pseudomonas*) species, for preliminary identification. For definitive identification, the organism is shown to ferment glucose only. A useful mnemonic to remember the fermentation patterns is G for glucose and gonococcus; and MG for maltose and glucose and meningococcus.

New transport media that support the survival and growth of gonococci have recently improved the recovery rate from specimens obtained in physicians' offices.[36]

Urine can be useful for the culture of gonococci in males, using a first-voided specimen.[41]

In women who have had hysterectomies, the urethra is the best area to culture.[47]

Serologic Diagnosis. Serum antibodies can be measured, but this is not yet practical or accurate for diagnostic purposes. There may be cross-reaction with meningococcal antibodies. Local urethral disease may occur without a detectable change in serum antibodies. Only in suspected gonococcal arthritis with negative cultures are serum antibodies likely to be useful in diagnosis.

Coagglutination. Colonies of oxidase-positive, gram negative diplococci on Thayer–Martin medium can be confirmed as gonococci using coagglutination with gonococcal antibody bound to staphylococci with protein A (Fig. 1-12)[38] This tests appears to be simple and accurate enough to make culture and identification of gonococci possible in the clinician's office laboratory.

Serologic Tests for Syphilis. The VDRL test for syphilis is usually indicated in patients with gonorrhea, because the two diseases can be transmitted at the same sexual contact.

Biologic Characteristics of Clinical Interest

Special Growth Requirements. Unless the laboratory is informed that gonococcus is suspected, the organism may not be recovered, since inoculation of special media is necessary.

Repeated Infection. This can occur in spite of detectable antibodies.

Animal Infections. Chimpanzees can be infected by intraurethral inoculation.[51] First signs (urethral discharge) and positive cultures appeared 4 days after inoculation, and the chimpanzees developed a rise in complement-fixing antibodies by 14 to 25 days after inoculation.

Beta-Lactamase Production. Gonococci highly resistant to penicillin were first detected in 1976. The mechanism of resistance is the production of a beta-lactamase, which can be transferred by a plasmid (see p. 50). This plasmid-mediated (R factor-mediated) resistance gene can be transferred via conjugation between two gonococci or between a gonococcus and an *E. coli.*

IgA-Cleaving Proteinase. As in the case of colonization with *N. meningitidis*, colonization, and perhaps disease, are enhanced in gonococcal strains that produce a proteinase which destroys IgA.[31]

Virulence Related to Pili, Colony Type, and Iron. Gonococcal strains are of variable virulence, in terms of disease produced in humans. The virulence is increased when laboratory strains are incubated with iron.[49] Differences in iron metabolism or iron concentrations in humans may explain variation of disease severity, although this hypothesis has not been tested.

Pili are protein-containing, hairlike appendages on bacterial cell surfaces. The presence of pili on gonococci is associated with increased virulence, apparently because pili are related to the adherence of the bacteria to urethral epithelium, to oral mucosal epithelium, or to sperm.[49]

Colony appearance can also be used to classify gonococci into types. The more virulent colony types can be identified by experienced observers, but so far this has not proved to be clinically practical information.

Circulating Immune Complexes. Gonococcal antigen–antibody complexes can be demonstrated in the blood of many patients with disseminated gonococcemia.[66]

Complement Deficiency.[50] Disseminated neisserial infections, both gonococcal and meningococcal, occur more frequently in individuals with a deficiency of a late component of complement (C7), which promotes bactericidal activity of normal serum.

Treatment of Gonorrhea

Penicillin. This is still the drug of choice. Probenecid, 1 g, should be given in addition.[50] Up until 1976, penicillin-resistant strains of gonococcus were only relatively resistant; only slightly higher concentrations were necessary to kill the organism. The penicillin concentrations required to inhibit the gonococcus increased between 1955 and 1970 from an average of about 0.2 units per ml to about 1.5 units per ml.[64] These concentrations were attainable with high doses (4.8 million units of procaine penicillin in adults). Failure rates for susceptible organisms are about two to four percent with this regimen.[43] Most treatment failures are probably due to reinfection, but there is a high frequency of treatment failure in patients with relatively resistant organisms.

This gradual increase in resistance to penicillin appears to have been a result of mutations which led to an altered gonococcal cell, less permeable to penicillin and other antibiotics.[61] However, a profound increase in resistance to penicillin has been a result of plasmid-mediated (R factor) acquisition of a penicillin-destroying enzyme.[61] Thus, in 1976, beta-lactamase producing strains of *Neisseria gonorrheae* were identified throughout the world, with one strain resistant to tetracycline first recognized in the Far East, and another strain more susceptible to tetracycline first recognized in West Africa.[55]

Unfortunately, some recent isolates of gonococci have been multiply resistant strains; they are resistant to penicillin, tetracycline, ampicillin, and spectinomycin. This suggests that a common plasmid (transferable DNA fragment) contains the genes to code for all of these resistance-producing enzymes.[56]

Because of possible beta-lactamase resistance of gonorrhea, follow-up cultures are recommended to confirm a cure.[52] If the clinical response is poor, spectinomycin is recommended. Isolates should be tested for beta-lactamase production. Contact tracing must be more aggressive now that therapy has a significant chance of being ineffective.

A single injection of cefoxitin, along with oral probenecid, is a possible alternative to spectinomycin, as it is effective against either penicillin-resistant or penicillin-sensitive gonococcal urethritis.

Treatment of Acute Salpingitis.[34] Aqueous crystalline penicillin G, 20 million units a day until improvement occurs, followed by ampicillin, 2 g daily, for a total course of 10 days is recommended.[34] Oral tetracycline, amoxicillin or ampicillin can be used for outpatients.

Treatment of Disseminated Gonococcal Infection.[34] Aqueous crystalline penicillin G, 10 million units a day until improvement, followed by ampicillin daily to complete 10 days was recommended in 1979. Oral tetracycline, amoxicillin, or ampicillin regimens can be used for outpatients, as described below.

Children. For children less than 50 kilograms, proportionally lower doses have been recommended.

Alternative Antibiotics.[34] Many antibiotics other than penicillin are also effective in penicillin-sensitive strains, including ampicillin, tetracycline, and erythromycin. However, effectiveness of oral therapy depends on patient cooperation in taking the drug.[34] Therefore, intramuscular spectinomycin is usually recommended when the patient is allergic to penicillin. Recommended dosages in adults are:[34]

Ampicillin 3.5 g or amoxicillin 3.0 g, with probenicid 1 g by mouth, if oral therapy is desired. Pharyngeal infection is difficult to treat and does not respond well to single dose ampicillin, so an additional 500 mg four times a day for 2 more days has been recommended.[39]

Spectinomycin 2 g intramuscularly for women or men, if the patient is allergic to penicillin, or if the organism is resistant to penicillin; this appears valuable for anorectal gonorrhea.

Tetracycline 0.5 g by mouth four times a day for a total of 10 g, if oral therapy is desired, for an individual who is allergic to penicillin, or if non-gonococcal urethritis may be the cause.

Trimethoprim–sulfamethoxazole is as effective as ampicillin for gonococcal urethritis.[58] Erythromycin was associated with persistent or recurrent gonococcal infection in one study of gonococcal urethritis.[32] Oral doxycycline had a failure rate of seven percent and was significantly inferior to oral ampicillin in another study of uncomplicated gonorrhea in males and females.[40]

Therapy for Incubating Syphilis

In a cooperative study, various antibiotic regimens for gonorrhea and a placebo were compared for efficacy in eradicating incubating syphilis.[59] Procaine penicillin, 2.4 to 4.8 million units, eradicated incubating syphilis, and patients so treated, with a negative serology at the time of therapy, do not need follow-up serologic tests for syphilis. However, tetracycline, and probably other broad spectrum antibiotics effective against the gonococcus, may only prolong the incubation period of syphilis, and patients so treated do require serologic follow up.[34,59] Spectinomycin is ineffective for incubating syphilis.

Prevention

Contact Investigation. Rapid investigation and treatment of contacts is useful to prevent spread of the disease. This has become much more important with the appearance of penicillin-resistant strains. Such a program is based on experimental evidence that most infections occur within a week of exposure. Therefore, contacts should be examined, cultured, and treated at once.

Prophylaxis in Newborns. Gonococcal conjunctivitis in newborns can be prevented by instillation of silver nitrate, penicillin, or erythromycin drops into the eyes after birth.[30,62] Recently, parents have begun requesting a delay in the use of eye drops to permit the newborn to see the mother, which may be important in mother–infant "bonding". Some states allow an hour after birth before eye drops are required. Data are lacking, but a delay of an hour to allow infant eye contact with the mother before instillation of eye drops probably does not increase the risk of gonococcal conjunctivitis.

Prophylaxis in Adults. Condoms are partially effective in prevention of transmission of gonorrhea from either sexual partner.[20] Antibiotic prophylaxis before or immediately after sexual intercourse has not been adequately studied in a controlled fashion to determine what oral dosages are sufficient.

REFERENCES

Meningococcus

1. Artenstein, M.S.: Prophylaxis for meningococcal disease. J.A.M.A., 231:1035–1037, 1975.

2. Baltimore, R.S., and Hammerschlag, M.: Meningococcal bacteremia. Clinical and serological studies of infants with mild illness. Am. J. Dis. Child., *131:*1001–1004, 1977.

3. Brandt, B.L., Smith, C.D., and Artenstein, M.S.: Immunogenicity of serogroup A and C *Neisseria meningiditis* polysaccharide vaccines administered together in humans. J. Infect, Dis., *137:*202–205, 1978.

4. Center for Disease Control: Meningococcal meningitis—Hawaii. Morb. Mort. Week. Rep., *17:*178, 1968.

5. ———Surveillance summary. Meningococcal disease—United States, epidemiological year 1970. Morb. Mort. Week. Rep., *19:*414–415, 1970.

6. Corbett, T.H., and Brody, J.A.: An epidemic in an Eskimo village due to Group B meningococcus. Part 2. Clinical features. J.A.M.A., *196:*388–390, 1966.

7. Dalldorf, F.G., and Jennette, J.C.: Fatal meningococcal septicemia. Arch. Pathol. Lab. Med., *101:*6–9, 1977.

8. Devine, L.F., et al.: Levels in serum and saliva and effect on the meningococcal carrier state. J.A.M.A., *214:*1055–1059, 1970.

9. Devine, L.F., and Hagerman, C.R.: Spectra of susceptibility of *Neisseria meningitidis* to antimicrobial agents in vitro. Appl. Microbiol., *19:*329–334, 1970.

10. Fass, R.J., and Saslaw, S.: Chronic meningococcemia. Possible pathogenesis role of IgM defiency. Arch. Intern. Med., *130:*943–946, 1972.

11. Feigin, R.D., San Joaquin, V., and Middlekamp, J.N.: Purpura fulminans associated with *Neisseria catarrhalis* septicemia and meningitis. Pediatrics, *44:*120–123, 1969.

12. Finley, R.A.: Prophylaxis against meningococcal disease. J.A.M.A., *236:*459–461, 1976.

13. Gold, R., et al.: Carriage of *Neisseria meningitidis* and *Neisseria lactamica* in infants and children. J. Infect. Dis., *137:*112–121, 1978.

14. Goldschneider, I., Gotschlich, E.C., and Artenstein, M.S.: Human immunity to the meningococcus. I. The role of humoral antibodies. II. Development of natural immunity. J. Exp. Med., *129:*1307–1326, 1327–1348, 1969.

15. Greenfield, S., Sheehe, P.R., and Feldman, H.A.: Meningococcal carriage in a population of "normal" families. J. Infect. Dis., *123:*67–73, 1971.

16. Hammerberg, S., Marks, M.I., and Weinmaster, G.: Reevaluation of the disk diffusion method for sulfonamide susceptibility testing of *Neisseria meningitidis.* Antimicrob. Agents Chemother., *10:*869–871, 1976.

17. Hoeprich, P.D.: Prediction of antimeningococcic chemoprophylactic efficacy. J. Infect. Dis., *123:*125–133, 1971.

18. Jacobson, J.A., Chester, T.J., and Fraser, D.W.: An epidemic of disease due to serogroup B *Neisseria meningitidis* in Alabama: report of an investigation and community-wide prophylaxis with a sulfonamide. J. Infect. Dis., *136:*104–108, 1977.

19. Jacobson, J.A., Filice, G.A., and Holloway, J.T.: Meningococcal disease in day-care centers. Pediatrics, *59:*299–300, 1977.

20. Kaiser, A.B., et al.: Seroepidemiology and chemoprophylaxis of disease due to sulfonamide-resistant *Neisseria meningitidis* in a civilian population. J. Infect. Dis., *130:*217–223, 1974.

21. Koppes, G.M., Ellenbogen, C., and Gebhart, R.J.: Group Y meningococcal disease in United States Air Force recruits. Am. J. Med., *62:*661–666, 1977.

22. Martin, G.I., and DeGrinney, J.T.: Intrafamilial infection with *Neisseria meningitidis,* group C. An intense intramural meningeal infestation in a family. Clin. Pediatr., *11:*538–540, 1972.

23. Meningococcal Disease Surveillance Group: Meningococcal disease. Secondary attack rate and chemoprophlaxis in the United States, 1974. J.A.M.A. *235:*261–265, 1976.

24. Mulks, M.H., and Plant, A.G.. IgA protease production as a characterisitc distinguishing pathogenic from harmless *Neisseriaceae.* New Engl. J. Med., *299:*973–976, 1978.

25. Nikoskelainen, J., et al.: Is group-specific meningococcal vaccination resulting in epidemics caused by other groups of virulent meningococci? Lancet, *2:* 403–405, 1978.

26. Peltola, H., et al.: Meningococcus group A vaccine in children three months to five years of age. J. Pediatr., *92:*818–822, 1978.

27. Reller, L.B., MacGregor, R.R., and Beaty, H.N.: Bactericidal antibody after colonization with *Neisseria meningitidis.* J. Infect. Dis., *127:*56–62, 1973.

28. Wilson, F.E., and Morse, S. R.: Therapy of acute meningococcal infections: early volume expansion and prophylactic low dose heparin. Am. J. Med. Sci., *264:*445–455, 1972.

Gonococcus

29. Abu-Nassar, H., Hill, N., Fred, H.L., and Yow, E.M.: Cutaneous manifestations of gonococcemia. Arch. Intern. Med., *112:*731–737, 1963.

30. Barsam, P.C.: Specific prophylaxis of gonorrheal ophthalmia neonatorum. A review. New Engl. J. Med., *274:*731–734, 1966: Corre-

spondence. New Engl. J. Med., 275:280–281, 1966.

31. Blake, M., Holmes, K.K., and Swanson, J.: Studies on gonococcal infection. XVII. IgA$_1$-cleaving protease in vaginal washings in women with gonorrhea. J. Infect. Dis., 139:89–92, 1979.

32. Brown, S.T., Pedersen, A.H.B., and Holmes, K.K.: Comparison of erythromycin base and estolate in gonococcal urethritis. J.A.M.A., 238:1371–1373, 1977.

33. Center for Disease Control: Morb. Mort. Week. Rep. Vol. 28. Annual Supplement. Summary, 1978, issued Sept. 1979.

34. ———CDC recommended treatment schedules for gonorrhea, 1979. Morb. Mort. Week Rep., 28:13–16,21, 1979.

35. Charles, A.G., Cohen, S., Kass, M.B., and Richman, R.: Asymptomatic gonorrhea in prenatal patients. Am. J. Obstet. Gynecol., 108:595–599, 1970.

36. Cross, R.C., Crecilius, H.G., and Counts, J.M.: Survival of Neisseria gonorrheae in the mail. Appl. Microbiol., 20:281, 1970.

37. Cunningham, F.G., et al.: Evaluation of tetracycline and ampicillin for treatment of acute pelvic inflammatory disease. New Engl. J. Med., 296:1380–1383, 1977.

38. Danielson, D., and Kronvall, G.: Slide agglutination method for the serological identification of Neisseria gonorrheae with antigonococcal antibodies adsorbed to Protein A-containing staphylococci. Appl. Microbiol., 27:368–374, 1974.

39. Di Caprio, J.M., et al.: Ampicillin therapy for pharyngeal gonorrhea. J.A.M.A. 239:1631–1633, 1978.

40. Enfors, W., and Eriksson, G.: Comparison of oral ampicillin and doxycycline in the treatment of uncomplicated gonorrhea. Br. J. Vener. Dis., 51:99–103, 1975.

41. Feng, W.C., Medeiros, A.A., and Murray, E.S.: Diagnosis of gonorrhea in male patients by culture of uncentrifuged first-voided urine. J.A.M.A., 237:896–897, 1977.

42. Hart, G.: Role of preventitive methods in the control of veneral disease. Clin. Obstet. Gynecol., 18:243–253, 1975.

43. Holmes, K.K., et al.: Single-dose aqueous procaine penicillin G therapy for gonorrhea: use of probenecid and cause of treatment failure. J. Infect. Dis., 127:455–460, 1973.

44. Holmes, K.K., Counts, G. W., and Beaty, H.N.: Disseminated gonococcal infection. Ann. Intern. Med., 74:979–993, 1971.

45. Keiser, H., Ruben, F.L., Wolinsky, E., and Kushner I.: Clinical forms of gonococcal arthritis. New Engl. J. Med., 279:234–240, 1968.

46. Kellogg, D.S., Jr., et al.: Neisseria gonorrheae. II. Colonial variation and pathogenicity during 35 months in vitro. J. Bacteriol., 96:596–605, 1968.

47. Klaus, B.D., Chandler, J.E., and Dans, P.E.: Gonorrhea detection in posthysterectomy patients. J.A.M.A., 240:1360–1361, 1978.

48. Klein, E.J., Fisher, L.S., Chow, A.W., and Guze, L.B.: Anorectal gonococcal infection. Ann. Intern. Med., 86:340–346, 1977.

49. Kolata, G.B.: Gonorrhea: more of a problem but less of a mystery. Science 192:244–247, 1976.

50. Lee, T.J., et al.: Familial deficiency of the seventh component of complement associated with recurrent bacteremic infections due to Neisseria. J. Infect. Dis., 138:359–368, 1978.

51. Lucas, C. T., Chandler, F., Jr., Martin, J.E., Jr., and Schmale, J.D.: Transfer of gonococcal urethritis from man to chimpanzee. An animal model for gonorrhea. J.A.M.A., 216:1612–1614, 1971.

52. McCormack, W.M.: Treatment of gonorrhea—is penicillin passé? New Engl. J. Med., 269:934–936, 1977.

53. Mahoney, J.F., Van Slyke, C.J., Cutler, J.C., and Blum, H.L.: Experimental gonococcal urethritis in human volunteers. Am. J. Syph. Gonor. Vener. Dis., 30:1–39, 1946.

54. Pariser, H., and Marino, A.F.: Gonorrhea: frequently unrecognized reservoirs. South. Med. J., 63:198–201, 1970.

55. Perine, P.L., et al.: Evidence for two distinct types of penicillinase-producing Neisseria gonorrhoeae. Lancet, 2:993–995, 1977.

56. Powell, J.T., and Bond, J.H.: Multiple antibiotic resistance in clinical strains of Neisseria gonorrheae isolated in South Carolina. Antimicrob. Agents Chemother., 10:639–645, 1976.

57. Reichert, J.A., and Valle, R.F.: Fitz-Hugh Curtis syndrome, a laparoscopic approach. J.A.M.A., 236:266–268, 1976.

58. Sattler, F.R., and Ruskin, J.: Therapy of gonorrhea. Comparison of trimethoprim-sulfamethoxazole and ampicillin. J.A.M.A., 240:2267–2270, 1978.

59. Schroeter, A.L., Turner, R.H., Lucas, J.C., and Brown, W.J.: Therapy for incubating syphilis. Effectiveness of gonorrhea treatment. J.A.M.A., 218:711–713, 1971.

60. Shore, W.B., and Winkelstein, J.A.: Nonvenereal transmission of gonococcal infections to children. J. Pediatr., 79:661–663, 1971.

61. Siegel, M.S., et al.: Penicillinase-producing Neisseria gonorrheae: results of surveillance in the United States. J. Infect. Dis., 137:170–175, 1978.

62. Thatcher, R.W., and Pettit, T.H.: Gonorrheal conjunctivitis. J.A.M.A., 215:1494–1496, 1971.

63. Thayer, J.D., and Martin, J.E., Jr.: Improved medium selective for cultivation of N. gonorrheae and N. meningitidis. Publ. Health Rep., 81:559–562, 1966.

64. Thayer, J.D., Samuels, S.B., Martin, J.E., Jr., and Lucas, J.B.: Comparative antibiotic susceptibility of *Neisseria gonorrhoeae* from 1955 to 1964. Antimicrob. Agents Chemother., *1964:* 433–436, 1965.

65. Wald, E.R.: Gonorrhea. Diagnosis by Gram stain in the female adolescent. Am. J. Dis. Child., *131:* 1094–1096, 1977.

66. Walker, L.C., Ahlin, T.D., Tung, K.S.K., and Williams, R.C., Jr.: Circulating immune complexes in disseminated gonorrhea infection. Ann. Intern. Med., *89:* 29–33, 1978.

67. Wheeler, J.K., Heffron, W.A., and Williams, R.C.: Migratory arthralgias and cutaneous lesions as confusing initial manifestations of gonorrhea. Am. J. Med. Sci., *260:* 150–159, 1970.

68. Wiesner, P.J., et al.: Clinical spectrum of pharyngeal gonococcal infection. New Engl. J. Med., *288:* 181–185, 1973.

MISCELLANEOUS BACTERIA

This chapter deals with the newly recognized bacillus of Legionnaires' disease and with other bacteria rarely encountered in human disease. Many of the bacteria in this chapter are normal flora of the human; they ordinarily colonize humans but rarely cause disease, except in compromised hosts. Other bacteria are soil or water bacteria, which also typically cause disease only in compromised individuals.

LEGIONNAIRES' DISEASE BACILLUS

Objectives

1. Describe two clinical patterns produced by the bacillus of Legionnaires' disease (Legionella pneumophilia).
2. Describe the currently available laboratory methods of documenting infection with Legionnaires' disease bacillus.

Definitions

Legionnaires' disease bacillus is a small, blunt bacterium (Fig. 6-1) which can be strained in tissues by the Dieterle silver impregnation stain, a stain designed for spirochetes (Fig. 6-2).[11] It can be Gram stained after laboratory passaging, and is a small gram-negative rod. It is tentatively named *Legionella pneumophilia* for the outbreak of pneumonia that occurred in Philadelphia in July, 1976 (during the Bicentennial festivities) among attendees at a state American Legion convention at the Bellvue Stratford Hotel.[19] There were about 200 cases and nearly 30 deaths in that outbreak. Using stored sera, several previously unexplained outbreaks of disease were recognized to have been caused by *Legionella pneumophilia*.

Frequency and Importance

This organism appears to account for a significant percentage (perhaps 10 percent) of "viral" pneumonias. It can be fatal, especially in the elderly, the immunosuppressed, or those with chronic pulmonary disease. Thus it is an important nosocomial (hospital-acquired) infection.

Past Outbreaks

The following outbreaks are often described by their location:[19]

1. *Legionnaires' disease (Philadelphia outbreak)*. The most famous outbreak occurred primarily in members of the American Legion attending a convention in a Philadelphia hotel in July and August 1976.
2. *Broad Street pneumonia*. This syndrome is defined as pneumonia that occurred in persons in the vicinity of the Philadelphia hotel during the summer of 1976.[19]
3. *Pontiac fever*. This outbreak occurred at the Pontiac, Michigan, health department in July 1968, and was recognized in retrospect by study of paired sera.[23] It may have been related to the air conditioning system.

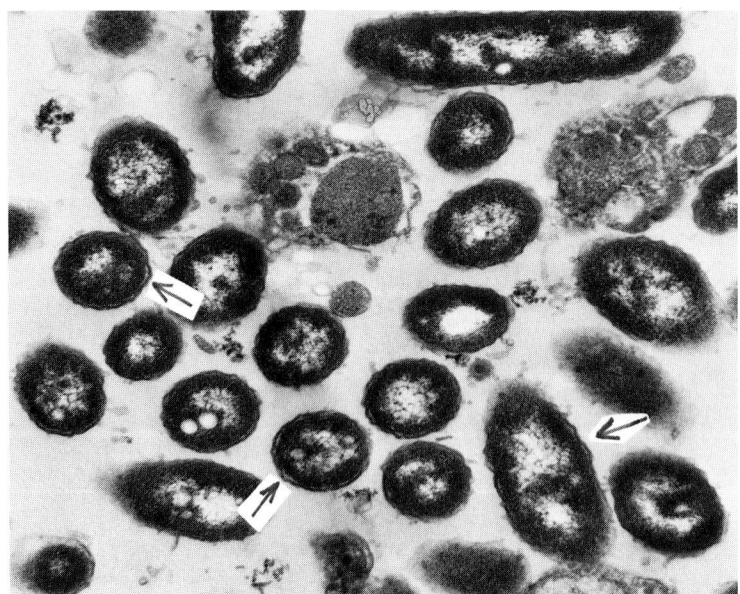

FIG. 6-1. Electron micrograph of Legionnaires' disease bacillus (Legionella pneumophilia) in a hen's egg yolk sac membrane. Note the enclosure of the organism by a double envelope (arrows). *(Photo from Dr. Francis W. Chandler)*

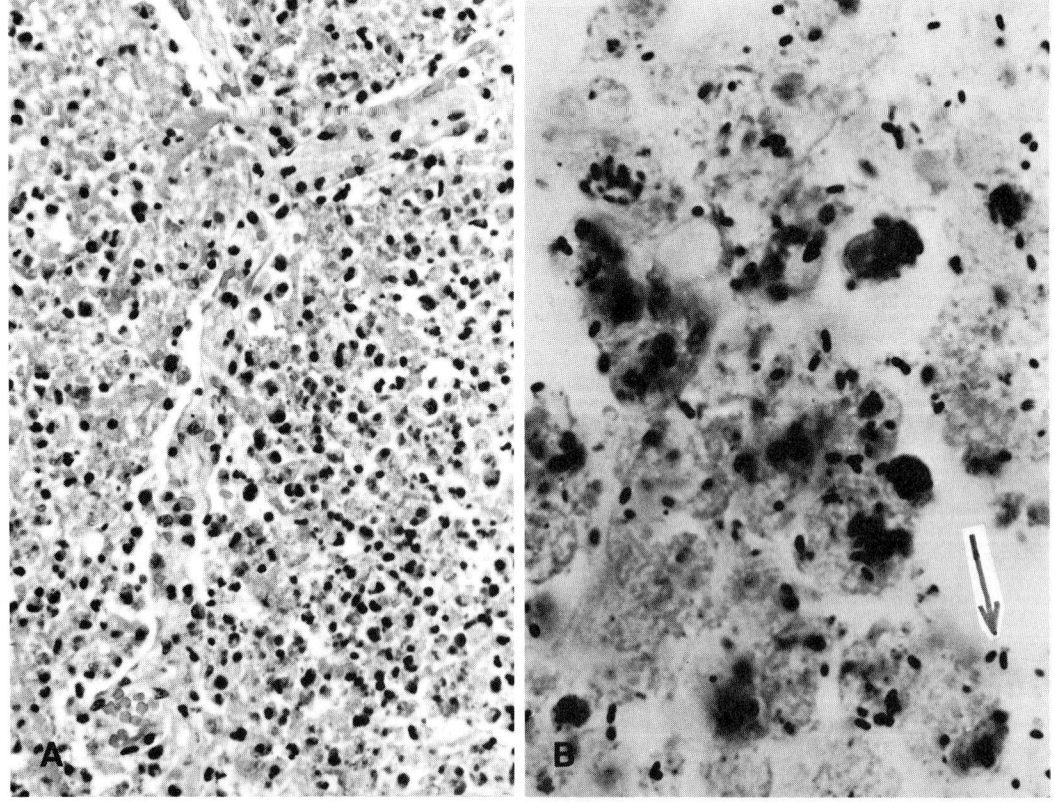

FIG. 6-2. (A) Lung tissue, showing severe disease, but no microorganisms by the conventional hematoxylin and eosin stain. (B) The same tissue section, stained with a silver-impregnation stain, showing Legionnaires' disease bacillus, a small, blunt pleomorphic rod, found both within and outside of phagocytes. *(Photo from Dr. F. W. Chandler)*

4. *St. Elizabeth's Hospital outbreak.* This occurred in Washington, D.C., in July 1965, and was also recognized in retrospect by study of paired sera.[42]

5. Sporadic outbreaks have been recognized frequently since serologic tests have become available.[9] Recent outbreaks occurred at the University of Indiana, Bloomington, Indiana; the Wadsworth Veterans Administration Hospital in Los Angeles, California; the New York City garment district; and in Vermont.

Clinical Patterns of Illness

Progressive Pneumonia. The 1976 Philadelphia outbreak was characterized by *progressive pneumonia*.[19] The pneumonia was typically bilateral, interstitial, and progressive. Later in the course there was focal consolidation, often nodular.[15] None in the 1976 outbreak had pleural effusion. However, analysis of subsequent confirmed cases indicates pleural effusion is common.[15] Cavitation and abscess formation occurs in some cases.

The onset is associated with a prodromal period with cough, fever, lethargy, and confusion.[19,29] Sputum reveals many neutrophils, but no bacteria by conventional stains (Fig. 6-1). The peripheral white blood count may or may not show a leukocytosis.[23,29] The sedimentation rate is elevated. The illness can be mistaken for a pulmonary infarction.[29] Capillary-alveolar diffusion block with hypoxemia (and some improvement with oxygen) is apparent early. The disease typically becomes progressively more severe. About 15 percent of all cases in the elderly have been fatal.[9]

Influenza-like Illness. The Pontiac, Michigan, outbreak was manifested by influenza-like illness with fever, headache, muscle aches, and no pneumonia and no fatalities.[9,23] This outbreak involved young, healthy adult visitors and personnel of a county health department building.

Complications. Fatalities due to complications are more common in the elderly or in immunosuppressed persons. Acute renal failure may occur. Confusion and disorientation (suggesting cerebral hypoxia) can be prominent.

Laboratory Diagnosis

Smear. The bacillus cannot be seen in an ordinary Gram stained smear. However, a Dieterle silver stain of sputum or transtracheal aspirate (or postmortem lung tissue) reveals the bacillus. The silver methenamine stain used to demonstrate *Pneumocystis carinii* does not stain the Legionnaires' disease bacillus. In frozen sections of lung, the Giménez stain is useful to demonstrate the organism. Recently, "half-a-gram" stain (crystal violet and iodine, without decolorizing) has been shown to demonstrate the bacillus in sputum.[14] A direct fluorescent antibody stain is also useful (Fig. 1-9), but several different type-specific antibody preparations are needed to detect all types.[7]

Serology. An increase in antibodies to Legionnaires' disease bacillus can be detected by several methods but usually the methods are available only in reference laboratories. One of the most sensitive methods is the ELISA test (see Fig. 13-6).[17] An indirect fluorescent antibody method is also useful (see Fig. 15-6).

A single high titer in convalescent serum is strongly suggestive of recent infection due to the organism. Significant antibody levels do not appear until about 3 weeks after the onset. Cold agglutinins are usually absent.

Culture. The organism was originally cultured from autopsy specimens of lung tissue inoculated into guinea pigs. After disease occurred, the animals' spleens were subcultured in yolk sacs of chick embryos (a technique used to culture rickettsiae).[34] Special enriched solid media will allow slow growth of the laboratory-passaged organism on subculture, and are being developed and improved for primary isolation from patient's specimens.[18] Some strains produce a water-soluble pigment.

Non-Specific Tests. The serum creatinine phosphokinase, transaminase, and lactic dehydrogenase enzymes are typically elevated. Low serum sodium may occur, apparently due to inappropriate antidiuretic hormone secretion.

Biologic Characteristics of Clinical Interest

Incubation Period. The symptoms begin about 2 to 10 days after exposure.[19]

Airborne Transmission. Major outbreaks appear to have involved airborne transmission. The disease is almost never transmitted to the patient's family, and isolation of the patient is not recommended. It has been suggested that the organism resides in soil, as excavation or new construction has been associated with outbreaks.[45] It has been recovered from creek water and air conditioning system water. Aerosolization of contaminated water has been suggested as an important route of transmission.

Serogroups. There appear to be several antigenically distinct serogroups of Legionnaires' disease bacillus, one of which is called the Togus strain, named after the Togus VA Hospital in Maine. The other types are numbered and named for the location where they were first recovered: serogroup 1, Knoxville, Philadelphia; serogroup 2, Togus; serogroup 3, Bloomington; and serogroup 4, Los Angeles. Serogroup 1 (Knoxville) appears to be the most common.

Legionella-like Bacteria. New techniques used to detect legionellae have recovered similar bacteria which differ sufficiently from *L. pneumophilia* so that legionella-like bacteria may eventually be assigned to different species or genus.

Electron Microscopy. This technique demonstrates the organism is a prokaryote, divides like a bacillus, and is engulfed by macrophages.[10]

Treatment and Prevention

Oxygen is important supportive treatment.

Erythromycin or rifampin prevent death in guinea pigs infected with the agent, but penicillin, chloramphenicol, tetracycline, or gentamicin show no significant effect.[20] Currently, erythromycin is usually recommended, and appeared to be effective when used in the Wadsworth Hospital outbreak.[29] Trimethoprim—sulfamethoxazole is also effective in vitro.

ACTINOBACILLUS AND HAEMOPHILUS APHROPHILUS

Actinobacillus actinomycetemcomitans is an aerobic, small gram-negative rod that rarely causes infections in humans.[8,37] Its cumbersome name refers to its concomitant growth with *Actinomyces* species in early studies of actinomycosis (Chap. 2).

Haemophilus aphrophilus is a small gram-negative rod closely resembling *A. actinomycetemcomitans*, and the two are frequently confused.[37] Aphrophilus means "carbon dioxide-loving". Both of these bacteria require CO^2 for primary isolation, but grow on blood agar without added growth factors. Both have been reported as a cause of endocarditis, brain abscess, and abscesses of soft tissues.[8,41]

AEROMONAS AND PLEISMONAS

Aeromonas is a genus of aerobic gram-negative rods with polar flagella which are classified in the same family as *Vibrio* species.[13] The usual species is *Aeromonas hydrophilia*.

Pleisomonas shigelloides is also called *Aeromonas shigelloides*.[13,16] *Aeromonas* species cause infections in fish and reptiles and rarely cause infections in humans. These bacteria are found in water and soil, and human infection can often be related to water exposure.[26]

A. hydrophilia has been reported to cause acute diarrhea, and has been shown to produce an enterotoxin.[31] It also can cause soft tissue abscesses, osteomyelitis, traumatic wound infections, and rarely endocarditis, meningitis, or sepsis in compromised hosts.[13,26,31] It is a cause of hospital-acquired infections.

Aeromonads are usually most susceptible to chloramphenicol, but are also often susceptible to aminoglycosides such as gentamicin.[13]

BRANHAMELLA

Neisseria catarrhalis was the former name for this gram-negative coccus now called *Branhamella catarrhalis*.[36] It has long been regarded as non-pathogenic normal flora of the upper respiratory tract, but recently has been recognized as an opportunistic pathogen, rarely causing endocarditis and meningitis, and apparently a rare cause of bronchitis or otitis media.

The organism is often susceptible to penicillin in vitro, but little information is available on treatment of clinical infections.

CALYMMATOBACTERIUM (DONOVANIA)

Calymmatobacterium (Donovania) granulomatis is a bipolar staining gram-negative rod which is the cause of granuloma inguinale, a rare venereal disease characterized by granulomatous ulcerations in genital areas.[13a] It has a capsule antigenically similar to klebsiellae.

Possibly the bowel is the source of infection, because of an apparent relation to anal contact. The disease responds to tetracycline.

CARDIOBACTERIUM

Cardiobacterium hominis is a small fastidious gram-negative rod which is a very rare cause of human infections.[21] It was formerly called Group II-D in the Center for Disease Control classification of organisms not yet named.

Evidence for its pathogenicity in humans was established by its repeated recovery in blood cultures of a few patients with bacterial endocarditis, so it was eventually called cardiobacterium.

It is part of the normal flora of the upper respiratory tract.

It appears to be susceptible in vitro and in vivo to penicillin or ampicillin.[21]

CHROMOBACTERIUM

Chromobacterium violaceum is a soil bacterium that produces violet pigmented colonies. It is a rare cause of human infection.[28] *Chromobacterium typhiflavum* is the most recent name for the organism, formerly called *Erwinia* species or *Enterobacter agglomerans*. This organism was involved in a nationwide outbreak of bacteremia due to contaminated commercial intravenous fluids. It is of special interest because of its ability to survive in the osmotic conditions of intravenous fluids.

EIKENELLA

Eikenella corrodens is a very rare cause of disease in humans.[6,27-29] Corrodens refers to the observation that the colony corrodes or eats into the agar, resulting in a visible pitting or depression, which allows a presumptive identification from colony morphology alone.

It is a small gram-negative rod which also has been called or confused with *Bacteroides corrodens* (because some strains are strictly anaerobic). It formerly was called HB-1, in the Center for Disease Control classification of bacteria with unassigned species names. Sometimes these "corroding bacilli" are classified as bacteroides if strictly anaerobic, and eikenella if facultatively anaerobic.[38] This has some clinical significance, since the *Bacteroides* strains are quite susceptible to clindamycin whereas the *Eikenella* strains are quite resistant.[38]

It is normal flora of the upper respiratory tract, and can produce cellulitis or osteomyelitis in human bites (or, more frequently, hand wounds contaminated by saliva, as when a fist strikes the opponent's teeth).[27] It can also cause meningitis, endocarditis, and abdominal abscesses.[6]

It appears to be susceptible to most antibiotics, and severe infections have been cured by penicillin alone.[22]

ERYSIPELOTHRIX

Erysipelothrix rhusiopathiae is an aerobic gram-positive rod that resembles diphtheroids on Gram stain, and *Listeria monocytogenes* in other laboratory characteristics (Chap. 2). It can cause skin infections in animals (especially swine). Typically, human disease has consisted of erysipeloid, (like erysipelas due to Group A streptococci), a purplish-red infection of the fingers seen in butchers, fish handlers, and others with similar occupational exposures. It has rarely caused bacteremia or endocarditis.[25,33]

Penicillin appears to be effective therapy, but high doses are required in endocarditis.

ERWINIA

This genus was named for Erwin Smith, a plant pathologist. These plant pathogens resemble enterobacter and serratia and ordinarily do not cause disease in humans.[5] In the 1970s, *Erwinia* species (also called *Enterobacter agglomerans*, now classified as *Chromobacterium typhiflavum*, discussed above), were contaminants in commercial intravenous solutions, and caused bacteremias and deaths in patients given such contaminated solutions.[35] These organisms may also be a cause of conjunctivitis.[32]

FUSOBACTERIUM

This is a genus of anaerobic bacteria which is normal flora of the human gastrointestinal tract, especially the mouth. In necrotizing ulcerative gingivitis, a noncontagious periodontal disease sometimes called trench mouth, fusobacteria may be found in increased numbers. However, dentists who have studied this disease regard fusobacteria (and also the oral spirochetes) as secondary invaders, but not the cause of the disease, which is poor hygiene.[1]

Fusobacterium species are members of the same family as *Bacteroides* species, and *Fusobacterium* species bacteremia secondary to human bites is described in Chapter 3.

LACTOBACILLUS AND BIFIDOBACTERIUM

These anaerobic bacteria are normal flora of the mouth and dental areas, bowel, and vagina.[3,24] *Lactobacillus acidophilus* may contribute to an acid environment in the vagina or periodontal area. *Lactobacillus bifidus* (now usually called *Bifidobacterium* species) is named for its bifurcated ends. These bifidobacteria appear to differ significantly from other lactobacilli.

Lactobacilli have rarely been associated with bacterial endocarditis and bacteremia complicating an abscess.[3]

Bifidobacteria have rarely been associated with disease, but appear to be a rare cause of empyema or lung abscess—particularly after aspiration.[24]

Lactobacilli are usually susceptible to penicillin, but very high doses may be required for endocarditis.[3]

Lactobacillus species are the major component of yogurt, which has been advocated for therapy of diarrhea during antibiotic therapy. However, yogurt has not been shown to be of medical value in controlled clinical studies.

PEPTOCOCCUS AND PEPTOSTREPTOCOCCUS

The genus *Peptococcus* can be conveniently regarded as the strictly anaerobic counterpart of the genus *Staphylococcus*. The genus *Peptostreptococcus* can be regarded as the strictly anaerobic counterpart of the genus *Streptococcus*.

Peptococci are normal flora of the mouth, vagina, and skin. The organism has been associated with severe infections of the feet in diabetics.[39]

Peptostreptococci are usually reported by the bacteriology laboratory as anaerobic streptococci, which are discussed in Chapter 1. These organisms are normal flora of the upper respiratory tract and bowel. They can produce aspiration pneumonia or lung abscess, brain abscess, synergistic gangrene of the skin, anaerobic myositis, and female pelvic infections, especially abscesses.[44]

Anaerobic gram-positive cocci are typically susceptible to penicillin in vitro and in vivo.

PROPIONIBACTERIUM

This "anaerobic diphtheroid," formerly called *Corynebacterium acnes*, is now called *Propionibacterium acnes*. It is normal flora of the skin, and hydrolyzes fatty acids of sebaceous glands, thus contributing to the pathogenesis of acne. It is often recovered from blood cultures as a contaminant from inadequately disinfected skin. Most strains are microaerophilic; that is, able to grow slowly and poorly in aerobic cultures.

It has rarely been demonstrated to be the cause of human infections, such as meningitis.[40] *P. acnes* is typically susceptible in vitro to penicillin and its derivatives.[43]

STREPTOBACILLUS

Streptobacillus moniliformis is a gram-positive rod which is a rare cause of human infection. It can be transmitted to humans by a rat bite or by drinking milk contaminated by rat excreta (Haverhill fever, named for Haverhill, Massachusetts.).[30] The organism is typically susceptible to penicillin.

VEILLONELLA

This genus of anaerobic gram-negative cocci was named for Veillon, a French bacteriologist who was one of the first to describe the typical species. Most species show red fluorescence under ultraviolet light, a characteristic which is useful for rapid detection in mixed cultures.[12]

Veillonella species are normal flora of the mouth and vagina which have rarely been associated with human disease, such as intraabdominal abscesses and osteomyelitis.[4]

REFERENCES

1. Baer, P.N., and Benjamin, S.D.: Periodontal disease in children and adolescents. Philadelphia, J. B. Lippincott. 1975. pp. 38–45.
2. Balows, A., and Fraser, D.W. (editors): International symposium on Legionnaires' disease. Ann. Intern. Med., 90:491–714, 1979.
3. Bayer, A.S., Chow, A.W., Betts, D., and Guze, L.B.: Lactobacillemia—report of nine cases. Important clinical and therapeutic considerations. Am. J. Med., 64:808–813, 1978.
4. Borchardt, K.A., Baker, M., and Gelber, R.: *Veillonella parvula* septicemia and osteomyelitis. Ann. Intern. Med., 86:63–64, 1977.
5. Bottone, E., and Schneierson, S.S.: *Erwinia* species: an emerging human pathogen. Am. J. Clin. Pathol., 57:400–405, 1972.
6. Brooks, G.F., et al.: *Eikenella corrodens*, a recently recognized pathogen. Medicine, 53:325–342, 1974.
7. Broome, C.V., et al.: Rapid diagnosis of Legionnaires' disease by direct immunofluorescent staining. Ann. Intern. Med., 90:1–4, 1979.
8. Burgher, L.W., Loomis, G.W., and Ware, F.: Systemic infection due to *Actinobacillus actinomycetemcomitans*. Am. J. Clin. Pathol., 60:412–415, 1973.
9. Center for Disease Control: Legionnaires' disease: diagnosis and management. Ann. Intern. Med., 88:363–365, 1978.
10. Chandler, F.W., et al.: Ultrastructure of the agent of Legionnaires' disease in the human lung. Am. J. Clin. Pathol., 71:43–50, 1979.
11. Chandler, F.W., Hicklin, M.D., and Blackmon, J.A.: Demonstration of the agent of Legionnaires' disease in tissue. New Engl. J. Med., 297:1218–1221, 1977.
12. Chow, A.W., Patten, V., and Guze, L.B.: Rapid screening of *Veillonella* by ultraviolet fluorescence. J. Clin. Microbiol., 2:546–548, 1975.
13. Davis, W.A., III, Kane, J.G., and Garagusi, V.F.: Human Aeromonas infections: a review of the literature and a case report of endocarditis. Medicine, 57:267–277, 1978.
13a. Davis, C.M.: Granuloma inguinale. JAMA, 211:632–636, 1970.
14. DeFreitas, J.L., Borst, J., and Meenhorst, P.L.: Easy visualization of *Legionella pneumophilia* by "half-a-gram" stain procedure. Lancet, 1:270–271, 1979.
15. Dietrich, P.A., Johnson, R.D., Fairbank, J.T., and Walke, J.S.: The chest radiograph in Legionnaires' disease. Radiology, 127:577–582, 1978.
16. Ellner, P.D., and McCarthy, L.R.: *Aeromonas shigelloides* bacteremia: a case report. Am. J. Clin. Pathol., 59:216–218, 1973.
17. Farshy, C.E., Klein, G.C., and Feeley, J.C.: Detection of antibodies to Legionnaires' disease organism by microagglutination and microenzyme-linked immunosorbent assay tests. J. Clin. Microbiol., 7:327–331, 1978.
18. Feeley, J.C., et al.: Primary isolation media for Legionnaires' disease bacterium. J. Clin. Microbiol., 8:320–325, 1978.
19. Fraser, D.W., et al.: Legionnaires' disease. Description of an epidemic of pneumonia. New Engl. J. Med., 297:1189–1198, 1977.
20. Fraser, D.W., et al.: Antibiotic treatment of guinea pigs with agent of Legionnaires' disease. Lancet, 1:175–178, 1978.
21. Geraci, J.E., et al.: *Cardiobacterium hominis* endocarditis. Four cases with clinical and laboratory observations. Mayo Clin. Proc., 53:49–53, 1978.
22. Geraci, J.E., Hermans, P.E., and Washington, J.A., III: *Eikenella corrodens* endocarditis. Report of cure in two cases. Mayo Clin. Proc., 49:950–953, 1974.
23. Glick, T.H., et al.: Pontiac fever. Am. J. Epidemiol., 107:149–160, 1978.
24. Green, S.L.: Case report. Fatal anaerobic pulmonary infection due to *Bifidobacterium eriksoni*. Postgrad. Med., 63:187–189, 1978.
25. Grieco, M.H., and Sheldon, C.: *Erysipelothrix rhusiopathiae*. Ann. N.Y. Acad. Sci., 174:523–532, 1970.
26. Hanson, P.G., Standridge, J., Jarrett, F., and Maki, D.G.: Freshwater wound infection due to *Aeromonas hydrophila*. J.A.M.A., 238:1053–1054, 1977.
27. Johnson, S.M., and Pankey, G.A.: *Eikenella corrodens* osteomyelitis, arthritis, and cellulitis of the hand. South. Med. J., 69:535–540, 1976.
28. Johnson, W.M., DiSalvo, A.F., and Steuer, R.R.: Fatal *Chromobacterium violaceum* septicemia. Am. J. Clin. Pathol., 56:400–406, 1971.
29. Kirby, B.D., Snyder, K.M., Meyer, R.D., and Finegold, S.M.: Legionnaires' disease: clinical features of 24 cases. Ann. Intern. Med., 89:297–209, 1978.
30. Lambe, D.W., Jr., McPhedran, A.M., Mertz, J., and Stewart, P.: *Streptobacillus moniliformis* isolated from a case of Haverhill fever. Am. J. Clin. Pathol., 60:854–860, 1973.
31. Ljungh, A., Popoff, M., and Wadström, T.: *Aeromonas hydrophila* in acute diarrheal disease: detection of enterotoxin and biotyping of strains. J. Clin. Microbiol., 6:96–100, 1977.

32. London, R., and Bottone, E.: *Erwinia* conjunctivitis in children. Pediatrics, *49:*931–932, 1972.

33. McCracken, A.W., Mauney, C.U., Huber, T.W., and McCloskey, R.V.: Endocarditis caused by *Erysipelothrix insidiosa.* J. Clin. Pathol., *59:*219–222, 1973.

34. McDade, J.E., et al.: Legionnaires' disease. Isolation of a bacterium and demonstration of its role in other respiratory disease. New Engl. J. Med., *297:*1197–1203, 1977.

35. Maki, D.G., Rhame, F.S., Mackel, D.C., and Bennett, J.V.: Nationwide epidemic of septicemia caused by contaminated intravenous products. Am. J. Med., *60:*471–485, 1976.

36. Ninane, G., Jolly, J., and Kraytman, M.: Bronchopulmonary infection due to *Branhamella catarrhalis:* 11 cases assessed by transtracheal aspiration. Br. Med. J., *1:*276–278, 1978. 1978.

37. Page, M.I., and King, E.O.: Infection due to *Actinobacillus actinomycetemcomitans* and *Haemophilus aphrophilus.* New Engl. J. Med., *275:*181–188, 1966.

38. Robinson, J.V.A., and James, A.L.: In vitro susceptibility of *Bacteroides corrodens* and *Eikenella corrodens* to ten chemotherapeutic agents. Antimicrob. Agents Chemother., *6:*543–546, 1974.

39. Sanderson, P.J.: Infection of the foot with *Peptococcus magnus.* J. Clin. Pathol., *30:*266–268, 1977.

40. Schlesinger, J.J.: *Propionibacterium acnes* meningitis in a previously normal adult. Arch. Intern. Med., *137:*921–923, 1977.

41. Sutter, V.L.: *Haemophilus aphrophilus* infections: clinical and bacteriologic studies. Ann. N.Y. Acad. Sci., *174:*468–487, 1970.

42. Thacker, S.B.: An outbreak in 1965 of severe respiratory illness caused by the Legionnaires' disease bacterium. J. Infect. Dis., *138:*512–519, 1978.

43. Wang, W.L.L., Everett, E.D., Johnson, M., and Dean, E.: Susceptibility of *Propionibacterium acnes* to seventeen antibiotics. Antimicrob. Agents Chemother., *11:*171–173, 1977. 173,1977.

44. Wilson, W.R., Martin, W.J., Wilkowske, C.J., and Washington, J.A., III: Anaerobic bacteremia. Mayo Clin. Proc., *47:*639–646, 1972.

7
Mycobacteria

MYCOBACTERIUM TUBERCULOSIS

Objectives

1. Describe the early childhood and adult patterns of pulmonary tuberculosis.
2. Describe the typical clinical pictures of miliary tuberculosis, and of tuberculous meningitis.
3. Discuss the interpretation of a smear that is positive for acid fast bacilli.
4. Discuss what type of specimens are appropriate for confirming the diagnosis of tuberculosis and describe how long it takes cultures to become positive.
5. Describe the typical histologic findings of tuberculosis.
6. List the advantages and major adverse effects of the following antituberculous drugs: INH, PAS, streptomycin, ethambutol, rifampin.
7. Define drug resistant *Mycobacterium tuberculosis* and describe under what conditions it may be found. Discuss the rationale for multiple drug therapy and give three examples.
8. Define the conditions under which a tuberculin test is indicated, the dose of tuberculin testing material used under various circumstances, the timing, the method of reading the interdermal test, and the interpretation of test.
9. Discuss the relationship of corticosteroids to tuberculosis, both in treatment and in predisposing patients to the disease.
10. Define preventive therapy of tuberculosis and give two situations in which such therapy would be indicated.
11. Discuss the indications for BCG vaccine.

Definitions

"Myco" refers to the fungus-like properties of mycobacteria, which grow in long branches and grow slowly. There are many species of *Mycobacterium*, but *M. tuberculosis* is the major pathogenic species of humans. Other mycobacteria, which include *M. leprae* (Hansen's bacillus, the cause of leprosy), and the atypical mycobacteria, are discussed in the next section.

Frequency and Importance

In the 1970s, about 30,000 to 40,000 new cases of tuberculosis were reported annually in the United States, with about 5,000 deaths annually. The case rate is highest in the South and Southwest, especially in Native Americans, Chicanos, blacks, and others from lower socioeconomic groups.

Tuberculosis remains an important disease in the United States, especially because transmission to others and serious disability or death in the patient can be prevented by prompt diagnosis and chemotherapy.

Clinical Patterns of Illness

Mycobacterium tuberculosis is a frequent cause of:

Peripheral Pneumonia With Hilar Lymphadenopathy (Primary Complex). If primary pulmonary tuberculosis in childhood produces any visible

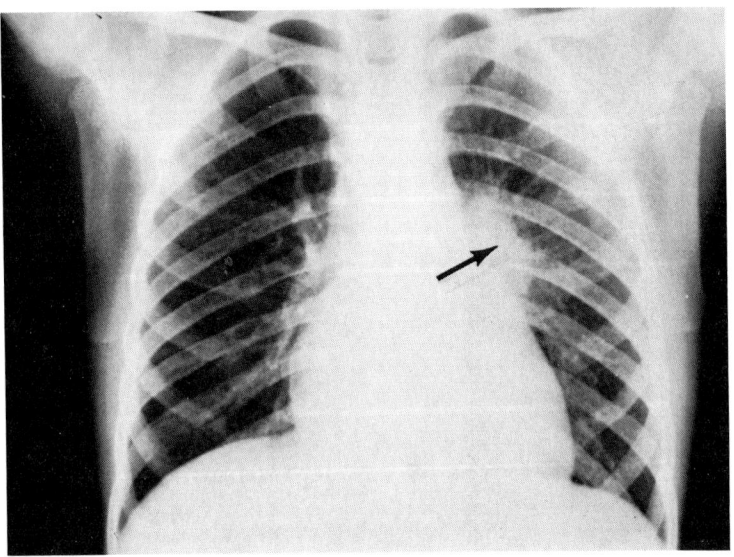

FIG. 7-1. *This child had primary pulmonary tuberculosis, with diffuse pneumonia and hilar adenopathy on the left (arrow). However, any pattern of pneumonia can be due to tuberculosis. (Radiograph from Dr. Justin Wolfson)*

radiologic abnormality, it is typically seen as a peripheral lesion in the lung with hilar adenopathy.[11,17] The peripheral opacities can be nodular, segmental, or subsegmental, but are rarely lobar (Fig. 7-1).

Apical Pneumonia. In adults, tuberculosis often occurs as apical pneumonia, in the apex of the upper lobes. Often the symptoms are minimal or absent, but may include cough, weight loss, and sweating at night.

Chronic Pneumonia. In adolescents or adults, persistent chronic pneumonia, sometimes with cavitation, is often due to *M. tuberculosis*. The mechanism can be either reinfection after new exposure or reactivation of old disease.[23,29]

Miliary or Nodular Pneumonia. Disseminated hematogenous tuberculosis typically results in miliary or nodular pneumonia with alveolar or small nodular infiltrates on chest film, and enlargement of the liver and spleen.[19,33] Often a severe viral pneumonia is suspected when there is acute severe respiratory failure with minimal radiographic findings.[19]

Nonpurulent Meningitis with Low Glucose. Tuberculous meningitis is a frequent cause of nonpurulent meningitis with low glucose. Tuberculous meningitis is most frequent in young infants and the elderly. Death or brain damage is still frequent, and chemoprophylaxis of exposed children is an important preventive measure.[33]

M. tuberculosis is an uncommon cause of osteomyelitis. Involvement of the spine (Pott's disease) or finger (dactylitis) is the most frequent form of tuberculous osteomyelitis.

M. tuberculosis is now a rare cause of primary peritonitis, pleural effusion, chronic pericarditis, granulomatous bowel disease, chronic skin ulcerations, retinitis or keratitis, chronic nephritis (typically manifested by pyuria with a negative urine culture), otitis media and mastoiditis.[15] In most of these situations the patient is already known to have a history of pulmonary tuberculosis or a positive tuberculin test.

Asymptomatic or Subclinical Infection

In the United States, the most frequent clinical pattern of tuberculosis, for most age groups in most locations, is absent or minimal symptoms. Affected individuals are typically detected by routine tuberculin testing, or by tuberculin testing individuals with minimal or nonspecific symptoms of fatigue, weight loss, or cough.

In some situations the chest film is negative.[14] The diagnosis may be proved by recovery of the organism from an individual with a tuberculin test which has recently converted to positive. In adults, active disease may be proved by culture in spite of no change in the chest film.[14]

Screening Programs

Detection of unrecognized tuberculosis is best done by routine tuberculin tests in high risk groups, such as immigrants from countries with much tuberculosis, health care personnel, and ethnic groups subjected to poverty, crowding, and infrequent medical care. Screening by annual chest roentgenograms has largely been abandoned.

Bovine Tuberculosis

Mycobacterium bovis was very rarely observed in cattle in the United States in the 1970s. Almost all cases in humans now occur in the elderly or in persons born and infected outside the United States or Canada.[39] The clinical presentation includes exposure to unpasteurized milk, a strongly positive standard tuberculin test, and any of the findings described above for *M. tuberculosis*. The organism is niacin-negative but produces disease in rabbits and guinea pigs.

Laboratory Approach

Smear for Acid-Fast Bacilli. Demonstration of acid-fast bacilli can provide a presumptive diagnosis, especially if a cavitary lesion is present. It is particularly useful when found in pus, sputum, secretions obtained by bronchoscopy, or spinal fluid smears. The demonstration of acid-fast bacilli is of less value when the specimen has been obtained from areas which have a normal flora containing, or easily contaminated by, saprophytic mycobacteria. For example, in the feces, sputum, or urine, mycobacteria other than *M. tuberculosis* can be seen on direct smear and later recovered from cultures, and may represent contamination in collecting the specimen, or urethral saprophytes. Nevertheless, some authorities emphasize the value of acid-fast stains of gastric aspirates in patients who cannot produce sputum.[25] The gastric aspirate contains a saprophytic mycobacteria in only one to three percent of normal individuals, but reveals acid-fast bacilli in about half of patients with *M. tuberculosis* eventually cultured from the sputum.[25] Other studies have indicated that smears of gastric aspirates result in too many false negatives and false positives to be reliable.[32]

Fluorescent Smears. The detection of *M. tuberculosis* in smears is easier if the specimen is stained with an auramine–rhodamine stain.[32] A microscope equipped with a special light source and special light filters is used to detect the mycobacteria, which appear fluorescent. It should be noted that this is not a fluorescent antibody technique, and no antigen–antibody reaction is involved, as is used in the detection of Group A streptococci (Chap. 1).

Culture. Specimens of urine or gastric contents should be sent promptly to the laboratory, because urine and gastric acid are toxic to mycobacteria. Specimens of bone marrow or spinal fluid should be refrigerated if prompt inoculation of solid media is not possible.

Preliminary laboratory diagnosis may be possible in 2 to 3 weeks, but about 8 weeks are required before the culture should be reported as negative.[22] *M. tuberculosis* typically appears buff-colored on solid media, whereas many of the atypical mycobacteria are yellow or orange. The most definitive laboratory characteristic to *M. tuberculosis* is the production of niacin.

Drug Susceptibility Testing. This can be done using microscopic examination of the inoculated plate for typical colonies, and allows susceptibility results to be available within 7 days.[24]

Histology. Histologic findings that suggest tuberculosis include caseation necrosis, and the production of tubercles with mononuclear and giant cells. The histologic lesions of atypical mycobacterial infection can be identical.

Serology. Serologic diagnosis of tuberculosis is investigational, but not yet practical.

Tuberculin Tests

Definitions. Conversion of the tuberculin test from negative to positive is useful for diagnosis of recent tuberculous infection. A person with a skin test which is known to have become positive in the past 2 years is called a convertor. One who has a positive tuberculin skin test of unknown duration or who had a negative test more than 2 years previously is called a reactor. Active disease in adults is often discovered as a result of detection of skin test convertors in childhood.[26]

Anergy. The tuberculin test is a model for the study of cell-mediated immunity, and a useful—but sometimes confusing—clinical tool.[9] It illustrates the concept of anergy, which is defined as confirmed tuberculosis with repeatedly negative tuberculin skin tests.[18] Anergy is sometimes associated with progressive, disseminated tuberculosis, but can occur in healthy persons.[18] Use of the second strength tuberculin test (250 tuberculin units) is recommended for the elderly or debilitated patient with a negative intermediate test (5 tuberculin units), but is not recommended for screening of healthy persons.

Infection Does Not Mean Disease. The tuberculin test also illustrates the concept that infection does not necessarily mean disease. Tuberculous infection is defined as the presence of the tubercle bacillus in the body sufficient to produce the delayed hypersensitivity of a positive tuberculin test. Tuberculous disease is defined as a clinical manifestation of tuberculosis, such as an abnormal chest radiograph.

The tuberculin skin test is suppressed by diseases that suppress cell-mediated immunity, such as wild (natural) measles infection, but only rarely is suppressed by attenuated virus vaccines.

Booster Effect of Tuberculin Testing. In populations repeatedly subjected to routine tuberculin testing, (such as hospital employees and medical students), conversion of the skin test occurs more frequently than is confirmed by detection of the disease.[34] In part, this is explainable by inexact reading of tuberculin tests. A conversion is best defined as an increase in induration (a firm feeling of the area due to edema and infiltration with lymphocytes) of at least 6 mm and increase in induration from less than 10 mm to 10 mm or more.[34] Technical requirements of a properly done tuberculin test include use of PPD (purified protein derivative) which is stabilized by Tween 80 and stored in the dark.

The booster effect may be observed when two tuberculin tests are given a week apart, in individuals who have tuberculin hypersensitivity that has waned, but is "boosted" by the first of the two tests. Recently, the use of two tuberculin tests, a week apart, has been recommended to determine the true tuberculin status of those who are routinely and repeatedly tested—but as yet is not widely used.[34]

The accidental injection of the antigen subcutaneously, rather than the proper intradermal injection, is not an important source of error. In one study of children with primary pulmonary tuberculosis, the extent of induration was not significantly different in concurrently administered intradermal and subcutaneous intermediate strength PPD tests.[37]

Biologic Characteristics of Clinical Interest

Acid-Fast Staining. This is due to the high lipid content of the cell. These organisms are difficult to stain, but once stained are not easily decolorized by acid, and so are called acid-fast.

Cording. This is the appearance of the clumps of acid-fast-stained tubercle bacilli, which resemble a rope composed of 10 to 20 strands of the bacilli. It is of value in the preliminary identification of the organism, and is closely correlated with virulence.

Obligate Aerobe. It is necessary to periodically loosen the caps of the tubes containing solid media in which the specimen is inoculated, to provide necessary oxygen for this obligate aerobe.

Slow Growth. Other mycobacteria, particularly the group called rapid growers, may grow to visible colonies in several days, but *M. tuberculosis* requires several weeks to produce visible colonies.

Contagiousness. Primary pulmonary tuberculosis in young children is usually regarded as noncontagious, because fewer organisms are present and cough is not prominent, although *M. tuberculosis* is present in the gastric aspirate, in a small proportion of cases.

In coughing adolescents or adults, the disease is contagious. In recent years, when most of the population have negative tuberculin tests, outbreaks in families and classrooms have been identified and well studied.[4] Most transmission to close family contacts has already occured before the disease is diagnosed.

Isolation Requirements.[15] Aerosol spread is the most important mechanism for spread of tuberculosis.[21] In hospitalized patients who are coughing, have not yet received chemotherapy, and have a positive sputum, the risk of spread by aerosol is significant. Hospitalization in a single room with 100 percent air exchange is essential.

Patients who have received chemotherapy for 3 to 6 weeks have a low risk of contagion even though the sputum may contain some acid-fast bacilli.[13]

At the present time, no gown or gloves are recommended, and special cleaning of the room after the patient leaves is unnecessary. Most masks provide no protection from an aerosol coughed by the patient. Instructing the patient to cover the mouth and nose when coughing is a reasonable approach.[15]

The fine particle aerosol (smaller than 5 millimicrons) produced by coughing is called a primary aerosol, can enter the alveoli, and is a potentially contagious aerosol. The coarse particles blown into the air from sweeping or blowing up the dust in a room is called a secondary aerosol, and has little, if any, contagious risk.

Hospitalization. Tuberculosis sanitariums have now been almost entirely abandoned. Most community hospitals have adequate facilities for initial isolation and treatment of patients with tuberculosis and the duration of hospitalization can usually be brief.[21] For those patients with advanced age, alcohol abuse, debilitation from other chronic disease, drug abuse, and failure to take medications, more prolonged hospitalization will probably be necessary.[5]

Chemotherapy

Principles of Multiple Drug Therapy. The purpose of the use of a combination of three drugs is to avoid the statistical chance of development of a resistant mutant strain. The more extensive the tuberculous disease, the larger the number of tubercle bacilli present, and the greater the risk of a random resistant mutant. Severe disease such as meningitis, miliary tuberculosis, or cavitary disease is usually treated with three drugs. Moderate disease such as non-cavitary pneumonia is treated with two drugs. A patient with a normal chest roentgenogram and a recently converted tuberculin test is usually treated with one drug.

Preventive Treatment. Patients who have no clinical or radiographic abnormality, but who have a positive tuberculin test (i.e. tuberculous infection without tuberculous disease) are at risk for developing tuberculous disease. Treatment of such patients is called preventive treatment, and usually consists of isoniazid for a year. Since isoniazid is associated with liver toxicity, the decision to give preventive treatment depends on which is the greater risk—development of tuberculous disease or the development of isoniazid toxicity.[1]

Groups with highest priority for preventive therapy, in order of priority, are:[1]

1. Household members and other close associates of persons with recently diagnosed tuberculous disease. If such close contacts are tuberculin negative, they should be given preventive treatment with isoniazid for 3 months, and then have another intermediate tuberculin test. The isoniazid need not be continued if the test is still negative after 3 months, and if the index case has been removed or treated sufficiently to be noncontagious. A full 12 months of isoniazid is recommended for any contact who has converted the skin test from negative to positive.
2. Positive tuberculin skin test reactors with radiologic evidence of non-progressive tuberculous disease, with negative sputum, and no history of antituberculous chemotherapy.
3. Tuberculin test convertors.
4. Tuberculin test reactors with special situations, such as corticosteroid therapy, immunosuppressive therapy, diseases associated with immunosuppression, silicosis, and after gastrectomy.
5. Positive reactors younger than 35, (except for pregnant women).
6. Positive reactors older than 35 who have not received prior antituberculous chemotherapy.[8]

First-Line Antituberculous Drugs.[36] Isoniazid (INH) is the most useful antituberculous drug and readily penetrates into cerebrospinal fluid (CSF) and caseous tissue.[36] The drug is readily absorbed by mouth, but can be given by injection if the patient is too sick to take oral medications.

In children with serious infections, the dose is 10 to 20 mg per kg per day, up to 500 mg. For prophylaxis, about 10 mg per kg per day, up to 300 mg daily, is used. In adults 300 mg daily is used for most infections, but is increased to 600 mg daily in meningitis.

Isoniazid is acetylated (and thus inactivated) by the liver. The rate of acetylation is genetically determined and patients can be classified as slow or rapid inactivators. Orientals are often rapid inactivators. About half of Americans and North Europeans are slow inactivators. Hepatitis is the most common adverse effect and is thought to be related to metabolic derivatives. Rapid acetylators, and persons who consume alcohol daily, might be more likely to develop isoniazid hepatitis.

INH toxicity typically occurs in the early weeks of therapy and consists of fever, myalgia, arthralgia, nausea or anorexia, and an elevation of the serum glutamic oxalacetic transaminase

(SGOT) levels.³ Isoniazid-associated hepatitis occurred in one percent of tuberculin reactions taking the drug in one series, with 0.1 percent having fatal hepatitis, compared to no such disease in matched, tuberculin-negative control patients who were not treated with INH.¹⁰

Streptomycin does not penetrate the CSF barrier unless there is meningitis. It should never be used as the only drug for tuberculosis, because of the rapid development of resistance. It must be given by injection. Streptomycin can produce vestibular and auditory damage.

Rifampin is a useful drug for tuberculosis and can be given orally.³⁵ Spinal fluid concentrations are about 20 percent of serum concentrations,⁶ and so rifampin is especially useful for tuberculous meningitis. It is relatively expensive. It makes the urine appear orange colored, and may stain contact lenses orange. It may affect the reliability of oral contraceptives. It also can cause hepatitis, which makes it inconvenient to use with isoniazid, since it is difficult to know which drug is causing the hepatitis.

Para-amino salicylic acid (PAS) was formerly used very frequently. It competes with INH for acetylation by the liver and increases the amount of free INH in the blood. It delays the development of resistance to streptomycin. It is associated with gastrointestinal intolerance, particularly vomiting.

Ethambutol is now usually used instead of PAS because it is much better tolerated, with less gastrointestinal side effects. It can produce decreased vision in high doses, but this rarely occurs with the currently recommended dose of 15 mg per kg per day. It is not recommended for children less than 13 years old, for whom testing for visual toxicity may be limited. Ethionamide is recommended instead of ethambutol in younger children.

Chemotherapy and Drug Resistance. Drug resistance in this context usually refers to resistance to one or more of the commonly used antituberculous drugs, particularly isoniazid, rifampin, streptomycin or ethambutol. Drug resistance is most commonly observed in patients with chronic or advanced pulmonary disease after treatment for many months. However, it is occasionally observed as primary drug resistance, without apparent development secondary to drug therapy.

It is important to try to culture M. tuberculosis from diseased patients, so that the isolate can be tested for drug susceptibility. Although susceptibility testing results may sometimes be available too late to aid in therapy of the index patient, recognition of a drug resistant strain may be life saving in the therapy of a contact of the index case who later develops disease.³⁰ Isoniazid prophylaxis is of no value in preventing disease if the organism is isoniazid resistant.⁷

In severe forms of tuberculosis, such as tuberculous meningitis, the risk of primary drug resistance may be so great that initial therapy should take this into account. For example, some authorities now use INH, streptomycin, ethambutol, and rifampin to treat tuberculous meningitis, because of the risk of primary drug resistance.³¹

In a continuing survey of primary drug resistance throughout the United States, an average of about nine percent of cultures obtained from patients without previous chemotherapy were resistant to one or more drugs.¹⁶ However, Asians and Hispanics had 15 to 20 percent average primary drug resistance. About four percent of the isolates were resistant to isoniazid, but less than one percent were resistant to rifampin or ethambutol.

Second-Line Tuberculous Drugs. These include viomycin, cycloserine, pyrazinamide, ethionamide, capreomycin, and thioacetazone.²

Other Therapy

Corticosteroids. The value of steroids in tuberculous meningitis is controversial. One prospective double-blind controlled study indicated no clear effect in preventing obstruction of spinal fluid flow by the high protein content, but suggested a beneficial effect on cerebral edema.²⁰ Steroids have also been recommended, by French physicians, for miliary tuberculosis with respiratory insufficiency, pleural effusion, and bronchial obstruction with segmental pneumonias, in order to prevent bronchiectasis.¹¹,¹²

Surgical Resection. This is rarely indicated in recent years; except, for example, in patients with persistent positive cultures from multiply resistant organisms or damaged areas of lung still present after prolonged chemotherapy.

Transfer Factor. Patients with defects of cellular immunity who have not responded to conventional chemotherapy have improved dramatically after treatment with dialyzable transfer factor obtained from leucocytes of tuberculin-positive donors.³⁸

Prevention

Treatment of Illness. This is important in prevention of spread of tuberculosis, since chemotherapy quickly renders the patient noncontagious.

BCG. An attenuated mutant of bovine tuberculosis (*M. bovis*), which has been carried through unfavorable conditions (bile-containing media) is known as Bacillus Calmette–Guerin or BCG. It can be used in immunization against tuberculosis. This vaccine appears to be effective, especially in areas where tuberculosis is frequent. However, BCG immunization results in a positive tuberculin test and makes the tuberculin test virtually useless to detect tuberculosis in immunized individuals,[28] unless quantitative measurements of tuberculin reactivity are possible.[27]

OTHER MYCOBACTERIA

Objectives

1. Name three mycobacteria species other than *M. tuberculosis* and describe the typical clinical diseases they cause.
2. Define how mycobacteria species other than *M. tuberculosis* are identified in the laboratory.
3. Discuss the use of skin test with atypical mycobacterial antigens.

There are many species of mycobacteria other than *Mycobacterium tuberculosis*.[43,55,60,67] *M. avium* causes tuberculosis of birds and rarely infects humans. *M. bovis*, which causes tuberculosis of cattle, is now rare in the United States and Canada, because of slaughtering of tuberculin-positive cattle, although it is still a rare cause of disease in older or foreign-born individuals.[53] *M. leprae* (Hansen's bacillus) is the cause of leprosy (Fig. 7-2). Most of the other mycobacterial species which had been called unclassified, anonymous, or atypical mycobacteria are now described by their formal species name.

Classification of Other Mycobacteria

The names and major characteristics of the other mycobacteria are summarized, following the grouping of Runyon.[67]

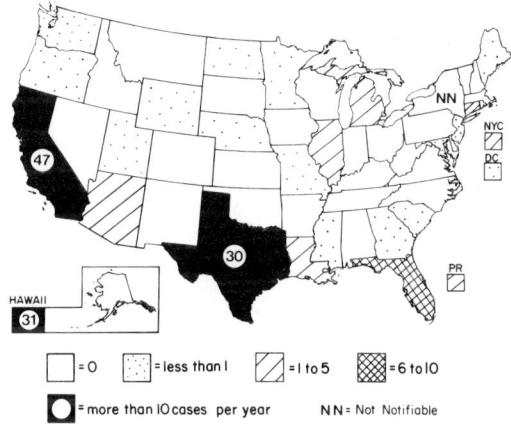

FIG. 7-2. *Leprosy: average annual reported cases, 1970–1972.*[42]

Group I. (Photochromogens). The colonies are yellow, but only when grown during exposure to light. *M. kansasii*, named after the state of Kansas, and *M. marinum*, a cause of disease in fish as well as humans, are the major members of the group.

Group II. (Scotochromogens). Scoto means dark, as in scotoma. The colonies are yellow or orange, whether grown in the light or the dark. *M. scrofulaceum* is the species responsible for most mycobacterial cervical adenitis in children. Many scotochromogens recovered by laboratories are contaminants, the so-called tap water scotochromogens, particularly *M. gordonae*, which is very common. *M. szulga*; may be a cause of chronic pulmonary or skin disease.[63]

Group III. (Nonchromogens). *M. intracellulare* (Battey bacillus) and *M. avium* are the major species in this group and are closely related. Battey is the name of a tuberculosis sanitarium in Rome, Georgia, where *M. intracellulare* was first recovered.

Group IV. (Rapid Growers). The colonies appear in a few days, instead of a few weeks, as in other mycobacteria. *M. fortuitum* is the major species in this group. *M. phlei* and *M. smegmatis* are also rapid growers and are normal flora of humans. *M. chelonei* is a rapid grower, often found in water. *M. xenopi* has been found as a contaminant of tap water and grows well in hot water (43 ° C). *M. rhodochrous* resembles nocardia or corynebacteria in initial laboratory studies.

Clinical Patterns of Illness

Mycobacterium species are *uncommon* causes of:

Cervical Adenitis in Children. *M. scrofulaceum* is the most frequent mycobacterial cause of cervical adenitis, but a few other species have been implicated.[55,57]

Chronic Skin Ulcers. Exposure of small skin wounds to sea water, swimming pools, or the water in fish tanks may result in chronic ulcerating skin lesions due to *M. marinum (M. balnei)*.[40,61] The swimming pool granuloma lesions are typically on the elbows or knees where the patient's skin is abraded by the edge of the pool.[61] *M. ulcerans* is a less common cause of skin ulcers.

Other Rare Mycobacterial Diseases. Mycobacterial species are a rare cause of chronic pneumonia,[45] cavitary pulmonary disease,[62] chronic osteomyelitis, nonpurulent meningitis with low CSF glucose, miliary pneumonias with hepatosplenomegaly,[64] and soil-contaminated wound infection.[31,59] Endocarditis also is rarely caused by mycobacteria.[56]

Disseminated atypical mycobacteremia may especially occur in patients with a pancytopenia, as does disseminated histoplasmosis.

Leprosy. About 100–150 new cases of leprosy are reported annually in the United States, most of which are imported.[50] It is first manifested by hypopigmented skin plaques. Lepromatous leprosy is a progressive, ultimately fatal disease producing waxy, thickened, or nodular skin lesions with ultimate destruction of peripheral nerves. Tuberculoid leprosy involves nerves and skin, producing anaesthetic hypopigmented plaques.

Armadillos in Louisiana are susceptible to natural and experimentally induced leprosy.[43]

Laboratory Approach

Smear. All mycobacteria are acid fast. The species cannot be identified by smear.

Culture. The species are identified by growth characteristics, particularly speed of growth, production of yellow or orange pigmented colonies, and other metabolic tests such as niacin and catalase production.[66]

Skin Tests. Pure protein derivative (PPD) can be prepared from various mycobacterial species, and standardized by the same methods used in the production of *M. tuberculosis* skin testing antigen. Patients with disease due to a mycobacterial species usually have a positive skin test to the PPD antigen prepared from their infecting organism. If standardized PPD antigens are used, the largest area of induration is made by the antigen most similar to the infecting organism. Usually, one antigen from each of Runyon's Groups I, II, and III are used. PPD-S refers to Standard *(M. tuberculosis)*. PPD-Y refers to Yellow *(M. kansasii*–Group I). PPD-G refers to Gause, a strain of scotochromogen (Group II). PPD-B refers to Battey, a Group III nonchromogen. PPD-A (Avian) is sometimes used.

Availability of atypical mycobacteria skin tests, which depends on federal testing and approval, was limited until 1979, when these skin testing antigens were withdrawn until they could be standardized, tested, and approved.

Serology. No satisfactory serologic tests are available.[48,49,54]

Recently, skin test antigens to *M. fortuitum* (PPD-F) and *M. chelonei* (PPD CG) have been prepared and shown to produce a positive skin test in patients with wound infections due to either of these mycobacteria.[52]

Biologic Features of Clinical Interest

Chronic Bacteremia in Leprosy.[47] Patients with lepromatous leprosy tolerate a chronic bacteremia from the skin lesions with few symptoms.

Normal Flora. Approximately 10 percent of normal adults have *Mycobacterium* species other than *M. tuberculosis* in mouth washings and sputum specimens, so that positive smears for acid-fast bacilli must be interpreted in terms of the clinical situation.[58] *M. phlei* and *M. smegmatis* are also normal flora.

Treatment

Chemotherapy. Rifampin appears to be effective therapy against cervical adenitis due to *M. scrofulaceum*.[57] Rifampin also appears to be effective against leprosy,[65] although dapsone is the traditional therapy.[51]

M. intracellulare isolates are typically more resistant to usual antituberculous drugs, and susceptibility testing is needed. Surgical removal of the diseased area may be necessary. *M. kansasii* is treated with drugs similar to those for *M. tuberculosis*.

M. fortuitum and other rapid growers are very resistant to chemotherapy, but amikacin is effective in vitro,[46] and may prove clinically effective.

Transfer Factor. Leprosy has been treated with transfer factor or whole lymphocytes from donors with delayed hypersensitivity to *M. leprae* antigens.[41]

REFERENCES

Mycobacterium Tuberculosis

1. American Thoracic Society: Preventive therapy of tuberculous infection. Am. Rev. Respir. Dis., *110:*371–374, 1974.
2. Byrd, R.B., Kaplan, P.D., and Gracey, D.R.: Critical review. Treatment of pulmonary tuberculosis. Chest, *66:*560–567, 1974.
3. Byrd, R.B., Nelson, R., and Elliott, R.C.: Isoniazid toxicity. A prospective study in secondary chemoprophylaxis. J.A.M.A., *220:*1471–1473, 1972.
4. Darney, P.D., and Clenny, N.D.: Tuberculosis outbreak in an Alabama high school. J.A.M.A., *216:*2117–2118, 1971.
5. Davis, H.L., White, W.G., and Sutliff, W.D.: Characteristics of hospitalized tuberculosis patients today. South. Med. J., *71:*1401–1403, 1978.
6. D'Oliveira, J.J.G.: Cerebrospinal fluid concentrations of rifampin in meningeal tuberculosis. Am. Rev. Respir. Dis., *106:*432–437, 1972.
7. Fairshter, R.D., Randazzo, G.P., Garlin, J., and Wilson, A.F.: Failure of isoniazid prophylaxis after exposure to isoniazid-resistant tuberculosis. Am. Rev. Respir. Dis., *112:*37–42, 1975.
8. Falk, A., and Fuchs, G.F.: Prophylaxis with isoniazid in inactive tuberculosis. Chest, *73:*44–48, 1978.
9. Freedman, S.O., and Kongshavn, P.L.: Immunobiology of tuberculin hypersensitivity. Chest, *68*(Suppl.):470–474, 1975.
10. Garibaldi, R.A., Drusin, R.E., Ferbee, S.H., and Gregg, M.B.: Isoniazid-associated hepatitis. Report of an outbreak. Am. Rev. Respir. Dis., *106:*357–365, 1972.
11. Gerbeaux, J.: Primary Tuberculosis in Childhood. Springfield, Illinois, Charles C. Thomas, 1970.
12. Gerbeaux, J., Baculard, A., and Couvrer, J.: Primary tuberculosis in childhood. Am. J. Dis. Child., *110:*507–518, 1965.
13. Gunnels, J.J., Bates, J.H., and Swindoll, H.: Infectivity of sputum-positive tuberculous patients on chemotherapy. Am. Rev. Respir. Dis., *109:*323–330, 1974.
14. Husen, L., et al.: Pulmonary tuberculosis with negative findings on chest x-ray films: a study of 40 cases. Chest, *60:*540–542, 1971.
15. Johnston, R.F., and Wildrick, K.H.: "State of the art" review. The impact of chemotherapy on the care of patients with tuberculosis. Am. Rev. Respir. Dis., *109:*636–664, 1974.
16. Kopanoff, D.E., et al.: A continuing survey of tuberculosis primary drug resistance in the United States: March 1975 to November 1977. Am. Rev. Respir. Dis., *118:*835–842, 1978.
17. Lincoln, E.M., and Sewell, E.M.: Tuberculosis in Children. New York, McGraw-Hill, 1963.
18. McMurray, D.N., and Echeverri, A.: Cell-mediated immunity in anergic patients with pulmonary tuberculosis. Am. Rev. Respir. Dis., *118:*827–834, 1978.
19. Murray, H.W., Tuazon, C.U., Kirmani, N., and Sheagren, J.N.: The adult respiratory distress syndrome associated with miliary tuberculosis. Chest, *73:*37–43, 1978.
20. O'Toole, R.D., Thornton, G.F., Mukherjee, M.M., and Nath, R.L.: Dexamethasone in tuberculous meningitis. Ann. Intern. Med., *70:*39–48, 1969.
21. Reichman, L.B.: Tuberculosis care: when and where? Ann. Intern. Med., *80:*402–406, 1974.
22. Reinisch, E.H., and Kaufmann, W.: A study of prolonged incubation of cultures for mycobacterium tuberculosis. Am. Rev. Tuberc., *84:*451–453, 1961.
23. Romeyn, J.A.: Exogenous reinfection in tuberculosis. Am. Rev. Respir. Dis., *101:*923–927, 1970.
24. Runyon, E.H.: A rapid mycobacterial drug susceptibility test. Am. J. Clin. Pathol., *59:*817–824, 1973.
25. Saslaw, S., and Perkins, R.L.: The gastric smear for acid-fast bacilli in the presumptive diagnosis of tuberculosis. Am. J. Med. Sci., *243:*470–474, 1962.
26. Sharpe, L.D., and Stead, W.W.: Epidemiologic significance of the sporadic case of tuberculosis. Clin. Pediatr., *12:*281–284, 1973.
27. Smith, D.T.: Diagnostic and prognostic significance of quantitative tuberculin tests. Ann. Intern. Med., *67:*919–946, 1967.
28. Special Panel. Public Health Service recommendations on the use of BCG vaccination

in the United States. Morb. Mort. Week. Rep., 15:350–351, 1966.

29. Stead, W.W.: Pathogenesis of the sporadic case of tuberculosis. New Engl. J. Med., 277:1008–1012, 1967.

30. Steiner, M., et al.: Primary drug-resistant tuberculosis. Report of an outbreak. New Engl. J. Med., 283:1353–1358, 1970.

31. Steiner, P., and Portugaleza, C.: Tuberculous meningitis in children. A review of 25 cases observed between the years 1965 and 1970 at the Kings County Medical Center of Brooklyn with special reference to the problem of infection with primary drug-resistant strains of *M. tuberculosis*. Am. Rev. Respir. Dis., 107:22–29, 1973.

32. Strumpf, I.J., Tsang, A.Y., Schork, M.A., and Weg, J.G.: The reliability of gastric smears by auramine–rhodamine staining technique for the diagnosis of tuberculosis. Am. Rev. Respir. Dis., 114:971–976, 1976.

33. Sumaya, C.V., et al.: Tuberculous meningitis in children during the isoniazid era. J. Pediatr., 87:43–49, 1975.

34. Thompson, N.J., Glassroth, J.L., Snider, D.E., Jr., and Farer, L.C.: The booster phenomenon in serial tuberculin testing. Am. Rev. Respir. Dis., 119:587–597, 1979.

35. Vall-Spinosa, A., and Lester, T.W · Rifampin: characteristics and role in the chemotherapy of tuberculosis. Ann. Intern. Med., 74:758–760, 1971.

36. Van Scoy, R.E.: Antituberculosis agents. Isoniazid, rifampin, streptomycin, ethambutol. Mayo Clin. Proc., 52:694–700, 1977.

37. Victoria, M.S., Steiner, P., and Rao, M.: The effect of intradermal and subcutaneous route of administration on variation in PPD sensitivity. Clin. Pediatr., 16:514–515, 1977.

38. Whitcomb, M.E., and Rocklin, R.E.: Transfer factor therapy in a ptient with progressive primary tuberculosis. Ann. Intern. Med., 79:161–166, 1973.

39. Wigle, W.D., Ashley, M.J., Killough, E.M., and Cosens, M.: Bovine tuberculosis in humans in Ontario. Am. Rev. Respir. Dis., 106:528–534, 1972.

Other Mycobacteria

40. Brown, J., Kelm, M., and Bryan, L.E.: Infection of the skin by *Mycobacterium marinum*: report of five cases. Can. Med. Assoc. J., 117:912–914, 1977.

41. Bullock, W.E., Fields, J.P., and Brandriss, M.W.: An evaluation of transfer factor as immunotherapy for patients with lepromatous leprosy. New Engl. J. Med., 287:1053–1059, 1972.

42. Center for Disease Control: Morb. Mort. Weekly Rep. Annual Suppl., 19–21:Summaries, 1970–1972.

43. ———: Leprosy-like disease in wild-caught armadillos—Louisiana. Morb. Mort. Weekly Rep. 28:18, 23, 1976.

44. Chapman, J.S.: Atypical mycobacterial infections. Med. Clin. North Am., 51:503–517, 1967.

45. Cooper, A.R., and Martin, R.S.: Pulmonary *Mycobacterium scrofulaceum* infection in a child. Pediatrics, 49:118–124, 1972.

46. Dalovisio, J.R., and Pankey, G.A.: Problems in diagnosis and therapy of *Mycobacterium fortuitum* infections. Am. Rev. Respir. Dis., 117:625–630, 1978.

47. Drutz, D.J., Chen, T.S.N., and Lu, W-S.: The continuous bacteremia of lepromatous leprosy. New Engl. J. Med., 287:159–164, 1972.

48. Edwards, L.B.: Current status of the tuberculin test. Ann. N.Y. Acad. Sci., 106:32–42, 1963.

49. Fogan, L.: PPD antigens and the diagnosis of mycobacterial diseases. Arch. Intern. Med., 124:49–54, 1969.

50. Grove, D.I., Warren, K.S., and Mahmoud, A.A.F.: Algorithms in the diagnosis and management of exotic diseases. XV. Leprosy. J. Infect. Dis., 134:205–210, 1976.

51. Hand, W.L., and Sandford, J.P.: *Mycobacterium fortuitum*—a human pathogen. Ann. Int. Med., 73:971–977, 1970.

52. Hoffman, P.C., Fraser, D.W., and Hinson, P.L.: Delayed hypersensitivity reactions in patients with *Mycobacterium chelonei* and *Mycobacterium fortuitum* infections. Am. Rev. Respir. Dis., 117:527–531, 1978.

53. Karlson, A.G., and Carr, D.T.: Tuberculosis caused by *Mycobacterium bovis*. Report of six cases: 1954–1968. Ann. Int. Med., 73:979–983, 1970.

54. Kendig, E.L., Jr.: Unclassified mycobacteria. Incidence of infection and cause of a false-positive tuberculin reaction. New Engl. J. Med., 268:1001–1002, 1963.

55. Lincoln, E.M., and Gilbert, L.A.: Disease in children due to mycobacteria other than *Mycobacterium tuberculosis*. Am. Rev. Respir. Dis., 105:683–714, 1972.

56. Lohr, D.C., Goeken, J.A., Doty, D.B., and Donta, S.T.: *Mycobacterium gordonae* infection of a prosthetic aortic valve. J.A.M.A., 239:1528–1531, 1978.

57. Mandell, F., and Wright, P.F.: Treatment of atypical mycobacterial cervical adenitis with rifampin. Pediatrics, 55:39–43, 1975.

58. Mills, C.C.: Occurrence of *Mycobacterium* other than *Mycobacterium tuberculosis* in the oral cavity and in sputum. Appl. Microbiol., 24:307–310, 1972.

59. Offer, R.C., Karlson, A.G., Spittell, J.A.: Infection caused by *Mycobacterium fortuitum*. Mayo Clin. Proc., *46:*747–750, 1971.

60. Owens, D.W.: General medical aspects of atypical mycobacteria. *67:*39–45, 1974.

61. Philpott, J.A., et al.: Swimming pool granuloma. A study of 290 cases. Arch. Derm., *88:*158–162, 1963.

62. Rauscher, C.R., Kerby, G., and Ruth, W.E.: A ten-year clinical experience with *Mycobacterium kansasii*. Chest, *66:*17–19, 1974.

63. Sybert, A., Tsou, E., and Garagusi, V.F.: Cutaneous infection due to *Mycobacterium szulgai*. Am. Rev. Respir. Dis., *115:*695–698, 1977.

64. Volini, F., Coulton, R., and Lester, W.: Disseminated infection caused by Battey type mycobacteria. Am. J. Clin. Pathol., *43:*39–46, 1965.

65. Waters, M.F.R., et al.: Rifampicin for lepromatous leprosy: nine years' experience. Br. Med. J., *1:*133–136, 1978.

66. Wayne, L.G., and Doubek, J.R.: Diagnostic key to mycobacteria encountered in clinical laboratories. Appl. Microbiol. *16:*925–931, 1968.

67. Wayne, L.G., and Runyon, E.H.: Mycobacteria: a guide to nomenclatural usage. Am. Rev. Respir. Dis., *100:*732–734, 1969.

Fungi

FUNGI IN GENERAL

Definitions and Classification

Fungi differ from mycobacteria and bacteria because they are eukaryotes. Bacteria are prokaryotes, which means they have a cell wall and a single strand of DNA. *Actinomyces* and *Nocardia* species are classified as bacteria because they are prokaryotes, although they are filamentous and resemble fungi (Table 8-1).

Fungi grow slowly compared to bacteria. Antibiotics that are effective against bacteria are usually not effective against fungi. Drugs that are effective against fungi are often toxic to humans.

It is useful to classify fungi into dimorphic fungi (which are in the yeast form at 37°C and the mycelial form at 25°C), yeast-like fungi (which grow as budding yeasts at 25°C in the laboratory and in the human host), and filamentous fungi (which are usually found in the mycelial form) (Table 8-1). Fungal infections are also classified as superficial (involving skin or mucous membranes) or deep (involving lung or CNS).

A mycetoma is a tumor-like mass produced by a fungus, especially found in the skin or lungs.

Some fungi are highly infectious to laboratory personnel, especially at room temperature, when the mycelial form can be inhaled as an aerosol.

Superficial Fungal Infections

Fungi that only infect the skin or nails (dermatophytes) produce infections called superficial dermatomycoses.[1,2] Tinea, the general term for a superficial fungal infection, is customarily followed by the location, for example, tinea pedis (athlete's foot), tinea cruris (jock itch), tinea corporis (body), and tinea capitis (scalp). Ringworm is a lay term for tinea. A kerion is an inflammatory pustular tinea capitis.[3]

Genera. Most tinea is caused by *Microsporum, Epidermophyton,* or *Trichophyton* species. Refractory foot, groin, or nail infections may be caused by *Candida albicans* (discussed later in this chapter).

Diagnosis. Many superficial mycoses show a green fluorescence under ultraviolet (Wood's) light. Culture is useful if the disease fails to respond as expected to treatment described below.

Treatment. Tolnaftate cream or powder is available without prescription. Miconazole and clotrimazole are very effective against both superficial mycoses and *Candida albicans.* Griseofulvin is given by mouth, but is very expensive for the long course required for refractory nail infections.

HISTOPLASMA CAPSULATUM

Objectives

1. Describe the typical clinical pattern of acute disseminated histoplasmosis.
2. Describe two other clinical patterns of histoplasmosis.
3. Describe the usual circumstances under which epidemic histoplasmosis occurs.

4. Describe methods used in the laboratory to determine infection with histoplasma, particularly the use of the histoplasmin skin test and serologic tests.

5. Discuss the circumstances in which chemotherapy might be indicated and indicate what chemotherapy is effective.

Definitions

Histoplasma capsulatum is a dimorphic fungus, which means it is found in two forms, as a yeast or as a fungus. The yeast phase is found at 37 ° C in human tissues and in incubated laboratory cultures. The mycelial (fungal) phase is found at 22 ° C in soil, bird manure, and cultures incubated at room temperature. The name histoplasma comes from an early belief that the organism was a protozoan. Different mycelial colony forms and different serotypes can be identified in the laboratory, but these variations have no known clinical significance.

Frequency and Importance

Histoplasmosis is most common in the Mississippi and Ohio River valleys of the United States (Fig. 8-1). In eight of these states, the majority of young adults have a strongly positive histoplasmin skin test.

Most individuals who acquire histoplasmosis have a mild or asymptomatic illness. Progressive fatal histoplasmosis is rare, but it is important to be aware of it, because chemotherapy with amphotericin B can be lifesaving.

TABLE 8-1. CLASSIFICATION OF MYCOTIC MICROORGANISMS.

Superficial fungi	*Filamentous fungi*
Microsporum	Aspergillus
Trichophyton	Allescheria
Epidermophyton	Phycomycetes
	Agents of mycetoma
Dimorphic fungi	*Yeast-like fungi*
Blastomyces	Cryptococcus
Coccidioides	Candida
Histoplasma	Torulopsis
Sporothrix	Geotrichum
Paracoccidioides	
	Fungi-like bacteria
	Actinomyces
	Nocardia
	Agents of mycetoma

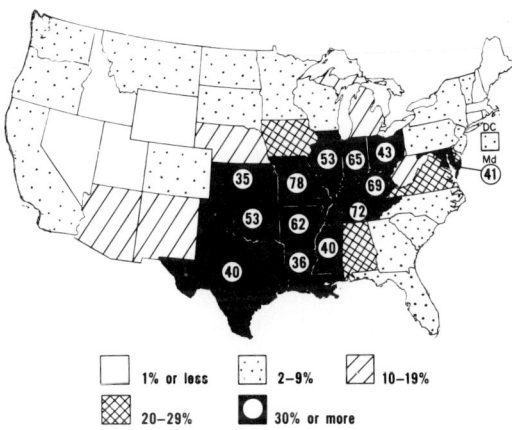

FIG. 8-1. *Histoplasmin skin tests in naval recruits with a lifetime one county, nonmetropolitan residence, 1958–1965.*[9]

Clinical Patterns of Illness

The spectrum of illnesses produced is similar to that of tuberculosis.[14] In endemic areas of the United States, *Histoplasma capsulatum* is a *rare* cause of:

Fever with hepatosplenomegaly is the usual presenting illness in acute disseminated histoplasmosis, and typically occurs in young infants or debiliated adults (Fig. 8-2).[14,16] The chest film is typically normal, but occasionally reveals miliary lesions. The histoplasmin skin test is typically negative. Hematologic abnormalities are usually present, and often are prominent enough to receive the major diagnostic attention. Less severe manifestations may occur in adults with chronic disseminated disease.[14] Immunosuppressed patients are also susceptible to chronic disseminated disease.[7]

Progressive hemolytic anemia with reticulocytosis, leukopenia, and thrombopenia is another rare occurrence.[16] Since fever and hepatosplenomegaly are often present, acute leukemia may be considered. The bone marrow does not reveal an abnormal number of blast cells, but usually does reveal the circular organisms of the yeast phase (Fig. 8-3). Acute histiocytosis, a malignant proliferation of histiocytes, can closely resemble this form of histoplasmosis, and the distinction is often difficult. Disseminated intravascular coagulation is a possible complication of disseminated histoplasmosis. Chronic disseminated histoplasmosis has similar clinical findings, but the course is prolonged.

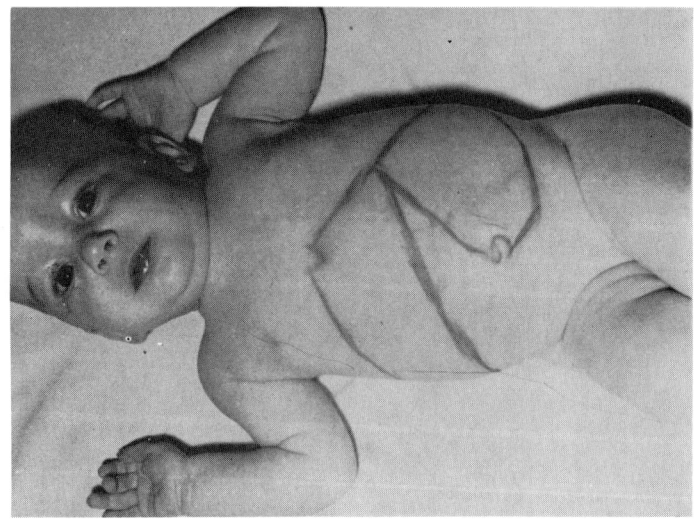

FIG. 8-2. *A 4-month-old with massive enlargement of liver and spleen due to disseminated histoplasmosis.*[18]

Acute Pneumonia. Various forms of acute pneumonia can be seen after a single unusually heavy exposure.[1] Acute pneumonia is the most common symptomatic illness. The incubation period is about 10–18 days after exposure.[14] The severity depends on the extent of exposure and the resistance of the individual. Influenza-like illness is the most frequent pulmonary presentation, with fever and cough, but little or no pulmonary infiltrates. Peripheral pneumonia with hilar adenopathy can also occur. Atypical pneumonia, with gradual onset and no response to antibiotics, may occur. Acute miliary pneumonia is typically related to a heavy exposure.[14] Giant cysts of the mediastinum, bronchiolithiasis, and compression of a major bronchus by an enlarged hilar node are other complications that can occur.[20]

Erythema Nodosum or Erythema Multiforme. These skin rashes may provide clues to the diagnosis of histoplasmosis.

Even in endemic areas, *Histoplasma capsulatum* is a *rare* cause of a number of other clinical pattern, including:

Non-Purulent Meningitis. In meningitis due to histoplasmosis the spinal fluid pleocytosis is usually lymphocytic, the CSF glucose may be low, and the protein elevated, as in tuberculous meningitis.[11,13] Histoplasmosis is usually listed as a possible cause of choroiditis and uveitis on the basis of indirect evidence, such as a positive skin test. However, the organism has not been cultured from human lesions, and so amphotericin B therapy is not justified in suspected, but unconfirmed, cases.[14]

Reinfection Pneumonia. A heavier exposure is necessary to produce a symptomatic illness in an individual with some existing immunity.[14] Chronic or progressive pneumonia can occur, with cavitation.[15]

Adrenal Insufficiency. In chronic disseminated disease due to histoplasmosis (as due to other disseminated fungi or tuberculosis), the adrenal glands may be involved, and adrenal insufficiency can occur, even years after successful treatment.[14]

Other diseases rarely due to histoplasmosis include pericarditis, endocarditis,[21] adenitis of mediastinal or mesenteric nodes, granulomatous

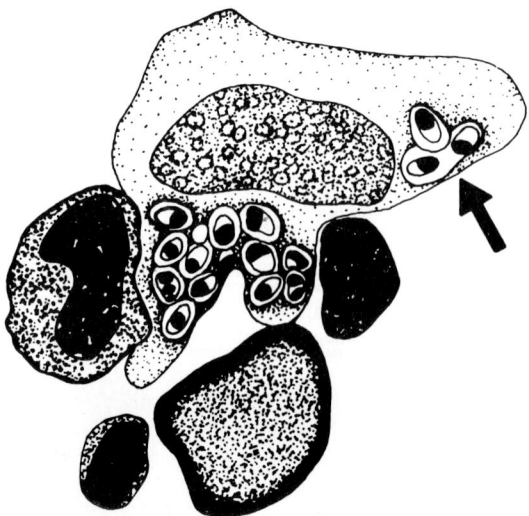

FIG. 8-3. *Histoplasma capsulatum appears as oval organisms in bone marrow.*

lesions of the colon, and ulcerations of the mucosa of the mouth, pharynx, or larynx.[14]

Skin lesions of histoplasmosis include chronic ulcers, chronic pustular lesions, nodules, or wound infection from accidental skin inoculation in the laboratory.[23] Skin lesions are rare and are typically secondary to disseminated disease occurring in a patient with depressed immunity or receiving steroid therapy.[22]

Asymptomatic or Subclinical Infection

Asymptomatic infection is the most common clinical pattern of infection, and is usually detected by a positive histoplasmin skin test. Asymptomatic miliary calcification of the lungs, liver or spleen is usually due to histoplasmosis.

Laboratory Approach

Culture. This is practical in only a few clinical situations, such as from sputum in pulmonary cavitation, or from bone marrow in disseminated disease. The organism is typically not recoverable from sputum of patients with acute pneumonia. Culture from bone marrow requires about 10 days, which may be too long to delay treatment in the disseminated form of the disease. Thus, a careful examination of a bone marrow smear should be done when disseminated histoplasmosis is suspected.

Smear. The organism can usually be identified by its typical morphology in Gomori stained smears of the bone marrow (Fig. 8-3) or liver biopsy. The clinician should tell the technician that histoplasmosis is expected, although an alert hematology technician may recognize the organism in a routine Wright's stain, even without a warning to look for it. Disseminated histoplasmosis in infancy often resembles acute leukemia, with thrombopenia, hepatosplenomegaly, anemia, leucopenia, but without excessive blast forms. The organisms should be looked for in all cases of suspected leukemia in which leukemia cannot be confirmed by bone marrow examination.

Fluorescent antibody stains of the bone marrow smear may be extremely useful if the disease is suspected,[16] but are unlikely to be available except in geographic areas where the disease is frequent.

Serologic Diagnosis. A variety of serologic tests are possible. Whole cells or extracts of either yeast or mycelial phase cultures may be used as an antigen.[4] The patient's serum can be tested for antibody against the various antigens. The antigen–antibody reaction can be detected by testing for complement fixation, observing a precipitin band in agar diffusion, indirect immunofluorescence, or agglutination tests using colloidion or latex particles as the carrier for histoplasmin (mycelial phase) antigen.

The histoplasmin latex agglutination test is inexpensive, commercially available, and can be done in small hospital laboratories. False positives in low titer can occur from tuberculosis or atypical mycobacteria.[8] False negatives are rare, but the titer must be 1:32 or higher to have significance in acute disease.[5] Most state health laboratories can do complement fixation tests using both yeast-phase and mycelial-phase antigen. An elevated yeast-phase antibody titer suggests recent infection.

Histoplasmin Skin Test. The skin test antigen is given as an intradermal injection of 0.1 ml. of a 1:100 dilution of U.S. Reference histoplasmin. The histoplasmin skin test is similar in most respects to the tuberculin test. The test usually becomes positive about 2 to 3 weeks after exposure and sometimes is positive at the time of the onset of symptoms (about 2 weeks after exposure). The major disadvantage of the skin test is that it can confuse the serologic diagnosis, as it may stimulate a rise in antibody titer, particularly if the patient has previously been infected.[6] Repetitive skin testing is even more likely to stimulate an antibody response, and should not be done.

Biologic Characteristics of Clinical Interest

Diphasic Morphology. The mycelial form, which is found at room temperature, is a potential hazard to laboratory personnel. The yeast form, which is found at body temperature, can be recognized in bone marrow smears in patients with disseminated disease, but is not contagious.

Geographic Localization. The organism persists in soil for many years, and is predominantly localized in the United States to the Mississippi and Ohio River valley states and the river valleys of the northwestern states (Fig. 8-1). Outbreaks can often be traced to a localized exposure to a house, or chicken coop, or cave.[16] Even outbreaks in cities can usually be traced to a small focus or point source.[12] One outbreak was a result of raking and sweeping a school yard that contained an old starling and blackbird roost, as a part of Earth Day activities.

Animal Infections. Birds, bats, dogs, and many other animal hosts can be infected. Bird manure appears to support or stimulate the growth of histoplasmosis. Pigeon, starling, or chicken droppings result in areas of high concentration of histoplasma growing in the soil.

Relation to Other Diseases. Disseminated histoplasmosis and active tuberculosis often occur together. Lymphoma is reported to occur slightly more frequently in patients with histoplasmosis. Host defects in cell-mediated immunity or blood leukocytes may be a factor in predisposing to disseminated disease.

Treatment

Observation without specific chemotherapy is the usual treatment. Amphotericin B, a potent and potentially toxic chemotherapeutic agent, is useful for chronic pulmonary histoplasmosis,[19] as well as for the disseminated or progressive form of the disease.[18] The drug must be given by the intravenous route, and regularly produces fever during administration. Phlebitis is common. Renal toxicity is a potential problem.

In experimental histoplasmosis in mice, therapy with rifampin and amphotericin B was more effective than either agent alone.[17]

Prevention

The disease is not usually regarded as contagious, but person-to-person transmission by the respiratory route has occasionally been suspected. Vaccines have been studied in animals.

COCCIDIOIDES IMMITIS

Objectives

1. List the four states in which coccidioidomycosis is most frequently found, and indicate the range of the incubation period.

2. Describe how coccidioidomycosis has been spread by indirect contact.

3. Describe the three most severe clinical patterns of coccidioidomycosis.

Definitions

The organism was originally named as a protozoan, because in tissue sections it appears as spherules containing circular endospores. It is dimorphic, with a yeast phase in tissue at 37 ° C, and a mycelial phase on cultivation at 25 ° C. The form of the disease with fever, arthralgia, bronchitis, and rash is often called valley fever by people who live in the southwestern United States, especially in the San Joaquin Valley of southern California.

Frequency and Importance

Coccidioidomycosis is a frequent, but usually mild disease in the southwestern United States. About 100,000 cases and 70 deaths occur annually.[26] More than 50 percent of young adults from many counties in California, Arizona, and Texas have positive skin tests (Figure 8-4). It is important because it rarely may cause severe meningeal or pulmonary disease, which respond to antifungal chemotherapy.

Clinical Patterns of Illness

In endemic areas in the United States, *Coccidioides immitis* is an *occasional* cause of:

Atypical pneumonia is manifested by cough, pleuritic chest pain, variable fever, and a somewhat circular pulmonary infiltrate which rapidly

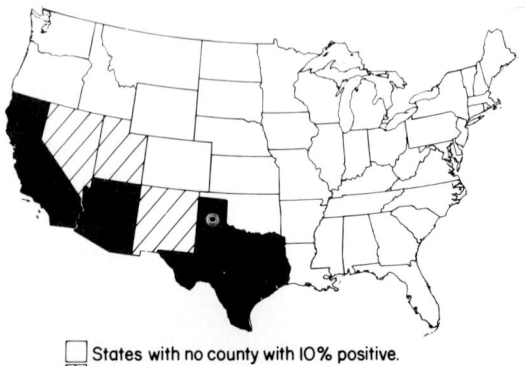

FIG. 8-4. *Coccidioidin skin tests of recruits with one county residence.*[27]

progresses and then regresses over about a 3 week period.[26] Pleural effusion may occur. Hilar adenopathy is common. There is no response to penicillin. A fleeting rash is present in 10 percent or more of the cases. Night sweats, weight loss, and a productive cough is not unusual. Eosinophilia of four to ten percent and an elevated sedimentation rate is observed in many of the symptomatic cases.[26] There may be a residual pulmonary nodule which does not calcify, as often occurs in histoplasmosis.

Influenza-like illness can occur, with fever, chills, cough, myalgia, and weakness.[26] A normal chest film is a common pattern.[33] Pleuritic chest pain is often a prominent symptom, but is usually gone in a few days. Physical exam rarely reveals rales or decreased breath sounds.

Rash. A nodular rash on the shins (erythema nodosum) or a generalized body rash with target and iris lesions (erythema multiforme) is the usual rash pattern, and is often the chief complaint. The rash can also be pruritic, or maculopapular, with peeling.[33] Arthralgias may accompany the rash.

Coccidioidomyces immitis is a rare cause of several other patterns. It can cause miliary pneumonia with hepatosplenomegaly, which resembles miliary tuberculosis or disseminated histoplasmosis. Disseminated coccidioidomycosis often is associated with nonpurulent meningitis, with a CSF lymphocytosis, elevated protein and low glucose.[24] Hypercalcemia may be present. Disseminated disease has been observed in the newborn period.[28] Other uncommon patterns include chronic cavitary pulmonary disease, laryngitis, pericarditis, urinary infection, chronic or recurrent cutaneous ulceration from inoculation of the organism, and inoculation osteomyelitis.[26,29,34]

Asymptomatic or Subclinical Infection

This is the most common pattern in skin test converters, whether laboratory workers, or new residents in the southwestern U.S.[26,29]

Laboratory Diagnosis

Coccidioidin and Spherulin Skin Tests. Skin tests are useful if the disease is not disseminated.[26] Spherulin is a new skin testing reagent which appears to be more efficient than coccidioidin. The intradermal skin test with coccidioidin or spherulin becomes positive very early in the illness, usually before antibodies are detectable, and usually by the time symptoms are present. There is probably little or no cross reaction with the histoplasmin skin test.[29] The skin test often reverts to negative when the individual moves away from the endemic area.

Culture. The laboratory should be informed if coccidioidosis is suspected, as the organism is extremely hazardous to laboratory personnel, because the mycelial spores are easily airborne and very infectious.[29] The organism is dimorphic, but the mycelial form is sometimes found at 37°C as well as at room temperature. The appearance of the arthrospores in the hyphae is diagnostic.

Serum Antibodies. A rise in antibody titer is diagnostic.[26] The complement fixation test is an index of disease activity.

Sputum. A Papanicolaou stain or potassium hydroxide preparation of sputum may reveal the typical spherules with endospores.

Lung Biopsy. This may be needed for rapid diagnosis in a compromised host.

Chest Radiograph. A variety of patterns can be observed, and none are typical.[26]

Biologic Characteristics of Clinical Interest

Geographic Localization. In the United States, the disease is found almost exclusively in the Southwest, where it lives in the soil. Inhalation of the spores in dust clouds after dry weather is the usual mechanism of acquiring the disease. However, Native American relics and fossils may contain spores, so indirect transmission to distant areas is possible.[31]

Contagion. The disease is not contagious by the airborne route, since the tissue (yeast) form is not spread by air or aerosol, in contrast to the mycelial arthrospores produced at room temperature, which are highly infective. Monkeys caged with experimentally infected monkeys did not get the disease when precautions were taken to prevent spores on the fur of the monkeys.[30] However, draining pus from skin or bone is infective if the organism is present in the pus.

Incubation Period. This is about 1 to 3 weeks after inhalation, on the basis of known exposures, and probably is proportional to the number of spores inhaled.[26,29,32]

Treatment

Amphotericin B is used for therapy of disseminated or severe cavitary disease.[24] Transfer factor has also been effective therapy.[32] Surgical excision of a persistent pulmonary cavity may be necessary. Ketoconazole is an oral, investigational drug which appears very promising.

Prevention

Inhalation of dust in endemic areas should be avoided if possible. Vaccines are experimental. Subcutaneous inoculation of small numbers of spores in animals protects against severe disease after later aerosol exposure, and different strains appear to have different virulence when injected.[25] Vaccines for humans are under investigation.

BLASTOMYCES DERMATIDITIS

Objectives

1. Describe three patterns of illness produced by *Blastomyces dermatiditis*.
2. Describe how the diagnosis can be confirmed in the laboratory.

Definitions

Disease produced by *B. dermatiditis* is sometimes called North American blastomycosis, but the organism has been observed outside of North America. Dermatiditis refers to the primary cutaneous ulcerations occasionally produced by this organism.

South American blastomycosis is due to *Paracoccidioides braziliensis*, and produces mucocutaneous lesions, but not respiratory disease. It is occasionally seen in the United States in people who come from South America.[36]

Frequency and Importance

Blastomycosis is a very rare disease in the United States. It is important because it responds to chemotherapy.

Clinical Patterns of Illness

Blastomyces dermatiditis is a rare cause of several syndromes.[37,42,45] Disseminated nodular pneumonia or chronic pneumonia may occur. Fatal progressive pneumonia has been observed, presumably as a primary infection, similar to the course which occasionally occurs with tuberculosis or the other systemic fungal disease.[41] Pleural effusion can occur. A marked leukocytosis (greater than 25,000) with neutrophilia is typical.

Disseminated blastomycosis with meningitis resembling tuberculous meningitis can occur.[38] Adrenal insufficiency is an occasional complication of disseminated blastomycosis.

Skin infection due to blastomyces usually appears as chronic ulcerating or verrucous lesion, but may appear as a circular erythematous plaque, with central scarring and scaling at the edges.

Chronic osteomyelitis is one of the patterns occasionally observed. Infection of the prostate, epididymis, and testes can occur, but involvement of the kidney is rare.

Pulmonary or disseminated disease in compromised hosts probably represents reactivation.[40]

Laboratory Diagnosis

Microscopic Examination. The organism is dimorphic, and like histoplasmosis, is a yeast at body temperature and a fungus at room temperature. The budding yeast forms usually can be found in a sputum specimen in patients with pulmonary disease (Fig. 8-5). The fresh sputum should be examined after addition of 20 percent potassium hydroxide.

Cultures. The organism grows well on standard media used for the isolation of fungi.

Serum Antibodies. Serologic tests are generally unsatisfactory. However, mycologists are often interested in obtaining serum on suspected or confirmed cases for development and evaluation of serologic tests, and some appear promising.[39]

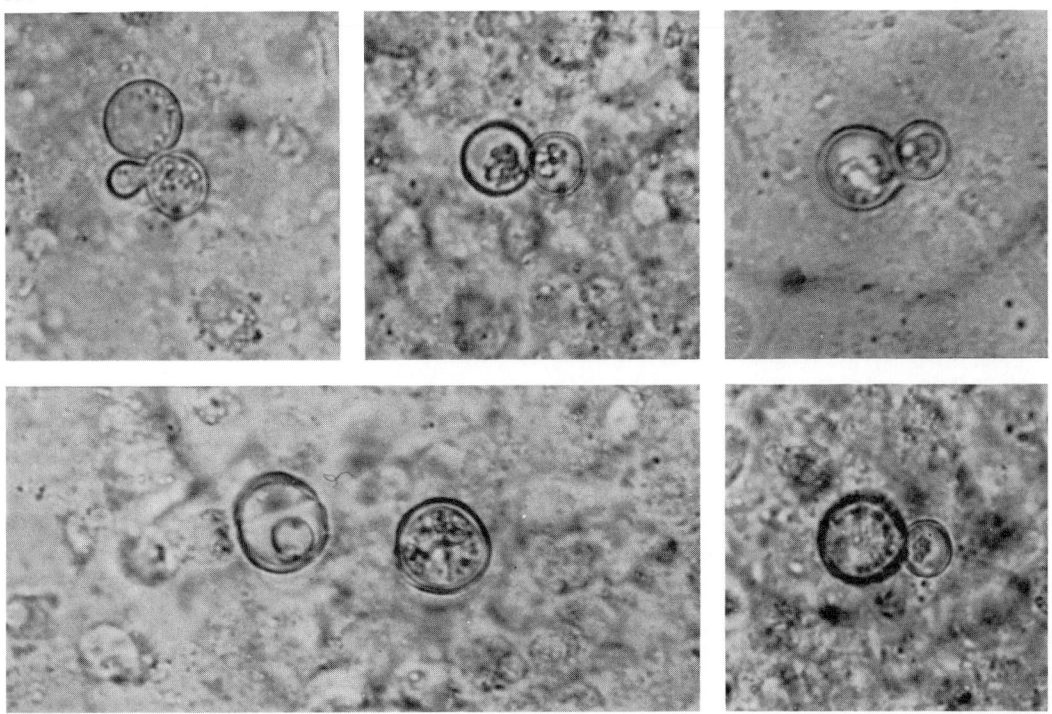

FIG. 8-5. *Blastomycosis. Yeast forms, including budding yeasts in unstained sputum, of a single patient. (Cryptococcus yeast forms have a very similar appearance in urine.)*

Skin Test. This has generally been regarded as of no value, because it was typically negative in patients with disseminated disease. However, in a recent common-source outbreak of acute pulmonary blastomycosis, 10 of 12 patients had a positive blastomycin skin test.[44]

Biologic Characteristics of Clinical Interest

Animal Infections. The dog is often naturally infected. Infection in horses has been observed.

Soil Sources. The organism has been found in the soil, and presumably it is a source.

Accidental Infection. A laboratory-acquired infection has been observed in which the onset of symptoms (cough, chills, fever) occurred 32 days after examination of a culture of *B. dermatiditis*.[35] Right hilar adenopathy developed, without radiologic pneumonia, and the organism was recovered from the patient's sputum. Serologic studies were negative.

Accidental inoculation of the pathologist's skin during performance of an autopsy has been observed.

Treatment

Hydroxystilbamidine is effective, and less toxic than amphotericin B, which is also effective.[37] In a recent common-source outbreak with seven severely ill patients, all recovered without specific antifungal therapy.[44] Miconazole therapy has been successful in patients who have relapsed after amphotericin B therapy.[43] South American blastomycotic pneumonia has been successfully treated with trimethoprim–sulfamethoxazole.[36]

SPOROTHRIX SCHENCKII

Objectives

1. Describe the typical clinical picture of cutaneous sporotrichosis.
2. Describe the methods for laboratory diagnosis and for treatment of sporotrichosis.

Definitions

Sporothrix schenckii is a dimorphic fungus, named after Schenck, who first recovered the

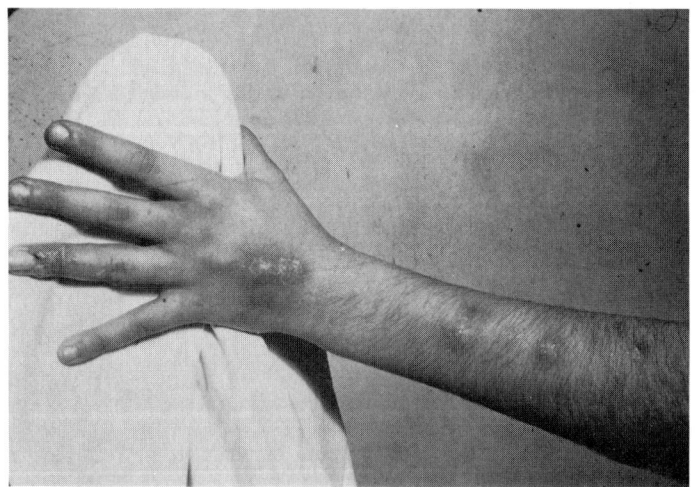

FIG. 8-6. *Typical appearance of sporotrichosis with superficial nodules along lymphatic drainage of a finger lesion. (Photo from Dr. Morris Leider; used by permission of Resident and Staff Physician and New York University School of Medicine Skin and Cancer Unit)*

organism from subcutaneous abscesses. *Sporotrichium* is the older genus name.

Frequency and Importance

The disease is rare, but is important because it responds to chemotherapy.

Clinical Patterns of Illness

Sporothrix schenckii is an *occasional* cause of:

Chronic Skin Ulcers With Nodular Lymphangitis. Typically, a relatively painless ulcer occurs on the hand or lower leg, in an individual exposed by gardening or forestry activities. There are typically several red nodules occurring up the arm or leg, along the lymphatic drainage, which appear within 2 weeks of the initial lesion (Fig. 8-6). The ulcerative lesions may drain for 6 to 8 weeks.[49] In a well-studied outbreak traced to sphagnum moss, about 25 percent of individuals handling the moss developed cutaneous sporotrichosis, with an incubation period of about 2 to 4 weeks.[48,49,53]

Sporothrix schenckii is a *rare* cause of several other syndromes. Chronic pneumonia with cavitation can occur.[46,52] Chronic joint effusions have also been observed.[52] Meningeal involvement is very rare.[52] Disseminated sporotrichosis is exceedingly rare.[54]

The organism has been recovered repeatedly over a period of 8 months from sputum cultures of an individual without clinical or radiologic evidence of lung disease, suggesting it can be a saprophyte.[50]

Laboratory Diagnosis

Culture. The organism can be recovered in about 3 to 5 days on fungal culture media.[51]

Smear. The organism is often missed on Gram stained smears of exudate from the skin lesion, but occasionally budding yeasts can be seen.[48,51]

Serum Antibodies. Several serologic methods are available, but are usually less practical than culture of the organism. A slide latex agglutination method appears to be simple and accurate, even in patients with localized skin disease.[47]

Skin Test. The antigen has been studied but is not commercially available.[48]

Biologic Characteristics of Clinical Interest

Occupational Exposures. Gardeners, florists, foresters, and others who handle roses, thorny bushes, or sphagnum moss are most likely to get the disease.

Growth in Iodide. The organism grows in 10 percent iodide, but iodide therapy is clearly effective in most patients with disease limited to the skin.[46]

Treatment

Iodides are effective treatment of the cutaneous disease and sometimes are effective in pulmonary sporotrichosis.[48] Intravenous amphotericin B may be necessary for effective treatment of extracutaneous disease.[46,52] Some patients with cavitary lesions have been treated with pulmonary resection.[52]

CANDIDA ALBICANS

Objectives

1. Describe three clinical situations that predispose to candidal infections.
2. Describe laboratory methods for confirming the diagnosis.
3. Describe the treatment of serious candidal infections.

Definitions and Species

Candida is a genus of dimorphic fungi in which both yeast and mycelial forms can be seen in the host or in laboratory cultures at 37°C. *C. albicans* is the most common species recovered, but *C. tropicalis* is occasionally found.[83] *C. albicans* forms a creamy, white yeasty-smelling colony on ordinary blood agar. This species is identified by its thick-walled spores called chlamydospores, from chlamydia (cloak). Monilia is an older name for the genus *Candida*, and the terms moniliasis and candidiasis are synonymous.

Geotrichium is another species of yeast that is rarely associated with human disease.

Frequency and Importance

Candida albicans is a very frequent cause of diaper rash in infants, and of chronic inflammation of the toes or inguinal area. Chronic infection of the mouth or the skin after infancy should raise the question of disorder of cell-mediated immunity, with possible adrenal insufficiency. Severe infections can occur, but respond to specific antifungal therapy.

Clinical Patterns of Illness

In normal persons, *Candida albicans* is a *common* cause of:

Stomatitis in infants, which is also called oral moniliasis or thrush. The mouth, especially the tongue, is coated with a white scum, and slight bleeding of the buccal mucosa or gums may occur.

Vaginitis has a discharge that is characteristically white and creamy, and budding yeasts are seen in a wet preparation. Candida vaginitis suggests the possibility of diabetes mellitus when it occurs in adults, but not in children.

Diaper Dermatitis. Candida is a frequent cause of "diaper rash". It has a rough, confluent, orange appearance, with satellite lesions on the abdomen, and to some extent on the trunk.[79]

Skin and Nail Infections. Tinea, (superficial fungus infection of the skin) can involve the toes, hands, or inguinal area. Nail infections are difficult to treat.

Cannula sepsis is a syndrome of fever and positive blood culture secondary to an intravenous cannula. *Candida albicans* is a frequent cause of this syndrome when the administered solution includes hypertonic glucose for hyperalimentation. Embolic eye infection (endophthalmitis) or infective endocarditis due to *Candida albicans* can occur as a complication of the candidal cannula sepsis.[62,66] Embolic skin lesions are papules or nodules, and contain candida.[57]

Urinary infection in normal individuals is usually secondary to an indwelling catheter, and typically responds to irrigation of the bladder with amphotericin B.

In immunologically compromised hosts, *C. albicans* is a *common* cause of:

Chronic mucocutaneous candidiasis is characterized by chronic or recurrent candida infection of the nails, skin, or mucous membranes.

Esophagitis is occasionally seen and has characteristic radiologic changes.[80]

In a compromised host, *Candida albicans* is an *occasional* cause of severe deep or disseminated fungal infections.[67] In normal individuals, these syndromes are rarely due to Candida.

Candida meningitis is rare.[71] The CSF glucose is low, the protein is high, and the cell count is usually less than 1000 white blood cells per cubic millimeter, predominately neutrophils.

Pneumonia is rare, but can occur in a host with malignancy or immunosuppression,[25] or in normal newborn infants.

Congenital infection with disseminated candidiasis has been reported.[61]

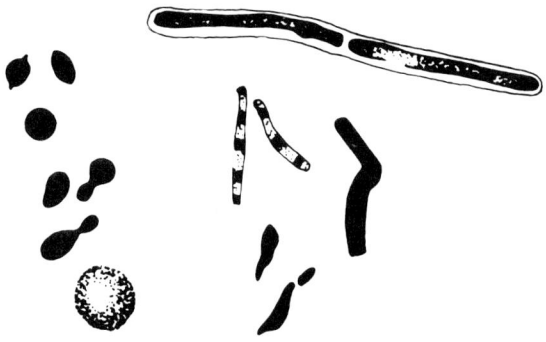

FIG. 8-7. Candida albicans *forms that might be seen in Gram stain of mouth, skin or intravenous catheter tip. Left, yeast forms and an erythrocyte. Right, mycelial forms. Only a few of these variants would be seen on a single specimen.*

Other rare disseminated infections include osteomyelitis[68] or arthritis[76] secondary to fungemia, and a verrucous skin lesion with plaques, sometimes associated with immunologic abnormalities.[74] Other complications include multiple mycotic aneurysms,[59] disseminated intravascular coagulation, nodular skin lesions and peritonitis.

Laboratory Diagnosis

Culture. *Candida* grows out on sheep blood agar plates in 24 to 48 hours, but more selective media are necessary when many other bacteria are present, as in the bowel.

Smear. Budding yeasts seen in specimens from the mouth are usually *C. albicans*. Fungemia has been recognized early on a smear of the peripheral blood.[78] Blood withdrawn from plastic intravenous cannula may show budding yeasts and mycelia[55] (Fig. 8-7).

Serology. Several tests are available in research laboratories,[60,70] and counterimmunoelectrophoresis (Fig. 4-2) appears useful.[70] Several antigen fractions are available.

Skin or Muscle Biopsy. Red papules or nodules sometimes are a manifestation of disseminated candidiasis and biopsy can be used for early diagnosis in patients at risk for this disease.[87]

Skin Test. This is useful as a test to evaluate cellular immunity, rather than as a test to diagnose candida infection. The frequency of positive individuals is related to age, with about 90 percent of individuals over a year old having a positive reaction.[82] The candida skin test is often applied concurrently with the tuberculin test to determine if the patient is anergic (Chap. 7).

Biologic Characteristics of Clinical Interest

Colonization as Normal Flora. *C. albicans* is often found in the mouth, vagina, and bowel in small numbers, but can become more numerous during antibiotic therapy.

Association with Endocrine Diseases. Hypoparathyroidism or hypoadrenalism (Addison's disease) are often found in patients with chronic mucocutaneous candidiasis. The endocrine disorders are not caused by the *Candida* infection, but candidiasis should always be thought of as a clue to consider endocrine defects.[58]

Association with Immunologic Diseases. Thymoma, thymic alymphoplasia, and congenital absence of the thymus and parathyroid glands are often associated with chronic mucocutaneous candidiasis. Patients with many other immune defects also often have superficial *Candida* infections, which should be regarded as a clue to consideration of such defects.[74]

Relation to Steroid Therapy. The mechanism is unknown, but presumably steroid suppression of immune mechanisms is involved.

Relation to Antibiotic Therapy. Apparently the suppression of the normal bacterial flora is the primary reason for the emergence of candidiasis during antibiotic therapy, although other mechanisms may be involved.[81]

Pathogenesis of Fungemia. Oral ingestion of *Candida albicans* has been studied prospectively in a normal individual, and the organism recovered twice from blood and urine several hours later.[72] *C. albicans* is a strict aerobe and blood culture bottles need venting to allow optimal detection.

Treatment

Correct Predisposing Cause. This usually means stopping antibiotic or steroid therapy.

Chemotherapy. Topical therapy with nystatin is sufficient for simple localized infections. Nystatin, named as a contraction of New York State, is only available as an oral preparation.

For disseminated candidiasis, parenteral amphotericin B is usually used. It is relatively toxic and is indicated only for severe disease.

Chronic mucocutaneous candidiasis has been successfully treated with transfer factor, in patients with non-lethal defects of cell-mediated immunity.[64] Miconazole or clotrimazole is effective treatment of chronic mucocutaneous candidiasis.[65,77]

Fluocytosine can be given by mouth and is usually effective therapy for disseminated candidiasis when given with amphotericin B, with which it is synergistic.[63,69] Rifampin also is often synergistic with amphotericin B in disseminated candidiasis.[56]

CRYPTOCOCCUS NEOFORMANS

Objectives

1. Describe the typical spinal fluid findings of cryptococcal meningitis.
2. Describe how infection can be detected in the laboratory by methods other than culture of the organism.
3. Discuss the presumed source of the disease in humans and the possible mode of transmission.

Name and Species

Cryptococcus neoformans is a yeast that was previously called *Torula histolytica*. The disease produced by this organism is called cryptococcosis or torulosis.[90] Other closely related organisms include *Torulopsis glabrata*, which typically infects only compromised hosts.[92] Rhodotorula is a red-colored cryptococcus which is a rare cause of disease.[91]

Frequency and Importance

Cryptococcal meningitis has been estimated to cause about 50 deaths per year in the United States, although this figure is very low because reporting is not required. The disease is not rare in immunosuppressed humans. Cryptococcosis is important because early treatment with amphotericin B is usually effective in preventing death or serious damage.

Clinical Patterns of Illness

Cryptococcus neoformans is an *occasional* cause of:

Nonpurulent Meningitis With Low Glucose. Cryptococcal meningitis typically resembles tuberculous meningitis, with 10 to 500 cells per cubic millimeter, predominantly lymphocytes, and increased protein and decreased glucose.[85,90] It can occur in normal hosts, as well as in individuals with malignancy or those receiving immunosuppressive drugs.

Cryptococcus neoformans is a *rare* cause of several syndromes. Pneumonia due to *Cryptococcus* usually occurs in patients with preexisting pulmonary disease.[98] It can be focal, interstitial, or miliary. Chronic pyelonephritis due to *Cryptococcus* can proceed meningeal involvement.[93] Other possible manifestations include skin ulcerations, prostatitis, adrenocortical insufficiency, lytic bone lesions, septic arthritis, and hepatic involvement.[84] Asymptomatic infection may occur, since antibodies can be found in individuals with no identifiable past clinical illness.[97] It is rarely found in the throat or between the toes of healthy individuals.[94]

Laboratory Approach

Smear. Budding yeasts can be found in an India ink preparation of the spinal fluid in about half of the patients with cryptococcal meningitis proved by culture. A higher proportion of positive results can be obtained if spinal fluid is filtered through a Millipore filter, and the filter examined, or if special centrifuging and special stains are used.[89]

It also can be found on silver stained smears of the bone marrow from patients with disseminated disease.[95]

Culture. The organism can be isolated on standard fungus media. Identification of *C. neoformans* is based on metabolic tests, or by demonstration of pathogenecity for mice, in which it produces hydrocephalus.

Serology. Serum antibodies against cryptococcus can be determined by several methods.[88,97] The antigen can be detected in the spinal fluid in cryptococcal meningitis by latex fixation, and this is a very useful and rapid method for laboratory diagnosis.[88] (Fig. 8-8).

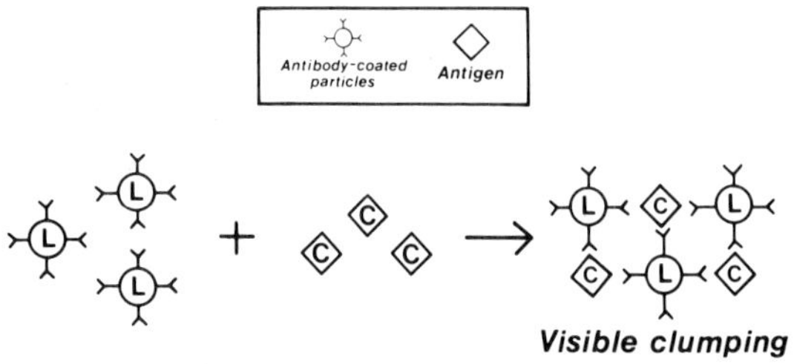

FIG. 8-8. *Latex fixation method for detection of Cryptococcus. L = latex particles coated with antibody to cryptococcus. C = cryptococci in spinal fluid. Visible agglutination of the latex particles is a positive result. Compare with Fig. 1-12.*

Skin Test. Antigen can be prepared and a positive reaction presumably indicates past infection.

Biologic Characteristics of Clinical Interest

Animal infection occurs naturally in a number of animals and birds. Birds may be involved in transmission. A study of pigeon fanciers indicated 22 percent were positive for cryptococcal antibodies compared to control individuals of whom three percent were positive.[97]

Therapy

Parenteral amphotericin B is effective, but intrathecal administration of amphotericin B, using a subcutaneous cerebrospinal fluid reservoir, may be necessary if the patient fails to respond.[86,90,96] Flucytosine is also effective. It is well absorbed when given by mouth, penetrates well into the spinal fluid,[87] and appears to be synergistic with amphotericin B.[96]

ASPERGILLUS SPECIES

Objectives

1. Describe at least two clinical diseases that can be produced by *Aspergillus* species.
2. Describe how the diagnosis can be confirmed in the laboratory.
3. Discuss the treatment of aspergillosis.

Definitions

Aspergillus is a genus of fungi named after the microscopic appearance of the spore cluster (rough head). Most disease in humans is due to *A. fumigatus*, but occasionally *A. niger* or other species are associated with disease.

Frequency and Importance

Aspergillus infection is rare, but occurs most frequently as a complication of asthma. Allergic aspergillosis is especially important because it often responds to steroids, which may prevent operative removal of part of the lung.

Clinical Patterns of Illness

Aspergillus species are an *occasional* cause of:

Pulmonary Infiltrate with Eosinophilia (PIE Syndrome). This syndrome, when due to aspergilli, is called allergic pulmonary aspergillosis. Wheezing may be prominent, and saccular upper lobe bronchiectasis may occur.[100] This pattern has been frequently recognized in Britain, and recently has been recognized in the United States.[105]

Aspergillus species are a *rare* cause of:

Acute Pneumonia. Typically, there is an underlying problem, such as steroid therapy, or a hematologic malignancy, but fulminating disease has been observed in apparently normal individuals.[100,107,109] There is typically necrotizing bronchopneumonia or a hemorrhagic pulmonary infarction. Occasionally, other patterns occur, including lobar pneumonia. The disease is typically acute, with a progressive necrotizing pneumonia, often with abscess formation. In two apparently normal children with fulminating cases, fever was not prominent, but marked leucocytosis was present.[107]

Other Diseases. Most patients with invasive aspergillosis have severe underlying disease, particularly cancer.[99] Fungus balls (aspergillomas) in lung cavities may occur secondarily to tuberculosis or other primary diseases.[108] Brain abscess may complicate pulmonary aspergillosis. Other diseases rarely due to aspergilli include otitis externa, myocarditis, and endocarditis.[101]

Laboratory Approach

Smear. Examination of a Gram stain of the sputum or tissue may reveal mycelia with septate hyphae.[105]

Culture. *Aspergillus* species usually grow out on simple media in a few days. Since the organism is frequently a contaminant, the significance of a positive culture must be interpreted in the light of the clinical situation.

Serum Antibodies. These can be measured in some reference laboratory, and may be useful in some cases. Normal individuals usually do not have precipitating antibodies, but individuals colonized without invasive disease also usually have such antibodies.[103]

Serum IgE. This is markedly elevated in patients with pulmonary allergic aspergillosis, and may be a guide to response to corticosteroid therapy.[104]

Skin Test. Antigens for skin testing may be available in some research laboratories.

Biologic Characteristics of Clinical Interest

Animal Hosts. Invasive pulmonary disease is produced in many species of birds, and may cause epidemics. It is a major cause of death of penguins in zoos, and can be a serious problem on chicken or turkey farms.

Thermophilia. *Aspergilli* thrive at elevated temperatures, which accounts for their ability to grow in birds and in decaying moldy hay.

Treatment

Corticosteroids are useful in allergic bronchopulmonary aspergillosis, which is characteristically associated with eosinophilia.[100]

Amphotericin B, by the intravenous route, is the best treatment for acute invasive aspergillosis.[99,108] Rifampin or 5-fluorocytosine is often synergistic with amphotericin B in vitro.[102]

Pulmonary resection may be indicated in patients with a fungus ball or localized, especially cavitary, disease.[106,108] It is important not to remove part of a lung for allergic bronchopulmonary aspergillosis since corticosteroid therapy would probably make any resection unnecessary.

OTHER INVASIVE FUNGI

The fungi discussed in this section are very rare causes of disease in humans. This section simply lists some fungi with relevant references in case the reader encounters them.

Phycomycosetes (Mucormycosis)

Phycomycosetes are a class of fungi characterized by nonseptate hyphae. The genera included in this class are *Mucor, Rhizopus, Basidiobolus, Absidia, Mortierrella,* and *Cunninghamella.*[118] Phycomycosis is also called mucormycosis, after the genus most often implicated.

Phycomycosis (also called zygomycosis) is very rare, and often is fatal. It typically occurs in individuals with debilitating conditions, particularly diabetes mellitus or malignancies. The most frequent pattern of illness is sinusitis, with bloody nasal discharge, facial swelling, proptosis, and invasion of blood vessels and the brain.[115,118] Black necrosis of the skin and mucosa is often seen. The disease also can involve the kidneys, gastrointestinal tract, or lungs.[121] *Rhizopus* species have been a contaminant of elastic bandages and a cause of wound infections.

Laboratory diagnosis is best made by histologic examination of the involved tissue. The typical broad, nonseptate branching can sometimes be seen immediately on Gram stain, as well as on stained tissue sections. Culture is definitive, and the organism can be tested for susceptibility to amphotericin B.[121]

Serologic methods of diagnosis are under investigation.[116]

Treatment with amphotericin B is sometimes lifesaving.[111,121,123] Surgical excision, particularly enucleation of the eye, is often necessary.

Allescheria (Maduromycosis)

Allescheria boydii is a fungus found in the soil. It is a rare cause of infections in immunosuppressed hosts, such as brain abscess or necrotizing pneumonia in leukemia.[120,125] The organism has an asexual phase called *Monosporium apiospermum.* It can produce a fungal tumor (mycetoma) called maduromycosis (madura foot), a progressive invasive infection of the hand or foot, which occurs after traumatic inoculation of the organism from the soil. (Maduromycosis is named for Madura, an Indonesian island near Java.) *Allescheria* species can also produce sinusitis, cavitary pneumonia, arthritis, eye infections and disseminated disease.

Recently the new name *Petriellidium* has been proposed as the genus name instead of *Allescheria*.

The hyphae may be seen in tissues with a silver stain.

Chemotherapy with amphotericin B (especially topically in the eye) or miconazole may be helpful.[120,125]

Cladosporium (Chromoblastomycosis)

This genus is very rarely recognized as a cause of disease in the United States, but is more common in South America. *Fonsecaea* and *Phialophora* are closely related genera. These fungi typically produce wart-like papules of the skin after traumatic inoculation. The fungi and the lesions are brown or black, so the infection is called chromoblastomycosis. Brain abscess or meningitis also can be produced by this group of fungi.[112] Fluocytosine may be of therapeutic value.

Other Miscellaneous Fungi

This section includes organisms likely to be mistaken for fungi and not easily classified elsewhere.

Cephalosporium is a genus that has been associated with meningitis in infants.[122]

Curvularia is a genus of saprophytic fungi that can cause lung or brain infection in the compromised human host.[117]

Dermatophilus is a genus commonly causing pustular skin lesions in animals, but has also been observed as a cause of subcutaneous nodular abscesses in an immunocompromised child.[110]

Fusarium is a genus of fungi, pathogenic to plants, that can infect burned or ulcerated skin and can produce disseminated infection in compromised hosts.[127] It can invade the plastic of hemodyalisis tubing, but can be cured by removal of the tubing.

Prototheca is a genus of saprophytic alga that resembles the yeast candida, and which can produce wound infection.[119]

Penicillium species are best known as the original source of penicillin. A *Penicillium* species has been recovered from the spleen of a patient with Hodgkin's disease.[113] The organism resembles the yeast phase of *Histoplasma capsulatum*. At least 14 human cases have been described.

Phialophora is a genus of fungi that has been associated with endocarditis on a prosthetic value.[124]

Rhodocrous is a genus of acid-fast organisms which is intermediate between rapid-growing mycobacteria and the bacterial genus *Nocardia*. It has been recovered from bone marrow and transtracheal aspirates of compromised hosts.[114]

Rodotorula and torulopsis are mentioned in this chapter in the section on cryptococcus.

Botryomycosis is a general name for a granulomatous process which resembles a mycetoma with granules, but can be caused by a variety of bacteria.[126] It can be caused by actinomyces, pseudomonas, staphylococci, or other bacteria.

REFERENCES

Fungi

1. Jacobs, P.H.: Fungal infections in childhood. Pediatr. Clin. North Am., 25:357–370, 1978.
2. Millikan, L.E.: Superficial and cutaneous fungal infections. Postgrad Med., 60:52–58, 1976.
3. Stocker, W.W., Richtsmeier, A.J., Rozycki, A.A., and Baughman, R.D.: Kerion caused by *Trichophyton verrucosum*. Pediatrics, 59:912–915, 1977.

Histoplasma Capsulatum

4. Bauman, D.S., and Smith, C.D.: Comparison of immunodiffusion and complement fixation tests in the diagnosis of histoplasmosis. J. Clin. Microbiol., 2:77–80, 1975.
5. Bennett, D.E.: The histoplasmin latex agglutination test: clinical evaluation and a review of the literature. Am. J. Med. Sci., 251:175–182, 1966.
6. Buechner, H.A., et al.: The current status of serologic, immunologic, and skin tests in the diagnosis of pulmonary mycoses. Chest, 63:259–270, 1973.
7. Davies, S.F., Khan, M., Sarosi, G.A.: Disseminated histoplasmosis in immunologically suppressed patients. Occurrence in a nonendemic area. Am. J. Med., 64:94–100, 1978.
8. DiSalvo, A.F., and Corbett, D.S.: Apparent false positive histoplasmin latex agglutination tests in patients with tuberculosis. J. Clin. Microbiol., 3:306–308, 1976.
9. Edwards, L.B., et al.: An atlas of sensitivity to tuberculin, PPD-B, and histoplasmin in the

United States. Am. Rev. Respir. Dis., 99 (Suppl.):1–132, 1969.

10. Edwards, L.B., Acquavita, F.A., and Livesay, V.T.: Further observations on histoplasmin sensitivity in the United States. Am. J. Epidemiol., 98:315–325, 1973.

11. Enarson, D.A., Keys, T.F., and Onofrio, B.M.: Central nervous system histoplasmosis with obstructive hydrocephalus. Am. J. Med., 64:895–896, 1978.

12. Furculow, M.L., et al.: The emerging pattern of urban histoplasmosis. New Engl. J. Med., 264:1226–1230, 1961.

13. Gerber, H.J., Schoonmaker, F.W., and Vazquez, M.D.: Chronic meningitis associated with *Histoplasma* endocarditis. New Engl. J. Med., 275:74–76, 1966.

14. Goodwin, R.A., Jr., and DesPrez, R.M.: State of the Art—Histoplasmosis. Am. Rev. Respir. Dis., 117:929–956, 1978.

15. Goodwin, R.A., Jr., et al.: Chronic pulmonary histoplasmosis. Medicine, 55:413–452, 1976.

16. Holland, P., and Holland, N.H.: Histoplasmosis in early infancy. Hematologic, histochemical, and immunologic observations. Am. J. Dis. Child., 112:412–421, 1966.

17. Kitahar, M., Kobayashi, G.S., and Medoff, G.: Enhanced efficacy of amphotericin B and rifampicin combined in treatment of murine histoplasmosis and blastomycosis. J. Infect. Dis., 133:663–667, 1976.

18. Moffet, H.L., Najjar, S., and Cramblett, H.G.: Successful therapy of disseminated histoplasmosis through the use of amphotericin B. J. Iowa Med. S., 49:625–631, 1959.

19. Parker, J.D., et al.: Treatment of chronic pulmonary histoplasmosis. New Engl. J. Med., 283:225–229, 1970.

20. Schwarz, J., Schaen, M.I., and Picardi, J.L.: Complications of the arrested primary histoplasmic complex. J.A.M.A., 236:1157–1161, 1976.

21. Segal, C., Wheeler, C.G., and Tompsett, R.: *Histoplasma* endocarditis cured with amphotericin. New Engl. J. Med., 280:206–207, 1969.

22. Studdard, J., Sneed, W.F., Taylor, M.R., Jr., and Campbell, G.D.: Cutaneous histoplasmosis. Am. Rev. Respir. Dis., 113:689–693, 1976.

23. Tesh, R.B., and Schenidan, J.D., Jr.: Primary cutaneous histoplasmosis. New Engl. J. Med., 275:597–599, 1966.

Coccidioides Immitis

24. Caudill, R., Smith, C.E. and Reinarz, J.: Coccidioidal meningitis. A diagnostic challenge. Am. J. Med., 49:360–365, 1970.

25. Converse, J.L., Pakes, S.P., Snyder, E.M. and Castleberry, M.W.: Experimental primary cutaneous coccidioidomycosis in the monkey. J. Bacteriol., 87:81–95, 1964.

26. Drutz, D.J. and Cantazaro, A.: State of the art. Coccidioidomycosis. Parts I, II. Am. Rev. Respir. Dis., 117:559–585, 727–771, 1978.

27. Edwards, P.Q. and Palmer, C.E.: Prevalence of sensitivity to coccidioidin, with special reference to specific and nonspecific reactions to coccidioidin and histoplasmin. Dis. Chest, 31:35–60, 1957.

28. Hyatt, H.W., Sr.: Coccidioidomycosis in a 3-week-old infant. Am. J. Dis. Child., 105:93–98, 1963.

29. Johnson, J.E., et al.: Laboratory-acquired coccidioidomycosis. A report of 210 cases. Ann. Intern. Med., 60:941–956, 1964.

30. Kruse, R.H., Green, T.D. and Leeder, W.D.: Infection of control monkeys with *Coccidioides immitis* by caging with inoculated monkeys. In Ajello, L.: Coccidioidomycosis. Tuscon, U. Ariz. Press, 1967. pp. 387–395.

31. Rothman, P.E., Graw, R.G., Jr., Harris, J.C., Jr., and Onslow, J.M.: Coccidioidomycosis—possible fomite transmission. A review and report of a case. Am. J. Dis. Child., 118:792–801, 1969.

32. Steele, R.W., et al.: Therapy for disseminated coccidioidomycosis with transfer factor from a related donor. Am. J. Med., 61:283–286, 1976.

33. Werner, S.B., Pappagianis, D., Heindl, I. and Mickel, A.: An epidemic of coccidioidomycosis among archeology students in Northern California. New Engl. J. Med., 286:507–512, 1971.

34. Winn, W.A.: Tuberculosis and coccidioidomycosis: a working classification of coccidioidal disease related to therapy based on differences between the two diseases. Am. Rev. Respir. Dis., 96:229–236, 1967.

Blastomyces Dermatiditis

25. Baum, G.L., and Lerner, P.I.: Primary pulmonary blastomycosis: a laboratory-acquired infection. Ann. Intern. Med., 73:263–265, 1970.

36. Bouza, E., Winston, A.J., Rhodes, J., and Hewitt, W.L.: Paracoccidioidomycosis (South American blastomycosis) in the United States. Chest, 72:100–102, 1977.

37. Duttera, M.J., and Osterhout, S.: North American blastomycosis: a survey of 63 cases. South. Med. J., 62:295–301, 1969.

38. Gonyea, E.F.: The spectrum of primary blastomycotic meningitis: a review of central nervous system blastomycosis. Ann. Neurol., 3:26–39, 1978.

39. Kaufman, L., McLaughlin, D.W., Clark, M.J., and Blumer, S.: Specific immunodiffusion test for blastomycosis. Appl. Microbiol., 26:244–247, 1973.

40. Laskey, W., and Sarosi, G.A.: Endogenous activation in blastomycosis. Ann. Int. Med., 88:50–52, 1978.
41. Palmer, P.E., and McFadden, S.W.: Blastomycosis. Report of an unusual case. New Engl. J. Med., 279:979–983, 1968.
42. Pfister, A.K., et al.: Pulmonary blastomycosis. Roentgenographic clues to the diagnosis. South. Med. J., 49:1441–1447, 1966.
43. Rose, H.D., and Varkey, B.: Miconazole treatment of relapsed pulmonary blastomycosis. Am. Rev. Respir. Dis., 118:403–408, 1978.
44. Sarosi, G.A., Hammerman, K.J., Tosh, F.E., and Kronenberg, R.S.: Clinical features of acute pulmonary blastomycosis. New Engl. J. Med., 290:540–543, 1974.
45. Witorsh, P., and Utz, J.P.: North American blastomycosis. A study of 40 patients. Medicine, 47:169–200, 1968.

Sporothrix Schenkii

46. Baum, G.L., et al.: Pulmonary sporotrichosis. New Engl. J. Med., 280:410–413, 1969.
47. Blumer, S.O., et al.: Comparative evaluation of five serological methods for the diagnosis of sporotrichosis. Appl. Microbiol., 26:4–8, 1973.
48. Chandler, J.W., Kriel, R.L., and Tosh, F.E.: Childhood sporotrichosis. Am. J. Dis. Child., 115:368–372, 1968.
49. D'Alessio, D.J., Leavens, L.J., Strumpf, G.B., and Smith, C.D.: An outbreak of sporotrichosis in Vermont associated with sphagnum moss as the source of infection. New Engl. J. Med., 272:1054–1058, 1965.
50. Lowenstein, M., Markowitz, S.M., Nottebart, H.C., and Shadomy, S.: Existence of *Sporothrix schenckii* as a pulmonary saprophyte. Chest, 73:419–421, 1978.
51. Park, C.H., Greer, C.L., and Cook, C.B.: Cutaneous sporotrichosis: recent appearance in northern Virginia. Am. J. Clin. Pathol., 57:23–26, 1972.
52. Parker, J.D., Sarosi, G.A., and Tosh, F.E.: Treatment of extracutaneous sporotrichosis. Arch. Intern. Med., 125:858–863, 1970.
53. Powell, K.E., et al.: Cutaneous sporotrichosis in forestry workers. Epidemic due to contaminated sphagnum moss. J.A.M.A., 240:232–235, 1978.
54. Satterwhite, T.K., et al.: Disseminated sporotrichosis. J.A.M.A., 240:771–772, 1978.

Candida Albicans

55. Anderson, A.O., and Yardley, J.H.: Demonstration of candida in blood smears. New Engl. J. Med., 286:108, 1972.
56. Beggs, W.H., Sarosi, G.A., and Walker, M.I.: Synergistic action of amphotericin B and rifampin against *Candida* species. J. Infect. Dis., 133:206–209, 1976.
57. Balandran, L., Rothschild, H., Pugh, N., and Seabury, J.: A cutaneous manifestation of systematic candidiasis. Ann. Intern. Med., 78:400–403, 1973.
58. Blizzard, R.M., and Gibbs, J.H.: Candidiasis: studies pertaining to its association with endocrinopathies and pernicious anemia. Pediatrics, 42:231–237, 1968.
59. Collins, G.J., Jr., et al.: Multiple mycotic aneurysms due to candida endocarditis. Ann. Surg., 186:136–139, 1977.
60. Dee, T.H., and Rytel, M.W.: Detection of candida serum precipitins by counterimmunoelectrophoresis: an adjunct in determining significant candidiasis. J. Clin. Microbiol., 55:453–457, 1977.
61. Dvorak, A.M., and Gavaller, B.: Congenital systemic candidiasis. Report of a case. New Engl. J. Med., 274:540–543, 1966.
62. Edwards, J.E., Jr., Foos, R.Y., Montgomerie, J.Z., and Guze, L.B.: Ocular manifestations of candida septicemia: review of seventy-six cases of hematogenous candida endophthalmitis. Medicine, 53:47–75, 1974.
63. Eilard, T., Alestig, K., and Wahlén, P.: Treatment of disseminated candidiasis with 5-fluorocytosine. J. Infect. Dis., 130:155–159, 1974.
64. Feigin, R.D., et al.: Treatment of mucocutaneous candidiasis with transfer factor. Pediatrics, 53:63–70, 1974.
65. Fischer, T.J., Klein, R.B., Kershnar, H.E., Borut, T.C., and Stiehm, E.R.: Miconazole in the treatment of chronic mucocutaneous candidiasis: a preliminary report. J. Pediatr., 91:815–819, 1977.
66. Haning, H.A.L., Johnston, R., Touloukian, R., and Margolis, C.Z.: Successfully treated candida endophthalmitis in a child. Pediatrics, 51:1027–1031, 1973.
67. Hart, P.D., Russell, E., Jr., and Remington, J.S.; The compromised host and infection. II. Deep fungal infection. J. Infect. Dis., 120:169–191, 1969.
68. Hirschmann, J.V., and Everett, E.D.: Candida vertebral osteomyelitis. Case report and review of the literature. J. Bone Joint Surg., 58-A:573–575, 1976.
69. Imbeau, S.A., Hanson, J., Langejans, G., and D'Alessio, D.: Flucytosine treatment of candida arthritis. J.A.M.A., 238:1395–1396, 1977.
70. Kozinn, P.J., et al.: Efficiency of serologic tests in the diagnosis of systemic candidiasis. Am. J. Clin. Pathol., 70:893–898, 1978.
71. Kozinn, P.J., et al.: Candida meningitis successfully treated with amphotericin B. New Engl. J. Med., 268:881–884, 1963.
72. Krause, W., Mathies, H., and Wulf, K.: Fungaemia and funguria after oral administration

of *Candida albicans*. Lancet, *1*:598–599, 1969.

73. Kressel, B., Szewczyk, C., and Tuazon, C.U.: Early clinical recognition of disseminated candidiasis by muscle and skin biopsy. Arch. Intern. Med., *138*:429–433, 1978.

74. Landau, J.W.: Chronic mucocutaneous candidiasis—associated immunologic abnormalities. Pediatrics, *42*:227–230, 1968.

75. Masur, H., Rosen, P.P., and Armstrong, D.: Pulmonary disease caused by *Candida* species. Am. J. Med., *63*:914–925, 1977.

76. Murray, H.W., Fialk, M.A., and Roberts, R.B.: Candida arthritis. A manifestation of disseminated candidiasis. Am. J. Med., *60*:587–595, 1976.

77. Pazin, G.J., Nagel, J.E., Friday, G.A., and Fireman, P.: Topical clotrimazole treatment of chronic mucocutaneous candidiasis. J. Pediatr., *94*:322–324, 1979.

78. Portnoy, J., Wolf, P.L., Webb, M., and Remington, J.S.: Candida blastospores and pseudohyphae in blood smears. New Engl. J. Med., *285*:1010–1011, 1971.

79. Robinson, R.C.V.: Cutaneous moniliasis in infants. J. Pediatr., *50*:721–723, 1957.

80. Rohrman, C.A., and Kidd, R.: Chronic mucocutaneous candidiasis: radiologic abnormalities in the esophagus. Am. J. Roentgenol., *130*:473–476, 1978.

81. Seelig, M.S.: Mechanisms by which antibiotics increase the incidence and severity of candidiasis and alter the immunologic defenses. Bacteriol. Rev., *30*:442–459, 1966.

82. Shannon, D.C., Johnson, G., Rosen, R.S., and Austen, K.F.: Cellular reactivity to *Candida albicans* antigen. New. Engl. J. Med., *275*:690–693, 1966.

83. Toala, P., Schroeder, S.A., Daly, A.K., and Finland, M.: Candida at Boston City Hospital. Arch. Intern. Med., *126*:983–989, 1970.

Cryptococcus Neoformans

84. Anon.: Cryptococcal infections (editorial). Br. Med. J., *1*:1008–1009, 1978.

85. Butler, W.T., Alling, D.W., Spickard, A., and Utz, J.P.: Diagnostic and prognostic value of clinical and laboratory findings in cryptococcal meningitis. New Engl. J. Med., *270*:59–67, 1964.

86. Diamond, R.D., and Bennett, J.E.: A subcutaneous reservoir for intrathecal therapy of fungal meningitis. New Engl. J. Med., *288*:186–188, 1973.

87. Fass, R.J., and Perkins, R.L.: 5-Fluorocytosine in the treatment of cryptococcal and candidal mycoses. Ann. Intern. Med., *74*:535–539, 1971.

88. Goodman, J.S., Kaufman, L., and Koenig, M.G.: Diagnosis of cryptococcal meningitis. Value of immunologic detection of cryptococcal antigen. New Engl. J. Med., *285*:434–436, 1971.

89. Jéquier, M., and Dufresne, J-J.: Diagnosis of cryptococcal meningitis (letter). New Engl. J. Med., *286*:285–286, 1972.

90. Littman, M.L., and Walter, J.E.: Cryptococcosis: current status. Am. J. Med., *45*:922–932, 1968.

91. Louria, D.B., Greenberg, S.M., and Molander, D.W.: Fungemia caused by certain nonpathogenic strains of the family Cryptococcaceae. Report of two cases due to *Rhodotorula* and *Torulopsis glabrata*. New Engl. J. Med., *263*:1281–1284, 1960.

92. Marks, M.I., Langston, C., and Eickhoff, T.C.: *Torulopsis glabrata*—an opportunistic pathogen in man. New Engl. J. Med., *283*:1131–1135, 1970.

93. Randall, R.E., Jr., et al.: Cryptococcal pyelonephritis. New Engl. J. Med., *279*:60–65, 1968.

94. Randhawa, H.S., and Paliwal, D.K.: Occurrence and significance of *Cryptococcus neoformans* in the oropharynx and on the skin of a healthy human population. J. Clin. Microbiol., *6*:325–327, 1977.

95. Robert, F., Durant, J.R., and Gams, R.A.: Demonstration of *Cryptococcus neoformans* in a stained bone marrow specimen. Arch. Int. Med., *137*:688–690, 1977.

96. Utz, J.P., et al.: Therapy of cryptococcosis with a combination of flucytosine and amphotericin B. J. Infect. Dis., *132*:368–373, 1975.

97. Walter, J.E., and Atchison, R.W.: Epidemiological and immunological studies of *Cryptococcus neoformans*. J. Bacteriol., *92*:82–87, 1966.

98. Warr, W., Bates, J.H., and Stone, A.: The spectrum of pulmonary cryptococcosis. Ann. Intern. Med., *69*:1109–1116, 1968.

Aspergillus Species

99. Aisner, J., Schimpff, S.S., and Wiernik, P.H.: Treatment of invasive aspergillosis: relation of early diagnosis and treatment to response. Ann. Int. Med., *86*:539–543, 1977.

100. Campbell, J., and Clayton, Y.M.: Bronchopulmonary aspergillosis; a correlation of the clinical and laboratory findings in 272 patients investigated for bronchopulmonary aspergillosis. Am. Rev. Respir. Dis., *89*:186–196, 1964.

101. Kammer, R.B., and Utz, J.P.: *Aspergillus* species endocarditis. Am. J. Med., *56*:506–521, 1974.

102. Kitahara, M., Seth, V.K., Medoff, G., and Kobayashi, G.S.: Activity of amphotericin B, 5-fluorocytosine, and rifampin against six clinical isolates of *Aspergillus*. Antimicrob. Agents Chemother., *9*:915–919, 1976.

103. Kurup, V.P., and Fink, J.N.: Evaluation of methods to detect antibodies against *Aspergillus fumigatus*. Am. J. Clin. Pathol., *69*:414–417, 1978.

104. Rosenberg, M., Patterson, R., and Roberts, M.: Immunologic response to therapy in allergic bronchopulmonary aspergillosis: serum IgE value as an indicator and predictor of disease activity. J. Pediatr., 91:914–917, 1977.

105. Slavin, R.G., Stancyzk, D.J., Lonigro, A.J., and Broun, G.O., Sr.: Allergic bronchopulmonary aspergillosis—a North American rarity. Am. J. Med., 47:306–313, 1969.

106. Soltanzadeh, H., et al.: Surgical treatment of pulmonary aspergilloma. Ann. Surg., 186:13–16, 1977.

107. Strelling, M.K., Rhaney, K., Simmons, D.A.R., and Thomson, J.: Fatal acute pulmonary aspergillosis in two children of one family. Arch. Dis. Child., 41:34–43, 1966.

108. Varkey, B., and Rose, H.D.: Pulmonary aspergilloma. A rational approach to treatment. Am. J. Med., 61:626–631, 1976.

109. Young, R.C., et al.: Aspergillosis. The spectrum of the disease in 98 patients. Medicine, 49:147–173, 1970.

Other Fungi

110. Albrecht, R., et al.: *Dermatophilus congolensis* chronic nodular disease in man. Pediatrics, 53:907–912, 1974.

111. Brown, J.F., Jr., Gottlieb, L.S., and McCormick, R.A.: Pulmonary and rhinocerebral mucormycosis. Successful outcome with amphotericin B and griseofulvin therapy. Arch. Int. Med., 137:936–938, 1977.

112. Crichlow, D.K., Enrile, F.T., and Memon, M.Y.: Cerebellar abscess due to *Cladosporium trichoides (bantianum)*. Am. J. Clin. Pathol., 60:416–421, 1973.

113. DiSalvo, A.F., Fickling, A.M., and Ajello, L.: Infection caused by *Penicillium marneffei*: description of first natural infection in man. Am. J. Clin. Pathol., 60:259–263, 1973.

114. Haburchak, D.R, Jeffrey, B., Higbee, J.W., and Everett, E.D.: Infections caused by Rhodochrous. Am. J. Med., 65:298–302, 1978.

115. Hale, L.M.: Orbital phycomycosis. South. Med. J., 63:886–890, 1970.

116. Jones, K.W., and Kaufman, L.: Development and evaluation of an immunodiffusion test for diagnosis of systemic zygomycosis (mucormycosis): preliminary report. J. Clin. Microbiol., 7:97–101, 1978.

117. Lampert, R.P., Hutto, J.H., Donnelly, W.H., and Shulman, S.T.: Pulmonary and cerebral mycetoma caused by *Curvularia pallescens*. J. Pediatr., 91:603–605, 1977.

118. Landau, J.W., and Newcomer, V.D.: Acute cerebral phycomycosis (mucormycosis). Report of a pediatric patient successfully treated with amphotericin B and cyclohexamide and review of the pertinent literature. J. Pediatr., 61:363–385, 1962.

119. Lee, W.-S., Lagios, M.D., and Leonards, R.: Wound infection by *Prototheca wickerhamii*, a saprophytic alga pathogenic for man. J. Clin. Microbiol., 2:62–66, 1975.

120. Lutwick, L.I., Galgioni, J.N., Johnson, R.H., and Stevens, D.A.: Visceral fungal infections due to *Petriellidium boydii (Allescheria boydii)*. In vitro drug sensitivity studies. Am. J. Med., 61:632–640, 1976.

121. Medoff, G., and Kobayashi, G.S.: Pulmonary mucormycosis. New Engl. J. Med., 286:86–87, 1972.

122. Papadatos, C., Pavlatou, M., and Alexiou, D.: *Cephalosporium* meningitis. Pediatrics, 44:749–751, 1969.

123. Sandler, R., Tallman, C.B., Keamy, D.G., and Irving, W.R.: Successfully treated rhinocerebral phycomycosis in well-controlled diabetes. New Engl. J. Med., 285:1180–1182, 1971.

124. Slifkin, M., and Bowers, H.M., Jr.: *Phialophora mutabilis* endocarditis. Am. J. Clin. Pathol., 63:120–130, 1975.

125. Winston, D.J., Jordan, M.C., and Rhodes, J.: *Allescheria boydii* infections in the immunosuppressed host. Am. J. Med., 63:830–835, 1977.

126. Wu, W.Q., Cattaneo, E.A., Lapi, A., and Halde, C.: Botryomycosis: first report of human brain involvement. South. Med. J., 71:1530–1533, 1978.

127. Young, N.A., et al.: Disseminated infection by *Fusarium moniliforme* during treatment for malignant lymphoma. J. Clin. Microbiol., 7:589–594, 1978.

9
Spirochetes

TREPONEMA PALLIDUM AND OTHER SPIROCHETES

Objectives

1. Describe some differences between treponema, leptospira, borrelia, and spirillum.
2. Describe the natural history of syphilis in its usual clinical pattern, including the relative length of time between various stages and the time required after exposure for the serology to become positive.
3. Define latent syphilis.
4. Describe two ways in which infection with syphilis may occur without venereal exposure.
5. Discuss the significance of a postive VDRL.
6. Define biologic false positive (BFP).
7. Name three diseases in which BFP reactions can occur.
8. Define some of the available confirmatory serologic tests for syphilis and indicate the circumstances in which they might be used.
9. Discuss the chemotherapy of syphilis in the following cases:
    a. A genital lesions resembles a chancre and a blood specimen for VDRL has just been obtained.
    b. A smear of a urethral discharge shows gram-negative diplococci, in addition to a. (above).
    c. VDRL is positive and the patient is allergic to penicillin.
10. Discuss the evidence that a particular regimen of antibiotics for syphilis can be demonstrated to be effective.
11. Discuss what antibiotic treatments of gonorrhea are also effective in eradicating incubating syphilis.

Classification of Spirochetes

There are three genera of spirochetes: treponema, leptospira, and borrelia.

Treponema. The name means turning thread, referring to the observation that the organism rotates rapidly about its long axis. *Treponema pallidum* is the spirochete that causes syphilis. Other treponemae include: *T. pertenue*, the cause of yaws (an ulcerating skin disease found in the tropics); *T. carateum*, the cause of pinta (a tropical skin disease manifested by increased pigmentation, followed by loss of skin pigment); and *T. microdentium* (one of many species of treponema found in the normal mouth).

Leptospira. The name means thin coil. The organism often appears to be hooked at one or both ends because a flagellum is attached near the tip, and its motion produces a hooking action. *Leptospira* species are discussed later in this chapter.

Borrelia. This genus was named after the French microbiologist Borel. *Borrelia* species can cause relapsing fever in humans. Headache, photophobia, and 2 to 7 day episodes of fever alternating with 2 to 4 day afebrile periods occur for about three episodes. The natural reservoir is rodents (e.g. a rat, squirrel, or chipmunk). Humans become infected by the bite of a tick or louse from the rodent. Approximately 10 to 15 cases are reported annually in the United States, almost always in people camping or hiking in the western states. Cases have been rarely observed east of the Mississippi River. The usual incubation period is about 7 days.

The diagnosis is best confirmed in the laboratory by finding spirochetes on microscopic examination of a Wright stained smear of peripheral blood or dark field examinations during the febrile period. A single oral dose of tetracycline or erythromycin is effective therapy.[20] Antibiotic therapy may produce a Jarisch–Herxheimer reaction, first described as a complication of the therapy of syphilis. The rapid lysis of the organism can produce chills, fever, rapid pulse, and a drop in blood pressure—findings that can be ameliorated by intravenous normal saline and hydrocortisone, if severe.[1]

Spirillum. This is a genus of spiral bacteria, and is not a spirochete. *Spirillum* species can be confused with some vibrios, which are curved gram-negative bacteria. The confusion is made worse by the fact that many species of *Spirillum* do not grow on artificial media, but require mouse inoculation. These organisms are a rare cause of bacteremia or meningitis.[9] Typically, human disease occurs after a rat bite, and is called rat bite fever.

Rat bite fever is a syndrome that may also be caused by *Streptobacillus moniliformis*, a gram-negative bacillus. Fortunately, rat bite fever due to either of these organisms responds well to penicillin therapy.[17]

Frequency and Importance

New cases of acquired syphilis were reported at frequency of about 21,000 cases per year in the United States in 1978. After an alarming increase from 1959–1962, the frequency seems to have leveled off (Fig. 9-1). The late effects of syphilis frequently caused serious brain disease and heart disease before 1940, but now are rare. Syphilis remains important because it is a treatable disease and the serious (though rare) complications of syphilis can be prevented if the infection is diagnosed early and properly treated.

Clinical Patterns of Illness

The primary and secondary stages of syphilis are often grouped together as early or infectious syphilis.[5,15]

Primary Syphilis: the Chancre. This hard-edged, painless ulcer usually appears on the external genitalia about two or three weeks after exposure

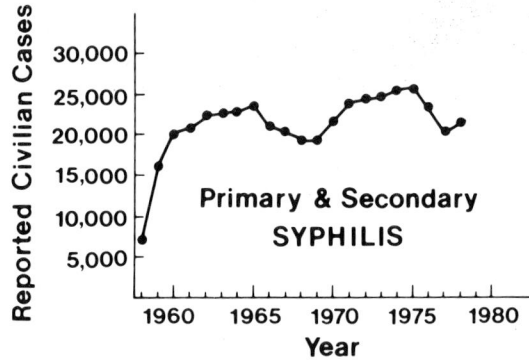

FIG. 9-1. *Primary and secondary syphilis: reported civilian cases in the United States, 1960–1977.*[3]

(Fig. 9-2). It is occasionally seen near the mouth or breast. The chancre lasts 3 to 12 weeks, and leaves a scar. The spirochetes spread to the regional lymph nodes, which appear enlarged about 3 to 4 days after the chancre appears. Lymphadenopathy is often marked in the inguinal area, but may be generalized. Syphilis is the most frequent cause of a chancre-like skin or mucosal ulcer in the genital area (Fig. 9-2). However, a similar lesion called soft chancre, or chanchroid, can be produced by *Haemophilus ducreyi* (Fig. 9-3).

Secondary Syphilis: the Rash. After the chancre appears, a generalized spirochetemia occurs. A generalized papular rash, and highly infectious mucosal ulcerations, appear about a month or two after the chancre, and last about 2 to 6 weeks. At the same time the rash is seen, secondary lesions are developing in many internal organs. Bone lesions, sudden deafness, or sudden blindness are rare complications of early syphilis. Secondary syphilis is now uncommon, but should still be considered by the physician in the differential diagnosis of skin or mucosal lesions.

Tertiary Syphilis: Ulcers, Gummas, and Others. Months to years after the chancre, tertiary lesions can develop. Tertiary syphilis is often latent. Latent syphilis is defined by the absence of clinical lesions and normal spinal fluid.[5] Early latent syphilis (less than 4 years duration) can relapse into the secondary stage and become infectious. Late latent syphilis (more than 4 years duration) is regarded as noncontagious.

Cerebrospinal fluid examination is mandatory in any patient with suspected symptomatic neurosyphilis, and desirable in other patients who have had syphilis more than a year to exclude asymptomatic neurosyphilis. The optimal treatment of syphilis of more than a year's duration

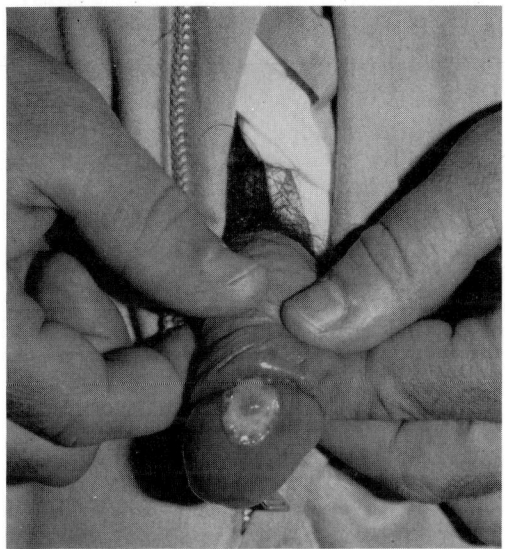

FIG. 9-2. *Syphilitic chancre. (Photo from Dr. Larry Lantis) Compare with Fig. 9-3.*

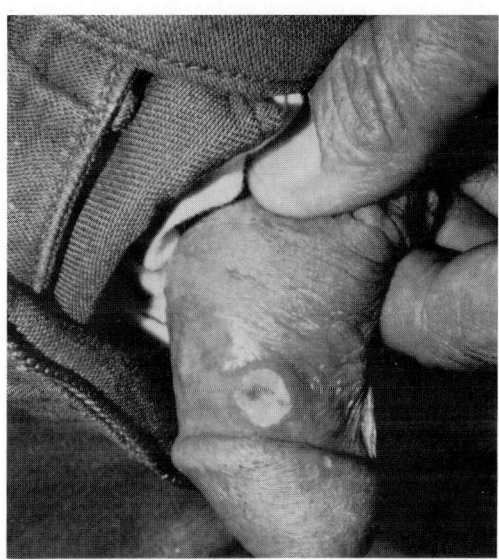

FIG. 9-3. *Chancroid, (soft chancre), due to Haemophilus ducreyi. This lesion may be impossible to distinguish by appearance from a syphilitic chancre. (Photo from Dr. Larry Lantis) Compare with Fig. 9-2.*

has not been clearly established, but higher doses and longer duration of therapy is recommended.

The lesions of tertiary syphilis include gummas, aneurysms, general paresis, and tabes dorsalis. Gummas are rubbery necrotic lesions which may be found in the skin and bones.[10] Aortic aneurysms are now rarely due to syphilis. General paresis is characterized by slow muscle movements and delusional fabrications. Tabes dorsalis is characterized by loss of position sense with resulting uncertain gait. All forms of tertiary syphilis are now very rare.

Transfusion Syphilis. No chancre occurs when syphilis is acquired by blood transfusion. The first sign of disease is usually the appearance of the secondary rash, which occurs about 1 to 4 months after the transfusion. Blood is routinely tested for syphilis before use, except in rare emergency circumstances. Storage of blood for 48 hours at refrigerator temperature (4 ° C) is sufficient to produce death of the organism.

Congenital Syphilis. During the primary or secondary stage of syphilis, spirochetes can spread from the pregnant woman to the fetus, via the blood, often resulting in abortion or stillbirth. The cogenital infection can result in no apparent disease, so a serologic test for syphilis should be done routinely during pregnancy, as described below.

Congenital syphilis can also produce clinically apparent disease varying in severity from delivery of an edematous, macerated infant to milder illness such as chronic nasal discharge, fissures about the mouth or anus, or pain and tenderness of an arm or leg due to infection of the bone.[4] Jaundice and enlargement of the liver and spleen may occur.

Congenital syphilis involving the CNS is indicated by white blood cells in the spinal fluid. When CNS involvement is present, the dose and duration of penicillin needed is greater than for the infant without neurologic infection,[11] as described later in this chapter.

Syphilis in Pregnancy. Congenital syphilis can be prevented by detection and treatment of syphilis during pregnancy. A syphilis serology is usually obtained at the pregnant woman's first prenatal visit in the first trimester, and also should be repeated later in pregnancy in the third trimester, if the woman has a high risk of exposure to syphilis again because of more than one sexual partner.

It was widely believed that syphilis was not transmitted to the fetus before the fifth month of pregnancy, because the Langhans layer of cells in the placenta was thought to be impermeable to spirochetes, but *T. pallidum* has recently been demonstrated by silver stain in fetal tissue during the first trimester.[8]

Laboratory Diagnosis

Serologic Tests for Syphilis (STS).

1. *VDRL.* The screening serologic test usually used is the VDRL, named after the Venereal Disease Research Laboratory at Atlanta, Georgia. The test is done on a glass slide, using undiluted patient's serum and standardized commercially available antigen called reagin. If positive, agglutination can be detected with a small hand lens. Twofold dilutions of a positive serum can be done to determine the patient's VDRL titer.[6,14,21]

A commerical antigen is now available which utilizes a plastic-coated card for the agglutination rather than a glass slide.[6] Charcoal particles are coated with reagin antigen and the particles agglutinate when antibody is present in this rapid plasma reagin test (RPR). The reagin screen test (RST) uses a lipid antigen, charcoal particles, and a blue dye for better visibiltiy.

The VDRL and similar floculation tests detect a nonspecific antibody called reagin, which appears about 4 weeks after exposure, and about a week after the chancre. If therapy is started before the appearance of the chancre, the patient's serum will usually not become reactive. Patients who receive treatment during the primary or secondary stage are likely to revert to seronegative within 6 to 24 months.[14]

2. *Confirmatory serologic tests.* Highly specific confirmatory tests are occasionally needed, particularly when a false positive VDRL is suspected. The reference laboratory usually chooses the most appropriate test or tests.

In the fluorescent treponemal antibody-absorbed (FTA-ABS) test, the antigen used is a strain of *T. pallidum*. The antibody is specific syphilitic antibody in the patient's serum. Fluorescent conjugated anti-human globulin is added to the slide, and if syphilitic antibody in the gamma globulin fraction has reacted with the treponema antigen, the anti-human globulin adheres and is detected by fluorescence. Recent infection or congenital infection can be detected by the IgM-FTA-ABS test, described below.

Nonpathogenic human treponemes, such as *Treponema microdentium*, can be a source of treponemal antibodies in normal individuals. These can be removed by using an extract of a treponema called Reiter protein to absorb out these antibodies from the serum to be tested.

The FTA-ABS test should not be used as a screening test, but rather reserved for confirmation of a positive VDRL test.

3. *IgM-specific antibody* (IgM-FTA-ABS). An infant born to a woman with a positive VDRL can have transplacentally transmitted IgG antibodies giving a positive VDRL test without infection being present in the newborn infant. Congenital syphilis can often be detected in this situation by testing the IgM fraction of the infant's serum for the presence of antibody using a properly standardized FTA-ABS test (see Fig. 15-6).[16]

Dark Field Examination. Spirochetes can often be seen in smears taken from a chancre or mucosal lesions by use of dark field microscopy. This procedure is not readily available, since experience is necessary for recognition of *Treponema pallidum* on dark field examination. Serologic procedures are more reliable and practical.

Culture. Of the spirochetes, only *Leptospira* species can be cultured on artificial media. Syphilis cannot yet be diagnosed by culture, although research in this area suggests that culture may soon be possible.

Biologic Characteristics of Clinical Interest

Survival of Treponema Pallidum. Outside the human host, survival of *T. pallidum* is limited to a few minutes.

Biologic False Positives (BFP). If the patient has nonsyphilitic antibodies that react with the VDRL antigen, this is called a biologic false positive VDRL. Acute diseases that can cause BFP reactions which disappear after a few months include primary atypical pneumonia, measles, infectious mononucleosis, tuberculosis, and malaria. Chronic diseases that produce persistent BFP reactions include lupus erythematosus, scleroderma, idiopathic thrombocytopenic purpura, and many carcinomas.[21,23] BFP reactions can occur in newborn infants as a result of transplacental transfer of antibody.[12]

At one time, there was no explanation for the observation that some individuals have persistently positive VDRL reactions which were thought to be biologic false positives. However, more sensitive specific tests have led to the conclusion that many unexplained chronic BFP reactors probably have had syphilis and should be managed as if they had syphilis.[23]

Treatment

Evidence for Efficacy of Chemotherapy. Many treatment regimens used for primary syphilis are judged to be effective on the basis of whether or not retreatment is needed. Some patients are reinfected, so large comparative groups are necessary. Continued rise in anti-syphilis antibody titer, or appearance of a rash of secondary syphilis after treatment of a chancre are criteria for retreatment.[10,19]

Antibiotics can also be studied for efficacy against syphilis by determining if individuals exposed to syphilis develop the disease. Thus, various antibiotic regimens for incubating syphilis can be tested during treatment of contacts. In one such study of patients exposed to syphilis within the preceding 30 days, 2.4 million units of benzathine penicillin G or 2.4 to 4.8 million units of procaine penicillin was 100 percent effective in aborting incubating syphilis, compared to placebo treatment, after which 30 percent developed syphilis.[18]

Current Recommendations for Treatment. Primary or secondary syphilis—or latent syphilis of less than a year's duration—can be treated with one injection of benzathine penicillin (2.4 million units), which provides effective treatment in a single visit.[11] If the patient is allergic to penicillin, oral tetracycline or erythromycin (500 mgm of either four times a day) are recommended for 15 days.[21] Spectinomycin, as used to treat gonorrhea is inadequate therapy for primary syphilis.[10] Tetracycline is avoided in pregnant women to avoid damage to fetal teeth or bones.

For syphilis of indeterminate or more than a years' duration, or cardiovascular or neurosyphilis, an injection of 2.4 million units of benzathine penicillin is recommended once a week for 3 successive weeks.[2] However, some clinicians prefer 2 to 4 million units of aqueous crystalline penicillin intravenously every 4 hours for 10 days for neurosyphilis, in order to achieve high brain and spinal fluid concentrations of penicillin.[2] The above daily oral doses of tetracycline or erythromycin are recommended for 30 days if the patient is allergic to penicillin, although the adequacy of this therapy has not been documented.

Newborn infants with congenital syphilis and an abnormal spinal fluid should be treated with aqueous crystalline penicillin (50,000 units per kg per day intravenously or intramuscularly in two divided doses for 10 days).[2,11] Although some have advocated benzathine penicillin for newborns with normal spinal fluid, it is more cautious to use the therapy recommended for newborns with abnormal spinal fluid, since benzathine penicillin does not achieve adequate spinal fluid levels, and neurologic involvement is difficult to exclude.[2]

The Jarisch–Herxheimer reaction may begin about 6 to 8 hours after starting treatment of syphilis. It is very frequent with treatment of secondary syphilis. This reaction consists of chills, fever, headache, rapid pulse, muscle aches, sore throat, and exacerbation of the local lesions. The reaction is believed to be due to the release of lipopolysaccharide (endotoxin) from the killed spirochetes, and endotoxin can be detected in the blood of such patients.[7] A study of this reaction in Ethiopian patients treated for borrelia indicated that intravenous saline, hydrocortisone, or acetaminophen may be helpful as described earlier in this chapter under Borrelia.

Prevention

Avoidance of exposure is effective but relatively impractical. Prophylactic antibiotics can be effective against incubating syphilis. Procaine penicillin 2.4 million units was 100 percent effective in aborting incubating syphilis in one study, whereas tetracycline 3 g was only about 50 percent effective.[18]

Contact finding, with therapy of contacts, can prevent spread by the contact. Premarital and prenatal serologic testing also result in some case findings.

Vaccines are in the experimental stage. Use of other treponemes might be worth studying as attenuated vaccines, as there appears to be considerable cross protection between syphilis, yaws, and pinta.[17]

LEPTOSPIRA SPECIES

Objectives

1. Describe how leptospira are distinguished from other treponema.
2. Describe two typical clinical presenting illness of leptospirosis.
3. Discuss the laboratory methods of confirmation of leptospira infection.

4. Discuss the animal reservoirs of leptospira and describe how humans usually come in common contact with the disease.
5. Discuss the treatment of leptospirosis.

Species

Leptospira are distinguished from other kinds of spirochetes by their corkscrew motion in dark field microscopy and by agglutination reactions. Most isolates are classified as serotypes of the species *Leptospira interrogans*. There are approximately 60 such serotypes, six of which account for most human infections. *L. interrogans* serotype *var.* (variety) *icterohemorrhagiae*, is, as the name implies, often associated with the pattern of illness of jaundice and bleeding (see below), and is usually related to a rodent source. Other species are *L. interrogans var. canicola*, usually acquired from dogs, and *L. interrogans var. pomona*, usually acquired from swine or cattle.

Frequency and Importance

About 100 cases of leptospirosis are reported annually in the United States, with about five deaths per year. Serious disease or death is rare. Many cases have a known exposure to rodents, dogs,[29] cattle, swine, or raccoons. Many patients have a history of wading or swimming in water possibly contaminated by the urine of such animals.

Clinical Patterns of Illness

Leptospirosis typically is a biphasic illness.[27,31] The first (septicemic) phase lasts about 4 to 7 days and is characterized by fever and leptospires in the blood. The second (immune) phase has variable manifestations, depending on the organ systems most prominently involved. A history of possible exposure to animal urine can often be elicited.

Leptospira species are a *rare* cause of:

Fever Without Localizing Signs. Chills, fever, headache, vomiting, abdominal pain, and myalgia are prominent in the early stages. Conjunctival effusion is often present, and in combination with the muscle pain and tenderness may suggest trichinosis, but there is no eosinophilia. Cough is not unusual. Lumbar puncture done in the first 5 days of fever usually reveals no pleocytosis. Vomiting or abdominal pain may be prominent. Occasionally a rash over the tibias is noted and this has been called pretibial fever.

Nonpurulent Meningitis. After 5 to 7 days of fever, stiff neck may appear. At this point, the spinal fluid usually shows about 50 to 500 white blood cells, with a normal glucose and protein. The CSF protein is sometimes slightly elevated and rarely the glucose is slightly depressed. Neutrophiles may predominate in the CSF at first, but later lymphocytosis is found. The CSF may be yellow if the patient is jaundiced.

Hepatitis and Nephritis (Weil's Disease). After a few days of fever, the patient may develop jaundice and an enlarged, tender liver, and is likely to be given a preliminary diagnosis of hepatitis. Hematuria, proteinuria, and pyuria in the second phase may suggest glomerulonephritis or pyelonephritis. The combination of nephritis, hepatitis and bleeding is called Weil's disease, and may occur after only a few days of fever in severe illnesses. Uremia, shock, or delirium may occur. Weil's disease was the first described clinical pattern of leptospirosis, but it is much less common than anicteric leptospirosis.

Kawasaki-like Illness. Kawasaki disease is the current name for a scarlet fever-like illness first reported in 1967 by Kawasaki. It was first called mucocutaneous lymph node syndrome. Typically a child develops fever, conjunctivitis, enlarged lymph nodes in the neck, an erythematious rash resembling scarlet fever that eventually desquamates. The illness rarely results in death related to coronary artery arteritis. Recently, a few children with illnesses resembling Kawasaki disease have been documented as having leptospirosis.[35] An erythematous or maculopapular rash which later desquamated was noted in some of these children. Gangrene of the hands or feet, cholecystitis and pancreatitis were also reported in this group.[35]

Leptospira species are an *extremely rare* cause of abdominal pain, pneumonia, uveitis, parotitis, orchitis, or arthritis.

Asymptomatic infection is rare, except in individuals with animal exposures, who may have leptospiral antibodies with no apparent history of disease.

The incubation period is usually 7 to 12 days, with a range of 2 to 20 days.

Laboratory Diagnosis

Serology. Agglutination tests are the most practical method.[30] Multiple leptospira antigens are commercially available, and provide screening that can be done in most laboratories. A positive agglutination test may appear as early as 4 days after the onset, but testing of both acute and convalescent sera is needed for accurate diagnosis.

There are two general types of serologic tests.[24] One type employs live antigens and is species specific, but is hazardous. The other is genus specific and uses a single antigen prepared from *L. biflexa*.[24] It is easy to perform and detects most cases regardless of the species.

Microscopic Exam of Tissue or Fluid. Leptospires can sometimes be found in dark field examination of urine or pleural fluid. Silver stains of tissue may be useful.

Isolation of the Organism. Animal inoculation (guinea pigs) was used frequently in the past, before culture on artificial solid media became available. Most routine hospital laboratories will not have culture media available, but the organism may survive a few days in anticoagulated blood, which can be sent to a reference laboratory.

Other Laboratory Studies. The VDRL is usually negative. The leucocyte count is usually not significantly elevated (usually less than 15,000), but there is often a shift to the left. The sedimentation rate is usually moderately elevated.

Biologic Characteristics of Clinical Interest

Animal Hosts.[32] Leptospirosis is common in animals, and most human cases have an animal exposure. Contact with water contaminated with animal (dog, rat, cattle) urine is frequently the source. Squirrel skinning or dressing has also been incriminated.[26] The portal of entry is usually minor cuts or abrasions of skin or mucous membranes.

Treatment

In experimental infections of animals, antibiotics are effective only when given during the prodromal stage, before the leptospires have reached the various tissues. In human infections, most observers believe that penicillin, streptomycin, or tetracycline do not appear to exert any appreciable effect on the course of the fever, jaundice, or renal involvement.[27] However, a controlled study of human infections in Malaya did indicate that tetracycline shortened the average duration of fever from 6 to 3 days, compared to ascorbic acid; penicillin or chloramphenicol did not shorten the course.[28,33]

Penicillin treatment has been recommended by some authorities, but may provoke a Herxheimer reaction as described in the sections on borrelia and syphilis.

Prevention

Avoid Exposure. Avoid swimming or wading in water contacted by animals. The use of rubber gloves in skinning or cleaning animals may prevent exposure to several animal diseases, including tularemia, and plague, as well as leptospirosis.[26]

Antibiotics. Prophylactic penicillin did not prevent disease in a laboratory worker exposed to leptospira.[25]

REFERENCES

Treponema Pallidum and Other Species

1. Butler, T., Jones, P.K., and Wallace, C.K.: *Borrelia recurrentis* infection: single dose antibiotic regimens and management of the Jarisch–Herxheimer reaction. J. Infect. Dis., 137:573–577, 1978.
2. Center for Disease Control: Syphilis—CDC recommended treatment schedules, 1976. J. Infect. Dis., 134:97–99, 1976.
3. ———Morbidity Mortality Weekly Rep., 26: Annual summary 1977, September, 1978. p. 2.
4. Curtis, A.C., and Philpott, O.S.: Prenatal syphilis. Med. Clin. North Am., 48:707–719, 1964.
5. Drusin, L.M.: The diagnosis and treatment of infectious and latent syphilis. Med. Clin. North Am. 56:1161–1174, 1972.

6. Dyckman, J.D., Wende, R.D., Gantenbein, D., and Williams, R.P.: Evaluation of reagin screen, a new serological test for syphilis. J. Clin. Microbiol., 4:145–150, 1976.

7. Gelfand, J.A., Elin, R.J., Berry, F.W., Jr., and Frank, M.M.: Endotoxemia associated with the Jarisch–Herxheimer reaction. New Engl. J. Med., 295:211–213, 1976.

8. Harter, C.A., and Benirschke, K.: Fetal syphilis in the first trimester. Am. J. Obstet. Gynecol., 124:765–711, 1976.

9. Kowal, J.: Spirillum fever. Report of a case and review of the literature. New Engl. J. Med., 264:123–128, 1961.

10. Lucas, J.B., and Price, E.V.: Cooperative evaluation of treatment for early syphilis: preliminary report with special reference to spectinomycin sulphate (actinospectacin). Br. J. Vener. Dis., 43:244–248, 1967.

11. McCracken, G.H., Jr., and Kaplan, J.M.: Penicillin treatment for congenital syphilis. A critical reappraisal. J.A.M.A., 228:855–859, 1974.

12. Miller, J.L., Meyer, P.G., Parrott, N.A., and Hill, J.H.: A study of the biologic falsely positive reactions for syphilis in children. J. Pediatr., 57:548–552, 1960.

13. Olansky, S.: Late benign syphilis (gumma). Med. Clin. North Am., 48:653–665, 1964.

14. ———Serodiagnosis of syphilis. Med. Clin. North Am., 56:1145–1150, 1972.

15. Pariser, H.: Infectious syphilis. Med. Clin. North Am., 48:625–636, 1964.

16. Rosen, E.U., and Richardson, N.J.: A reappraisal of the value of the IgM fluorescent treponemal antibody absorption test in the diagnosis of congenital syphilis. J. Pediatr., 87:38–42, 1975.

17. Roughgarden, J.W.: Antimicrobial therapy of ratbite fever. A review. Arch. Intern. Med., 116:39–54, 1965.

18. Schroeter, A.L., Turner, R.H., Lucas, J.B., and Brown, W.J.: Therapy for incubating syphilis. Effectiveness of gonorrhea treatment. J.A.M.A., 218:711–713, 1971.

19. Schroeter, A.L., Lucas, J.B., Price, E.V., and Falcone, V.H.: Treatment of early syphilis and reactivity of serologic tests. J.A.M.A., 221:471–476, 1972.

20. Southern, P.M., Jr., and Sanford, J.P.: Relapsing fever, A clinical and microbiological review. Medicine, 48:129–149, 1969.

21. Sparling, P.F.: Diagnosis and treatment of syphilis. New Engl. J. Med., 284:642–653, 1971.

22. Thatcher, R.W.: The search for a vaccine for syphilis. An epidemiological approach. Br. J. Bener. Dis., 45:10–12, 1969.

23. Tuffanelli, D.L., Wuepper, K.D., Bradford, L.L., and Wood, R.M.: Fluorescent treponemal antibody absorption tests. Studies of false-positive reactions to tests for syphilis. New Engl. J. Med., 276:258–262, 1967.

Leptospira Species

24. Andrew, E.D., and Marrocco, G.R.: Leptospirosis in New England. J.A.M.A., 238:2027–2028, 1977.

25. Broom, J.C., and Norris, T.St.M.: Failure of prophylactic oral penicillin to inhibit a human laboratory case of leptospirosis. Lancet, 1:721–722, 1957.

26. Diesch, S.L., Crawford, R.P., McCullouch, W.F., and Top, F.H.: Human leptospirosis acquired from squirrels. New Engl. J. Med., 276:838–842, 1967.

27. Edwards, G.A., and Domm, B.M.: Human leptospirosis. Medicine, 39:117–156, 1960.

28. Fairburn, A.C., and Semple, S.J.G.: Chloramphenicol and penicillin in the treatment of leptospirosis among British troops in Malaya. Lancet, 1:13–15, 1956.

29. Feigin, R.D., Lobes, L.A., Anderson, D., and Lickering, L.: Human leptospirosis from immunized dogs. Ann. Intern. Med., 79:777–785, 1973.

30. Galton, M.M.: Methods in the laboratory diagnosis of leptospirosis. Ann N.Y. Acad. Sci., 98:675–685, 1962.

31. Heath, C.W., Jr., Alexander, A.D., and Galton, M.M.: Leptospirosis in the United States. Analysis of 483 cases in man, 1949–1961. New Engl. J. Med. 273:857–864, 915–922, 1965.

32. Reinhard, K.R.: Leptospirosis. In Hull, T.G.: Diseases Transmitted from Animals to Man. Springfield, Thomas. 1963. pp. 624–651.

33. Russell, R.W.R.: Treatment of leptospirosis with oxytetracycline. Lancet, 2:1143–1145, 1958.

34. Turner, L.H.: Leptospirosis. Br., Med. J., 1:231–235, 1969.

35. Wong, M.L., et. al.: Leptospirosis: a childhood disease. J. Pediatr., 90:532–537, 1977.

10

Chlamydia, Rickettsia, Mycoplasma

CHLAMYDIA SPECIES

Objectives

1. Describe how chlamydia are different from viruses or rickettsia.
2. Describe the typical clinical pictures of psittacosis, lymphogranuloma venereum, neonatal inclusion conjunctivitis, and infantile chlamydial pneumonia.
3. Describe how the diagnosis of psittacosis, lymphogranuloma venereum or chlamydial conjunctivitis or pneumonia can be confirmed in the laboratory.
4. Discuss the treatment of psittacosis, lymphogranuloma venereum, inclusion conjunctivitis, and chlamydial pneumonia.
5. Discuss the prevention of psittacosis by control of birds.
6. Discuss the relationship of trachoma to inclusion conjunctivitis.
7. Describe the typical clinical picture of lymphogranuloma venereum.

Definitions

Chlamydia means "cloak" and refers to the mantle of carbohydrate that surrounds the inclusion body. They are now classified as bacteria. Chlamydia have a common group antigen with a unique growth cycle.[7,13] Other names that have been used for this group include bedsoniae, after Bedson, who isolated the agent of psittacosis; and miyagawanella, after Miyagawa, who described the elementary bodies of lymphogranuloma venereum.

Classification and Species

The order *Chlamydiales* has one genus, *Chlamydia*, and two species, *C. psittaci* and *C. trachomatis*. *C. trachomatis* is susceptible to sulfonamides and produces inclusion bodies that stain with iodine, neither of which occurs with *C. psittaci*.

Chlamydia trachomatis is the cause of trachoma, a blinding eye disease rare in the United States, and inclusion conjunctivitis, which is frequent in the United States, in newborns and adults, due to genital to eye transmission. TRIC agents, using TR from trachoma, and IC from inclusion conjunctivitis, was an early name for this microorganism. A few types can cause lymphogranuloma venereum, and many types can cause other genital disease, described in more detail below.

Psittacus is the Greek word for parrot, and psittacosis is sometimes called parrot fever. Psittacosis in humans is usually related to exposure to psittacine birds (parrots, parakeets). Ornithosis is the term used to describe the same disease in extrapsittacacine birds (pigeons, ducks, turkeys, lovebirds), and is sometimes used to describe the disease in humans.

Types

C. trachomatis has many serotypes.[13] Types A, B, Ba, and C are associated with trachoma. Types D through K are associated with inclusion conjunctivitis, nongonococcal urethritis, genital disease of both sexes, and infantile chlamydial pneumonia. Types L-1, L-2, and L-3 are associ-

149

ated with lymphogranuloma venereum, and these types are more invasive in mice and cell cultures.

C. psittaci also has many unidentified serotypes. Possibly there are two major groups in birds—one causing abortions and the other causing arthritis and conjunctivitis.[13]

Frequency and Importance

Chlamydia trachomatis is now believed to be the most frequent cause of venereal disease (particularly nongonococcal urethritis) in the United States. Nongonococcal urethritis is more common than gonorrheal urethritis in many venereal disease clinics. Proper diagnosis and antibiotic therapy is helpful. About 600 cases of lymphogranuloma venereum, an uncommon manifestation of *C. trachomatis*, are reported annually in the United States, although it is undoubtedly underreported.

Infection of the uterine cervix is probably very frequent, as high as 20 per cent of women in some clinics. About half of the babies born to infected women develop inclusion conjunctivitis. Infantile chlamydial pneumonia is frequent—now more frequent than whooping cough in the United States—but accurate statistics are not yet available.

Trachoma is the most frequent cause of blindness in the world, particularly in areas of poverty. About 1000 cases of trachoma occur annually in the United States, mostly in Native Americans in the Southwest.

Psittacosis may be common in birds, but it is rare in humans. About 50 to 100 cases were reported in the United States each year from 1970 through 1978. The disease is rarely fatal, but usually is improved by proper antibiotic therapy.

Clinical Patterns of Illness of C. Psittaci

C. psittaci is a *rare* cause of:

Pneumonia. Typically, there is a history of exposure to birds. Headache is frequent and typically very severe.[14,15] Fever is usually present at this point. Vomiting can occur. Other findings include chills, sweating, body aches, and arthralgias. This nonrespiratory onset is followed by cough and pulmonary involvement of variable severity. Typically, there is a patchy infiltrate and no leucocytosis. Occasionally, there is an acute lobar or segmental pneumonia.[14] Rarely, there may be multiple nodules on chest film. Fulminating pneumonia can occur and may be associated with myocarditis, pancreatitis, or renal failure.[5] Confusion and delerium are common in such severe cases.

Influenza-like illness with no radiologic evidence of pneumonia is a common pattern. The illness typically lasts 1 to 3 weeks. Anemia, eosinophilia, albuminuria, elevated transaminase and alkaline phosphatase are sometimes observed.[14] The white blood count is usually 10,000 to 20,000 with a mild neutrophilia, but occasionally is less than 5,000.[14,15]

Hepatitis. Jaundice with pneumonia should suggest psittacosis. Liver involvement may be the most prominent manifestation.[19]

Rashes. Erythema nodosum was described in one outbreak, but the patients did not have exposure to birds.[12]

Clinical Patterns of Illness of C. Trachomatis

C. trachomatis is a *common* cause of:

Bilateral Interstitial Pneumonia in Infants. Often the infant has had persistent conjunctivitis during the newborn period.[2] The illness is chronic, without fever, and begins with persistent cough and rapid breathing as early as the second or third week of life, and typically lasts several weeks. In some infants, the illness resembles pertussis, except that neither a whoop nor lymphocytosis is observed. *C. trachomatis* has been recovered from biopsy of lung tissue in infants with this disease. The illness is self limited, but may be improved by erythromycin or sulfisoxazole therapy.[3]

Nongonococcal Urethritis in Men. There is typically pain on urination. Unlike gonorrheal urethritis, where there often is a spontaneous purulent discharge from the penis, the discharge typically is not spontaneous, requiring penile stripping, and is mucoid. Postgonococcal urethritis is usually due to chlamydiae, which have persisted after penicillin has eradicated the gonococcus.

Acute "Idiopathic" Epididymitis. According to needle aspiration cultures for chlamydia, gonococci, and other bacteria, *C. trachomatis* is the major cause of acute "idiopathic" epididymitis, when gonococcal studies of the urethra are negative.[4]

Acute Salpingitis. In a Swedish study of acute salpingitis using diagnostic laparoscopy to confirm the diagnosis and obtain specimens for culture, C. trachomatis was frequently recovered from the cervix and infected fallopian tube.[11]

Cervicitis in Women. Cervical erosion or discharge can be caused by C. trachomatis. This is the source of chlamydial conjunctivitis in newborn infants. Chlamydial cervicitis can cause a slightly abnormal Papanicoloau smear, which reverts to normal after tetracycline therapy.[6] Genital chlamydial infection may be more frequent than gonorrhea in some populations.[9] Infants born to women with chlamydia in the cervix have a high risk of developing conjunctivitis or later pneumonia.

Conjunctivitis in the Newborn. This is called inclusion conjunctivitis, after the inclusions seen in cells scraped from the conjunctiva (Figs. 10-1 and 10-2). The disease develops about 3 to 14 days after exposure at birth.

Sporadic Conjunctivitis. Outside of the newborn age group, chlamydial sporadic conjunctivitis can be acquired by eye contact with vaginal or urethral secretions via fingers.

Trachoma. C. trachomatis can produce a very severe conjunctivitis called trachoma (rough tumor), named after the rough pebbly appearance of the conjunctivae, which may proceed to scarring. Trachoma occurs primarily in underdeveloped nations in conditions of very poor hygiene, but also occurs in Native Americans in the

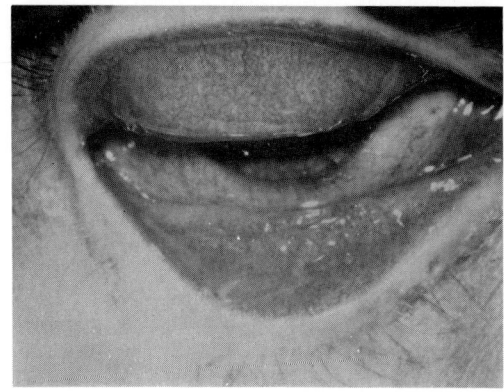

FIG. 10-1. *Inclusion conjunctivitis due to* Chlamydia trachomatis, *showing typical pebbly appearance on lower lid. (Photo from Dr. Fred Brightbill)*

southwestern United States. Trachoma appears to occur in individuals sensitized by previous contact, while inclusion conjunctivitis is the result of a single exposure.[8,17]

Sporadic inclusion conjunctivitis after the newborn period presumably represents a first exposure to the agent. A blind volunteer infected with this agent developed the clinical and laboratory pattern of inclusion conjunctivitis on the first experimental exposure, and the clinical and laboratory pattern of trachoma when exposed a year later.[8]

Inguinal Adenitis.[10] Classic lymphogranuloma venereum is characterized by a genital vesicle or ulcer, followed by inguinal adenitis, which suppurates to form a fluctuant abscess or bubo. Fistulae or strictures can occur in the rectal, vaginal or urethral areas.

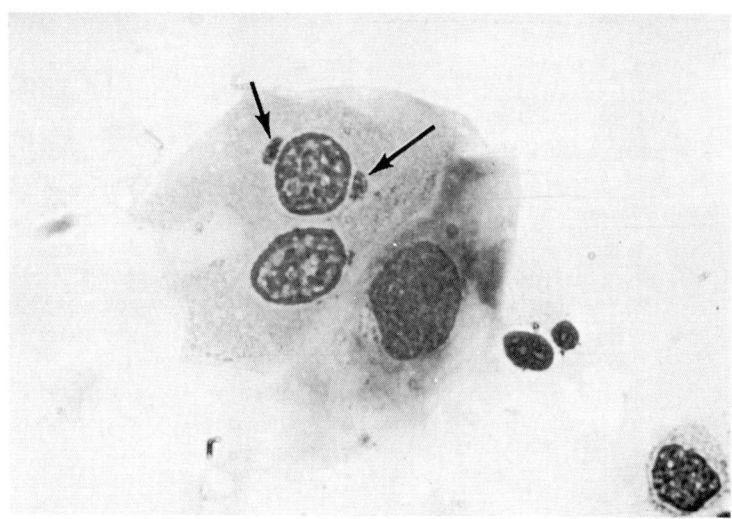

FIG. 10-2. *Two inclusions (arrows) on each side of the nucleus of a cell obtained from conjunctival scraping. (Photo from Dr. Fred Brightbill)*

Other Diseases. *C. trachomatis* has been suspected as a possible cause of postpartum endometritis, urethritis in the female, proctitis, infection of Bartholin's gland, and otitis media.

Laboratory Approach

Serum Antibodies. Serologic study of paired sera is the best method of laboratory detection of chlamydial infections, especially psittacosis. The complement fixation test uses an antigen that is shared by all chlamydia, so cross reactions occur. *C. trachomatis* antibodies can be detected by an indirect immunofluorescence test (Fig. 15-6) or by the ELISA method (Fig. 13-7). The first serum often is suggestive of the diagnosis, since the patient may have been sick for a week or more before seeking medical care, and a low titer antibody may already be present. Infants with chlamydial interstitial pneumonitis often have very high titer of chlamydia antibodies and IgM antibody.

Culture. *C. trachomatis* can be recovered from swabs from the urethra or conjunctivae by inoculation of the specimen into McCoy cell cultures, which have been treated with cycloheximide to stop cell proliferation. The inclusion cells can be seen in the cells after staining.[18]

C. psittaci can be cultured by inoculation of McCoy cell cultures, which are more sensitive than mouse inoculations.

Smear. *Chlamydia trachomatis* inclusion bodies are best detected by a specially stained smear of a conjunctival scraping, done with a sterile spatula, rather than a cotton swab.

Special stains of psittacosis-infected tissues may demonstrate dense particles which are typical, and distinguishable from the appearance of rickettsiae. Smear of pus from an inguinal bubo may reveal the elementary bodies of the lymphogranuloma venereum agent.

Skin Tests. An intradermal skin test, called the Frei test, becomes positive 1 to 6 weeks after the onset of the adenitis in lymphogranuloma venereum. The antigen used is usually made from pus of another patient's bubo. Lygranum is a skin testing antigen made from a psittacosis agent, and can occasionally be helpful in the retrospective diagnosis of psittacosis.

Biologic Characteristics of Clinical Interest

Animal Infections. Other chlamydia species can cause pneumonia in mice and other mammals.[16] *C. psittaci* from psittacine birds may be more virulent for man than the strain from other birds.[16]

Contagiousness. Psittacosis very rarely can be spread from human to human, apparently by coughing out an aerosol of infected droplets.[1] Lymphogranuloma venereum is spread by sexual contact.

Growth Cycle. The pattern of replication of chlamydiae is unique, and involves binary fission. The organism must grow in cells. Two forms exist in the cycle: a small, filterable form, and a large cell which is noninfectious, has no cell wall, and divides by binary fission.[7,13]

Susceptibility to Antibiotics. A tetracycline is usually effective in eradication of the organism, and usually results in a prompt clinical response. Tetracyclines are often given to birds in pet shops or on turkey farms, and reduce mortality, but may not eliminate the carrier state.

Cell cultures can be used to test chlamydial cultures for susceptibility to various antibiotics.

Latency. Both *C. trachomatis* in humans and *C. psittaci* in birds tend to produce chronic, latent subclinical infections.

Treatment

Chemotherapy. Tetracycline is generally considered the treatment of choice for psittacosis. A tetracycline such as doxycycline is also effective in chlamydial urethritis, and should also be given to the sexual partner. Sulfonamides are effective for lymphogranuloma venereum and do not mask syphilis. A sulfonamide or doxycycline is effective therapy against trachoma. Spectinomycin (sometimes used to treat gonorrhea) is ineffective against genital *C. trachomatis*. Erythromycin may improve the course of infantile chlamydial pneumonia.[3]

Supportive Therapy. Oxygen and other respiratory supportive therapy may be indicated for psittacosis. Aspiration of buboes may be indicated for lymphogranuloma venereum.

Prevention

Chlamydia psittaci can be eliminated from pet birds by use of tetracycline. Treatment of cervicitis due to *C. trachomatis* in pregnant women is under study to try to prevent inclusion conjunctivitis of the newborn. Discharges containing *C. trachomatis* are infectious, and isolation techniques are required.

No vaccines are available.

RICKETTSIA

Objectives

1. Describe the typical clinical picture of Rocky Mountain spotted fever.
2. Describe the serologic methods readily available in most hospital laboratories to detect rickettsial infection.
3. Discuss the reservoirs and treatment for Rocky Mountain spotted fever.
4. Discuss the approximate frequency of other rickettsial diseases in the United States, particularly murine typhus, louse-borne (epidemic) typhus, and Q fever.

Definitions

Rickettsiae are intracellular parasites which are intermediate in size between viruses and bacteria. *Rickettsiae* can be seen as coccobacilli with a light microscope, after special staining.

Species

Rickettsia rickettsi is the agent that causes Rocky Mountain spotted fever (RMSF). It was named for Howard Taylor Ricketts, who did fundamental experiments in the transmission of the disease from humans to animals and between animals and ticks. He died of typhus after being bitten by an infected louse in his laboratory. Rocky Mountain spotted fever occurs much more frequently in Maryland, Virginia and the Carolinas than in the Rocky Mountain states (Fig. 10-3). RMSF is also called tick typhus to distinguish it from louse-borne typhus and flea-borne typhus, which are other rickettsial spotted fevers.

"Typhus" is derived from a Greek work for "hazy" or "smoky", which refers to the confusion or stupor sometimes seen in typhus and

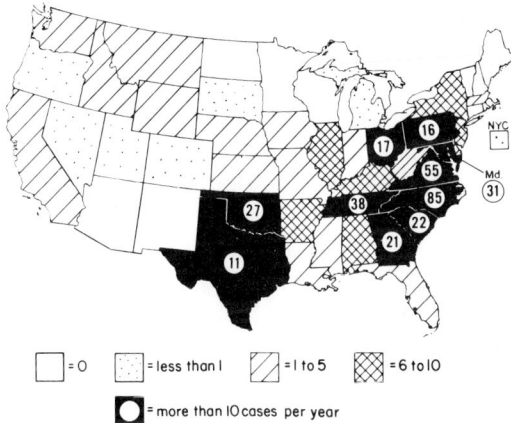

FIG. 10-3. *Rocky Mountain spotted fever: average annual reported cases, 1968–1972.*[22]

typhoid fever. The name Q fever comes from Query fever, as this rickettsial disease was originally observed as a fever of unknown cause. This rickettsia is named *Coxiella burnetti*, after Cox and Burnet who discovered the organism.

Frequency and Importance

RMSF is the most frequent rickettsial disease in the United States and has been increasing in frequency since 1960.[22] In 1977, over 1000 cases were reported, of which about 10 percent were fatal (Fig. 10-4). More than half of the reported cases occur in the South Atlantic states, but cases have been reported from most sections of the United States (Fig. 10-3). It is the only important rickettsial disease in the United States. Early diagnosis and proper treatment may be lifesaving.

Murine typhus, (flea-borne typhus due to *R. mooseri*) is reported at an average of about 30 cases per year in the United States, with no deaths. In recent years, the majority of cases have occurred in a few counties in southern Texas. The disease was often suspected and laboratory tests done because the cases occurred in a known endemic area. The typical clinical pattern was an influenza-like illness, with fever, chills, headache, and myalgia, with only one-fourth of the patients having a petechial rash.[33]

Q fever (caused by the rickettsia, *Coxiella burnetti*) is an optionally reported rickettsial disease, with 25 cases reported from California and Colorado in 1970. *Rickettsialpox* is presumed to be a rare disease, but is mild and probably often unrecognized, so its true frequency is unknown. Scrub typhus (caused by *R. tsutsugamushi*) is

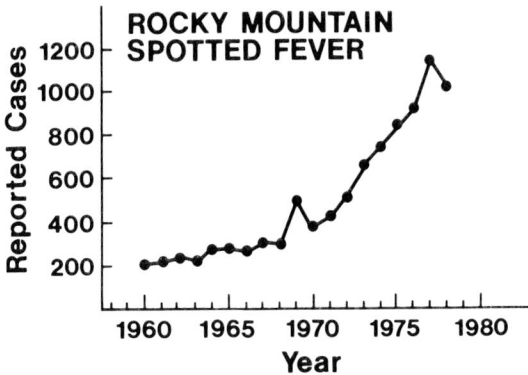

FIG. 10-4. Rocky Mountain spotted fever. Annual reported cases, 1960–1978.[23]

occasionally seen in individuals returning to the United States from Southeast Asia.[37]

Classical typhus (endemic or epidemic typhus) due to R. typhi is not found in the United States except as a laboratory acquired infection.

Clinical Patterns of Illness

In endemic areas in the United States, R. rickettsia is an *occasional* cause of:

Fever and Petechial Rash. Rocky mountain spotted fever usually has a typical clinical pattern of exposure to wood ticks or dog ticks, with headache, chills, and high fever.[21,28,30,31] At this point, the diagnosis is very difficult if the exposure is not known and the rash has not yet appeared. Muscle aches, vomiting, abdominal pain, and diarrhea are not unusual.[31] A macular and petechial rash occurs on about the fourth day of illness, and is first noted on the extremities. Edema of the eyes and extremities, and delerium are common. Shock and renal failure may occur in severe cases.

Atypical measles (Chap. 13) is often misdiagnosed as RMSF.

Thrombocytopenia and low serum sodium are common.[31] Papilledema, exudates, and tortuous blood vessels may be observed in the retina. In fatal cases, myocraditis is common.

In experimental infections in beagle dogs, the severity of illness was proportional to the number of rickettsia inoculated.

In a case transmitted by a blood transfusion, fever occurred in the recipient 3 days after the transfusion, chloramphenicol was begun a day later, and the patient recovered completely.[39] The donor developed fever 3 days after donating blood and died on the seventh day of illness. The rickettsiae had survived for 9 days before transfusion into the recipient.

The incubation period from tick bite to fever is 2 to 5 days in severe disease and 3 to 12 days in mild disease.

Rare Rickettsial Diseases

Rickettsialpox is also a fever–rash disease.[20] It is transmitted from mice to humans by a mite. The diagnosis may be suspected when a 1 cm black eschar is noted on an extremity at the site of the bite of the mite. The rash usually includes some vesicles and crusts, and so is sometimes mistaken for chickenpox.

Scrub typhus is also a fever–rash disease associated with a black eschar at the site of the bite of the chigger, but the disease is not seen in the United States, except in individuals recently arrived from Asia or Australia.

Q fever is a rare cause of atypical pneumonia. The typical clinical illness in humans usually includes an inhalation exposure to cattle, sheep, goats, exotic pets such as pythons, or laboratory exposure to the organism.[27] Infected milk does not seem to be a source of the disease.[29] Fever and chills and headache are prominent early, but the cough does not appear until several days after the onset. Physical signs of pneumonia are minimal, but segmental or lobar consolidation may be seen on radiograph. Occasionally, fever, and liver enlargement with hepatitis and slight jaundice are the only manifestations of Q fever. It is also a cause of endocarditis with negative cultures.

Laboratory Diagnosis

Serum Antibodies. Serology is the best way to confirm a rickettsial infection. Most hospital laboratories can do a Weil–Felix test, a screening test which will detect antibodies to certain proteus antigens, which are also found in RMSF, murine typhus, and epidemic typhus. The titer may be high enough to be suggestive as early as 5 or 6 days after the onset. More specific serologic tests for all rickettsiae can be done by reference laboratories.[35]

Culture. Isolation of rickettsia is difficult, and carries a significant risk of laboratory infection.

Histology. Histologic diagnosis can be done by fluorescent antibody staining of the organism in skin obtained at biopsy.[38]

Biologic Characteristics of Clinical Interest

Endothelial Lesions. The endothelium of small blood vessels is the site of multiplication of most rickettsia, and a diffuse vasculitis is produced.[26] This accounts for the rash, and the renal and cerebral manifestations.

The kallikrein–kinin system is activated, also contributing to the disseminated intravascular coagulation, the hemorrhagic rash, edema, and shock.[40]

Weil-Felix Reaction. This is the agglutination of proteus antigens by antibodies produced by patients infected with some rickettsial species. Rickettsia and proteus apparently have some identical antigens, and this coincidental relationship is the basis for a simple diagnostic serologic test.

Reservoirs. Transovarian transmission of the rickettsia of RMSF can occur in the tick, and allows the perpetuation of the organism. However, the human body louse is eventually killed by the rickettsia of epidemic typhus, and humans are the only reservoir.

Recrudescence. Epidemic typhus can occur as a mild fever–rash illness in an individual infected many years previously, and this is called Brill–Zinsser disease.[32] A person with recrudescent typhus could be contagious to others if he or she is infested with lice.

Treatment

Antibiotic therapy of RMSF with either a tetracycline or chloramphenicol is effective, and usually modifies the course of the illness dramatically, if begun soon enough. Tetracycline and chloramphenicol appear to be equally effective in RMSF experimental infections in monkeys, but erythromycin is of less value.[36] Sulfonamides make RMSF worse.[30] Antibiotic therapy should be begun on the basis of clinical findings, without delaying for a confirmed laboratory diagnosis.

Tetracycline is more effective than chloramphenicol for scrub typhus.[37] Q fever endocarditis has been successfully treated with trimethoprim-sulfamethoxazole.[25]

Corticosteroids appear to have a beneficial effect in seriously ill, "toxic" patients with RMSF, presumably by protective effects on the endothelium, or perhaps by interfering with antigen–antibody reactions.

Prevention

Vaccines using killed organisms are used to protect against epidemic typhus and RMSF. An older RMSF vaccine failed to protect one laboratory worker, who was apparently exposed by aerosol.[24] However, multiple boosters appear to protect exposed laboratory workers,[34] and a new vaccine developed in the late 1970s appears to be more potent and effective than older vaccines.

Insecticides, particularly DDT, have been effective in preventing the spread of epidemic typhus. However, DDT-resistant lice have been observed. Tick repellents, long clothing, and inspection of the body for ticks are recommended.

Prophylaxis with tetracycline delayed development of the disease in monkeys exposed by aerosol, but did not prevent death.[36]

MYCOPLASMA PNEUMONIAE AND OTHER MYCOPLASMAS

Objectives

1. Define mycoplasmas and distinguish them from L forms, spheroplasts, protoplasts, and pleuropneumonia-like organisms (PPLO).
2. Describe some differences between *Mycoplasma pneumoniae* and other mycoplasmas, as defined in the laboratory.
3. List the mycoplasmas found in the urogenital tract and the diseases with which they are thought to be associated.
4. Describe the typical clinical findings in pneumonia produced by *Mycoplasma pneumoniae*.
5. Describe how an early presumptive laboratory diagnosis of mycoplasmal pneumonia might be made and how the laboratory diagnosis of infection with *Mycoplasma pneumoniae* can be confirmed.
6. Discuss the evidence that antibiotics are effective in pneumonia due to *Mycoplasma pneumoniae*.

Definitions

Mycoplasmas. The word mycoplasma refers to the plasticity of the morphology of these organisms without cell walls.[56] Most mycoplasmas

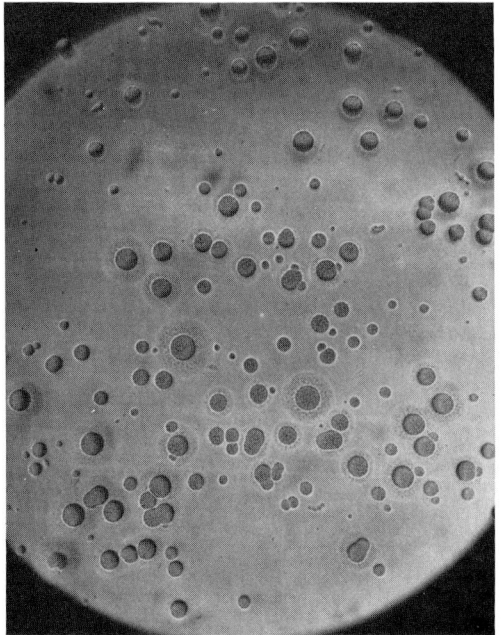

FIG. 10-5. Mycoplasma salivarium *colonies, showing "fried egg" appearance of some colonies due to surrounding surface growth. (U.S. Navy photograph)*

are slightly smaller than bacteria, and some are smaller than large viruses, but do not require cells for growth and can grow on solid media. Mycoplasmas were originally called pleuropneumonia-like organisms (PPLO). *Mycoplasma mycoides,* the cause of bovine pleuropneumonia, was the first PPLO isolated. Mycoplasma colonies on agar plates often resemble fried eggs, and are tiny granules under the microscope (Figure 10-5). Mycoplasma should be distinguished from L forms, spheroplasts, and protoplasts, all of which are bacteria without cell walls.

L Forms. Named for the Lister Institute in England, where they were first studied, L forms are bacteria that have lost their cell wall. L forms produce colonies on agar which resemble fried eggs and which resemble mycoplasma colonies. Unstable L forms, also called L variants, can revert to the bacterial form. Stable L forms cannot revert, but can be identified by knowing their origin or composition.

Spheroplasts. These are bacteria that have a defective cell wall and so assume an irregular spherical or globular shape. Spheroplast is best defined as a bacteria in which some of the cell wall components might still be present.

Protoplasts. These are bacteria which have had the cell wall permanently removed, as by penicillin therapy or enzymatic digestion, and cannot revert to the original bacteria.

Human Mycoplasma Species

Mycoplasma Pneumoniae. This organism is a frequent cause of the atypical pneumonia syndrome. In the past, it was called Eaton agent, named after the physician who recovered this filterable agent from patients with atypical pneumonia and showed that it produced pneumonia in rats.

Pharyngeal Mycoplasmas. The following mycoplasma can be regarded as normal flora of the mouth or pharynx: *M. salivarium, M. orale* type 1 (*M. pharyngis*), *M. orale* type 2, *M. hominis* type 1, and *M. hominis* type 2.

M. hominis type 1 can produce exudative pharyngitis in experimental infections in volunteers, but apparently does not cause naturally occurring pharyngitis.[59,62]

Genital Mycoplasmas. The following mycoplasma are often found in the urogenital tract: *M. hominis, M. fermentans,* and *T* mycoplasmas (T strains), which are now called *Ureaplasma ureolyticum*.[60] T strains refer to the tiny colonies on agar. Such mycoplasmas have been recovered from patients with urethritis, cervicitis, or vaginitis, and self-inoculation experiments have been done to produce disease. Recovery of such mycoplasmas from Bartholin's gland abscess, abscesses in the newborn,[67] or from blood of women with septic abortion or endometritis[77] indicates these diseases can occasionally be produced by these mycoplasma.

Association of *M. hominis* and *Ureaplasma ureolyticum* with infertility, spontaneous abortion, stillbirth and prematurity has been suggested, but remains unproved.[51,77] However, recovery of these microorganisms from the urethra or cervix increases in direct relation to the number of sexual partners. The most valuable data about genital disease therefore come from studies of people with very few sexual partners.

Ureaplasmas produce a urease that aids in their detection. There are several serotypes, some of which may be more pathogenic to humans.

Frequency and Importance

M. pneumoniae is one of the most frequent causes of pneumonia in school age children and young adults, but this has only been recognized since the 1960s.[41,47,59]

Genital mycoplasma are suspected to be a frequent and important cause of maternal and perinatal disease.

Clinical Patterns of Illness

M. pneumoniae is a *frequent* cause of:

Atypical Pneumonia. Atypical features of pneumonia are defined as the opposite of the typical findings of pneumococcal lobar pneumonia.[47] Atypical manifestations include gradual onset, prominent extrapulmonary signs and symptoms, radiologic findings not accurately predicted by physical examination, absence of leukocytosis, and failure to improve after penicillin therapy.

Atypical pneumonia due to *Mycoplasma pneumoniae* (mycoplasmal pneumonia) typically occurs in school age children and young adults, and is associated with cold agglutinins, a relatively long incubation period (2 to 3 weeks), and slow spread within families.

Lobar pneumonia[41] or pneumonia with effusion or residual pleural thickening can occur.[63]

Acute pneumonia with a severe oxygen diffusion block can also occur, responding to oxygen and corticosteroids.[65]

Bullous myringitis, with painful red tympanic membranes and blisters or bullae on the ear drum, are observed in about five to 10 percent of patients with mycoplasmal pneumonia.[48,66] Bullous myringitis also occurs in some of the individuals experimentally infected with *M. pneumoniae.*[70]

M. pneumoniae is an *occasional* cause of:

Influenza-like Illness. Cough, headache, fever, and myalgia can be due to *M. pneumoniae.*[45] Pharyngitis, which is often exudative, occurs in experimental infections,[70] and in naturally occurring infections of adults.[48]

Headache and fever may be the only signs of illness.[70] In experimental infections, headache, malaise and mild fever may last for only one or two days.

Rashes. In children, associated rashes are usually maculopapular, but can be urticarial or vesicular.[43] Erythema multiforme exudativum (Stevens–Johnson syndrome) is sometimes associated with *M. pneumoniae* infection.[58]

M. pneumoniae is a *rare* cause of toxic psychosis,[42] hemolytic anemia,[76] nonpurulent meningitis,[64] encephalitis,[64] polyradiculitis (Guillain–Barre syndrome),[64] and polyarthritis.[55]

Mycoplasma hominis is an *occasional* cause of:

Postpartum Fever.[77] Fever after giving birth is sometimes due to infection of the uterus (endometritis) with blood cultures positive for *Mycoplasma hominis* or *Ureaplasma urealyticum* (mycoplasmemia).[77]

Mycoplasmemia after urethral instrumentation in males has recently been recognized.[69]

Acute Pyelonephritis. *M. hominis* has been recovered more frequently from the upper urinary tract in patients with acute pyelonephritis, compared to a control group.[75]

Cervicitis and Vaginitis. Both *U. urealyticum* and *M. hominis* appear to cause cervicitis and vaginitis,[60] although this is difficult to document since these microorganisms can be found frequently in the female genital tract.

Pharyngitis. Exudative pharyngitis can be produced by experimental infection of volunteers with *M. hominis,* but rarely, if ever, occurs naturally.[62]

Ureaplasma urealyticum is a *possible* cause of:

Reproductive Failure.[52] Infertility,[52] abortion,[9] prematurity, and fetal death have all been attributed to maternal infection with *Ureaplasma urealyticum.*[57] A tetracycline, such as doxycycline, can eradicate the ureaplasma carrier state, but has not been shown to result in increased fertility,[52] although some studies support an etiologic role in infertility for this agent.[57]

Non-gonococcal Urethritis. Although *Chlamydia trachomatis* appears to be the usual cause, *Ureaplasma urealyticum* appears to be a possible cause, because inoculation of ureaplasmas into the male urethra results in urethritis, which usually responds to tetracycline.[73]

Laboratory Diagnosis

Cold Agglutinins. This is an IgM antibody which appears early in most severe illnesses due to *M. pneumoniae*, but also appears in many viral pneumonias, especially in adenoviral pneumonia. It also is found in children with lower respiratory disease due to RS virus or parainfluenza virus.[72]

Serum Antibodies. Serologic diagnosis, using paired sera, is the most practical method of laboratory diagnosis of *M. pneumoniae* infection. Many methods of serodiagnosis are possible, but the complement fixation test is usually the most available.[49] If a serum is obtained when the patient first sees a physician, it often will have a low titer of *M. pneumoniae* antibodies, which is specific evidence of early or recent infection.

Countercurrent immunoelectrophoresis (CIE) can demonstrate serum precipitating antibodies early in *M. pneumoniae* infection (Fig. 4-2).[61]

Culture. This is sometimes practical, but is not as likely to be available. Some species of mycoplasmas have colonies which have a typical "fried egg" appearance (Fig. 10-5).

Various species of mycoplasmas are defined by neutralization tests which are done by a method similar to antibiotic susceptibility testing with paper discs. The organism is spread over an agar plate and discs impregnated with species-specific antiserum are placed over it. The disc that inhibits the growth around it identifies the species (Fig. 10-6). *M. pneumoniae* hemolyzes sheep erythrocytes, a characteristic which distinguishes it from all other mycoplasma species.

Smear. Mycoplasmas in tissues and upper respiratory secretions can be detected and defined by species in experimental animal infections, using a direct ELISA method.[54] This method can be used for detecting Herpes simplex virus in vesicles (Fig. 11-3). The method appears potentially useful for diagnosis in humans.

Biologic Characteristics of Clinical Interest

Animal Diseases. Mycoplasma-like organisms (PPLO) cause naturally occurring pneumonias in cattle, rodents, and swine.

Long Incubation Period. The incubation period was about 2 to 3 weeks in family exposures,[49] and about 13 days in a college fraternity outbreak.[48]

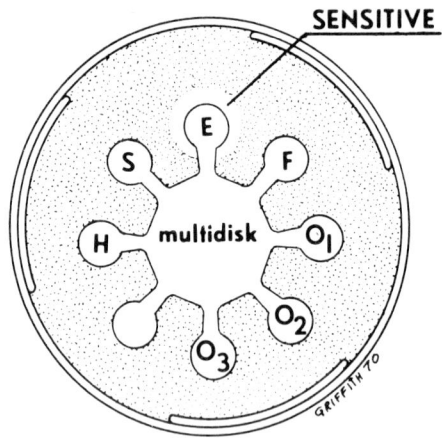

FIG. 10-6. *Identification of* Mycoplasma *species by inhibition by species-specific antiserum in paper disc. Inhibition of growth of the isolate around the* **E** *disc identifies the isolate as* M. pneumoniae. *(U.S. Navy photograph)*

This suggests that there is a delay of several days after natural exposure before colonization or before infection occurs.

Cell Culture Contaminants. Mycoplasmas are important and frequent contaminants of cell cultures, and produce cytopathic effects occasionally ascribed to a "new virus".[53]

Tumors. Mycoplasmas have been cultured from various human malignancies, but are generally regarded as having been cell culture contaminants, rather than a cause of tumors.[53]

Cold Agglutinins. *Mycoplasma pneumoniae* infection is often associated with the production of cold agglutinins in the infected patient. Cold agglutinins are not specific for *M. pneumoniae* infection, but suggest a nonbacterial cause of pneumonia. *M. pneumoniae* antigens apparently cross react with the I antigen of human erythrocytes, so that antibodies to the *Mycoplasma* produce agglutination of the patient's red blood cells when a tube of the citrated whole blood specimen is put into ice.[46]

Autoimmune Phenomena. Atypical pneumonia is frequently associated with antilung antibodies, eosinophilia, false positive syphilis reactions, urticaria, or erythema multiforme; suggesting induced autoimmune phenomena.[74]

Second Attacks. Naturally acquired immunity to *M. pneumoniae* appears to be of limited duration and second attacks have been clearly documented.[50]

Treatment

Mycoplasmas are resistant to penicillin and to polymyxin B, which act on cell walls, but are susceptible to tetracyclines and erythromycin. If tetracycline or erythromycin therapy is begun early enough in the illness, the patient's temperature and clinical findings improve faster than a control group.[68,70] However, in most situations, antibiotic therapy seems to be of little value.[71] The organism may persist in the pharynx in spite of antibiotic therapy or serum antibodies.[67] Sulfonamides, which are usually effective in chlamydial urethritis, are not effective against *M. hominis* or *U. urealyticum*.

Prevention

Killed Vaccine. Volunteers who produced antibodies after receiving killed vaccine were protected from illness when challenged. However, volunteers who did not produce antibodies after receiving killed vaccine developed more severe illness than in control volunteers.[70]

REFERENCES

Chlamydia

1. Barrett, P.K.M., and Greenberg, M. J.: Outbreak of ornithosis. Br. Med. J., 2:206–207, 1966.
2. Beem, M.O., and Saxon., E.M.: Respiratory tract colonization and a distinctive pneumonia syndrome in infants infected with *Chlamydia trachomatis*. New Engl J. Med., 296:306–310, 1977.
3. Beem, M.O., Saxon, E., and Tipple, M.A.: Treatment of chlamydial pneumonia in infancy. Pediatrics, 63:198–203, 1979.
4. Berger, R.E., et al.: *Chlamydia trachomatis* as a cause of acute "idiopathic" epididymitis. New Eng. J. Med., 293:301–304, 1979.
5. Byrom, N.P., Walls, J., and Mair, H.J.: Fulminant psittacosis. Lancet, 1:353–356, 1979.
6. Carr, M.C., Hanna, L., and Jawetz, E.: Chlamydiae, cervicitis, and abnormal Papanicolaou smears. Obstet. Gynecol., 53:27–30, 1979.
7. Grayston, J.T., and Wang, S-p.: New knowledge of chlamydiae and the diseases they cause. J. Infect. Dis., 132:87–105, 1975.
8. Jones, B.R., and Collier, L.H.: Inoculation of man with inclusion blennorrhea virus. Ann. N. Y. Acad. Sci., 98:212–228, 1962.
9. McCormack, W.M., et al.: Fifteen month follow-up study of women with *Chlamydia trachomatis*. New Engl J. Med., 300:123–125, 1979.
10. McLelland, B.A., and Anderson, P.C.: Lymphogranuloma venereum. Outbreak in a university community. J.A.M.A. 235:56–57, 1976.
11. Mårdh, P.-A., Ripa, T., Svenson, L., and Weström, L.: *Chlamydia trachomatis* infection in patients with acute salpingitis. New Engl. J. Med., 296:1377–1379, 1977.
12. Sarner, M., and Wilson, R.J.: Erythema nodosum and psittacosis. Report of five cases. Br. Med. J., 2:1469–1470, 1965.
13. Schacter, J.: Chlamydia infections. New Engl. J. Med., 298:428–435, 490–495, 540–549, 1978.
14. Schaffner, W., Drutz., D.J., Duncan, G.W., and Koenig, M.G.: The clinical spectrum of endemic psittacosis. Arch. Intern. Med., 119:433–443, 1967.
15. Seibert, R.H., Jordan, W.S., and Dingle, J.H.: Clinical variations in the diagnosis of psittacosis. New Engl. J. Med., 254:925–930, 1956.
16. Shaughnessy, H.J.: Psittacosis (Ornithosis), In Hull, T.G. (ed.): Diseases Transmitted from Animals to Man. Springfield, Thomas 1963, p. 350–373.
17. Tarizzo, M.L., Nataf, R., and Nabli, B.: Experimental inoculation of thirteen volunteers with agent isolated from inclusion conjunctivitis. Am. J. Ophthalmol., 63:1120–1128, 1967.
18. Thomas, B.J., Evans, R.T., Hutchinson, G.R., and Taylor-Robinson, D.: Early detection of chlamydial inclusions combining the use of cyclohexamide-treated McCoy cells and immunofluorescence staining. J. Clin. Microbiol., 6:285–292, 1977.
19. Yow, E.M.: The pathology of psittacosis. A report of two cases with hepatitis. Am. J. Med., 27:739–749, 1959.

Rickettsia

20. Barker, L.P.: Rickettsialpox: clinical and laboratory study of twelve hospitalized cases. J.A.M.A., 141:1119–1123, 1949.
21. Bradford, W.D., and Hawkins, H.K.: Rocky Mountain spotted fever in childhood. Am. J. Dis. Child., 131:1228–1232, 1977.
22. Center for Disease Control: Morb. Mort. Weekly Rep. Annual Supplement, Summary 1970. 19:1–60, 1971.
23. ———: Morb. Mort. Weekly Rep. Volumes 17–26: Annual supplements. Summaries 1968–1978.

24. Calia, F.M., Bartelloni, P.J., and McKinney, R.W.: Rocky Mountain spotted fever. Laboratory infection in a vaccinated individual. J.A.M.A., 211:2012–2014, 1970.

25. Freeman, G., and Hodson, M.E.: Q fever endocarditis treated with trimethoprim and sulphamethoxazole. Br. Med. J., 1:419–420, 1972.

26. Hand, W.L., Miller, J.B., Reinarz, J.A., and Sanford, J.P.: Rocky Mountain spotted fever. A vascular disease. Arch. Int. Med., 125:879–882, 1970.

27. Johnson, J.E., and Kaduli, P.J.: Laboratory-acquired Q fever. A report of fifty cases. Am. J. Med., 41:391–403, 1966.

28. ———: Rocky Mountain spotted fever acquired in a laboratory. New Engl. J. Med., 277:842–847, 1967.

29. Krumbiegel, E.R., and Wisniewski, H.J.: Q fever in Milwaukee. II. Consumption of infected raw milk by human volunteers. Arch. Environ. Health., 21:63–65, 1970.

30. Linnemann, C.C., Jr., and Janson, P.J.: The clinical presentations of Rocky Mountain spotted fever. Comments on recognition and management based on a study of 63 patients. Clin. Pediatr., 17:673–675, 678–679, 1978.

31. Middleton, D.B.: Rocky Mountain spotted fever: gastrointestinal and laboratory manifestations. South. Med. J., 71:629–632, 1978.

32. Murray, E.S., et al.: Brill's disease. I. Clinical and laboratory diagnosis. J.A.M.A., 142:1059–1066, 1955.

33. Older, J.J.: The epidemiology of murine typhus in Texas, 1969. J.A.M.A., 214:2011–2017, 1970.

34. Oster, C.N., et al.: Laboratory-acquired Rocky Mountain spotted fever. The hazard of aerosol transmission. New Engl. J. Med., 297:859–863, 1977.

35. Philip, R.N., et al.: A comparison of serologic methods for diagnosis of Rocky Mountain spotted fever. Am. J. Epidemiol., 105:56–67, 1977.

36. Saslaw, S., Carlisle, H.N., and Wolfe, G.L.: Antibiotic prophylaxis and therapy of Rocky Mountain spotted fever initiated in monkeys via the respiratory route. Antimicrob. Agents Chemother., 1964:652–655, 1965.

37. Sheehy, T.W., Hazlett, D., and Turk, R.E.: Scrub typhus. A comparison of chloramphenicol and tetracycline in its treatment. Arch. Int. Med., 132:77–80, 1973.

38. Walker, D.H., Cain, B.G., and Olmstead, P.M.: Laboratory diagnosis of Rocky Mountain spotted fever in immunofluorescent demonstration of *Rickettsia rickettsii* in cutaneous lesions. Am. J. Clin. Pathol., 69:619–623, 1978.

39. Wells, G.M., Woodward, T.E., Fiset, P., and Hornick, P.B.: Rocky Mountain spotted fever caused by blood transfusion. J.A.M.A. 239:2763–2765, 1978.

40. Yamada, T., et al.: Activation of the kallikrein–kinin system in Rocky Mountain spotted fever. Ann. Intern. Med., 88:764–768, 1978.

Mycoplasma

41. Alexander, E.R., et al.: Pneumonia due to *Mycoplasma pneumoniae*. New Engl. J. Med., 275:131–136, 1966.

42. Balassanian, N., and Robbins, F.C.: *Mycoplasma pneumoniae* infection in families. New Engl. J. Med., 277:719–725, 1967.

43. Cherry, J.D., Hurwitz, E.S., and Welliver, R.C.: *Mycoplasma pneumoniae* infections and exanthems. J. Pediatr., 87:369–373, 1975.

44. Crawford, Y.E.: A laboratory guide to the mycoplasmas of human origin. Great Lakes, Illinois, Naval Medical Research Unit No. 4, 1966.

45. Cordero, L., Cuadrado, R., Hall, C.B., and Horstmann, D.M.: Primary atpyical pneumonia: an epidemic caused by *Mycoplasma pneumoniae*. J. Pediatr., 71:1–12, 1967.

46. Costea, N., Yakulis, V.J., and Heller, P.: The mechanism of induction of cold agglutinins by *Mycoplasma pneumoniae*. J. Immunol., 106:598–604, 1971.

47. Denny, F.W., Clyde, W.A., Jr., and Glezen, W.P.: *Mycoplasma pneumoniae* disease: clinical spectrum, pathophysiology, epidemiology, and control. J. Infect. Dis., 123:74–92, 1971.

48. Evatt, B.L., Dowdle, W.R., Johnson, M., and Heath, C.W., Jr.: Epidemic *Mycoplasma pneumoniae*. New Engl. J. Med., 285:374–377, 1971.

49. Foy, H.M., et al.: Epidemiology of *Mycoplasma pneumoniae* infection families. J.A.M.A., 197:859–866, 1966.

50. ———Second attacks of pnemonia due to *Mycoplasma pneumoniae*. J. Infect. Dis., 135:673–677, 1977.

51. Frieberg, J.: Genital mycoplasma infections. Am. J. Obstet. Gynecol., 132:573–578, 1978.

52. Harrison, R.F., deLouvois, J., Blades, M., and Hurley, R.: Doxycycline treatment and human infertility. Lancet, 1:605–606, 1975.

53. Hayflick, L., and Chanock, R.M.: *Mycoplasma* species of man. Bacteriol. Rev., 29:185–220, 1965.

54. Hill, A.C.: Demonstration of mycoplasmas in tissue by the immunoperoxidase method. J. Infect. Dis., 137:152–154, 1978.

55. Jones, M.C.: Arthritis and arthralgia in infection with *Mycoplasma pneumoniae*. Thorax, 25:748–750, 1970.

56. Kenny, J.F.: Role of cell-wall-defective microbial variants in human infections. South. Med. J., 71:180–190, 1978.

57. Koren, Z., and Spigland, I.: Irrigation technique for detection of *Mycoplasma* intrauterine infection in infertile patients. Obstet. Gynecol., 52:588–590, 1978.

58. Ludlam, G.B., Bridges, J.B., and Benn, E.C.: Association of Stevens–Johnson syndrome with antibody for *Mycoplasma pneumoniae*. Lancet, 1:958–959, 1964.

59. Purcell, R.H., and Chanock, R.M.: Role of mycoplasmas in human respiratory disease. Med. Clin. North Am., 51:791–802, 1967.

60. McCormack, W.M., et al.: The genital mycoplasmas. New Engl. J. Med., 288:78–89, 1973.

61. Menonna, J., Chmel, H., Menegus, M., Dowling, P., and Cook, S.: Precipitating antibodies in mycoplasma infection. J. Clin. Microbiol., 5:610–612, 1977.

62. Mufson, M.A.: *Mycoplasma hominis* I in respiratory tract infections. Ann. N.Y. Acad. Sci., 174:798–808, 1970.

63. Mufson, M.A., Sanders, V., Wood, S.C., and Chanock, R.M.: Primary atypical pneumonia due to *Mycoplasma pneumoniae*. Report of a case with a residual pleural abnormality. New Engl. J. Med., 268:1109–1112, 1963.

64. Murray, H.W., Masur, H., Senterfit, L.B., and Roberts, R.B.: The protean manifestations of *Mycoplasma pneumoniae* infection in adults. Am. J. Med., 58:229–242, 1978.

65. Noriega, E.R., Simberkoff, M.S., Gilroy, F.J., and Rahal, J.J., Jr.: Life-threatening *Mycoplasma pneumoniae* pneumonia. J.A.M.A., 229:1471–1472, 1974.

66. Rifkind, D., et al.: Ear involvement (myringitis) and primary atypical pneumonia following inoculation of volunteers with Eaton agent. Am. Rev. Resp. Dis., 86:479–489, 1962.

67. Sacker, I., Walher, M., and Brunell, P.A.: Abscess in newborn infants caused by mycoplasma. Pediatrics, 46:303–304, 1970.

68. Shames, J.M., et al.: Comparison of antibiotics in the treatment of mycoplasmal pneumonia. Arch. Intern. Med., 125:680–684, 1970.

69. Simberkoff, M.S., and Toharsky, B.: Mycoplasmemia in adult male patients. J.A.M.A., 236:2522–2524, 1976.

70. Smith, C.B., Chanock, R.M., Friedewald, W.T., and Alford, R.H.: *Mycoplasma pneumoniae* infections in volunteers. N.Y. Acad. Sci., 143:471–483, 1967.

71. Stevens, D., et al.: Mycoplasma pneumoniae infections in children. Arch. Dis. Child., 53:38–42, 1978.

72. Sussman, S.J., Magoffin, R.L., Lennette, E.H., and Schieble, J.: Cold agglutinins, Eaton agent, and respiratory infections of children. Pediatrics, 38:571–577, 1966.

73. Taylor-Robinson, D., Csonka, G.W., and Prentice, M.J.: Human intraurethral inoculation of ureaplasmas. Q. J. Med., 46:309–325, 1976.

74. Thomas, L.: Circulating autoantibodies and human disease. With a note on primary atypical pneumonia. New Engl. J. Med., 270:1157–1159, 1964.

75. Thomsen, A.: Occurrence of mycoplasmas in urinary tracts of patients with acute pyelonephritis. J. Clin. Microbiol., 8:84–88, 1978.

76. Turtzo, D.F., and Ghatak, P.K.: Acute hemolytic anemia with *Mycoplasma pneumoniae* pneumonia. J.A.M.A., 236:1140–1141, 1976.

77. Wallace, R.J., et al.: Isolation of *Mycoplasma hominis* from blood cultures in patients with postpartum fever. Obstetr. Gynecol., 51:181–185, 1978.

11
Major DNA Viruses

HERPESVIRUS HOMINIS (HERPES SIMPLEX VIRUS)

Objectives

1. Name five syndromes that can be caused by *Herpesvirus hominis* virus. Indicate which type is most likely to be involved in each syndrome.
2. Define latent infection and give the evidence that latent infection due to herpes simplex virus can occur in humans.
3. Describe the laboratory methods used to detect *Herpesvirus hominis* infection.
4. Discuss the evidence that some illnesses in humans due to herpes simplex virus can be effectively treated with chemotherapy.
5. Discuss the evidence that herpes simplex virus is related to cervical carcinoma.

Species and Types

Herpesvirus hominis is also called herpes simplex virus, which is often abbreviated HSV. The word herpes is related to the Greek word for creep or crawl.

Herpes simplex virus should be distinguished from Herpes zoster virus (*Herpesvirus varicellae*), the cause of zoster and chickenpox. In common usage, "herpes" usually refers to herpes simplex virus or disease due to this virus.

Many animals have one or more herpesviruses which infect that species. Examples include the monkey herpesvirus (*Herpesvirus simiae*), which can cause ascending paralysis in man, and the canine herpesvirus, which can be an experimental model in dogs of neonatal and disseminated herpes.[7] Humans can be infected with four herpesviruses: *H. hominis, H. varicellae, H. cytomegalovirus*, and the Epstein–Barr virus.

There are two types of *Herpesvirus hominis*. Type 1 (oral) is the commonest and is the type usually found in gingivostomatitis. Type 2 (genital) is the type usually found in lesions of the penis or cervix, and in disseminated herpes of the newborn.

Frequency and Importance

Herpes simplex virus is best known as a frequent, recurrent infection of the lips, herpes labialis, called cold sores or fever blisters by lay people. It is also important as a cause of fatal, or brain-damaging encephalitis, although its frequency is controversial. Antiviral chemotherapy is of value in herpetic eye infections, and possibly of value in herpetic encephalitis. Herpes simplex virus is a suspected causal factor in carcinoma of the uterine cervix, as discussed later.

Clinical Patterns of Illness

Herpes simplex virus is the *usual* cause of:

Recurrent Lip or Facial Vesicles.[6] Recurrent vesicles near the lips or nose are well known manifestations of herpes simplex virus, which can be easily cultured if fluid from the lesion in inoculated into cell cultures.

Gingivostomatitis is characterized by shallow ulcers of the buccal mucosa (stomatitis), tongue (glossitis), or gums (gingivitis), which bleed easily. Acute gingivostomatitis in children is usually due to Herpes simplex virus, which is readily recovered in virus cultures.

Herpes simplex virus is a *frequent* cause of:

Ulcerative Pharyngitis. There are ulcers visible in the oropharynx. When due to herpes simplex virus, superficial ulcerations are also usually found on gums, tongue, lips, or buccal mucosa, so the diagnosis of gingivostomatitis is made and herpes simplex virus suspected. Occasionally, the lesions of the tongue or gums are missed on a single examination. Like the other *H. hominis* infections of the oral area, ulcerative pharyngitis is usually due to the type 1 virus, but rarely is due to type 2 virus.[17]

Vesicular Skin Lesions. The skin lesions of herpes simplex virus are typically groups of vesicles, which evolve into pustules, ulcers, and crusts. When located in the genital area, the shallow ulcer may be mistaken for a syphilitic chancre. Genital herpes is often called herpes progenitalis. When the virus infects burns or eczematous skin, the purulent exudate often leads to a mistaken diagnosis of a bacterial infection. When the virus infects a finger of a doctor or nurse who touches an infected patient's lesions, the lesion also resembles that of a deep bacterial infection. In a newborn infant infected from the mother's cervix, the vesicles may be seen on the scalp.

Herpes simplex virus can rarely produce bullous lesions, as well as the target or iris lesions of erythema multiforme.[11]

Genital Herpes (Herpes Genitalis or Progenitalis). Ulcerative or necrotizing lesions of the uterine cervix, or vulvar area can be produced by herpes simplex virus. Venereal transmission of the virus occurs. Penile ulcerations are painful and tend to be recurrent.

Chronic or Recurrent Corneal Ulceration. Keratitis, is characterized by a branching ("dendritic") ulceration of the cornea, which can be seen after straining the cornea with fluoroscein (Fig. 11-1). Herpes keratitis is an important, treatable cause of blindness in the United States.

Herpes simplex virus is an *occasional* cause of:

Acute Encephalitis.[14] This syndrome is defined by a severe and nontransient disturbance of consciousness, with 50 to 500 white blood cells (predominantly lymphocytes) in the spinal fluid. Herpes simplex virus is the most common viral

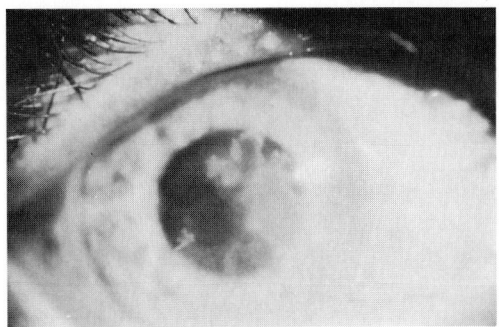

FIG. 11-1. *Herpes simplex keratitis. Staining of the cornea with fluorescein reveals branching (dendritic) ulceration as well as nonspecific geographic ulceration. (Photo from Dr. Fred Brightbill)*

cause of sporadic acute encephalitis in the United States, although the majority of patients with acute encephalitis do not have a specific etiology found. In herpes simplex encephalitis, the patient is usually an adult. Bizarre behavior and acute hallucinations of taste or smell may occur at the onset. Temporal lobe involvement is common, and many of the nonfatal patients who do not die have severe brain damage. Convulsions or very high fever is common.

Other neurologic manifestations include urinary retention, myelitis with paralysis of the legs, or neuralgia of the hips and legs secondary to genital herpes, tic douloureux (neuralgia of the facial nerve area) due to reactivation of latent herpesvirus in the trigeminal ganglia, and recurrent abdominal pain ascribed to virus in ganglia serving this area.

Suspected Neonatal Sepsis. Disseminated herpes of the newborn may produce an illness often mistaken for septicemia.[5] The infant infected in utero may be born with extensive lesions. The infant not exposed until the time of delivery or when in the newborn nursery is more likely to develop mild infection than disseminated disease.

Congenital Malformations. Microcephaly, eye malformations, or diffuse brain damage may result from intrauterine herpes infections.[10]

Herpes simplex virus is a *suspected* cause of:

Prostatitis. Nonbacterial prostatitis has been associated with cultures positive for herpesvirus more frequently than normals, but a causal relationship has not been proved.

Carcinoma of the cervix is statistically more frequent in patients with serologic evidence of previous Herpes simplex type 2 viral infection. However, a causal relationship has not been clearly established.[1,10]

Psychosis might be a manifestation of herpes infection of the brain, although this is not certain.

Asymptomatic or Subclinical Infection

On the basis of antibody studies, it is clear that asymptomatic or unrecognized infection is very frequent.

Laboratory Diagnosis

Culture. This virus is relatively easy to grow on most cell cultures of human tissue (Fig. 11-2). Preliminary identification can be made on the basis of the appearance of the cytopathic effect and the particular cell lines which support the growth of the virus. A conclusive identification can be done by fluorescent antibody methods. Separation of type 1 from type 2 can be done by quantitative neutralization, using type-specific antiserum, by fluorescent antibody methods, by ELISA methods (Fig. 11-3), by plaque formation, or by differential temperature sensitivity testing.[13]

Efficacy of various antiviral drugs can be tested in cell cultures.

Serum Antibodies. Complement fixation or neutralizing antibodies can be measured in paired sera. Recurrent herpes labialis may not result in an antibody rise.

Histologic Diagnosis. A smear of the base of a vesicle can be made and stained with Giemsa stain (the Tzanck test),[3] but the typical histology seen with herpes simplex virus cannot be distinguished from that of varicella–zoster virus. A Papanicolaou smear of the cervix (for cancer cells) also will often detect HSV infected cells. Similarly, the inclusion bodies in histologic sections and the viral particles as seen on electron microscopy, cannot be distinguished from those of varicella–zoster virus (see Fig. 11-6).

Herpes simplex virus antibody can be conjugated with fluorescein, and used for specific identification of herpes simplex virus in tissues, (Fig. 1-9). Herpes simplex virus antibody can also be conjugated with peroxidase and used for direct detection of the virus from skin or vesicular lesions, using the ELISA method shown in Figure 11-3, and explained in more detail in the section on cholera in Chapter 3.

Biologic Characteristics of Clinical Interest

Chronic or Latent Infection. *Herpesvirus hominis* is the most common recurrent or chronic virus infection of humans. The first infection with herpes simplex virus in a patient without anti-

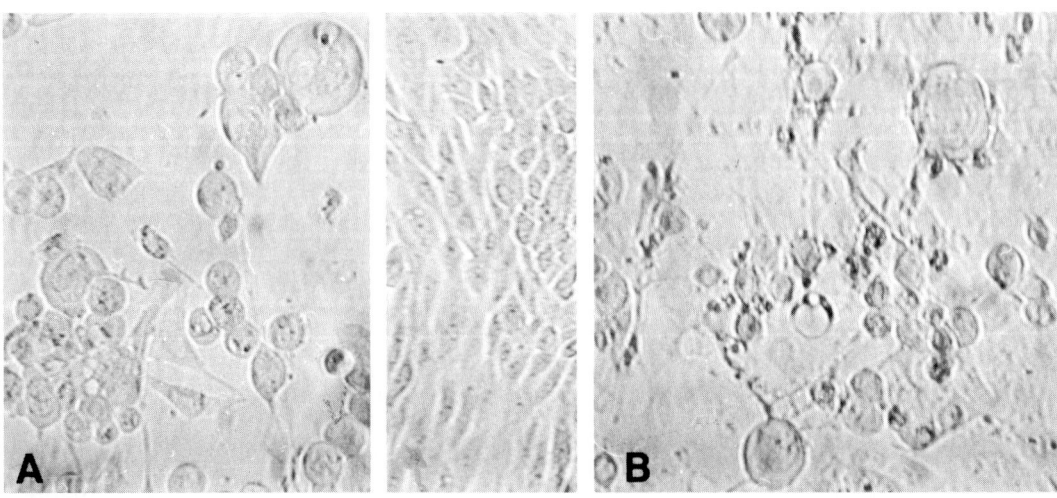

FIG. 11-2. (A) *Herpes simplex and* (B) *varicella-zoster viruses have very similar cytopathic effect on human embryo kidney (HEK) cell cultures. Normal cell cultures are shown in the center.*

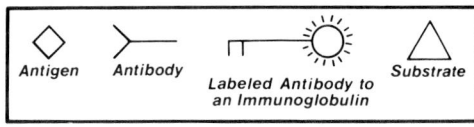
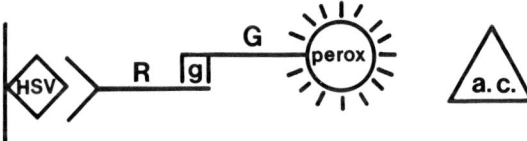

FIG. 11-3. *ELISA method to detect or identify antigen of herpes simplex virus.*[13] *HSV = antigen to be identified (vesicular fluid or virus in cell culture). R = rabbit antibody to HSV-1 or HSV-2. G = peroxidase-labeled goat anti-rabbit IgG antibody. g = IgG portion of rabbit antibody. a.c. = aminoethyl carbozole substrate, used to detect the peroxidase.*

bodies is called a primary infection. Subsequent episodes of typical herpes illnesses are called recurrences, as in recurrent herpes labialis. Recovery of the virus is sometimes possible in the absence of any illness, although it is often looked for at the time of respiratory symptoms. The term chronic implies that the virus is present, and culturable, between episodes of clinical disease, as is sometimes the case with eye infections. The term latent means that the virus is present, but modified in some way, so that it cannot be detected. Neutralizing antibodies in the serum usually do not protect against recurrent infection, consistent with the view that the virus is intracellular.

The virus has recently been recovered at autopsy from trigeminal ganglia of normal humans, indicating that sensory ganglia may be a site of persistence of the virus.[2]

Contiguous Spread of Infection. Most herpetic lesions are focal and spread from cell to cell. This accounts for the relative ineffectiveness of antibody in modifying the disease. The virus has been shown to spread via nerves in experiments in animals.

Heat Sensitivity. In canine herpesvirus infections, only young pups with low body temperatures (35°C–37°C) get disseminated experimental or natural infections.[7] Experimental *Herpesvirus hominis* infections in mice also seem to indicate heat sensitivity of this virus, particularly type 2.[10]

Treatment

Chemotherapy. Idoxuridine (IDU) is effective when applied directly to the cornea of infected rabbits. In an experimental model using type 2 infection in newborn mice, IDU had no effect on mortality, but did reduce viremia and replication of the virus in the lung, but the virus spread to the brain via the nerves.[9] This may explain in part the lack of efficacy in encephalitis.

Adenine arabinoside (ara-A, vidarabine) is a DNA antagonist which has antiviral properties and is being recommended for herpes simplex encephalitis.[16]

Vaccination. Repeated vaccination with smallpox vaccine has been used in the past to try to prevent herbes labialis. This is not only of unproved value, but is contraindicated because of the risks of the complications of smallpox vaccine. Various herpes simplex virus vaccines have been tried for recurrent herpes, and have had brief periods of apparent success. One double blind study of recurrent herpes labialis showed no difference between placebo and vaccine, with both having about a 70 percent effectiveness.[8] This high rate of spontaneous improvement probably accounts for preliminary reports of success with past vaccines.

Phototherapy. In this therapy, the vesicles are opened, and light-absorbing dye is applied. The lesion is then exposed to a cool, white, 15 watt fluorescent light at 6 inches for 15 minutes.[12] This therapy is of disputed effectiveness.[15] It might result in induction of malignancy, as some cases of carcinoma of the penis have occurred following dye–light treatment for herpes genitalis.[4]

Local Application of Chemicals. Ether has been used as local therapy because it inactivates herpesviruses in vitro by dissolving their lipid coat. Ether was ineffective in preventing recurrences of penile herpes in a prospective study. Other topical virucidal drugs have also been ineffective against herpes labialis or herpes genitalis when studied in a controlled fashion. For example, adenine arabinoside was unsuccessful for herpes labialis when subjected to a carefully controlled trial. For genital herpes, 2-deoxy-D-glucose, an inhibitor of viral glycosylation, is under investigation, and appears promising.

VARICELLA–ZOSTER VIRUS

Objectives

1. Describe the evidence that the viruses of chickenpox and zoster are the same.
2. Describe the typical clinical disease of chickenpox and of zoster.
3. Describe two complications of each disease.
4. Describe methods for confirmation of the diagnosis in the laboratory.
5. Discuss the method of production and the indications for the use of varicella–zoster immune globulin (VZIG).

Definitions

Varicella (chickenpox) and herpes zoster (shingles) are caused by the same virus, varicella-zoster (V–Z) virus. Zoster (Greek for belt) and shingles (Latin cingula, for belt) both refer to the sensory nerve dermatome distribution of herpes zoster. The virus is also called *Herpesvirus varicellae*.

Previous chickenpox does not prevent an individual from getting zoster; in fact, patients with zoster have had chickenpox at some time in the past, often very early in life.[43]

Several kinds of evidence indicate these two viruses are identical. Chickenpox crusts or zoster vesicle fluid are antigenically identical by standardized serologic procedures. Injection of zoster fluid into chickenpox susceptibles produces localized chickenpox.[20] Chickenpox may occur in chickenpox-susceptible individuals after exposure to zoster.[20,21]

Frequency and Importance

Approximately 200,000 cases of chickenpox are reported annually in the United States. It is generally regarded as a benign disease of childhood. Some complications can be serious, as described below. Chickenpox can be fatal in children with malignancies or who are immunosuppressed.

Zoster is not a notifiable disease. It is relatively more frequent in older individuals, and in patients of all ages with malignancies. It appears to be a recrudescence of the virus in individuals with impaired cellular immunity.

Clinical Patterns of Illness

Varicella–zoster virus is the *usual* cause of:

Chickenpox-like Illness. Typical chickenpox is usually a mild illness, with minimal respiratory symptoms, and mild to moderate fever. The characteristic lesions are a clear vesicle with erythema around it, small pustules, and small crusted ulcers. These lesions appear in crops, and several stages of the lesion are present at the same time. Itching is prominent, and often physicians are consulted primarily because of the itching. Lymphadenopathy also is prominent, particularly in nodes draining the scalp, or areas of scratched lesions. The mouth and other mucous membranes are often involved. (Fig. 11-4).

Zoster-like Illness. The typical zoster rash occurs in the distribution of a sensory nerve dermatome, especially on the trunk, or in the distribution of the fifth cranial nerve. Typically it ends abruptly at the midline. In adults, pain usually precedes the rash and persists for many days, but zoster usually is not painful in children.[21] The rash evolves through the same vesicle, pustule, crust pattern as chickenpox.

Complications of Chickenpox

Pneumonia. Severe interstitial pneumonia can occur in adults with chickenpox.[40] However, routine chest radiographs of healthy young adults with chickenpox indicated only 16 percent had radiologic evidence of pneumonia, and few had cough or dyspnea.[40] The white blood count is usually less than 10,000; recovery typically occurs in 6 to 10 days without antibiotics or steroids.[42] A diffusion block with hypoxemia can occur, and oxygen therapy is usually helpful.

Secondary Bacterial Skin Infection. This can be of variable severity, and typically occurs as boils or an abscess, usually due to *Staphylococcus aureus*. Severe streptococcal skin infection can also occur as a complication.[39]

Progressive Disseminated Varicella. Chickenpox may be progressive and fatal in patients with malignancies or receiving immunosuppressive agents or steroids. Asthmatics receiving low doses of steroids typically do well when they get chickenpox.[24] Even in children with leukemia, the prognosis is not invariably bad, with a mortal-

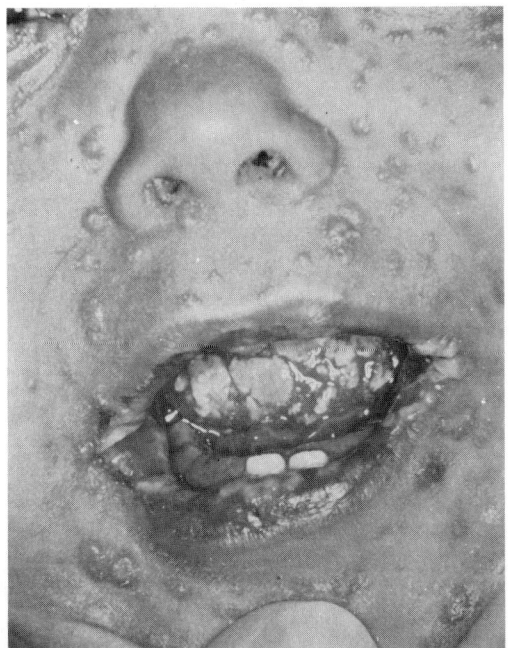

FIG. 11-4. *This child with moderately severe chickenpox, presented with most skin lesions in the pustular stage, and mucosal lesions of the mouth.*

ity rate of about seven percent.[25] Pneumonia is the usual cause of death in such cases.

Varicella Bullosa. Large hemorrhagic bullae (blisters) sometimes occur with chickenpox. The bullous lesions of chickenpox must be distinguished from secondary staphylococcal infection, with bulla formation.[38]

Congenital Chickenpox. If a woman gets chickenpox early in pregnancy, there is a small risk that congenital anomalies such as skin scarring, eye defects, and hypoplastic limbs may occur.[27] When the pregnant woman gets chickenpox late in pregnancy, there may be widespread visceral involvement of the infant, with an especially high mortality rate, if the baby is born within 4 days of the onset of chickenpox in the mother.[35] If a newborn infant gets chickenpox in the first 10 days of life, it was infected before birth, and this is considered congenital infection.

Neonatal Chickenpox. This is defined as chickenpox in the first month of life, with exposure occurring after birth. However, relatively few newborns get chickenpox when exposed, because of maternal antibodies.[29] In the first 6 months of life, when protection from maternal antibodies is variable, the disease is typically mild. Thus, exposure of the newborn to chickenpox is a much less serious risk than exposure of the fetus to maternal infection.

Encephalitis. This is rare, but can be severe.[33] The CSF typically reveals a normal glucose and protein, and 10 to 200 white blood cells, predominately monuclear.[33] Cerebellar ataxia or transverse myelitis also may occur.[34] Reye's syndrome, a rare and severe encephalopathy with liver failure, sometimes follows chickenpox.[32] It is characterized by vomiting, and a severe change of consciousness, often progressing to coma. Early clues to the diagnosis of Reye's are an elevated serum transaminase and elevated blood ammonia. Early diagnosis and early supportive treatment may decrease the mortality rate. The cause of Reye's syndrome is unknown, and it may occur after other viral infections, such as influenza.

Rare Complications. These include purpura fulminans with gangrene, and arthritis.[37] Acute glomerulonephritis has been reported following chickenpox, but is probably due to unrecognized streptococcal infection.

Complications of Zoster

Neuralgia. This painful complication of zoster occurs almost exclusively in older adults.

Eye Involvement. Zoster can involve the ophthalmic division of the trigeminal nerve, and may produce conjunctivitis or keratitis (Fig. 11-5).

Disseminated Zoster. As in the case of chickenpox, dissemination may occur in patients with malignancy or in patients receiving immunosuppressive drugs.[22]

Laboratory Diagnosis

Viral Cultures. Isolation of the virus is possible by culture of vesicles or pustule fluid. The specimen should be inoculated without freezing on cell cultures, such as human embryo kidney cells (Fig. 11-2). The cytopathic effect is typical, and resembles that of herpes simplex virus, with large intranuclear inclusions.

Smear of a Vesicle. If the base of a vesicle is scraped, put on a slide and stained with Giemsa

168 MAJOR DNA VIRUSES

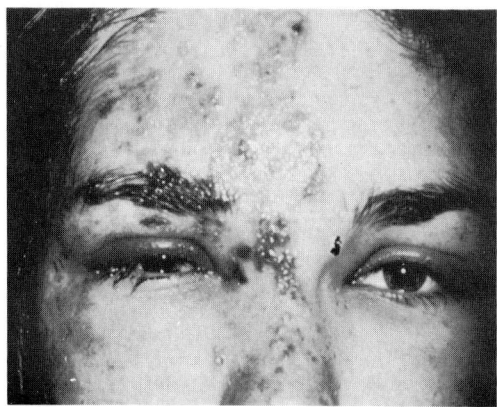

FIG. 11-5. *Herpes zoster involving the eye (zoster opthalmicus). The skin lesions stop slightly beyond the midline, and are in the distribution of the ophthalmic branch of the trigeminal nerve. (Photo from Dr. Fred Brightbill)*

stain, multinucleated giant cells with intranuclear inclusions may be seen.[19] However, these cells cannot be distinguished from those seen in the base of a vesicle due to herpes simplex virus. Electron microscopy of vesicular fluid can be used to distinguish chickenpox virus from smallpox virus. However, herpes simplex virus and varicella–zoster virus are identical on electron microscopy (Fig. 11-6).

Serum Antibodies. Serologic studies using paired sera may be available in a research or reference laboratory.[26].

Antigen Identification. An agar gel diffusion method appears to be useful in rapidly distinguishing chickenpox virus from smallpox virus, by using crusts or vesicular fluid as the source of antigen.[41]

Biologic Characteristics of Clinical Interest

Incubation Period and Period of Infectivity. Contagion of chickenpox has best been observed in contagious disease hospitals, when chickenpox was inadvertently introduced into scarlet fever wards.[31] The range of the incubation period is 10 to 21 days (usually 13 to 17 days). The patient with chickenpox is rarely contagious more than 24 hours before the rash or after the fifth day of the rash.[31] The virus can be recovered from the lesions of zoster for as long as 10 days, but is usually recovered only if the lesions are vesicular.[30] The incubation of varicella after exposure to zoster is about 18 to 21 days.[20]

Intracellular Localization. The virus is very difficult to grow unless intact infected cells are used as the inoculum. For this reason, V–Z virus once was classified in Subgroup B of the herpesviruses, along with the cytomegaloviruses.

Latency. Varicella–zoster virus is believed to remain latent after the chickenpox illness, at least

FIG. 11-6. *Herpes simplex virus and varicella–zoster virus have a similar appearance on electron microscopy of tissue.* Left: *Herpesvirus hominis.* Right: *Herpesvirus varicellae.*

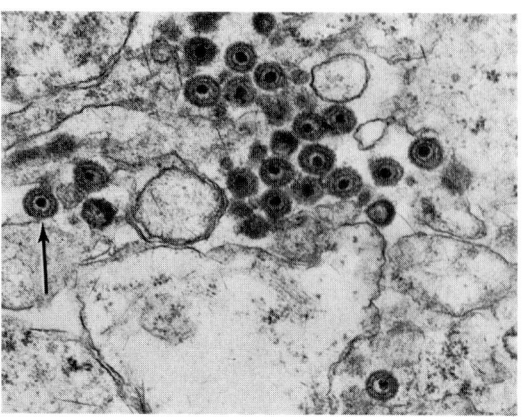
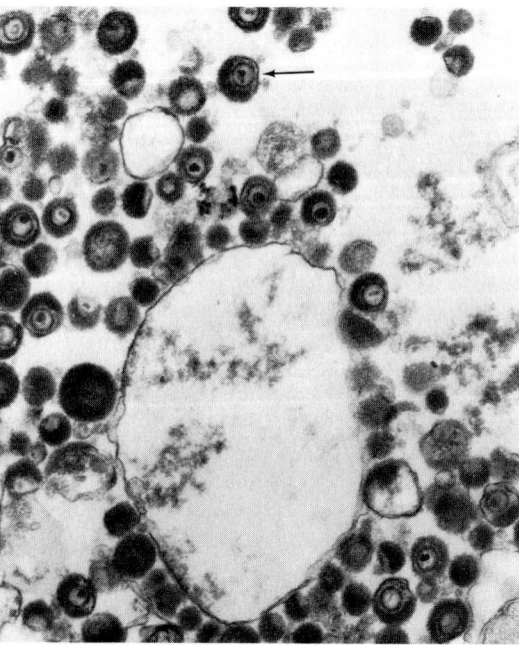

in those individuals who later develop zoster.[36] The virus can remain latent in dorsal nerve root ganglia.

Treatment

Chemotherapy. Antiviral drugs, such as cytosine or adenine arabinoside, are still being investigated as treatment for disseminated varicella or zoster.[22]

Corticosteroids. In adults steroids are effective in the relief of pain and do not result in dissemination of the disease.[23] In children, pain is rare and steroids are not indicated.

Convalescent Plasma or Hyperimmune Globulin. Antibody therapy is not effective in the treatment of disseminated zoster, as these patients already have antibodies.

Prevention

Gamma Globulin. Ordinary human immune globulin given to exposed susceptible children makes chickenpox milder, with fewer skin lesions, but does not prevent the disease. Its effectiveness depends on the presence of some antibodies to chickenpox in the general population used as the source of the gamma globulin.

Varicella–Zoster Immune Globulin (VZIG, Formerly Called ZIG). This is a hyperimmune gamma globulin made from serum obtained from individuals convalescing from zoster or chickenpox, usually obtained from blood bank blood. Its antibody content can be quantitated and the dose calculated so that the preparation can be used to prevent chickenpox in exposed susceptibles who are receiving immunosuppressive drugs, or who have a serious underlying illness, such as leukemia. It is also used for prevention of severe chickenpox in newborn infants whose mothers developed chickenpox 4 days or less before delivery.[35]

Zoster Immune Plasma (ZIP), obtained from patients with recent chickenpox or zoster is also effective in modifying or preventing chickenpox, if given to susceptible compromised hosts shortly after exposure.[18] However, plasma has a higher risk of hepatitis than does hyperimmune globulin.

Exposure While Receiving Steroids. Children with malignancies who are receiving corticosteroids should probably have the drug tapered to physiologic levels, in addition to receiving convalescent plasma or VZIG.[24] Children receiving low dose steroids for severe chronic asthma do not have more severe chickenpox than such children not receiving steroids.[24]

Vaccine. A live attenuated varicella vaccine is under investigation for possible use in susceptible children with malignancies or receiving immunosuppressive therapy.[28]

CYTOMEGALOVIRUS

Objectives

1. Discuss the origin of the name of this virus and other names used for this virus.
2. Describe the classical clinical picture of congenital cytomegalovirus infection.
3. Describe two syndromes of acquired cytomegalovirus infection in the adult.
4. Discuss methods used in the laboratory to detect cytomegalovirus infection.
5. Discuss the chemotherapy of cytomegalovirus infections.

Definitions

Cytomegalovirus was named for the disease called cytomegalic inclusion body disease.[55,63] The virus also has been called salivary gland virus, because typical inclusion cells are frequently found in the salivary glands.

Species and Types

Cytomegaloviruses are closely associated with cells and are difficult to separate from the cell. There is apparently some antigenic variation among human cytomegaloviruses, but types are not clearly defined as yet.[63] Many animal species have their own types of cytomegalovirus, which closely resemble the human type.

Frequency and Importance

About 80 percent of adults have antibodies to the virus, and presumably have had asymptomatic or unrecognized infections. Serious acquired infections are rare, and occur primarily in immunosuppressed hosts.

Congenital infection as detected by asymptomatic excretion of the virus in the urine, occurs in one to two percent of newborn infants, and rarely causes serious congenital disease, including deafness and retardation.

Clinical Patterns of Illness

Cytomegalovirus infections can be congenital or acquired. The virus and its most severe manifestations were first recognized and studied in newborn infants.[55]

In newborn infants, congenital cytomegalovirus infection is an *occasional* cause of:

Chronic Congenital Infection Syndromes. A severe form may occur, with neonatal jaundice, with hepatosplenomegaly and purpura. Isolated hepatosplenomegaly or transient petechiae also can be caused by cytomegalovirus.

Neurologic or eye defects also suggest a chronic congenital infection. Microcephaly, spastic paralysis, psychomotor retardation, cerebral calcifications, or chorioretinitis may occur as isolated findings after infection of the fetus. However, a typical severe pattern might include jaundice, purpura, microcephaly, and chorioretinitis.

The presence of serum antibody to CMV does not necessarily protect the infant from infection.

In normal individuals, cytomegalovirus is an *occasional* cause of:

Infectious Mononucleosis-like Syndrome. Fever, splenomegaly, and atypical lymphocytes are present in cytomegalovirus mononucleosis, but the heterophil test is negative, and the serum transaminase is normal.[50]

In unusual hosts, cytomegalovirus is an *occasional* cause of:

Postperfusion Syndrome. This consists of fever, lymphocytosis with atypical lymphocytes, and splenomegaly, occurring about 3 to 6 weeks after open heart surgery, where fresh blood has been used in the extracorporeal circulation. Cytomegalovirus has been recovered from the blood of such patients.[52] Transfusion of whole blood can be associated with seroconversion and can be a source of infection with cytomegalovirus.[60]

Disseminated Disease. In individuals with an immunologic deficiency, disseminated disease may occur, manifested by chronic pneumonia, acquired hemolytic anemia, or chronic hepatitis. Disseminated disease has been observed in patients with renal transplantation, malignancies, and congenital immunologic deficiencies.[58]

Chronic Pneumonia occasionally occurs as an isolated complication in compromised hosts, especially after renal transplantation.

Asymptomatic Infection

About 80 percent of adults over 35 years of age have serologic evidence of previous infection. Most patients with renal transplantation develop cytomegalovirus viruria, without any evidence of resulting renal disease.[45] About one to two percent of normal newborn infants have congenital cytomegalovirus infection as indicated by excretion of cytomegalovirus in the urine, but usually do not develop any sequelae.[51] However, the frequency of deafness or mental retardation is increased after asymptomatic congenital infection.[48]

Laboratory Diagnosis

Urine Cytology. Renal tubule cells containing large inclusions may be found by examination of urine after concentration of the sediment by filtration[57] (Fig. 11-7). A casual examination of the urine sediment by an untrained observer is not a satisfactory way to exclude cytomegalovirus infection. However, rapid microscopic detection of typical cytomegalic cells may be possible in severe cases of congenital disease. Electron microscopy also can be used to detect virus particles in urine sediment and is more sensitive than viral culture.[54]

Viral Cultures. Culture of the urine or saliva is the most sensitive method to detect infection. The virus grows in human fibroblasts, and usually requires several weeks to produce the typical cytopathic effect (Fig. 11-8). In severe congenital infections, however, the cytopathic effect may be observed as early as 24 to 48 hours after inoculation of the urine specimen.

Serum Antibodies.[47] Serologic diagnosis is difficult, because it is hard to grow the virus to high enough titer to make a good antigen, and because of some antigenic variation. However, absence of antibody in the newborn infant usually excludes fetal infection.

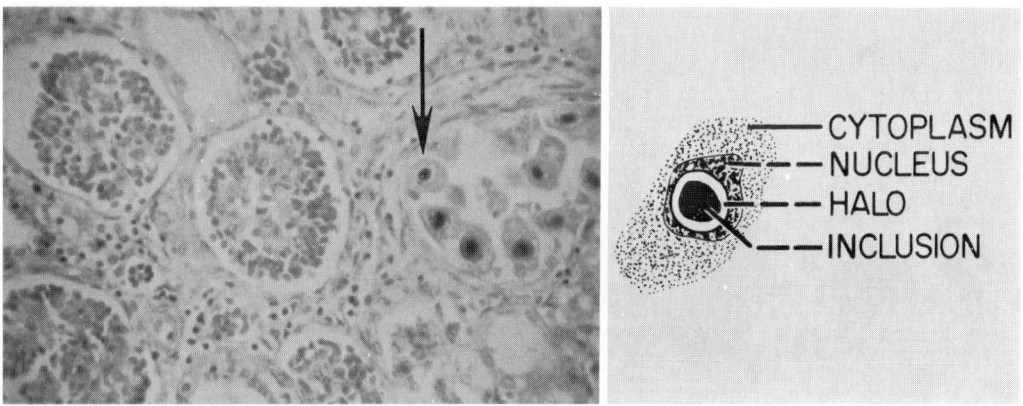

FIG. 11-7. (A) *Cytomegalovirus inclusions in renal tubular cells (250×). (B) A diagram of the cell as seen in the urine. (Photo from Dr. Enid Gilbert)*

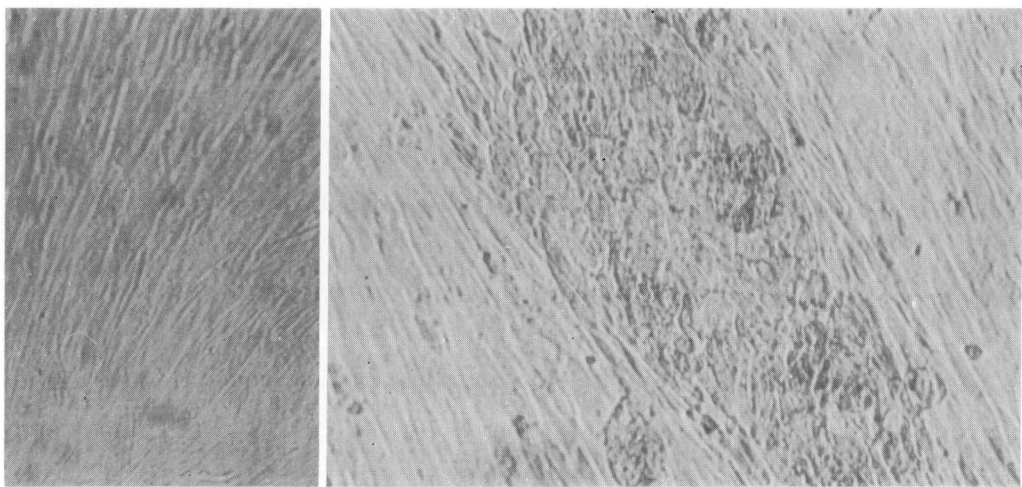

FIG. 11-8. *Cytomegalovirus cytopathic effect (CPE) on WI-38, a human diploid cell line. Left, normal. Right, CPE.*

Fluorescent Antibody Methods. Elevated IgM levels are often found in infants with congenital cytomegalovirus infection, as with congenital infection with syphilis, rubella, or toxoplasmosis. Using an indirect fluorescent antibody method, the infant's maculoglobulin fraction can be tested to determine if the infant has cytomegalovirus-specific IgM antibody which would indicate fetal infection.[61]

Biologic Characteristics of Clinical Interest

Latent or Chronic Infection.[63] Infection often persists in spite of neutralizing antibodies. In some cases, viruria may persist for years, without clinical disease. Reactivation can occur.

Species Specificity. Many animal species have their own strain of cytomegalovirus, which will grow only in its own cell lines.[63] As is the case of EB virus infections in humans (discussed later in this chapter), CMV infection of mice inhibits T cell function.

Transmission. The virus has been recovered from human milk,[49] semen[53], and from the cervix during pregnancy.[56] Transmission from urine occurs in institutions with poor sanitation, and transfusion of fresh blood is clearly a source, with an incubation period of 3 to 5 weeks.[49] Reactivation of a latent infection acquired at birth may be more important.[63] Venereal transmission of cytomegalovirus mononucleosis apparently occurs,[46] apparently with an incubation period of 4 or more weeks (like Epstein–Barr virus).

Treatment

Adenine arabinoside or phosphonacetic acid is effective antiviral therapy in CMV infection of mice.[59] Adenine arabinoside or transfer factor therapy are under investigation for human disease and appear to be of some benefit.

Prevention

Control of Exposures. The virus is excreted in high concentration in the saliva and urine in infants with congenital infections. Pregnant women should avoid exposure to these infants, if possible, although transmission of the virus from the baby to nursery contacts has not been adequately studied. Unlike congenital rubella, which is highly contagious, congenital cytomegalovirus appears to be of little risk to attendants exercising the usual isolation precautions.

Avoid Subsequent Pregnancies. Congenital infection can occur in more than one pregnancy,[62] and the risk of subsequent spontaneous abortions appears to be increased.[44]

Avoid Venereal Exposure. Cervical cytomegalovirus infection appears to be correlated with venereal exposure, although this is not yet conclusively proved.[46]

Vaccine. A live attenuated CMV vaccine is currently under investigation, and may prove useful in preventing CMV disease after renal transplantation.

EPSTEIN–BARR VIRUS

Objectives

1. Describe how Epstein–Barr virus is detected in the laboratory.
2. List the chronic diseases that frequently have of higher than normal EB virus antibody titers.
3. Describe the evidence that EB virus is the cause of heterophile-positive infectious mononucleosis.

Definitions

Epstein–Barr virus (EBV) was first described by Epstein, Barr, and Achong,[66] and was named after the EB cell culture line.[69] The virus was discovered in cells cultured from biopsy specimens of Burkitt's lymphoma, a tumor of the jaw found primarily in African children. Burkitt is a British physician who recognized the unusual frequency of this lymphoma in certain regions in Africa, and postulated an infectious etiology for the tumor.

Frequency and Importance

The virus has a worldwide distribution. Many individuals have serologic evidence of Epstein–Barr infection by 2 years of age, but this varies with geographic region and socioeconomic group. Only those individuals without EBV antibodies are likely to develop heterophile-positive infectious mononucleosis.[68] EBV has additional importance because of its association with Burkitt's lymphoma and other rare malignancies.

Clinical Patterns of Illness

Epstein–Barr virus is the *usual* cause of:

Infectious mononucleosis syndrome has several of the following clinical findings: fever, pharyngitis, generalized lymphadenopathy, and splenomegal, and often swollen eyelids (Fig. 11-9). The peripheral blood smear often shows many atypical lymphocytes and the serum transaminase is often elevated.

The patient's serum often contains antibodies called infectious mononucleosis heterophile antibodies that agglutinate sheep erythrocytes and which are absorbed by guinea pig antigen.

The clinical findings of infectious mononucleosis and atypical lymphocytes, with a negative heterophile, is called the heterophile-negative infectious mononucleosis syndrome. EBV is the most frequent cause of this syndrome, but cytomegalovirus and toxoplasmosis are other uncommon causes.

The first clue that Epstein–Barr virus was related to infectious mononucleosis resulted from the observation that a virology technician working with the virus developed EBV antibodies concurrently with infectious mononucleosis.[69] Subsequently, it was found that leukocyte cultures from patients with infectious mononucleosis regularly contain EBV.[71]

Retrospective studies of stored sera indicated that of 94 entering college students with EBV antibody, none developed infectious mononucleosis in college.[68] Of 268 entering college students without EBV antibody, about 15 percent developed infectious mononucleosis, indicating

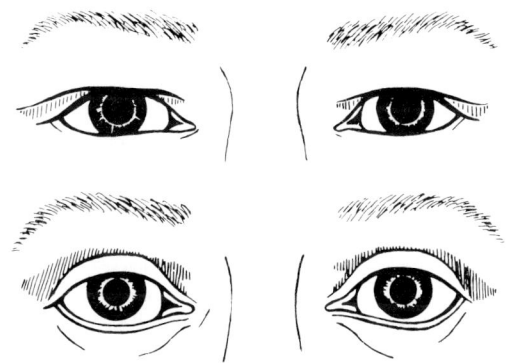

FIG. 11-9. Top, *Periorbital edema in infectious mononucleosis.* Bottom, *normal appearance of eyelids.*

EBV antibody is associated with immunity to infectious mononucleosis. Missing evidence needed to confirm this relationship is production of the disease by inoculation of the virus into antibody-negative humans.[71] Because of the association of EBV with malignancies, this is not likely to be done in humans, but can be done in nonhuman primates.

Epstein–Barr virus is a *rare* cause of several syndromes, but it is customary to refer to these patterns as complications or unusual manifestations of infectious mononucleosis if the heterophile test is positive. These include:

Severe acute complications of pharyngeal airway obstruction or ruptured spleen.[81]

Neurologic Presentations. Encephalitis, cranial nerve paralysis, ascending paralysis (Guilain–Barré syndrome) and transverse myelitis can occur.

Hematologic Complications. Hemolytic anemia, thrombocytopenia, and pancytopenia rarely occur.

Other Rare Complications. Arthritis, agammaglobulinemia, myocarditis, pericarditis, acute liver failure, acute renal failure, and interstitial pneumonia are extremely rare.[81]

Asymptomatic or Subclinical Infection

There is a gradual increase in the frequency of EBV antibody with increasing age.[79] The peak frequency of typical infectious mononucleosis is in children and young adults. However, it can occur at any age, even in adults over 60 years of age, in whom it may be confused with leukemia or hepatitis.[65]

Most individuals with antibody do not give a history of a recognized infectious mononucleosis-like illness. Children younger than 10 years of age rarely have a typical infectious mononucleosis-like illness, and rarely have a positive heterophile. It is likely that a number of syndromes can be caused by EBV, but the spectrum of illness has not yet been fully defined.

Laboratory Approach

Serum Antibodies. EBV antibodies can be detected by an indirect fluorescent antibody (FA) method.[79] Burkitt tumor cells containing the virus are used as antigen, since the virus has not yet been grown free of these cells. The specific EBV antibody in the patient's serum adheres to the antigen in the cells after the rest of the serum has been washed off. Fluorescein-labeled antihuman globulin is then added, and adheres to the antibody, and the fluorescein is then detected under a fluorescent microscope. (see Fig. 15-6). The EBV antibody in the patient's serum can be titered. This EBV antibody is an IgG immunoglobulin which is transplacentally transmitted, but is gone by 6 to 7 months of age,[82] in contrast to the infectious mononucleosis heterophile antibody, which is an IgM antibody, and is not transplacentally transmitted.

The indirect fluorescent EBV antibody has usually reached high levels by the time the patient with infectious mononucleosis sees a physician, and often does not rise further. Levels of EBV antibody remain high for months. Therefore, EBV antibody is less useful in the serologic diagnosis of infectious mononucleosis than is the heterophile antibody, which is readily detectable in simple slide tests (such as the Mono-Spot test). A positive slide test is virtually diagnostic of recent or current infection when detected. EBV-specific IgM antibodies to the viral capsid or nuclear antigen can be measured,[72] but are unlikely to become as readily available as the heterophile slide tests.

A variety of animal erythrocytes can be used for detecting heterophile antibody (antibody to cells of other species). The horse erythrocyte agglutination test is slightly more sensitive than the beef erythrocyte hemolysin test, which is in turn slightly more sensitive than the sheep erythrocyte agglutination test, using EBV IgM antibody as the standard.[67]

Culture of the Virus. The virus grows in explant cultures of tumor and in leukocytes, but only a small proportion of the cells appear to be infected, by electron microscopy. Recovery of the virus in cell cultures is not yet a practical method for diagnosis of EBV infection in patients, but is useful in studies of the virus.

Throat washings from individuals recovering from infectious mononucleosis can be inoculated into fresh human leukocytes from cord blood. The EB virus can be detected by its transformation of the leukocytes so that they continue to proliferate, with detection of EBV antigen by CF or FA techniques applied to the transformed cells.[78] These studies indicate that the leukocyte-transforming agent often persists for many months after recovery from infectious mononucleosis.[78] However, leukocyte-transforming agents were also recovered from the pharynx of almost 20 percent of patients in a general outpatient clinic, so this test cannot be taken as conclusive evidence for EBV excretion.

Smear. Electron microscopy of infected cells is not practical for diagnosis, except in research studies of an individual patient (Fig. 11-10).[76]

IgE. Also elevated in allergic disorders, the IgE level in the serum rises sharply early in the course of infectious mononucleosis.[64]

Biologic Characteristics of Clinical Interest

Transformation. Leukocyte cultures of normal individuals with no EBV antibody ordinarily do not undergo spontaneous lymphoblastic transformation (ability to multiply in cell cultures indefinitely). However, Epstein–Barr virus can produce transformation in such cells. Many healthy individuals with EBV antibodies have leukocytes which can eventually be established as a lymphoblastoid cell line.[70]

Acquired Agammaglobulinemia. In some families, males have developed agammaglobulinemia with a severe infectious mononucleosis-like clinical illness after an EBV infection.[80] Peripheral B lymphocytes, the source of immunoglobulins, were quantitatively normal, but did not respond to antigenic challenges by production of antibodies. EBV virus appears to infect B lymphocytes preferentially, and perhaps produces proliferation of T lymphocytes (which are the atypical

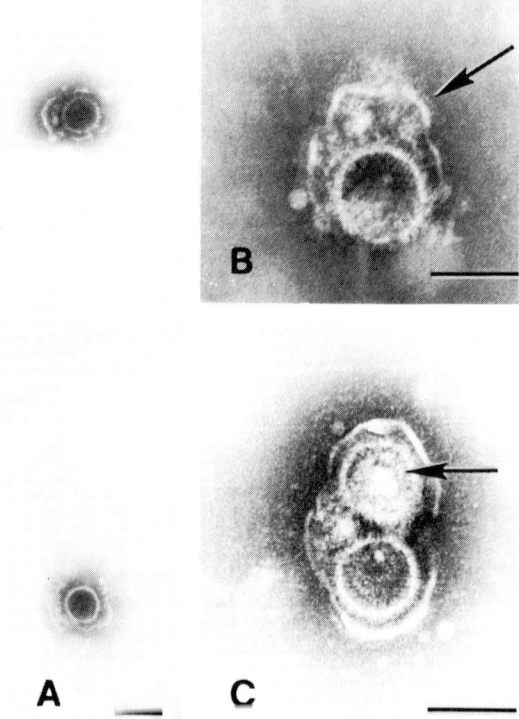

FIG. 11-10. Electron micrograph of EB virus in specimens from the pharynx.[76] All phosphotungstic acid stained virus particles shown in this figure were found in the throat washing from a renal transplant patient. (A) Two herpes-like virus particles in the same field. Both have intact capsids and fragmented envelopes. (B) At higher magnification capsid detail and envelope projections (arrow) are evident. (C) Two capsids surrounded by a single envelope. The negative stain has penetrated one of the capsids and revealed the viral core as well as the inner capsid layer (arrow). Bars represent 100 nanometers. (Lipman, M., Andrews, L., Niederman, J., and Miller, G.: Direct visualization of enveloped Epstein–Barr herpesvirus in throat washing with leukocyte-transforming activity. J. Infect. Dis., 132: 520–523, 1975; Reproduced with permission of the University of Chicago Press)

lymphocytes seen), which may attack and destroy the B lymphocytes.

Contagiousness. Individuals without EBV antibody exposed to a member of the family with infectious mononucleosis develop EBV antibodies only in about 10 to 15 percent of cases.[74] Antibodies to Epstein–Barr virus are acquired more frequently and at an earlier age than are antibodies to herpes simplex virus and cytomegalovirus; about 60 to 80 percent of individuals have EBV immunofluorescence antibody by about 2 years of age.[79] Patients from lower socioeconomic groups acquire EBV antibody slightly more frequently than do those from higher socioeconomic groups.[79]

The virus can be detected in throat garglings of patients with infectious mononucleosis as long as 16 months after the onset of the clinical illness, so they presumably are contagious for many months.[78]

Because of the persistence of the virus in saliva, and the early observation that West Point cadets got the disease about a month after vacations, but rarely infected their roommates, it has been postulated that kissing is an important mode of transmission.

Heterophile. Children younger than 10 years of age rarely have a positive heterophile, although infectious mononucleosis-like illness occurs in such young children. The heterophile antibody of infectious mononucleosis is defined as a sheep erythrocyte agglutinin which is not absorbed by guinea pig kidney antigen. There is as yet no explanation for the observation that adolescents and young adults regularly produce the heterophile antibody with EBV infection, while young children rarely do.

Association with Tumors. Epstein–Barr virus is associated with Burkitt's lymphoma[75] and nasopharyngeal carcinoma.[77] Herpes-like virus particles can be seen in cell cultures derived from Burkitt's lymphoma, and the EBV genome can be detected.

The Epstein–Barr virus genome can be demonstrated to occur in a plasmid or episomal form (Fig. 11-11). It is found intracellularly in the chromosomes of Burkitt's lymphoma, nasopharyngeal carcinomas, and the B lymphocytes of persons with infectious mononucleosis. The EBV plasmid is a closed circular, supercoiled, double-stranded DNA molecule.[78A] Its biological significance may be that it is the molecular form taken by herpes-group virus genomes when they are latent in cells.

All patients with Burkitt's lymphoma and nasopharyngeal carcinoma have a high titer of antibodies to EBV. Other B cell lymphoproliferative tumors possibly associated with EBV include B-cell sarcomas and plasmacytomas.

Association with Other Diseases. Patients with sarcoidosis or leprosy have high titers of antibodies to EBV, compared to normal individuals.[73] A defect in cell-mediated immunity may be involved. Immunosuppressed patients may develop reactivation of EBV, such as occurs in other herpesviruses.

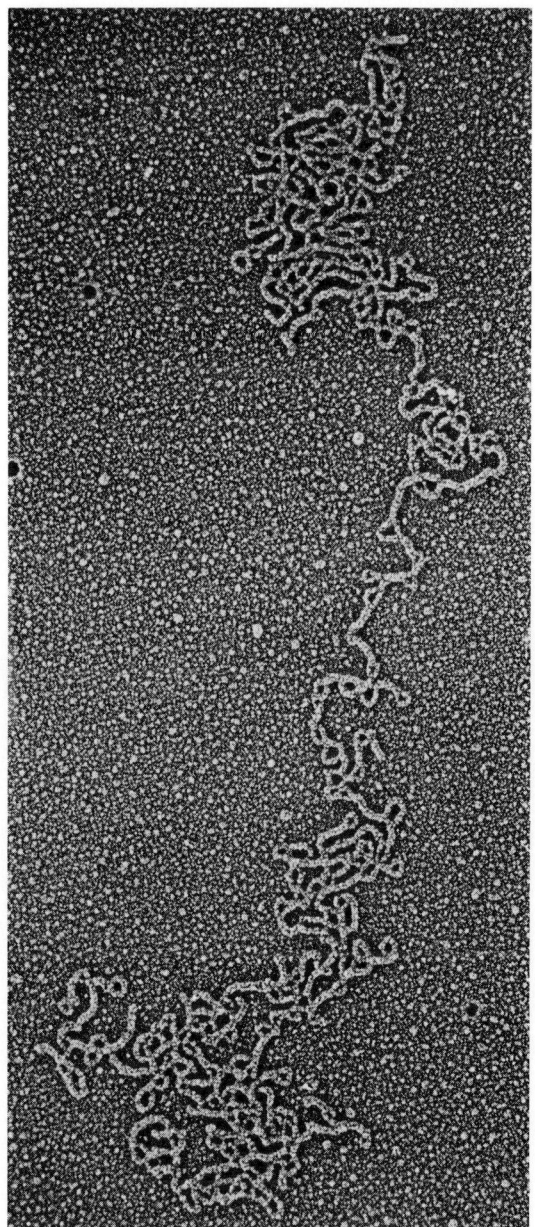

FIG. 11-11. *Electron micrograph of the EB virus genome in its plasmid or episomal form, as found in the chromosomes of human tumor cells of Burkitt's lymphoma or nasopharyngeal carcinoma, and the B lymphocytes of persons with infectious mononucleosis. (Pagano, J. S.: The Epstein–Barr virus plasmid. In ICNA–UCLA Symposium on Extrachromosomal DNA. New York, Academic press [In press])*

Transmission by Blood Transfusion. Like serum hepatitis virus and cytomegalovirus, EBV can be transmitted by blood transfusion.[70] Most patients have no apparent illness after transfusion infection with EBV. The incubation period appears to be about 6 to 8 weeks.

Treatment and Prevention

No specific treatment is available, but ara-A (adenine arabinoside) is under investigation. Vaccines are not likely to be developed soon because of the association of EBV with malignancy.

ADENOVIRUSES

Objectives

1. Describe four syndromes frequently produced by adenoviruses.
2. Describe how adenovirus infections can be detected in the laboratory.

Definitions and Types

Adenoviruses are named after adenoids, from which they were first grown in 1953. Other older names for this group of viruses include AD (Adenoid Degeneration) agents and APC (Adenoidal-Pharyngeal-Conjunctival) viruses. The definitive characteristic of an adenovirus is the presence of a common complement-fixation antigen.

Approximately 30 types of adenoviruses have been recovered from humans. Immunity is type-specific; infection with one type does not protect the individual from future infection with other types.

Frequency and Importance

Adenoviruses are a frequent cause of respiratory illnesses in children and young adults. Adenoviruses can produce tumors when injected into animals, and have investigational importance in this area.

Clinical Patterns of Illness

Adenoviruses are a *common* cause of:

Nonstreptococcal Pharyngitis. Pharyngitis with a throat culture negative for beta-hemolytic streptococci is often due to an adenovirus, especially in preschool children.[102]

Nonpurulent Conjunctivitis. In children and young adults, nonpurulent conjuctivitis is often due to an adenovirus. Sometimes the cornea is also involved (keratoconjunctivitis), producing a foreign body sensation (Fig.11-12).

Pharyngo-conjunctival Fever. This syndrome has been observed in outbreaks in military recruits and civilians, and is typically due to an adenovirus.[83]

Adenoviruses are an *occasional* cause of:

Influenza-like illness is also called Acute Respiratory Disease (ARD) or acute febrile respiratory illness, and often occurs in outbreaks in military recruits.[95] Pharyngitis and conjunctivitis are frequently present in this syndrome, which also occurs sporadically in children and young adults.

Atypical Pneumonia. This syndrome is often due to adenovirus.[87] Pharyngitis and rhinitis are usually present. The white blood count is usually normal with a neutrophilia, but can be as high as 30,000.[87]

Bronchiolitis. This syndrome of acute lower respiratory obstruction in young infants is sometimes caused by an adenovirus. During a period of 5 years of observation at Children's Memorial Hospital in Chicago, an adenovirus was recovered from 15 percent of 117 infants with bronchiolitis, compared to six percent of 286 normal controls.

Adenoviruses are a *rare* cause of:

Bronchiectasis. In Canada, an adenovirus type 3 outbreak in infants was associated with a severe pneumonia, often with permanent lung damage, including bronchiectasis.[97]

Progressive Fatal Pneumonia. Adenovirus has been recovered from the lung, and typical inclusion bodies have been seen in the lung in infants and in healthy young recruits with this syndrome.[90] These patients usually have a leukopenia, and the adenovirus recovered from the lung is usually Type 7.[90]

Adenoviruses are a *suspected* cause of several syndromes associated with recovery of the virus from patients with the disease, but alternative etiologies have not been adequately excluded. These include:

Pertussis-like Syndrome. Adenoviruses have been recovered from patients with illnesses resembling

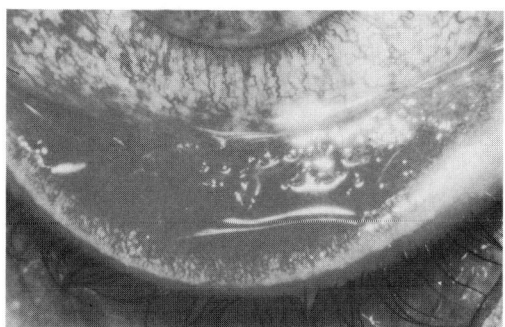

FIG. 11-12. *Severe conjunctivitis due to adenovirus. (Photo from Dr. Fred Brightbill)*

pertussis (whooping cough), including some patients with extreme lymphocytosis, and some with extreme eosinophilia.[104] The etiologic role of adenoviruses in pertussis-like illnesses is not yet established.

Unilateral Hyperlucent Lung. This is a complication of segmental collapse and compensatory overinflation, and has been observed after adenovirus pneumonia.[100]

Severe Infantile Diarrhea. This syndrome is associated with recovery of an adenovirus more frequently than controls.[101] However, experimental infection with adenovirus in adults does not produce diarrhea.[89]

Mesenteric Adenitis. Adenoviruses have been recovered from the mesenteric nodes at operation for suspected appendicitis, with a postoperative diagnosis of mesenteric adenitis.[85]

Intussusception. Adenoviruses have been recovered from mesenteric nodes obtained at operation for intussusception.[88] Adenovirus-like virions and inclusion bodies have been observed in an appendix removed incidentally at an operation for intussusception.[106]

Aseptic Meningitis Syndrome. An adenovirus has rarely been recovered from the spinal fluid.[98] Typically there is fever, headache, and pharyngitis, with usually a predominance of lymphocytes in the CSF.

Hemorrhagic Cystitis. This disease is characterized by gross hematuria and bladder symptoms such as suprapubic pain and tenderness, with pain on voiding.[103] Gross hematuria has many other causes, such as bacterial infection, trauma or post-streptococcal glomerulitis. However, adenoviruses have been recovered in this syndrome with the more frequent causes excluded.

Asymptomatic or Subclinical Infection

Neutralizing antibodies against various adenovirus types are found with increasing frequency from infancy on. Adenoviruses are rarely recovered from the throat of normal individuals, but are recovered from the feces of about five percent to ten percent of normal children.[101] Serial studies indicate that many infants may excrete an adenovirus in the stool intermittently for months after an infection.[93] For this reason, fecal recovery of an adenovirus may be unrelated to the patient's current illness. Recovery of an adenovirus from the throat is much more likely to be etiologically related to the patient's current illness.

Laboratory Diagnosis

Viral Cultures. Culture of adenovirus is relatively easy in primary or continuous cell cultures, such as human embryo kidney or HEp-2. The cytopathic effect of adenoviruses in cell culture is usually distinctive (Fig. 11-13), so that a reliable presumptive laboratory diagnosis of an adenovirus can often be made on the basis of cytopathic effect. Confirmation as an adenovirus can be done by identifying the type by CPE neutralization tests, or by demonstrating that the isolate can act as an adenovirus complement fixation antigen.

Serum Antibodies. Serologic diagnosis is usually based on the complement fixation (CF) test on paired sera. However, if cultures are negative for adenovirus, the CF test is very unlikely to detect an infection, in an acute illness.[102] In addition, the CF test fails to detect a considerable number of infections detected by the more cumbersome type-specific neutralization tests.

Histologic Diagnosis. Electron micrographs of adenoviruses are quite distinctive, showing a regular crystal-lattice like pattern (Fig. 11-14). Intranuclear inclusions are characteristic but not absolutely diagnostic. Adenoviruses have been detected in throat swab specimens using immune electron microscopy.[105] Rapid diagnosis of adenovirus infections has been achieved by FA staining of throat scrapings.[99]

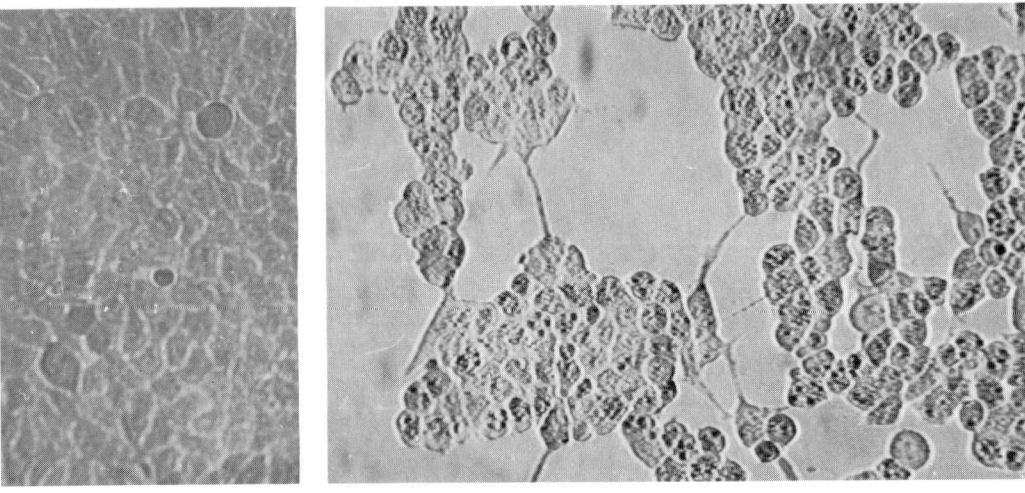

FIG. 11-13. *Adenovirus type 3: cytopathic effect on Hep-2 cell cultures. Normal cells at left.*

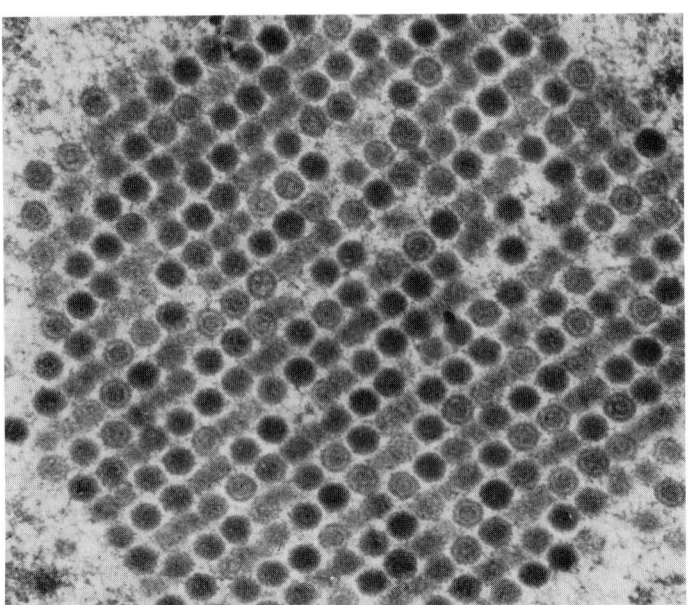

FIG. 11-14. *Adenovirus. Electron micrograph showing crystal-like arrangement.*

Biologic Characteristics of Clinical Interest

Tumor Production. Sarcomas have been produced in hamsters by intramuscular injection of some adenovirus types.[96] T antigens (tumor antigens) were first discovered in adenoviruses. However, antibodies to T antigens have not been found in humans with malignancies.[94]

Adenovirus-Associated Viruses. These are not adenoviruses, but are defective viruses which cannot grow in tissue culture unless an adenovirus is also present. Serologic evidence suggests that these viruses are associated with respiratory disease.[86]

Biologic Variation in Adenovirus Types. Type 4 is almost never recovered from children, but is a common cause of severe outbreaks of acute respiratory disease in military recruits. Type 7 is the usual type recovered in fatal adenovirus pneumonia.

Conjunctival Route of Infection. Inoculation of the conjunctivae in susceptible volunteers regularly produced conjunctivitis, with nasal obstruction and discharge, sore throat, and occasionally cough and headache; whereas inoculation of the nasopharynx with the same adenoviruses produced rhinitis in a small proportion of volunteers which was not significantly different from the rate of rhinitis in uninoculated controls.[84]

Treatment

DNA inhibitors, such as idoxuridine, are of theoretical value for therapy of a life-threatening pneumonia due to adenovirus, but are of no practical value, because of the difficulty in proving the diagnosis. Idoxuridine was of no value in virologically proven adenovirus conjunctivitis, when compared to control patients.[92]

Prevention

Live attenuated adenovirus vaccines have had clinical investigational trials in military recruits, for whom acute respiratory disease due to adenovirus is a major cause of hospitalization.[91] However, the importance of prevention of illness due to adenovirus infection in other populations has not been established. Adenovirus Type 4 vaccine has been effective in reducing disease in recruits, but disease due to Type 7 then became more frequent. At present, combined Type 4 and 7 vaccine appears to be effective in preventing respiratory disease, when given orally.[91]

SMALLPOX AND VACCINIA VIRUSES

Objectives

1. Discuss the reasons for the discontinuation of routine immunization against smallpox in the United States in 1970.
2. Describe the three most frequent complications of smallpox vaccination.
3. Describe what laboratory facilities might be available to identify smallpox virus and distinguish it from chickenpox virus and discuss the importance of this maneuver.
4. Discuss the indications, the sources, and the availability of vaccinia immune globulin.

Definitions

The word vaccination is derived from *vacca*, the Latin word for cow. Vaccinia virus is the virus used in smallpox vaccination. The official name is *Poxvirus officinale*, and the present strain now used as vaccine is of uncertain origin.[107] It is not the same virus as wild cowpox virus, and is probably highly attenuated smallpox virus. Other poxviruses include *P. variolae* (smallpox) and *P. bovis* (cowpox).

Importance

Smallpox has not occurred in the United States since 1949. Its importance is based on the massive public health effort which would be called forth if a single case should occur in the United States. An example of this type of effort occurred in 1965 when a patient entered Washington D.C. from Ghana and was thought to have smallpox, as described later in this section.

Frequency and Geographic Distribution of Smallpox

Worldwide eradication programs based on smallpox vaccination have led to a steady reduction of worldwide cases since 1960. In 1963, about 132,000 cases were reported, compared to about 31,000 cases in 1970, when routine smallpox immunization was discontinued in the United States.

Brazil was the last country in the Western hemisphere in which smallpox was endemic, and reported aboout 7,000 cases in 1969. Bangladesh, India, Pakistan, Somalia, Kenya, and Ethiopia were important endemic areas in the early 1970s.

Eradication of smallpox in isolated villages in the mid-1970s was aided by paying a large cash reward for reporting the disease.[119] By 1977, only about 3200 cases were reported for the year. In 1978, no cases had been reported for more than a year.

A laboratory-associated case of fatal smallpox occurred in a medical photographer in a medical school in England in 1978, presumably by airborne transmission. She worked in the same building, one floor above the virology laboratory. Over 200 of her contacts were placed under surveillance. Her mother was the only contact to develop smallpox. This incident has led to stricter controls on the availability of the virus for laboratory study.

Clinical Patterns of Smallpox Virus Infection

A description of the variety of smallpox illnesses is useful, even through the physician in the United States will probably never encounter the disease. Individuals entering the United States may have been exposed in endemic areas, or in smallpox-free areas where smallpox has recently been introduced, as in London in 1973 and 1978 and Yugoslavia in 1972.

Classical Smallpox (Variola Major). Most commonly, the illness begins about 12 days after exposure, with fever, prostration and headache.[109] The eruption begins about 3 days after the onset of fever. The lesions progress through the stages of macule, papule, vesicle, pustule, eschar and scar. Confluence of the lesions on the face and arms, secondary fever, or laryngeal lesions occur in the four severe forms. Absence of confluence or secondary fever, and hard, pearly lesions, with abortion of some lesions, occurs in the three milder forms, which have a mortality of zero to two percent.

Typically, the rash is denser on the upper face, back and dorsal surfaces of hands feet. The lesions are all in the same stage at the same time. History of vaccination does not exclude the diagnosis, and disease of any severity may occur in the vaccinated person.

Overall mortality rates for variola major in an unvaccinated population are estimated to be 20 percent to 50 percent.[109]

Variola Minor (Alastrum). This form of smallpox is due to a permanent variant of the smallpox virus, definably by laboratory studies or epidemiologic observations. Any degree of severity may occur, but typically the illness is milder, with fewer lesions.

Dixon has objected to the terminology of "classical" or "atypical" smallpox.[109] He has classified smallpox cases into nine types on the basis of secondary fever, laryngeal lesions, mental symptoms, and the extent and appearance of the rash, and has illustrated this classification with excellent pictures. His two extreme types (fulminating and *variola sine eruptione*) are described below.

Fulminating Hemorrhagic Smallpox. In this form of the disease, anxiety is prominent, hemorrhages appear early, especially in mucous membranes, and the rash is often absent, or soft and velvety, with death occurring in a few days, before pustulation of the rash occurs.[109]

Inapparent Infection (Variola Sine Eruptione). Sometimes conjuctivitis is the only manifestation of illness.[111]

Clinical Patterns of Vaccinia Virus Infection

Vaccinia virus and smallpox virus infection have many similarities. It is useful to be familiar with the course of primary infection with vaccinia virus, since smallpox disease resembles hundreds of primary vaccinations over the entire body.

Primary Reaction to Smallpox Vaccination. After inoculation of the skin with vaccinia virus for the first time, a small papule or vesicle appears in about 3 days, and progresses to a pustule, then to a black crusted ulcer, then to a scar. The peak of the reaction is about 7 to 10 days after inoculation, when the flat pustule is about 1 cm in diameter, often surrounded by an area of tense erythema about 4 cm in diameter. At this point, the patient may have a temperature of about 104°F. Gradual improvement and decreased fever typically occurs after the crust appears on the pustule.[119]

Major and Equivocal Reactions to Smallpox Vaccination. If the patient has had a previous smallpox vaccination, a revaccination can produce a major reaction, such as an area of definite palpable induration. There may be a small pustule and crusted pustule. Fever is minimal or absent, and the evolution of the lesion is accelerated so that the lesion is dry in about 5 days. The older terms "accelerated reaction" and "immune reaction" have been abandoned.

An equivocal reaction is anything other than a major reaction.[119]

Complications

In the United States, vaccinia virus was an occasional cause of several diseases.[110,113]

Vaccinia-Infected Eczema. Occasionally a child with atopic dermatitis (eczema) is exposed to someone who has been recently vaccinated, and the vaccinia virus invades the eczematous skin (eczema vaccinatum). High fever and purulent exudate over the entire area of the dermatitis, as a primary reaction to the virus, occurs. This reaction is often not recognized as a virus infection, and is often thought to be a bacterial infection. Eczema vaccinatum is occasionally fatal.

Herpes simplex virus also can infect eczematous skin, and the resulting purulent skin infection usually cannot be distinguished from that produced by vaccinia virus. Eczema infected by either virus is sometimes called Kaposi's varicelliform eruption.

Purulent Conjunctivitis. Accidental inoculation of the eye from the primary inoculation site may produce purulent conjunctivitis, which may appear very severe. Often a typical skin lesion can be found near the eye.

Pustular-Ulcerative Skin Lesions. Accidental vaccination of skin or mucosa can occur, by transfer of the virus from the site of primary inoculation, usually by the patient's fingers. The diagnosis is not difficult if the patient is the source, but may not be considered if the patient was exposed to a vaccination on someone else.

Other Complications of Vaccinia Virus Infection. Encephalitis is very rare, with a 30 percent mortality rate and brain damage in about 20 percent.[113] Progressive vaccinia occurs in immunocompromised hosts. The inoculated site does not heal, but progresses, with necrosis, with a fatality rate of about 20 percent.[110,113]

Laboratory Approach

When smallpox is suspected in a smallpox-free country, rapid and exact laboratory diagnosis is urgent. Public health authorities should be informed early, and specimens are usually flown to the most expert reference laboratories available.[115] Isolation of the patient *at home* has been recommended to avoid transmission of the diseases if a hospital does not have extremely effective contagion facilities.[115].

In the rare situation of suspected smallpox, the usual diagnostic problem is to distinguish smallpox virus from chickenpox virus. In the situation of suspected vaccinia virus infection, the usual diagnostic problem involves distinguishing vaccinia virus from herpes simplex virus in patients with infected eczema or conjunctivitis.

Electron Microscopy. Poxviruses (smallpox, vaccinia) can be distinguished from herpesviruses (herpes simplex, varicella–zoster) on the basis of size and shape under electron microscopy (Fig. 11-15).[112] The method is rapid and sensitive for detection of virus in pus or crusts. Recovery of the virus by culture and identification by neutralization with specific antiserum is necessary to distinguish between vaccinia and smallpox virus.

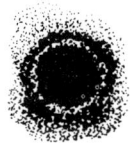

HERPESVIRUS   **POXVIRUS**

FIG. 11-15. *Comparison of size and appearance of smallpox virus (a poxvirus) and chickenpox virus (a herpesvirus) on electron microscopy.*

Virus Culture. Vaccinia virus and smallpox virus grow readily in cell cultures. The cytopathic effect of vaccinia virus resembles that of herpes simplex virus, but herpes simplex virus does not grow well on rhesus monkey kidney cells, as vaccinia virus does.

Identification of the virus isolates can be confirmed by neutralization of the virus by specific antisera.

Serum Antibodies. Serologic tests are not necessary for diagnosis of vaccinia, since virus is easily isolated from pustules or crusts. If smallpox is suspected, serum should be obtained for possible serologic studies.

Problems of Identification. Occasionally, the laboratory identification of smallpox virus may be very difficult, as illustrated by the following example.[108] In May, 1965, in Washington, D.C., a 31-year-old woman arrived from Ghana and developed a disease resembling smallpox. Blood and vesicular fluid from this patient were flown to the Center for Disease Control in Atlanta, Georgia. The fluorescent antibody tests were positive for smallpox, and negative for herpes simplex and chickenpox. The complement fixation serologic tests were inconclusive. Isolation attempts in chick embryos were suspicious for smallpox virus.

In the next week, 100 public health workers tracked down 118 direct contacts, 665 indirect contacts, and 246 secondary contacts, vaccinated them, and placed them under surveillance. However, the pocks on the eggs were later found to be nonspecific. The virus was grown on cell culture and shown to be chickenpox (varicella–zoster) on the basis of CPE neutralization, comparing varicella–zoster virus antiserum to smallpox virus antiserum. The patient developed an antibody response of 1:8 to 1:256 against chickenpox virus and no antibody rise against vaccinia virus.

This incident illustrates the occasional clinical and laboratory confusion of chickenpox with smallpox, and the massive public health vaccination program which may be done when a suspected smallpox case is recognized in the United States.

Biologic Characteristics of Clinical Interest

Stability. Dry crusts which have fallen to the floor may contain live virus after a year at room temperature. For this reason, hospital rooms which have had a patient with eczema vaccinatum should not be used for patients with eczema unless extremely thorough cleaning has been done, and environmental cultures for vaccinia virus are negative.

Animal Poxviruses. Vaccinia virus is closely related antigenically to monkeypox, rabbitpox, and mousepox viruses. Monkeypox virus can produce an illness closely resembling smallpox in unvaccinated humans exposed to monkeys.

Killed by Antiviral Agent. Methisazone, which prevents maturation of the virus particles, can kill vaccinia and smallpox virus in animal models, and has had a limited use in humans.[116]

Treatment

Vaccinia Immune Globulin (VIG) can be used to treat accidental vaccination of the eye and eczema vaccinatum. However, since smallpox vaccination is now done so rarely, the preparation is rarely needed, and production of VIG was discontinued by the commercial producer in 1977. VIG had been manufactured by separating out the gamma globulin fraction from plasma obtained from volunteers (often servicemen) immediately after a primary reaction. It was distributed without charge from about nine regional Center for Disease Control Public Health Service Quarantine Centers, after approval by the regional medical consultant.[119] VIG is effective in preventing or modifying illness in susceptible individuals exposed to smallpox.[113]

Methisazone. This drug has been used to treat progressive vaccinia.[116]

Prevention

Vaccine. Smallpox vaccination is effective in preventing smallpox. The Center for Disease Control, Atlanta, Georgia, maintains a stock for emergencies.

Hyperimmune Globulin. Vaccinia immune globulin can prevent smallpox after exposure,[113] but is in short supply.

Chemotherapy. Methisazone can prevent smallpox after exposure, and has been used to treat progressive vaccinia.[116]

Recommendations for Smallpox Vaccination. Routine smallpox vaccination of children was discontinued in the United States in 1970.[114,118] In 1976, because of the decrease in smallpox in the world, routine smallpox vaccination of hospital and health personnel in the United States was discontinued.[117] In May 1978, smallpox vaccination was required for travelers entering the United States only if smallpox was ocurring in the country from which they came, but by August 1978, such vaccination was not required for any arriving traveler.

Smallpox vaccination is still necessary for travelers to those few foreign countries that require it for entry. It is also indicated for laboratory personnel who work with the virus, although such work should be limited to a few centers which use extreme precautions.

At the present time, a letter from a physician stating that the traveller should not be given smallpox vaccine "for medical reasons" is recommended instead of vaccine, if the traveller is going to a country with an obsolete requirement for the vaccination. The physician's official letterhead stationery should be used.

OTHER DNA VIRUSES

Monkeypox

The monkeypox virus is closely related to the smallpox virus. It occurs naturally in primates in Africa, and occasionally infects humans. It can be mistaken for smallpox.[120]

Molluscum Contagiosum

Molluscum contagiosum is a very common poxvirus of humans, and occurs primarily in children.

The lesion consists of a number of flesh-colored papules, which may become umbilicated. The lesion has a pulpy white core which is often scratched out by the patient. Occasionally, the papule is surrounded by a peculiar red halo.[121]

The virus can be grown in cell culture, but usually the diagnosis is made on clinical appearance.[123]

Orf

Orf virus is a poxvirus that causes a skin disease of young sheep or goats, but is occasionally transmitted to humans. The virus persists in soil from season to season.

In humans the lesions characteristically occur on the hands, and consist of a red, raised, slightly tender nodule with an ulcerated surface.[122] The lesion may be mistaken for a skin malignancy. Acute and convalescent sera can be shown to have a rise in titer of orf complement fixation antibodies in a reference laboratory.

REFERENCES

Herpes Simplex Virus

1. Anon.: Genital herpes and cervical carcinoma (editorial). Br. Med., J. *1*:807, 1978.
2. Baringer, J.R., and Swoveland, P.: Recovery of herpes-simplex virus from human trigeminal ganglions. New Engl. J. Med., *268*:648–650, 1973.
3. Barr, R.J., Herten, J., and Graham, J. H.: Rapid method for Tzanck preparations. J.A.M.A., *237*:1119–1120, 1977.
4. Berger, R.S., and Papa, C.M.: Photodye herpes therapy—Cassandra confirmed? J.A.M.A., *238*:133–134, 1977.
5. Francis, D.P., et al.: Nosocomial and maternally acquired *Herpesvirus hominis* infections. A report of four fatal cases in neonates. Am. J. Dis. Child., *129*:889–893, 1975.
6. Haynes, R.E.: The spectrum of herpes simplex virus infections in children. South. Med. J., *69*:1069–1078, 1976.
7. Huxsoll, D.L., and Hemelt, I.E.: Clinical observations of canine herpesvirus. Am. J. Vet. Med. Assoc., *156*:1706–1713, 1970.
8. Kern, A.B., and Schiff, B.L.: Vaccine therapy in recurrent herpes simplex. Arch. Dermatol., *89*:844–845, 1964.
9. Kern, E.R., Overall, J.C., Jr., and Glasgow, L.A.: *Herpesvirus hominis* infection in newborn mice. I. An experimental model and therapy with iododeoxyuridine. J. Infect. Dis., *128*:290–299, 1973.
10. Nahmias, A.J., and Roizman, B.: Infection with herpes simplex viruses 1 and 2. New Engl. J. Med., *289*:667–674, 719–725, 781–789, 1973.
11. MacDonald, A., and Feiwel, M.: Isolation of herpes virus from erythema multiforme. Br. Med. J., *2*:570–571, 1972.
12. Melnick, J.L., and Wallis, C.: Photodynamic inactivation of herpes simplex virus: a status report. Ann. N.Y. Acad. Sci., *284*:171–181, 1977.
13. Mills, K.W., et al.: Serotyping herpes simplex virus isolates by enzyme-linked immunosorbent assays. J. Clin. Microbiol., *7*:73–76, 1978.
14. Olson, L.C., Buescher, E.L., Artenstein, M.S., and Parkman, P.D.: Herpesvirus infections of the human central nervous system. New Engl. J. Med., *277*:1271–1277, 1967.
15. Sabin, A.B.: Misery of recurrent herpes: what to do? (editorial) New Engl. J. Med., *293*:986–988, 1975.
16. Whitley, R.J., et al.: Adenine arabinoside therapy of biopsy-proved herpes simplex encephalitis. New Engl. J. Med., *297*:289–294, 1977.
17. Young, E.J., et al.: Acute pharyngitis caused by herpesvirus type 2. J.A.M.A., *239*:1885–1886, 1978.

Varicella-Zoster Virus

18. Balfour, H.H., Jr., et al.: Prevention or modification of varicella using zoster immune plasma. Am. J. Dis. Child., *131*:693–696, 1977.
19. Blank, H., et al.: Cytologic smears in diagnosis of herpes simplex, herpes zoster, and varicella. J.A.M.A., *146*:1410–1412, 1951.
20. Brodkin, R.H.: Zoster causing varicella. Current dangers of contagion without isolation. Arch. Dermatol., *88*:322–324, 1963.
21. Brunnell, P.A., Miller, L.H., and Lovejoy, F.: Zoster in children. Am. J. Dis. Child., *115*:432–437, 1968.
22. Dolin, R., Reichman, R.C., Mazur, M.H., and Whitley, R.J.: NIH Conference. Herpes zoster-varicella infections in immunosuppressed patients. Ann. Intern. Med., *85*:375–388, 1978.
23. Eaglstein, W.H., Katz, R., and Brown, J.A.: The effects of early corticosteroid therapy on the skin eruption and pain of herpes zoster. J.A.M.A., *211*:1681–1683, 1970.
24. Falliers, C.J., and Ellis, E.F.: Corticosteroids and varicella. Six years experience in an asthmatic population. Arch. Dis. Child., *40*:593–599, 1965.

25. Feldman, S., Hughes, W.T., and Daniel, C.B.: Varicella in children with cancer: seventy-seven cases. Pediatrics, 56:388–397, 1975.
26. Forghani, B., Schmidt, N.J., and Dennis, J.: Antibody assays for varicella-zoster virus: comparison of enzyme immunoassay with neutralization, immune adherence hemagglutination, and complement fixation. J. Clin. Microbiol., 8:545–552, 1978.
27. Frey, H.M., Bialkin, G., and Gershon, A.A.: Congenital varicella: case report of a serologically proved long-term survivor. Pediatrics, 59:110–112, 1977.
28. Gershon, A.A.: Varicella-zoster virus. Prospects for active immunization. Am. J. Clin. Pathol., 70(Suppl.):170–174, 1978.
29. Gershon, A.A., et al.: Antibody to varicella-zoster virus in partuient women and their offspring during the first year of life. Pediatrics, 58:692–696, 1976.
30. Gold, E.: Serologic and virus-isolation studies of patients with varicella or herpes zoster infections. New Engl. J. Med., 274:181–185, 1966.
31. Gordon, J.E., and Meader, F.M.: The period of infectivity and serum prevention of chickenpox. J.A.M.A., 93:2013–2015, 1929.
32. Huttenlocher, P.R.: Reye's syndrome: relation of outcome to therapy. J. Pediatr., 80:845–850, 1972.
33. Johnson, R., and Milbourn, P.E.: Central nervous system manifestations of chickenpox. Can. Med. Assoc. J., 102:831–834, 1970.
34. McCarthy, J.T., and Amer, J.: Postvaricella acute transverse myelitis: a case presentation and review of the literature. Pediatrics, 62:202–204, 1978.
35. Meyers, J.D.: Congenital varicella in term infants: risk reconsidered. J. Infect. Dis., 129:215–217, 1974.
36. Miller, L.H., and Brunell, P.A.: Zoster: reinfection or activation of latent virus? Observations on the antibody response. Am. J. Med., 49:480–483, 1970.
37. Priest, J.R., Urick, J.J., Groth, K.E., and Balfour, H.H., Jr.: Varicella arthritis documented by isolation of virus from joint fluid. J. Pediatr., 93:990–992, 1978.
38. Saslaw, S., and Prior, J.A.: Varicella bullosa. J.A.M.A., 173:1214–1217, 1960.
39. Smith, E.W.P., et al.: Varicella gangrenosa due to Group A beta-hemolytic streptococcus. Pediatrics, 57:306–310, 1976.
40. Triebwasser, J.H., Harris, R.E., Bryant, R.E., and Rhoades, E.R.: Varicella pneumonia in adults. Report of seven cases and a review of literature. Medicine, 46:409–423, 1967.
41. Uduman, S.A., Gershon, A.A., and Brunell, P.A.: Rapid diagnosis of varicella-zoster infections by agar-gel diffusion. J. Infect. Dis., 126:193–195, 1972.
42. Weber, D.M., and Pellecchia, J.A.: Varicella pneumonia. J.A.M.A., 192:572–573, 1965.
43. Winkelmann, R.K., and Perry, H.O.: Herpes zoster in children. J.A.M.A., 171:876–880, 1959.

Cytomegalovirus

44. Berenberg, W., and Nankervis, G.: Long-term follow-up of cytomegalic inclusion disease of infancy. Pediatrics, 46:403–410, 1970.
45. Betts, R.F., Freeman, R.B., Douglas, R.G., Jr., and Tolley, T.E.: Clinical manifestations of renal allograft derived primary cytomegalovirus infection. Am. J. Dis. Child., 131:759–763, 1977.
46. Chretien, J.H., McGinniss, C.G., and Muller, A.: Venereal causes of cytomegalovirus mononucleosis. J.A.M.A., 238:1644–1645, 1977.
47. Hanshaw, J.B.: Congenital cytomegalovirus infection: laboratory methods of detection. J. Pediatr., 75:1179–1185, 1969.
48. Hanshaw, J. B., et al.: School failure and deafness after "silent" congenital cytomegalovirus infection. New Engl. J. Med., 295:468–470, 1976.
49. Hayes, K., Danks, D.M., Gibas, H., and Jack, I.: Cytomegalovirus in human milk. New Engl. J. Med., 287:177–178, 1972.
50. Jordan, M.C., et al.: Spontaneous cytomegalovirus mononucleosis. Clinical and laboratory observations in nine cases. Ann. Intern. Med., 79:153–160, 1973.
51. Kumar, M.L., Nankervis, G.A., and Gold, E.: Inapparent congenital cytomegalovirus infection. A follow-up study. New Engl. J. Med., 288:1370-1372, 1973.
52. Lang, D.J., and Hanshaw, J.B.: Cytomegalovirus infection and the postperfusion syndrome. New Engl. J. Med., 280:1145–1149, 1969.
53. Lang, D.J., and Kummer, J.F.: Cytomegalovirus in semen: observations in selected populations. J. Infect. Dis., 132:472–473, 1975.
54. Lee, F.K., Nahmias, A.J., and Stagno, S.: Rapid diagnosis of cytomegalovirus infection in infants by electron microscopy. New Engl. J. Med., 299:1266–1270, 1978.
55. Medearis, D.N.: Cytomegalic inclusion disease. An analysis of the clinical features based on the literature and six additional cases. Pediatrics, 19:467–480, 1957.
56. Montgomery, R., Youngblood, L., and Medearis, D.N., Jr.: Recovery of cytomegalovirus from the cervix in pregnancy. Pediatrics, 49:524–531, 1972.
57. Naib, Z.M.: Cytologic diagnosis of cytomegalic inclusion body disease. Am. J. Dis. Child., 105:153–159, 1963.
58. Nankervis, G.A., and Kumar, M.: Diseases produced by cytomegaloviruses. Med. Clin. North Am., 62:1021–1035, 1978.

59. Overall, J.C., Jr., Kern, E.R., and Glasglow, L.A.: Effective antiviral chemotherapy in cytomegalovirus infection of mice. J. Infect. Dis., 133(Suppl.): A237–A244, 1976.

60. Prince, A.M., Szmuness, W., Millian, S.J., and David, D.S.: A serologic study of cytomegalovirus infections associated with blood transfusions. New Engl. J. Med., 284: 1125–1131, 1971.

61. Robertson, P.W., Kertesz, V., and Cloonan, M.J.: Elimination of false-positive cytomegalovirus immunoglobulin M-fluorescent antibody reactions with immunoglobulin M serum fractions. J. Clin. Microbiol., 6: 174–175, 1977.

62. Stagno, S., Reynolds, D.W., Lakeman, A., Charamella, L.J., and Alford, C.A.: Congenital cytomegalovirus infection: consecutive occurrence due to viruses with similar antigenic compositions. Pediatrics, 52: 788–794, 1973.

63. Weller, T.H.: The cytomegaloviruses: ubiquitous agents with protean clinical manifestations. New Engl. J. Med., 285: 203–214, 267–274, 1971.

Epstein–Barr Virus

64. Bahna, S.L., Horowitz, C.A., Fiala, M., and Heiner, D.C.: IgE response in heterophil-positive infectious mononucleosis. J. Allergy. Clin. Immunol., 62: 167–173, 1978.

65. Carter, J.W., Edson, R.S., and Kennedy, C.C.: Infectious mononucleosis in the older patient. Mayo Clin. Proc., 53: 146–150, 1978.

66. Epstein, M.A., Barr, Y.M., and Achong, B.G.: Studies with Burkitt's lymphomas. Wistar Institute Symp. Mono., 4: 69–82, 1965.

67. Evans, A.S., et al.: A prospective evaluation of heterophile and Epstein–Barr virus-specific IgM antibody tests in clinical and subclinical infectious mononucleosis: specificity and sensitivity of the tests and persistence of antibody. J. Infect. Dis., 132: 546–554, 1975.

68. Evans, A.S., Niederman, J.C., and McCollum, R.W.: Seroepidemiologic studies of infectious mononucleosis with EB virus. New Engl. J. Med., 279: 1121–1127, 1968.

69. Henle, G., Henle, W., and Diehl, V.: Relation of Burkitt's tumor-associated Herpes-ytpe virus to infectious mononucleosis. Proc. Natl. Acad. Sci. U.S.A., 59: 94–101, 1968.

70. Henle, W., et al.: Antibody responses to the Epstein–Barr virus and cytomegaloviruses after open-heart and other surgery. New Engl. J. Med., 282: 1068–1074, 1970.

71. Henle, W., and Henle, G.: Epstein–Barr virus and infectious mononucleosis. New Engl. J. Med., 288: 263–264, 1973.

72. Henle, W., Henle, G., and Horwitz, C.A.: Epstein–Barr virus specific diagnostic tests in infectious mononucleosis. Hum. Pathol., 5: 551–565, 1974.

73. Hirshaut, Y., et al.: Sarcoidosis, another disease associated with serologic evidence for herpes-like virus infection. New Engl. J. Med., 283: 502–506, 1970.

74. Joncas, J., and Mitnyan, C.: Serological response of the EBV antibodies in pediatric cases of infectious mononucleosis and in their contacts. Can. Med. Assoc. J., 102: 1260–1263, 1970.

75. Judson, S.C., Henle, W., and Henle, G.: A cluster of Epstein–Barr Virus-associated American Burkitt's lymphoma. New Engl. J. Med., 297: 464–468, 1977.

76. Lipman, M., Andrews, L., Niederman, J., and Miller, G.: Direct visualization of enveloped Epstein–Barr *Herpesvirus* in throat washing with leukocyte-transforming activity. J. Infect. Dis., 132: 520–523, 1975.

77. Miller, D., Goldman, J.M., and Goodman, M.L.: Etiologic study of nasopharyngeal cancer. Arch. Otolaryngol., 94: 104–108, 1971.

78. Niederman, J.C., et al.: Infectious mononucleosis. Epstein–Barr Virus shedding in saliva and the oropharynx. New Engl. J. Med., 294: 1355–1359, 1976.

78A. Pagano, J.S.: The Epstein-Barr Virus plasmid. ICN-UCLA Symposium on Extrachromosomal DNA. New York, Academic Press, in press (1980).

79. Porter, D.D., Wimberly, I., and Benyesh-Melnick, M.: Prevalence of antibodies to EB virus and other herpesviruses. J.A.M.A., 208: 1675–1679, 1969.

80. Provisor, A.J., et al.: Acquired agammaglobulinemia after a life-threatening illness with clinical and laboratory features of infectious mononucleosis in three related male children. New Engl. J. Med., 293: 62–65, 1975.

81. Rapp, C.E., Jr., and Hewetson, J.F.: Infectious mononucleosis and the Epstein–Barr virus. Am. J. Dis. Child., 132: 78–86, 1978.

82. Shapiro, L.R., Hirshaut, Y., Kanef, D.M., and Glade, P.: Epstein–Barr virus in infancy. J. Pediatr., 80: 1025–1026, 1972.

Adenovirus

83. Bell, J.A., et al.: Pharyngoconjunctival fever. Epidemiologic studies of a recently recognized disease entity. J.A.M.A., 157: 1083–1092, 1955.

84. ———: Studies of adenoviruses (APC) in volunteers. Am. J. Pub. Health, 46: 1130–1146, 1956.

85. Bell, T.M., and Steyn, J.H.: Viruses in lymph nodes of children with mesenteric adenitis and intussusception. Br. Med. J., 2: 700–702, 1962.

86. Blacklow, N.R., et al.: A seroepidemiologic

study of adenovirus-associated virus infection in infants and children. Am. J. Epidemiol., 94:359–366, 1966.

87. Bryant, R.E., and Rhoades, E.R.: Clinical features of adenoviral pneumonia in Air Force recruits. Am. Rev. Respir. Dis., 96:717–723, 1967.

88. Clarke, E.J., Phillips, I.A., and Alexander, E.R.: Adenovirus infection in intussusception in children in Taiwan. J.A.M.A., 208:1671–1674, 1969.

89. Couch, R.B., et al.: Aerosol-induced illness resembling the naturally occurring illness in military recruits. Am. Rev. Respir. Dis., 93:529–535, 1966.

90. Dudding, B.A., et al.: Fatal pneumonia associated with adenovirus type 7 in three military trainees. New Engl. J. Med., 286:1289–1292, 1972.

91. ———: Acute respiratory disease in military trainees. The adenovirus surveillance program. 1966–1971. Am. J. Epidemiol., 97:187–198, 1973.

92. Dudgeon, J., Bharbava, S.K., and Ross, C.A.C.: Treatment of adenovirus infections of the eye with 5-iodo-2'-deoxyuridine. Br. J. Ophthalmol., 53:530–533, 1969.

93. Fox, J.P., Hall, C.E., and Cooney, M.K.: The Seattle virus watch. VII. Observations of adenovirus infections. Am. J. Epidemiol., 105:362–386, 1977.

94. Gilden, R.V., et al.: Serologic surveys of human cancer patients for antibody to adenovirus T antigen. Am. J. Epidemiol., 91:500–509, 1970.

95. Ginsberg, H.S., et al.: Etiologic relationship of the RI-67 agent to acute respiratory disease (ARD). J. Clin. Invest., 34:820–831, 1955.

96. Girardi, A.J., Hilleman, M.R., and Zwickey, R.E.: Tests in hampsters for oncogenic quality of ordinary viruses including adenovirus type 7. Proc. Soc. Exp. Biol. Med., 115:1141–1150, 1964.

97. Herbert, F.A., Wilkinson, D., Burchak, E., and Morgante, O.: Adenovirus type 3 pneumonia causing lung damage in childhood. Can. Med. Assoc. J., 116:274–276, 1977.

98. Kelsey, D.S.: Adenovirus meningoencephalitis. Pediatrics, 61:291–293, 1978.

99. McCormick, D.P., Galapon, Q., and Berling, C.: Pathology of exfoliated oropharyngeal epithelial cells infected with wild-type adenovirus. Appl. Microbiol., 24:389–397, 1972.

100. MacPherson, R.I., Cumming, G.R., and Chernick, V.: Unilateral hyperlucent lung: a complication of viral pneumonia. J. Can. Assoc. Radiol., 20:225–231, 1969.

101. Moffet, H.L., Doyle, H.S., and Burkholder, E.: Etiology of severe infantile diarrhea. J. Pediatr., 72:1–14, 1968.

102. Moffet, H.L., Siegel, A.S., and Doyle, H.S.: Non-streptococcal pharyngitis. J. Pediatr., 73:51–60, 1968.

103. Mufson, M.A., and Belshe, R.B.: A review of adenovirus in the etiology of acute hemorrhagic cystitis. J. Urol., 115:191–194, 1976.

104. Nelson, K.E., et al.: The role of adenoviruses in the pertussis syndrome. J. Pediatr., 86:335–341, 1975.

105. Valters, W.A., Boehm, L.G., Edwards, E.A., and Rosenbaum, M.J.: Detection of adenovirus in patient specimens by indirect immune electron microscopy. J. Clin. Microbiol., 1:472–475, 1975.

106. Yunis, E.J., and Hashida, Y.: Electron microscopic demonstration in appendix vermiformis in a case of ileocecal intussusception. Pediatrics, 51:566–570, 1973.

Smallpox and Vaccinia Viruses

107. Baxby, D.: The origins of vaccinia virus (editorial). J. Infect. Dis., 136:453–455, 1977.

108. Communicable Disease Center: Presumptive smallpox. Change of diagnosis. Morb. Mort. Week. Rep., 14:169–170, 177–178, 277–278, 1965.

109. Dixon, C.W.: Smallpox. J. and A. Churchill, Ltd. London, 1962. 512 pp.

110. Goldstein, J.A., Neff, J.M., Lane, J.M., and Koplan, J.P.: Smallpox vaccination reactions, prophylaxis, and therapy of complications. Pediatrics, 342–347, 1975.

111. Heiner, G.G., et al.: A study of inapparent infection in smallpox. Am. J. Epidemiol., 94:252–268, 1971.

112. Long, G.W., et al.: Experience with electron microscopy in the differential diagnosis of smallpox. Appl. Microbiol., 20:497–504, 1970.

113. Kempe, C.H.: Studies on smallpox and complications of smallpox vaccination. Pediatrics, 26:176–189, 1960.

114. ———: Commentaries. The end of routine smallpox vaccination in the United States. Pediatric, 49:489–492, 1972.

115. Koplan, J.P., and Hicks, J.W.: Smallpox and vaccinia in the United States—1972. J. Infect. Dis., 129:244–226, 1974.

116. McLean, D.M.: Methisazone therapy in pediatric vaccinia complications. Ann. N.Y. Acad. Sci., 284:118–221, 1977.

117. Meiklejohn, G.: Smallpox: is the end in sight? J. Infect. Dis., 133:347–353, 1976.

118. Public Health Service Advisory Committee on Immunization Practices: Vaccination against smallpox in the United States. A reevaluation of the risks and benefits. Morb. Mort. Week. Rep., 20:339–345, 1971.

119. Public Health Service Advisory Committee on Immunization Practices: Smallpox vaccine. Morb. Mort. Week. Rep., 27:156–158, 163–164, 1978.

Other DNA Viruses

120. Cho, C.T., and Wenner, H.A.: Monkeypox virus. Bacteriol. Rev., *37:*1–18, 1973.
121. Rockhoff, A.S.: Molluscum dermatitis. J. Pediatr., *92:*945–947, 1978.
122. Leavell, U.W., Jr., et al.: Orf. Report of 19 human cases with clinical and pathological observations. J.A.M.A., *204:*657–664, 1968.
123. Neva, R.A.: Studies on molluscum contagiosum. Observations on the cytopathic effect of molluscum suspensions in vitro. Arch. Intern. Med., *110:*720–725, 1962.

12
Helical RNA Viruses

INFLUENZA VIRUS

Objectives

1. Define the term "influenza-like illness" as used by the Center for Disease Control.
2. Describe the typical clinical picture of classical influenza disease in school age children.
3. Explain the abbreviations used in the definitions of particular types or strains of influenza virus; for example: A/Japan/305/57 ($H_2 N_2$), and A/Russia/1977 ($H_1 N_1$).
4. Describe how influenza virus infection can be detected in the laboratory.
5. Discuss the importance to the community of laboratory confirmation of influenza.
6. Describe what statistics are used to determine if an outbreak of influenza is occurring.
7. Describe the probable role of the hemagglutinin and neuraminidase in the pathogenesis of influenza infection.
8. Describe how an influenza virus isolate is studied in order to determine if it is antigenically different from previous viruses.

Definitions

Influenza is a word used in several different ways. Patients may refer to many kinds of gastrointestinal illnesses as influenza or stomach flu or intestinal flu, although this kind of illness is not caused by influenza virus. Occasionally, a patient refers to any febrile or respiratory illness as the flu, and the physician may sometimes have to accept this usage to communicate with the patient. However, in scientific communications, the physician should restrict the use of the word influenza to refer to influenza virus, influenza-like illness (defined later), or *Haemophilus influenzae*.

Types and Strains

Influenza virus is classified into three types: Type A, Type B, and Type C. Each type has a distinctive internal nucleocapsid antigen. In the past, influenza type A was divided into three subtypes called A (or A-0), A-1, and A-2, but this subtype classification has been abandoned.

Recently revised nomenclature for influenza A viruses now includes the hemagglutinin (H) and neuraminidase (N) antigens (Table 12-1). For example, the current naming of influenza viruses is illustrated by the Russian strain of 1977, which represented a return to the earlier hemagglutinin (H_1) and neuraminidase (N_1) antigens and is called A/USSR/77 ($H_1 N_1$). Types B and C have no subtypes.

TABLE 12-1. FAMILIAR NAMES AND DESCRIPTIVE STRAIN NAMES OF SOME INFLUENZA A VIRUSES.

Familiar name	Strain name
1957 Asian influenza	A/Japan/305/57 ($H_2 N_2$)
1968 Hong Kong influenza	A/Hong Kong/8/68 ($H_2 N_2$)
1972 English influenza	A/England/42/72 ($H_3 N_2$)
1976 Swine influenza	A/New Jersey/76 ($Hsw_1 N_1$)
1977 Russian influenza	A/USSR/77 ($H_1 N_1$)

The hemagglutinin–neuraminidase designation also allows accurate description of laboratory-grown hybrids and animal isolates of similar antigenic structure. For example, A/New Jersey/1976 ($Hsw_1 N_1$) was the swine influenza virus of 1976 which had the same hemagglutinin antigen as a previously defined swine strain.

Influenza virus strains are defined by using antisera prepared in animals (Table 12-1). When a strain of influenza virus is recovered for the first time in an area, it is numbered and named for the state, city, or country. The Asian influenza strain of 1957, now called A/Japan/1957 ($H_2 N_2$), was originally called A-2/Japan/305/57—where A-2 was the subtype classification then used, 305 was the strain number, and 57 was the year it was recovered.

Minor antigenic changes occur within subtypes about every year. Major antigenic changes resulting in a different subtype occur once or twice a decade and can result in worldwide epidemics.

Frequency and Importance

Influenza A virus is associated with periodic world wide epidemics (pandemics) which are associated with many deaths. Attempts to prevent influenza by the use of vaccine has been an important medical effort for many years. However, vaccines have been generally unsuccessful against any very marked antigenic changes. Sufficient knowledge about the mechanisms of antigen changes might make it possible to have an effective vaccine against a very antigenically changed virus before it spreads throughout the world. Thus, study of the antigenic variation of the virus, development of better vaccines, and surveillance are extremely important in order to attempt to prevent the severe pandemics which periodically have occurred.

Influenza B virus has only minor antigenic variations and does not cause worldwide epidemics. The clinical illness is similar to that of influenza A, and the virus may rarely cause death in compromised hosts. Influenza virus infection is very rarely followed by Reye's syndrome, a very rare encephalopathy which is often fatal.

Influenza C virus has little if any antigenic variation and appears to cause less severe illness that the other two influenza subtypes. Its frequency is not clearly defined because it grows only in eggs, and requires constant refrigeration to detect hemagglutination. It appears to be an occasional cause of mild respiratory illness in children.

Clinical Patterns of Illness

During outbreaks, influenza virus is a *common* cause of:

Classic influenza-like illness in school-age children is manifested by chilliness, fever, headache, myalgia, extraordinary fatigue or weakness, cough, sore throat, and occasionally by substernal pressure or a congested feeling.[27] Outbreaks of classic influenza-like illness in civilians are likely to be due to influenza virus. Sporadic cases are more likely to be due to infection with an adenovirus or parainfluenza virus.

Fever is often as high as 104 ° F (39 ° C), even in adults, but the patient typically does not appear critically ill. In infants and young children, diarrhea and vomiting are common.[33]

Sometimes the muscle involvement is quite prominent, with severe pains in the gastrocnemius and soleus muscles of the calf, elevated serum levels of muscle enzymes, and influenza virus-like particles in muscle biopsy specimens.[6,9] Influenza virus has been recovered from muscle and liver specimens in Reye's syndrome, which is a very rare complication of influenza virus infection.[35]

Minor Respiratory Illnesses. A modified illness may occur in adults or older children, if they have had past infection with very closely related strains. These minor respiratory illnesses often resemble the common cold. Laryngitis or bronchitis without fever may occur. Such afebrile minor respiratory illnesses are usually not classified as influenza-like illness, unless myalgia and fatigue are prominent.

During outbreaks, influenza virus is an *occasional* cause of:

Bronchiolitis or laryngitis may be severe in infants or young children, and apnea may be noted.[33,34]

Asthmatic Episodes. Influenza virus infection can produce exacerbations of refractory wheezing in patients with asthma or chronic pulmonary disease.[19]

Pneumonia. Some individuals with a typical influenza-like illness also have clinical and radiologic pneumonia. Usually it is patchy and not severe. Bacterial pneumonia may occur as a complication, so even mild pneumonias are usually treated with antibiotics. Chronic pulmonary complications, such as pulmonary fibrosis, bronchiectasis or obliterative bronchiolitis, rarely have occurred.[20]

During outbreaks, influenza virus is a *possible* cause of these *rare* syndromes:

Fulminating pneumonia is a rare pattern of influenza virus infection.[28,33] The lungs are hemorrhagic and edematous. This pattern may be associated with the presence of a bacterial pathogen, especially *Staph. aureus*, or *Pseudomonas aeruginosa*.[21,40] In many fulminating cases, the patient has an underlying problem, such as a debilitating heart or lung disease.[21] Sometimes there may be a temporarily increased susceptibility, due to extraordinary fatigue or undernutrition.

Brain Involvement. Encephalitis or encephalopathy or postinfluenzal demyelinating disease is rare.[28] Febrile convulsions, however, are not rare in younger children, and may be associated with a mild lymphocytic CSF pleocytosis and recovery of the virus from the spinal fluid.[33]

Acute Renal Failure. Severe influenza A virus infection is occasionally associated with acute renal failure. Acute myoglobulinuria or disseminated intravascular coagulation which may be the mechanisms of the renal failure.[41] Profound muscular weakness, inability to walk, severe respiratory distress, and dark urine typically precede the oliguria.

Experimental Infections

Wild influenza virus has been used to infect human volunteers and subhuman primates. It is often difficult to find antibody-free volunteers. Influenza A $(H_3 N_2)$ was administered to ten healthy young adults and compared to the effects of attenuated influenza A virus, in order to study the effect the infection on T lymphocytes.[3] Mild illness was observed in the volunteers. Both wild and attenuated viruses produced a reduction in T lymphocytes.

The 1976 New Jersey swine influenza virus was inoculated into six volunteers.[1] The virus was excreted in the nasal washings for 3 to 8 days, and produced only mild clinical illness, compared to the frequent production of moderate to severe clinical illnesses in previous volunteer experiments with other influenza A viruses in the same laboratory. These experiments also differ from the severity of illness due to swine influenza virus in the soldiers in Fort Dix, New Jersey, where four of the 13 confirmed cases had radiologic evidence of pneumonia and one soldier died after a training march.[8]

Influenza A virus $(H_3 N_2)$ infections from three different years (1968, 1972, and 1974) have been compared in volunteer experiments.[37] Virus shedding in nasal secretions was observed as early as a day after intranasal inoculation, and peaked on the second through the fourth day after inoculation, concurrent with maximal clinical findings and interferon production. Since natural interferon production occurs so quickly after the onset of clinical illness, it seems unlikely to be of value for therapy—although these experimental results may not be applicable to severe clinical illnesses.

In squirrel monkeys, infection with both influenza A virus and the pneumococcus resulted in more severe pneumonia than that with either organism alone.[3]

Asymptomatic or Subclinical Infection

During outbreaks of influenza, many individuals develop serologic evidence of infection without symptoms. In one study of an influenza Type B outbreak in a children's home, about 25 per cent of the children in the home had asymptomatic infection.[27] When the older children and adolescents were interviewed after the outbreak, some admitted to having had minor respiratory symptoms during the outbreak, but the majority did not.[27]

Laboratory Diagnosis

Serum Antibodies. Serologic diagnosis is a useful method for laboratory diagnosis of influenza. The hemagglutination inhibition (HI) test is commonly used. A current strain of dead influenza virus, which agglutinates human erythrocytes, is used as the antigen. The highest dilution of serum which prevents the agglutination is the hemagglutination inhibition (HI) titer of the serum.

At the onset of an illness due to influenza virus, the patient's serum typically will have a low HI titer, less than 1:8. Three weeks after the illness, the patient's serum will have an increase in HI antibodies, usually to a dilution of about 1:64.

Viral Cultures. Isolation of influenza virus is done by inoculation of the patient's throat swabbing onto monkey kidney cells or into the amniotic sac of chicken eggs. Freezing the specimen decreases the frequency of recovery of influenza virus.[42] The virus can be detected after about a week, by demonstrating hemadsorption to the kidney cells or by demonstrating hemagglutinating activity of the amniotic fluid. Some types of influenza virus produce a cytopathic effect in some kinds of cell cultures.

Identification of the strain of influenza virus can be done by hemagglutination, using antisera produced by injection of animals (usually chickens) with various known virus strains (Table 12-2). A new influenza isolate can be quantitatively related to known strains by serologic studies which test the antisera of the old strains against the newly isolated virus. A table can be made of these reactions, and usually shows that recently recovered strains usually have slight antigenic differences from the older strains. This gradual change in antigen is called antigenic drift.

For example, in Table 12-2, the A/England/72 virus strain requires a 1:160 dilution of antiserum prepared against the A/Hong Kong/68 virus to inhibit hemagglutination, while the much more dilute concentration of 1:640 of Hong Kong antisera inhibits the Hong Kong virus. This indicates a moderate antigenic shift, and implies that a vaccine prepared with the Hong Kong antigen would be slightly less effective against the England strain than a vaccine prepared with the England strain. On the other hand, the 1973 Georgia strain is only slightly different from the 1972 England strain, and the 1973 Georgia and the 1973 Oregon strains are nearly identical.

Histologic Diagnosis. Influenza virus antigen can be identified in tissues by staining the patient's tissue with specific flourescent-labeled antibody (Fig. 1-9). Otherwise the histologic findings in the lung, for example, are not specific. Recognition of a virus particle the size and shape of influenza using electron microscopy of the tissue would be strongly suggestive, but not conclusive, since other viruses can resemble influenza virus. Such virus particles are usually extremely difficult to find.

New Laboratory Methods. Rapid identification of influenza virus subtypes can be done using the coagglutination method (Fig. 1-12), or by immunodiffusion methods (Fig. 4-2). Serum antibodies can be measured by an ELISA method, as described for rubella antibodies (Fig. 13-7).

Biologic Characteristics of Clinical Interest

Hemagglutination and Hemadsorption. An important laboratory characteristic of the virus is its ability to attach to red blood cells, producing hemadsorption or hemagglutination. Hemagglutination is the clumping of red blood cells by the virus. Hemadsorption is the attachment of erythrocytes to cell cultures infected by the virus. Thus, the virus can be detected by using erythrocytes in two ways. If the patient's throat swabbing is inoculated onto monkey kidney cells, the presence of the virus is detected about a week later by adding guinea pig erythrocytes, which attach (hemadsorb) to the monkey kidney cells (Fig. 12-1). Alternatively, if the patient's specimen is inoculated into the amniotic sac of a developing chick embryo, the virus is detected by taking some amniotic fluid, putting it in a tube, adding erythrocytes and seeing hemagglutination, the clumping of the added red blood cells.

Hemagglutinins and Neuraminidase. The process of infection with influenza virus has been studied experimentally and shown to involve attachment of the spikes (hemagglutinins) on the surface of the virus with host mucoprotein (Fig. 12-2). The host's respiratory secretions contain mucopro-

TABLE 12-2. INFLUENZA VIRUS ANTIGENIC DRIFT

Virus strains	Antisera			
	A/HK/68	A/Eng/72	A/Georgia/3/73	A/Oregon/4/73
A/Hong Kong/68	640	113	320	320
A/England/72	160	*320*	—	640
A/Georgia/3/73	113	160	*640*	—
A/Oregon/4/73	113	226	640	*905*

Antigenic drift within a H_2N_2 subtype as shown by gradual titer changes. Read table from top down. The titers are the reciprocal of the mean hemagglutination inhibition titers. Italicized titers are homologous (virus vs. own antiserum). Some titers are omitted for simplification.[4]

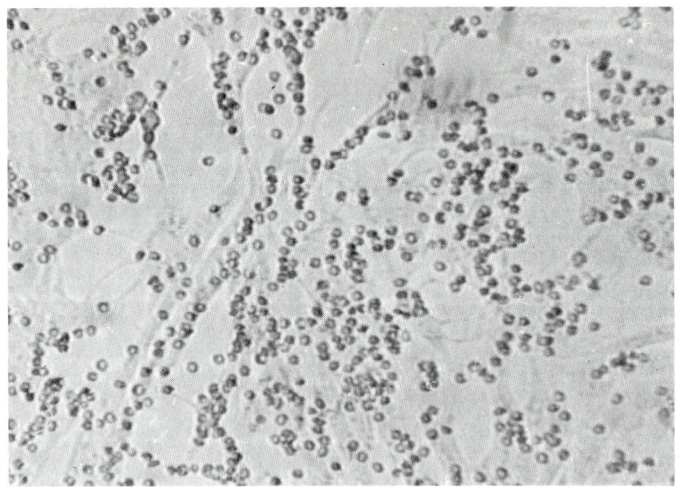

FIG. 12-1. *Positive hemadsorption (HA). Note the erythrocytes absorbed to the cell monolayer (African green monkey kidney cells). The tube, which has the monolayer growing along one side, is turned upside down to let unattached erythrocytes wash off.*

teins which combine with the virus and perhaps could prevent infection, but neuraminidase from the virus hydrolyzes the mucoprotein and prevents its acting as an inhibitor. The virus is classified as a myxovirus, because of its ability to combine with mucoprotein (myxo means mucus).

Protection against influenzal illness is related to antibodies against the hemagglutinin and the neuraminidase.[7] Antibodies against the hemagglutinin are more important in prevention of infection.[29] Antibodies against the neuraminidase also provide some protection against infection, perhaps by preventing the spread of virus from cell to cell.[31]

Antigenic Variation. The most important epidemiologic feature of influenza virus is its antigenic variation, which is the basis for the occurrence of epidemics. When a new, major change in the antigenic structure of the virus occurs, especially the hemagglutinin antigen, this new variant of the virus may spread over the entire world. Most of the exposed individuals become infected, because their antibodies from previous infections are not protective.

"Antigenic drift" is a term used to describe minor antigenic change, and probably is due to point mutations in the polypeptide chain of the hemagglutinin.[23,24] These mutations can be speeded by growing the virus in the presence of specific antisera so that eventually all possible mutants might be grown in the laboratory before they appear in nature. The gradual nature of antigenic drift can be seen in cross reactions between different virus strains of the same subtypes and the antisera produced against each strain (Table 12-2).

Recombination. Major antigenic changes in influenza A virus can occur through recombination.[23,24] This process can occur in cell cultures in the laboratory or in the human host when there is concurrent infection with two different subtypes. The discreet units of RNA from core of the virion (Fig. 12-2) are exchanged and recom-

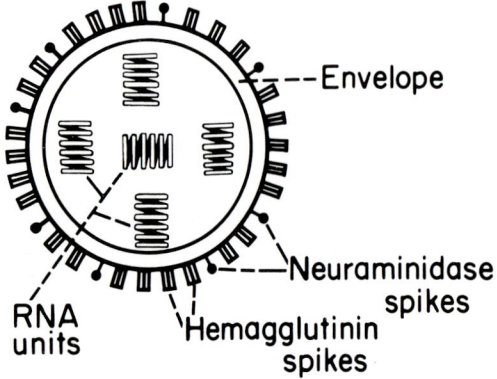

FIG. 12-2. *Diagram of influenza virus. Hemagglutinin spikes are more numerous than neuraminidase spikes. RNA is found in about eight discreet units.*

bined into hybrids as the two different viral types replicate. Such hybrids produced in the laboratory are being investigated as vaccines.

Animal Models. Influenza viruses can produce acute respiratory disease in animals and birds. Swine influenza virus and A/Equine/Miami/63 are examples of subtypes of Influenza Type A virus that have been found in animals.

Toxicity. An interesting, but poorly explained, clinical feature of the virus is sometimes called toxicity. The toxicity of human influenza infections is often striking, producing fatigue or prostration far beyond that expected from the other clinical signs. Dead influenza virus is toxic to cells and produces fever in experimental animals. A somewhat similar effect may occur after administration of killed vaccine to humans.[13]

Contagion. Influenza virus infection is typically very contagious. The incubation period is only about 2 days, as determined by illness occurring in individuals with a single known exposure.[27] Modified influenza is probably much less contagious than classical influenza; at least, the virus is less likely to be recovered from such cases.[27]

Treatment

No specific viral chemotherapy has been demonstrated to be effective in animals or humans. Amantadine therapy, which is under investigation, is described later in this section.

Oxygen is important supportive therapy in severe pneumonia.

Clinical Surveillance

Surveillance for influenza is done by public health authorities because early recognition of influenza virus by surveillance in one geographic area may allow time for people in another area to be protected by a vaccination program. Surveillance can be subdivided into clinical surveillance (detection of disease probably due to influenza virus) and laboratory surveillance (recovery of influenza virus from patients in an outbreak, and identification of the strain involved).

Clinical surveillance, which often includes telephone surveys to key areas, allows early detection of an outbreak and estimation of its severity. Laboratory studies should be done as soon as possible to confirm that the outbreak is caused by an influenza virus, so that the strain can be characterized antigenically and the proper vaccines used in geographic areas not yet affected. When an outbreak of influenza-like disease occurs, it is useful to recover the virus from some patients in order to provide evidence that the entire outbreak is probably caused by influenza virus.

The clinical methods of surveillance for influenza include:

Absenteeism Rates. School or industrial absenteeism provides early information. Influenza is practically the only febrile respiratory disease that causes extremely high absentee rates from schools.

Pneumonia and Influenza Mortality Rates. These rates are routinely reported to the Center for Disease Control in Atlanta, Georgia, from about 100 United States cities with populations over 100,000. On the basis of expected mortality rates from the previous 5 years, excessive pneumonia and influenza mortality rates, defined by expected rates called the epidemic threshold, are strongly indicative of influenza. The peak in excess mortality occurs about 3 to 5 weeks later than the peak in illness so the excess mortality is useful primarily as a retrospective method.

Influenza-Like Illness. At the present time, this is defined by the Center for Disease Control as fever, headache, and one or more of the following: cough, malaise, myalgia, or rhinorrhea. This definition of influenza-like illness is used on questionnaire surveys to determine if there is an increase in clinical disease compatible with influenza at sentinel stations.

Civilian and Military Sentinel Areas. Sentinel physicians in widely scattered geographic areas in the United States are used to look for clinical influenza-like illness. The United States Air Force has a surveillance program in 11 overseas and 11 continental installations which submit both clinical and laboratory information for worldwide surveillance of influenza.

Comparison of Pre-Illness and Convalescent Sera. In this method, approximately ten acute serum specimens are collected from individuals who have not yet had any clinical illness. At the same time, ten or more convalescent serum specimens are obtained from patients 2 or 3 weeks after an influenza-like illness. The mean titer of antibody is determined for the individuals who have not been sick and compared to the mean titer for the convalescent group. If there is a significant difference between the two groups, the outbreak of illness is probably due to influenza virus.

Animal Surveillance. After the appearance of swine influenza in humans in New Jersey in 1976, surveillance of virus isolates from swine and birds has been done.

Age-Specific Attack Rates. This information is useful to help determine if an outbreak is related to a significant antigenic change. For example, if the frequency of illnesses decreases sharply in individuals over 21 years of age, then most people over 21 have probably had infection with a similar strain in the past.

Prediction of Epidemics

As the results of clinical and laboratory surveillance become available, public health experts attempt to predict the spread and the virulence of a new strain, which led to an unprecedented national vaccination effort against swine flu in 1976. In the past ten years, such predictions have often been inaccurate, producing much debate and criticism, and suggestions that predictions be abandoned. Perhaps the most rational and scientific explanation of the problems involved has been made by Kilbourne.[18]

The reappearance of $H_1 N_1$ influenza virus (Russian influenza) in 1977 rekindles the issues of early prediction about a new influenza virus.[10] The well-documented severity of the worldwide epidemics (pandemics) of 1889, 1917, 1957, and 1968 is a good reason for the surveillance and attempts at predictions. Whereas minor antigenic drift occurs about every year and results in local or sporadic outbreaks, significant nationwide or worldwide epidemics are theoretically possible when a major change in the hemagglutinin or neuraminidase antigens occurs, as from $H_3 N_2$ of the 1968 to 1977 strains to $H_1 N_1$ in the case of Russian influenza.[10]

Prevention

Commercial Killed Vaccines. Killed influenza virus vaccine is effective in reducing the frequency of illness in susceptible individuals only if the current virus strain is antigenically closely related to the vaccine strain. Killed virus vaccines have been the only vaccines available for general use through 1979, but live attenuated vaccines have been under investigation for many years. A classification of investigational and commercially available new vaccines is shown in Table 12-3.

Monovalent killed vaccines contain only one influenza virus strain, such as A/New Jersey/76 ($Hsw_1 N_1$). Bivalent vaccines contain two virus strains, such as bivalent influenza A vaccine, containing A/New Jersey/76 and A/Victoria/1975. Another bivalent vaccine contained A/Victoria/75 and B/Hong Kong/72.

A subunit killed vaccine is made from highly purified suspensions of a killed influenza virus that has been disrupted (as by some vibration) or "split", and purified into various components by differential centrifugation.[12,43] Subunit vaccines are also called subvirion or split vaccines.

Subunit vaccines are less likely to cause fever in children than whole virus vaccines. However, two doses are necessary to produce an adequate antibody response.

Since 1978, viral antigen in a vaccine has been expressed as micrograms (μg) of the hemagglutinin protein per dose, rather than the older method of quantitation of virus by chick cell agglutination (CCA) units.

The 1979–1980 influenza vaccine contained 7 μg each of A/Brazil/78 ($H_1 N_1$), A/Texas/77 ($H_3 N_2$), and B/Hong Kong/72 (no H or N antigens in the B subtype) for adults. The vaccine for youths (adults less than 27 years of age) and children—who had no prior contact with $H_1 N_1$ antigen—was the same, except that two doses were used for children less than 13 years of age. The subunit (subvirion) form of the vaccine was more highly purified, but two doses a month apart were recommended because the subunit vaccine was less immunogenic.

Investigational Killed Vaccines. A recombinant vaccine using the neuraminidase antigen of a current wild influenza virus strain and an equine hemagglutinin antigen was found to be somewhat less protective than the killed vaccine prepared from the current wild strain.[32] This is an example of a neuraminidase-specific vaccine (Table 12-3).

Recent Vaccine Recommendations. The available killed vaccine and recommendations for its use changes every season. In the 1979–1980 season, a polyvalent vaccine containing two strains of A ($H_1 N_1$ and $H_3 N_2$) and a recent strain of B was recommended for older individuals, and all persons with chronic diseases such as heart or lung

TABLE 12-3. INVESTIGATIONAL AND NEW INFLUENZA VACCINES.

Kind of vaccine	Characteristics	Example
Killed virus		
Subunit	Purified suspensions of disrupted virus	1978–1979 "split-virus" vaccine for children
*Recombinant	Composed partly of new strain, and partly of an old easily grown stock strain	
Attenuated, live virus		
*Temperature sensitive (ts)	Grows in nose, but not in lungs	
*Cold-adapted	Similar principle to ts vaccines	
*Serum-inhibitor-resistant recombinant strain	See text	A/England/42/72 ("Alice")

*Investigational vaccine

disease.[36] The value of vaccine in immunosuppressed patients is still under investigation, but it appears to be useful, if it can be done at a time of minimal immunosuppression.[11]

Investigational Live Vaccines. Temperature sensitive (ts) live attenuated influenza virus vaccines have been under investigation since about 1972. Ts viruses replicate well in the cooler temperature of the human nose (32–34 ° C), but do not replicate adequately at the temperature of the lower respiratory tract (37 ° C).[30]

Cold adapted vaccines are different from temperature sensitive vaccines. Cold adapted vaccines are often recombinants of viruses that have been adapted to grow at 25°C.[5]

Serum inhibitor-resistant recombinant strains of influenza (attenuated, live influenza A vaccines) are under investigation. The "Alice" strain has the antigenic characteristics of A/England/42/72 (H_3 N_2) and the growth characteristics of A/PR/8/34.[26]

Hybrid vaccines have an advantage for commercial production since the antigenic component can be derived from the current strain, and the other part of the hybrid selected for rapid growth in the laboratory so that the new hybrid can be produced in large quantities.[17]

Toxicity of Killed Vaccines. Guillain–Barré syndrome occurred as a complication of swine flu immunization.[25] The reasons for this association are unknown, but apparently this syndrome can occur as a complication of any of the killed virus vaccines.[25]

Systemic toxicity, as manifested by fever, chills, and myalgia, is not related to impurities such as endotoxin, but rather appears to be caused by an intrinsic toxicity of the viral antigen.[13] Treatment of influenza B antigen by prolonged exposure to formalin reduces the vaccine toxicity.

Children with asthma have a transient impairment of expiratory flow after killed influenza vaccine, but have significant protection from subsequent natural influenza.[2]

Hospital-Acquired Infections

Influenza is very contagious and can easily spread within a hospital or any other closed group. By 1979, the United States Public Health Service had not issued any formal recommendation for controlling nosocomial influenza.

High risk patients can be protected to some extent by several methods.[16] Prior administration of vaccine before hospitalization would be advisable, as most older or compromised patients may have some side effects of vaccine after hospitalization which might confuse issues in their care. Chemoprophylaxis with amantadine is practical for some hospitalized patients, as described in the following section.

Immunization of all hospital employees is not practical, and immunization of all hospital visitors implies universal immunization without regard to the individual's need.

Protective isolation of unimmunized patients at risk and strict isolation of patients with suspected influenza is also not practical or known to be effective. Reports of isolation and other control measures being effective are likely to reflect only the waning of the outbreak. Closing large sections of hospitals or banning visitors has not

Chemoprophylaxis and Chemotherapy

Amantadine. Influenza is the only virus for which an oral prophylactic drug is available. This chemical agent is amantadine, a name derived from "adamant", which means hard, so named because this particular chemical structure is very resistant to degradation. The drug can be given by mouth and can prevent influenza in a statistically significant number if taken everyday in advance of exposure.[14,15,38] Amantadine appears to act by blocking the penetration of influenza virus into cells.[14] It must be taken daily for the duration of potential exposure, often for one to two months. It is not a substitute for influenza immunization.

Amantadine has also been used for treatment of adults who have already had the onset of symptoms of influenza. In one study of previously healthy adults, the duration of fever was reduced, but no other benefit was noted.[22] No studies have yet established any definite value in the prevention or treatment of pneumonia or other complications after symptoms have begun.

Preexisting antibodies do not interfere with amantadine's effect, and amantadine does not interfere with immunization with killed virus.[15]

Side effects of amantadine occur more frequently in the elderly and include blurred vision, confusion, hallucinations, headache, slurred speech, dizziness, and sleep disturbances. The drug is teratogenic in animals and contraindicated during pregnancy.

PARAINFLUENZA VIRUS

Objectives

1. Discuss the similarities between parainfluenza virus and influenza virus.
2. Describe three clinical illnesses that parainfluenza viruses can cause.
3. Describe what tests might be done to confirm the diagnosis of parainfluenza virus infection.

Definitions and Types

Parainfluenza viruses were so named because of some resemblances to influenza viruses, including the production of respiratory disease and the property of hemadsorption. Parainfluenza viruses were first called hemadsorption agents, because they were first detected in the laboratory by the method of hemadsorption. There are four types of parainfluenza virus. Type 2 was originally called croup-associated virus because it was first recovered from children with croup (laryngitis).

Frequency and Importance

Parainfluenza viruses are the most frequent cause of laryngitis in children, and are a moderately common cause of respiratory infections in older children and adults.

Clinical Patterns of Illness

In children, parainfluenza viruses are a *frequent* cause of:

Laryngitis. Parainfluenza viruses account for at least 50 percent of laryngitis in children.[47,53]

Bronchitis, Bronchiolitis, and Bronchopneumonia. In young children, these syndromes are frequently caused by parainfluenza viruses.[53]

Parainfluenza virus are an *occasional* cause of:

Fever Without Localizing Signs. In one study of hospitalized children with new episodes of fever, parainfluenza virus was frequently associated with fever alone.[50] In an outbreak in an orphanage, fever with minimal respiratory findings was a frequent pattern.[48]

Common Cold Syndrome. In adults or children, reinfection with parainfluenza viruses may occur, in spite of the presence of neutralizing antibodies, with the resulting illness resembling the common cold syndrome.[52] This syndrome is characterized by nasal discharge and obstruction, sneezing, dry cough, and red nasal and pharyngeal mucous membranes. Inoculation of adult volunteers with parainfluenza virus produces the common cold syndrome, even though the volunteers have prechallenge neutralizing antibody to the virus.[49]

Parotitis. Parainfluenza virus is an occasional cause of parotitis, and is a likely explanation for parotitis in a child who has previously had mumps vaccine, or past mumps virus infection.

Laboratory Diagnosis

Virus Culture. Inoculation of the patient's nasal washing onto rhesus monkey kidney cell cultures is the most sensitive method for primary isolation of parainfluenza virus. Parainfluenza viruses are detected by hemadsorption on monkey kidney cell cultures, as described in the section on influenza virus. Parainfluenza viruses usually produce little cytopathic effect, so that hemadsorption of the inoculated cell cultures is necessary for detection.

The interpretation of a positive hemadsorption is made difficult by the frequent presence in monkey kidney cell cultures of monkey viruses (simian viruses). These simian viruses may be latent in monkey kidney cell cultures and may produce a hemadsorption effect which is indistinguishable from that produced by parainfluenza or influenza viruses. Thus, neutralization of the hemadsorption effect by type-specific antiserum, or detection of parainfluenza antigen by addition of specific fluorescent antibody (Fig. 1-9)[51] is necessary to identify the virus as a parainfluenza virus and to exclude the possibility of a simian virus.

Serum Antibodies. Parainfluenza virus infection can be demonstrated using paired sera, one from early in the illness and one obtained 2 to 4 weeks later. An increase in complement fixing or hemadsorption inhibition antibodies indicates infection. However, there is some cross reaction between the four parainfluenza virus types and, to a minor degree, mumps virus. The newly developed ELISA test for antibody detection appears very useful (Fig. 13-7).

Smear. Parainfluenza virus infection has been diagnosed by recognition of the typical virion in electron micrographs of throat swabbings which have been concentrated by ultra centrifugation.[46] This method of diagnosis is not practical or available, but is of interest because one might not expect it to be possible. Parainfluenza virus closely resembles other paramyxoviruses in size and shape, but might be distinguishable from influenza virus (Fig. 12-3).

Biologic Characteristics of Clinical Interest

Organ Culture. The virus has been cultivated in organ cultures of human respiratory tract tissue.[45] The technique allows for the observation of virus cytopathology independent of host defenses.

No Antigenic Variation. Parainfluenza viruses do not undergo antigenic variation, as in the case of influenza virus, and major epidemics of parainfluenza virus infections do not occur. However, the virus is a cause of many respiratory infections every year in the United States.

Animal Infections. Several animals have respiratory disease due to viruses which closely resemble the human parainfluenza viruses. A virus closely resembling parainfluenza virus type 3 produces a febrile respiratory illness called shipping fever in cattle. Simian virus Type 5 (SV-5) has been recovered from monkeys with respiratory disease and from dogs with kennel cough.[44]

Treatment

No specific antiviral drug is available to treat parainfluenza virus. Respiratory insufficiency, particularly due to upper respiratory obstruction

FIG. 12-3. *RNA viruses that can cause respiratory diseases, showing their relative size. Picornaviruses (rhinoviruses, coxsackie, and echoviruses) are about 20 to 30 nanometers, while paramyxoviruses are about 120 to 250 nanometers in diameter.*

Picornaviruses (Coxsackie, ECHO)   Orthomyxoviruses (Influenza)   Paramyxoviruses (Parainfluenza, Mumps, Respiratory Syncytial, Measles)

in laryngitis, should be anticipated and treated early. Tracheotomy may be necessary in severe cases.

Prevention

Killed vaccines were studied in the 1960s, but in the 1970s there appears to be no vaccine that is likely to be used in the near future.[54]

MUMPS VIRUS

Objectives

1. Discuss the difference between parotitis and mumps virus infection.
2. Describe a classical case of parotitis due to mumps virus infection.
3. Discuss possible interpretations of a second episode of parotitis.
4. Describe three common complications of mumps virus infection.
5. Discuss the significance of the serum amylase in patients with parotitis.
6. Discuss the possible sources of mumps hyperimmune globulin and the evidence that it prevents or modifies mumps virus infection.
7. Discuss the possible explanations for an adult male giving a history of never having had mumps as a child, and discuss what should be done in such a situation.

Definitions

Parotitis is defined by the clinician as acute enlargement of the parotid gland, usually with fever. The word mumps was originally used to refer to the swelling seen in epidemic parotitis, and later became the name of the virus that was the usual cause of parotitis. However, parotitis is occasionally caused by other viruses, such as parainfluenza or influenza virus, and mumps virus often produces infection without parotitis. Therefore, the word mumps should not be used as if parotitis and mumps virus were identical. The best diagnostic phrasing is "parotitis, probably due to mumps virus". Outbreaks of parotitis, or cases of parotitis that can be related to exposure to another person with parotitis, are almost always due to mumps virus.

Types

There is only one type of mumps virus, and infection is believed to be followed by lifelong immunity, whether or not parotitis occurs on one or both sides or not at all. There are no definite laboratory-documented cases of a clinical recurrence of parotitis due to mumps virus, and second episodes of parotitis are presumed to be due to another virus. However, parotid swelling in mumps infection may increase or decrease rapidly over a period of a few days, as the parotid ducts become obstructed and then open up. Recurrent parotid swelling may be due to a number of causes, but is usually due to narrowed parotid ducts which are easily obstructed by an inflammatory process which would ordinarily not produce obstruction.

Frequency and Importance

Mumps virus infection formerly occurred in almost every individual during childhood or early adult life, and frequency statistics were not kept. Shortly after mumps vaccine became available in 1967, mumps (reported as parotitis) became a notifiable disease. Since mumps vaccine is not given to everyone in childhood, parotitis due to mumps virus remains a frequent disease, with about 70,000 cases reported in 1972, and 21,000 cases reported in 1977.[61]

Fatal or permanently damaging complications due to mumps virus were rare complications of a common disease, but have now assumed a greater importance since mumps has become a preventable disease. The role of mumps virus infection in producing congenital anomalies remains unsettled.

Clinical Patterns of Illness

In susceptible individuals who have not had previous parotitis or mumps vaccine, mumps virus is a *frequent* cause of:

Parotitis. Classical mumps virus infection is painful parotitis, with enlargement of the parotid glands, and fever. Inside the mouth, edema of the openings of Stensen's ducts can often be seen. The sublingual or submaxillary glands may also be enlarged. Edema of the neck or presternal area can occur in severe cases. Headache and vomiting may be very severe. Other clinical patterns seen in association with parotitis are discussed under complications.

Fever and Vomiting Without Parotitis. Many children develop a febrile illness about 3 weeks after exposure to presumed mumps parotitis in a sibling. Because of the known exposure, these patients are often correctly diagnosed as having mumps virus infection, and often a brief or subtle parotid enlargement noted because it is looked for carefully.

In susceptible individuals, mumps virus is an *occasional* cause of:

Croup, bronchiolitis, or pneumonia apparently can be caused by mumps virus.[65]

Pharyngitis or Cervical Adenitis. One might expect that the recovery of mumps virus from patients with these syndromes might be explained by errors in physical examination, especially the mistaking of salivary gland enlargement for lymph node enlargement. However, mumps virus is recovered from patients with these diagnoses often enough to make it an etiologic consideration in such syndromes.[79]

Persistent Vomiting. This is occasionally associated with recovery of mumps virus.

Asymptomatic or Subclinical Infection

Asymptomatic mumps virus infection is very frequent. In outbreaks of mumps, about 25 percent of infections are inapparent.[77,80] Before mumps vaccine was used, about 80 percent of children had had mumps by 10 years of age.[68] In one well-studied outbreak in Florida, about one half of adults without a history of parotitis had serologic evidence of immunity, and did not develop mumps infection when exposed.[77] Of 59 medical students without a history of mumps, 88 percent had detectable neutralizing antibody in their serum.[58] Some of these adults may have had asymptomatic mumps virus infection, but probably some of the unrecognized mumps virus infections were symptomatic illnesses with fever or vomiting.

Complications

The frequency of various complications of mumps has seldom been adequately studied in a prospective fashion. Some reported serious complications may in reality have been coincidental, since mumps was a frequent disease before vaccine was available. The definite complications include:

Orchitis. Testicular pain and swelling occurs in about 10 to 20 percent of adult males with mumps, according to the very few outbreaks of mumps that have been studied prospectively for orchitis.[72,80] Long term prospective studies of sterility have not been done. Before live vaccine was available for prevention of mumps, and when mumps hyperimmune globulin was expensive and of doubtful value, uncontrolled studies and editorials emphasized the infrequency of mumps as a cause of testicular atrophy or sterility.[56] After live attenuated mumps vaccine became available, the danger of orchitis was emphasized as an important indication for the use of the vaccine.[64]

Mumps orchitis rarely results in male infertility. In one study of 105 males with infertility due to a testicular cause, bilateral testicular biopsies were done.[90] Only two men had histologic findings compatible with mumps as a cause, and a past history of mumps orchitis was needed to corroborate the histologic diagnosis. Although mumps is a socially acceptable explanation for male infertility (compared to Klinefelter's syndrome, for example), it is rarely the medical cause.

Oophoritis. Involvement of the ovary may occur, but atrophy apparently does not occur, presumably because the ovary is not surrounded by an inelastic tunica albuginea, as is the testis. In one outbreak, pelvic pain and tenderness, presumably oophoritis, was observed in four (seven percent) of 59 female cases.[82]

Deafness. Although exceedingly rare, deafness apparently does occur, but there are not adequate data on the frequency of this complication.[76,86]

Nonpurulent Meningitis (Aseptic Meningitis Syndrome). This is defined by stiff neck and a moderate spinal fluid pleocytosis. Parotitis may not be present, in which case mumps may not be suspected. The patient may have severe headache, and transient delirium, but clinical evidence of brain involvement (encephalitis) or permanent brain damage is extremely rare.[57,76,83]

Encephalitis. This is defined as a severe and persistent disturbance of consciousness, with a CSF lymphocytosis, and is rare.[57,76,83] Death due to mumps virus infection is sometimes reported on death certificates, and might occur in encephalitis; but death due to mumps rarely has been documented by virus isolation, with exclusion of other causes. From 1969–1975, about 20 deaths due to mumps were reported annually, usually as a complication of encephalitis.[61]

Pancreatitis. This is probably much less frequent than is commonly believed on the basis of serum amylase determinations, as discussed below under laboratory approach.

Other Rare Complications. These complications are very rare or possibly coincidental. Endocardial fibroelastosis has been associated with positive mumps skin tests in some studies of infants with this heart disease, but not in others, and a casual relationship remains unproved. Experimental mumps in chick embryos results in heart lesions of myocarditis or endocardial fibroelastosis.[84] Prospective studies of viral infections in pregnancy have not indicated any adverse effects except increased risk of abortion.

Hydrocephalus, secondary to aqueductal stenosis has been reported in experimental infections in rodents, and might occur in humans.[71] Other rare complications include arthritis,[67] thyroiditis,[63] myocarditis,[70] facial paralysis, transient psychosis, thrombocytopenic purpura,[73] and transverse myelitis.[88]

Diabetes. Some studies have suggested a statistical correlation of mumps virus infection with later diabetes mellitus.[87] Some of the support for this proposed relationship is based on the presumed frequent occurrence of pancreatitis, as discussed below.

Laboratory Diagnosis

Viral Culture. The virus is relatively easily isolated in laboratories equipped for virus isolation. The virus produces both hemadsorption and a typical cytopathic effect on rhesus monkey kidney cells (Fig. 12–4).[79] The virus can be detected in cell cultures in a few hours to 3 days, using fluorescent antibody methods.[74]

Serum Antibodies. Paired sera should be obtained. The complement fixation test is usually done. The testing of a single serum for complement fixing antibodies using an S (soluble) antigen and a V (virus) antigen is not as useful as testing paired sera. In one study of experimental mumps virus infection in institutionalized children, both anti-S and anti-V antibodies were usually elevated by the time clinical disease was evident, and the titers were similar or identical in some cases.[69] Cross-reactions occur between mumps and parainfluenza virus complement fixation tests.[75]

Neutralizing antibody measurement is a reliable test of susceptibility, but is not readily available. Newer antibody techniques are more sensitive and appear to have promise for susceptibility testing.[78]

Mumps Skin Test. Antigen prepared in research laboratories was shown to be a reliable test of immunity in the 1940s, but the commercially available skin testing antigen appears to be very unreliable.[59]

Serum Amylase. This is elevated in patients with parotitis, and may be useful as a laboratory guide

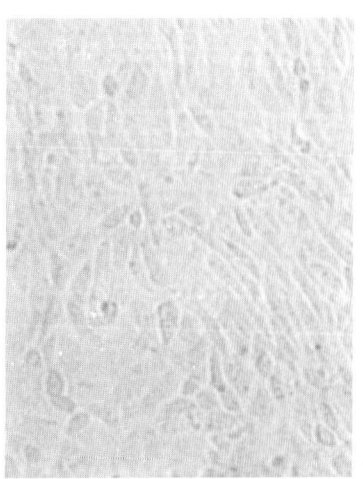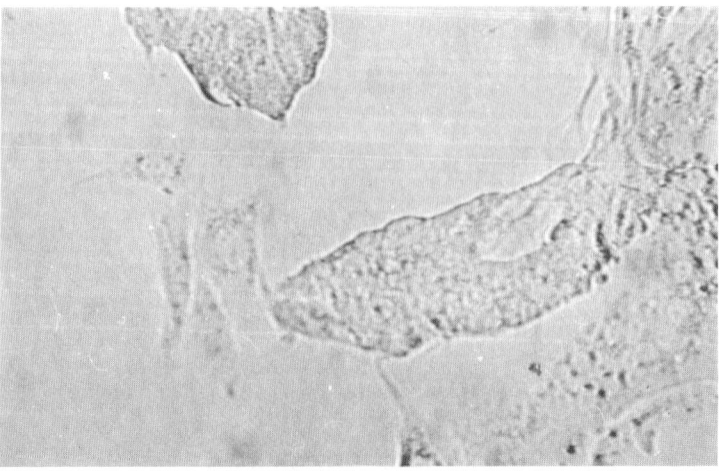

FIG. 12-4. *Right, mumps virus: cytopathic effect on rhesus monkey kidney cell cultures. Left, normal cells.*

to distinguish parotitis from cervical adenitis or other masses in the neck, but does not help distinguish between various causes of parotitis.[60] Amylase elevation persists in about one-third of patients after the parotid swelling disappears, so elevated amylase can indicate recent parotid gland disease.[60] Serum amylase elevation should not be interpreted as indicating pancreatitis, since it is proportional to the severity of parotitis, but not to abdominal pain.[89] Lipase elevation is similarly proportional to vomiting but not to abdominal pain, and has been attributed to infection involving the lipase secreting intestinal glands.[89]

Perhaps fractionation of serum amylase into salivary and pancreatic isoenzymes will clarify this problem.[85]

Biologic Characteristics of Clinical Interest

Contagion and Incubation Period. Daily cultures of experimentally infected children and naturally exposed children indicate the mumps virus is present in the saliva up to 5 to 7 days before and after the time parotitis is first noted. Fever often precedes the parotid enlargement. Parotitis and excretion of the virus can occur as early as 14 days after experimental exposure,[69] but the average incubation period (between exposure and parotid enlargement) is usually about 18 days. Usually the patient has already exposed contacts for several days before parotitis is recognized, but the patient is often kept from school for about 9 days after the onset of parotitis or for the duration of swelling, to allow a margin of safety.

Hemadsorption. This is the characteristic of the mumps virus used for detection of the virus in tissue culture. The hemadsorption appears the same as that of parainfluenza or influenza viruses, so hemadsorption inhibition with specific antiserum is necessary for conclusive identification.

Animal Infections. Accidental natural infection of dogs may occur. Cats and primates can be infected experimentally.

Reinfection. Mumps virus infection with parotitis has been documented in patients with low antibody titers, and this has been interpreted as reinfection, but also may be explained by the fact that the low antibody titers may not have been specific for mumps.[62]

Treatment

There is no specific treatment, and little symptomatic treatment of value for mumps parotitis. It is customary to emphasize bed rest to prevent complications. Although bed rest is not of proved value, complications that occur in patients allowed activity are usually blamed on the activity and whoever allowed it.

In one controlled study of mumps orchitis, steroids were of no value in relief of pain, swelling, or tenderness, and apparently did not influence testicular atrophy on evaluation 6 months after infection.[72] Aspirin or codeine, and support of the scrotum provide some relief.

Prevention

Live Attenuated Mumps Virus Vaccine. This vaccine appears to be safe and effective.[81]

Hyperimmune Mumps Globulin. This currently available preparation is not indicated for prevention of mumps in exposed adult males.[81] The hyperimmune globulin obtained from individuals convalescing from the disease was the preparation originally available and appears to be more effective than the currently available preparation obtained from volunteers who have been hyperimmunized with killed vaccine. In one early study using hyperimmune mumps globulin derived from plasma of patients convalescing from the disease, the frequency of orchitis was reduced to 8 percent, compared to 27 percent in controls.[66] However, current commercially available preparations of hyperimmune mumps globulin are derived from volunteers hyperimmunized with killed mumps virus vaccine. Such a preparation was not particularly effective in preventing orchitis, which occurred in 20 percent of 15 exposed men who received the hyperimmune globulin, and in 27 percent of 44 exposed men who did not receive it.[82] Fortunately, most adults presumed to be susceptible have probably had asymptomatic mumps virus infection. Thus, hyperimmune mumps globulin is not currently recommended. Instead, live attenuated vaccine is recommended if it can be given soon after exposure.[81]

MEASLES VIRUS

Objectives

1. Describe the typical clinical picture of classical measles.
2. Describe the circumstances under which modified (milder) measles may occur. Define atypical measles and indicate the circumstances under which it occurs.
3. Describe the clinical picture of giant cell pneumonia and subacute sclerosing panencephalitis, and their relationship to measles virus infection.
4. Describe the major complications of measles and their approximate frequency.
5. Describe the laboratory methods that might be used to confirm the diagnosis of measles virus infection.
6. Discuss the evidence that measles virus may be related to multiple sclerosis or optic neuritis.

Definitions

Measles was formerly called by the older name, rubeola, but this term should be avoided because it sounds like "rubella". More descriptive names, often used by lay people, are "the hard measles", "the red measles", or "the seven-day measles". The word measles without further modification, should always refer to "red measles", and should not refer to rubella (German measles, three-day measles). It is sometimes useful to distinguish between measles virus and measles-like illness, but in common usage measles can mean either.

Frequency and Importance

Before measles vaccine was available, measles virus infection occurred in nearly 100 percent of individuals during childhood. In the early 1960s, about 400,000 cases were reported annually in the United States.[95] In the late 1970s, more than ten years after the live vaccines were introduced in 1963, between 20,000 and 60,000 cases were reported annually (Fig. 12-5).

In the 1970–71 outbreak in St. Louis, Missouri, there were 130 children hospitalized with measles and its complications.[96] Of these, 66 had pneumonia with six deaths, and six had encephalitis with three requiring custodial care because of residual neurologic damage.[96]

If the rate of measles immunization in the United States continues to be only 70 to 90

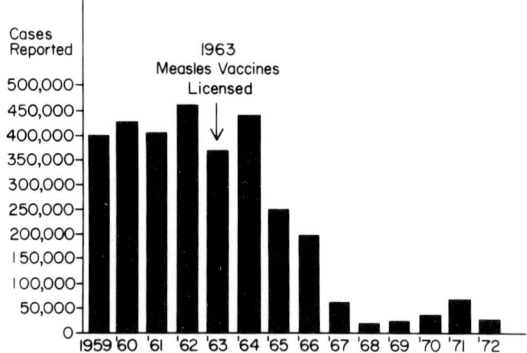

FIG. 12-5. *Measles: frequency in the United States, 1959–1972.*

percent of children, measles may continue to be a frequent disease of children.

Measles is important because of its serious complications of encephalitis and pneumonia. Measles virus has also been demonstrated to cause a rare, fatal disease called subacute sclerosing panencephalitis (SSPE or SSE), an average of seven years after the original illness. Measles virus antibodies are found in high titer in the spinal fluid of patients with multiple sclerosis or optic neuritis, and measles virus has a suspected, but unproved, relationship to these diseases (see p. 249).

Clinical Patterns of Illness

In susceptible children, measles virus is a *frequent* cause of:

Severe Measles-like Illness. Classical measles consists of high fever, cough, and rhinitis for several days, with red eyes (bulbar and palpebral conjunctivitis), and a measly (maculopapular) rash beginning on the face and spreading downward to involve the entire body and lasting about a week or longer. It usually becomes confluent. The lymphocytes are often suppressed, with a peripheral white blood count of less than 5,000 with less than 40 percent lymphocytes.

A mnemonic for the findings in measles is CCCK, for Cough, Conjunctivitis, Coryza (rhinitis) and Koplick spots.

Classical measles is highly contagious, but gamma globulin modified measles is much less contagious.[94]

Mild measles-like illness is now seen very frequently. Measles can be modified (made milder or atypical) by three circumstances. First, maternal (transplacental) antibodies are present in the infant's blood for approximately the first 6 months of life, if the mother has had measles. These antibodies make measles milder to a variable degree. The effect of transplacentally acquired antibodies from mothers who have had vaccine measles, rather than wild measles, is as yet unknown.

Second, gamma globulin (immune serum globulin) can be given by the physician to modify measles in an exposed susceptible child. Third, measles vaccine, particularly the killed virus vaccine, will occasionally allow the patient to have an atypical or milder measles, rather than produce the intended result of complete prevention of the disease. Atypical measles following killed measles virus vaccine appears to be influenced by delayed hypersensitivity from the vaccine, and the rash may have petechial, vesicular, or urticarial components.[100]

Atypical measles. This clinical pattern occurs in individuals who have previously received killed measles vaccine, or "live" measles vaccine that is really dead.[100] The diagnosis is difficult. The patient may have some features suggesting measles, such as cough and conjunctivitis, but the rash is atypical. The rash often begins on the extremities, rather than descending from the face to the trunk to the legs. It may be petechial or pustular, rather than maculopapular and confluent, as in classic measles. Atypical measles is often mistaken for Rocky Mountain spotted fever, which is petechial and begins on the extremities. Pneumonia may be prominent.

Atypical measles can also occur in children who have previously received live attenuated measles vaccine.[97]

Complications

Complications can occur in classic unmodified measles. These include:

Encephalitis. This occurs in about 1 of 1000 reported cases of measles and has a mortality rate of about 10 percent.[109] It typically occurs when the rash is still present. Measles encephalitis should not be confused with subacute sclerosing encephalitis, discussed below.

Pneumonia. Radiographic pulmonary densities are frequent in measles. Pneumonia is the usual cause of death in young children with measles, and encephalitis is the most frequent cause of death in teenagers.[117] *H. influenzae* is sometimes recovered from the blood of patients with pneumonia complicating measles.[117]

Fatal Giant Cell Pneumonia. In patients with severe host defects, such as leukemia, there may be no rash, with giant cells found at autopsy (giant cell pneumonia).[99] (Fig. 12-6). This complication occasionally occurs in adults who are receiving immunosuppressive therapy. It also occurs in children with acute leukemia in remission.[115]

Subacute Sclerosing Panencephalitis (SSPE or SSP). This was formerly called Dawson's encephalitis. It is an exceedingly rare disease which occurs about 2 to 10 years after the initial measles virus infection. About 10 to 15 cases are recognized each year in the United States. Patients with Dawson's encephalitis have extremely high antibody titers against measles virus,[105] especially measles-specific IgM.[106] Measles virus has been

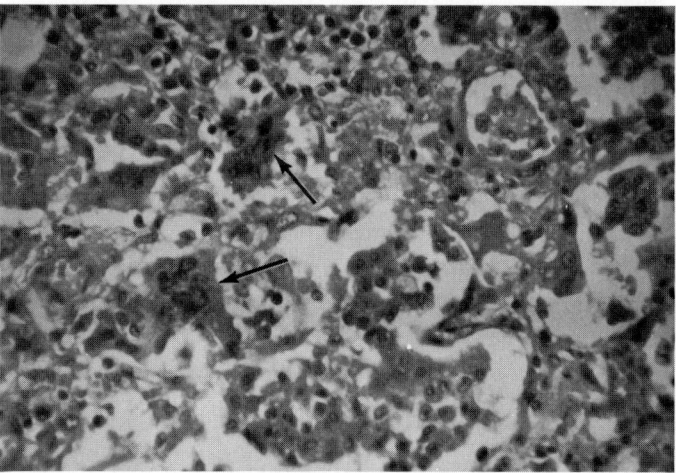

FIG. 12-6. *Lung section of a patient with fatal measles pneumonia showing multinucleated giant cells* (**arrows**). *(Photo from Dr. Enid Gilbert)*

recovered from the brain of patients with this disease.[114] The measles virus appears to be incomplete and its recovery requires the technique of cocultivation, using live brain cells from the patient grown in the same tubes as cell cultures that support the growth of the virus. The reason that some patients develop this disease is unknown, but subtle host immunologic abnormalities may be involved.[102] The possible risk of measles vaccine virus for producing this complication is unknown, but is less than that of wild virus infection. SSPE has been decreasing in frequency since about 5 to 10 years after the introduction of measles vaccine.

Encephalopathy in Children with Malignancies.[112] Children with leukemia or neuroblastoma can develop a chronic progressive encephalopathy which appears to be caused by measles virus.

Other Complications. Croup may occur and is sometimes severe. Otitis media is usually nonbacterial and usually responds to nasal vasoconstrictors, but antibiotics are often used. Activation and dissemination of latent tuberculosis may occur.[91]

Laboratory Diagnosis

Serum Antibodies. Testing for measles hemagglutination inhibition (HI) antibodies in paired sera is the most reliable method of laboratory diagnosis.

Viral Culture. Isolation of the virus is difficult and less certain than serologic studies.

Histology. Multinucleated giant cells with inclusion bodies may be found at autopsy in pneumonia complicating measles.[107] The virus particles can sometimes be seen by electron microscopy in the brain of patients with subacute sclerosing panencephalitis. Using fluorescent antibody techniques, nasopharyngeal cells can be stained for measles antigen, providing a rapid method to diagnose measles in difficult cases.[101]

Biologic Characteristics of Clinical Interest

Giant Cell Formation. Multinucleated giant cells may be found in nasal secretions, sputum, and lymphoid tissue of the gastrointestinal tract. Similar giant cells are seen in cell culture as a cytopathic effect (Fig. 12-6).

Interference with Delayed Hypersensitivity. The tuberculin test often becomes negative or less positive in patients infected with measles. This occurs with wild measles and with attenuated measles vaccine, and lasts several weeks. This is evidence of a suppression of cell-mediated immunity by measles virus infection, and probably accounts for the occasional activation or dissemination of tuberculosis after measles.

Animal Diseases Resembling Measles. Human measles virus cannot be transmitted to animals. However, rinderpest virus of cattle, and distemper virus of dogs and cats are serologically related to measles. Measles vaccine protects dogs against canine distemper.[92]

Multiple Sclerosis. Patients with multiple sclerosis or optic neuritis have high antibody titers against measles virus.[104,110] The significance of this is unknown.

Chromosome Breaks. This occurs in leukocytes infected with measles. The significance is not clear.

Virus in the Rash. Measles virus is present in the rash[113] and in Koplik's spots, as seen in electron micrographs.[116]

Measles in Adolescents and Adults. Immunization of children has changed the age distribution of measles virus infection. In 1967, about 99 percent of 18-year-olds were immune to measles on the basis of past infection. In the late 1970s, measles was appearing more frequently in adolescents and young adults, who had escaped exposure to natural disease and had not been immunized or who were not protected by the vaccine. Outbreaks are becoming more frequent in colleges. The exposure of pregnant women who might be susceptible to measles has become a more frequent problem. Accurate practical tests of measles susceptibility need to be more readily available, as is the case for rubella.

Prevention

Vaccine. Live attenuated measles virus vaccine is safe and effective. The vaccine is not recommended for infants until 15 months of age unless there is an outbreak of measles and exposure is likely. In the first year of life, some transplacentally acquired antibodies may be present and interfere with vaccine virus infection. It is currently recommended that individuals who were vaccinated in the first year of life and those who received killed vaccine should be revaccinated, but other individuals should not be revaccinated.[111] There may be some risk to a second inoculation of attenuated measles virus, so revaccination should not be done unless it is needed.

Gamma Globulin. Ordinary immune serum globulin (ISG) can be used to prevent or modify measles, depending on the dose used. Measles should usually be prevented in a patient with a serious resistance defect. However, most exposed susceptibles should be given a modifying dose. A modifying dose of gamma globulin clearly prevents encephalitis in exposed susceptibles in almost all cases.[103] The possibility that administration of live measles vaccine to prevent measles after exposure to wild measles is currently under investigation.

Avoid Exposure. Infants 6 to 15 months of age should be protected by preventing exposure, if possible.

RESPIRATORY SYNCYTIAL VIRUS

Objectives

1. Discuss the origin of the name respiratory syncytial (RS) virus.
2. Describe the illnesses that can be produced by RS virus in young infants and in adults.
3. Describe how infection with RS virus can be confirmed by laboratory studies.

Definitions and Types

Respiratory syncytial virus is an important cause of respiratory disease and produces a characteristic and diagnostic syncytial effect in cell cultures. It was originally recovered from chimpanzees with rhinitis, and was first called chimpanzee coryza agent. There are several antigenic variants of the virus, but these are regarded as a single serotype. Lack of lasting secretory antibody after infection, not antigenic variation, appears to be responsible for recurrences of RS virus infections.[118]

Frequency and Importance

RS virus is a frequent cause of serious lower respiratory infections in young infants, and occasionally causes pneumonia in elderly adults. However, fatalities due to this virus are quite rare.

Clinical Patterns of Illness

In young children, RS virus is a *frequent* cause of:

Bronchiolitis and pneumonia are the most serious illnesses produced by this virus, and typically occur in infants less than 2 years of age. Bronchiolitis is characterized by expiratory prolongation and low diaphragms in an infact less than 2 years of age. Typically, the infant has had cough and rhinitis for a few days before developing a rapid respiratory rate (60 to 80 per minute). Pneumonia is usually suspected, but chest film usually reveals minimal or no pulmonary infiltrate.

Pneumonia without bronchiolitis also can be caused by RS virus, but bronchiolitis is more common.[135]

Apnea and Sudden Death. In infants less than 6 months of age with RS virus infection, apnea was observed in about 20 percent. It has been suggested that RS virus may be one of the causes of sudden infant death syndrome.[120]

In older children and adults, RS virus is an *occasional* cause of:

Common Cold Syndrome. Mild respiratory illnesses, particularly the common cold, are occasionally due to RS virus. It can also produce minor respiratory illnesses in the newborn period,[135] and can be quite contagious in a neonatal intensive care unit. In a study of RS infections in families, serologic evidence of infection was observed during the surveillance year in 18 percent of children 1 to 9 years of age, 14 percent in children 10 to 19 years of age, and five percent of adults 20 to 39 years of age.[133]

The common cold syndrome in adults is typically a reinfection that occurs in spite of the presence of neutralizing antibodies.[130] Reinfection apparently occurs frequently in children as young as 3 to 5 years of age, usually with mild clinical illnesses resembling the common cold.[119]

Exacerbation of Asthma and Chronic Bronchitis. Pulmonary function studies on healthy young adults indicate increased pulmonary resistance and are consistent with the observation that RS virus infection exacerbates asthma and chronic bronchitis.[129]

Heart Block. RS virus infection has been associated with cardiac arrhythmias, especially heart block.[125]

Pneumonia in the Elderly. An outbreak of RS virus infection in a nursing home was associated with pneumonia in some elderly patients.[123]

Asymptomatic Infection

Seroconversions without illness can occur, but usually there is lower respiratory disease.[133]

Laboratory Approach

Viral Cultures. The virus is detected in cell cultures by a syncytial cytopathic effect (CPE), which occurs about a week after inoculation. The syncytial CPE is typical (Fig. 12-7). The RS virus does not survive well after freezing, so the throat swabbing should be inoculated onto cell cultures immediately after it is obtained, without freezing.

An adequate specimen must be obtained for optimal recovery of the virus. This is done by squirting about 5 ml of phosphate-buffered saline solution into a nostril and immediately sucking it back, using a rubber bulb syringe.[127]

Serum Antibodies. Serologic diagnosis can be done using paired sera. However, young infants or adults may have symptoms associated with the isolation of the RS virus, without showing an antibody rise.[130]

Smear. RS virus has been detected in smears from patient specimens by using an indirect immunofluorescent technique.[131] The smear of the patient's pharyngeal secretions is fixed in acetone and then covered with anti-RS virus bovine antiserum. The slide is then washed and stained with fluorescein-conjugated antibovine antiserum and examined for fluorescence.[131]

Biologic Characteristics of Clinical Interest

Animal Models. Calves have naturally occurring respiratory syncytial virus-caused febrile respiratory disease.

Epidemic Occurrence. RS virus infections are often observed in outbreaks lasting just a few months in one area, during which time RS virus is responsible for a large proportion of the lower respiratory diseases in infants and children.[124] Outbreaks, which occur almost every year, can give the impression that bronchiolitis is usually due to RS virus. However, bronchiolitis is a syndrome that has many possible causes, including allergy, and occurs sporadically throughout

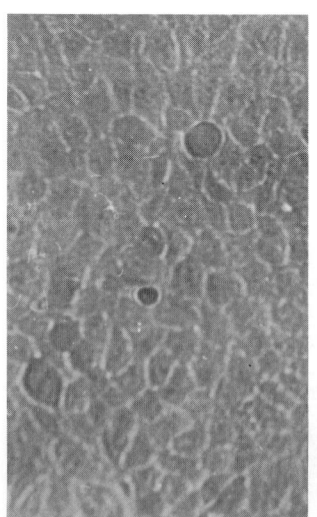
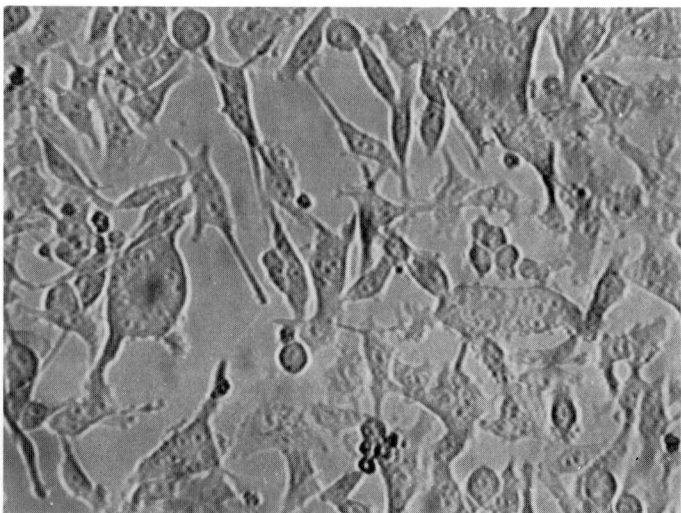

FIG. 12-7. *Respiratory syncytial virus, showing syncytial cytopathic effect on HEp-2 cell culture. Normal HEp-2 cells are shown at the left.*

the year. For example, during a 5 year period at Children's Memorial Hospital, RS virus was recovered from only 21 percent of 117 infants with bronchiolitis compared to less than one percent of 286 normal individuals. Thus, RS virus is not recovered in the majority of cases, although it is the most frequent single cause.

Repeated Symptomatic Infections. These infections are apparently unrelated to antigenic variation.[118] Reinfection occasionally is as severe as the first infection, but usually subsequent infection is associated with a milder illness.[118]

Severity Worse After Killed Virus Vaccine. RS virus disease is worse if infant has received killed virus vaccine.[132] This suggests that severe manifestations may be in part an antigen–antibody reaction. However, infants with transplacentally acquired IgG antibodies do not have more severe disease than infants without antibodies.[121]

Contagion. Inoculation of volunteers indicates an incubation period of 5 days.[130] The virus appears to spread among individuals with antibodies, resulting in modified illnesses, and older children or even adults are the usual source of serious illnesses in young infants.[126,133] Infants often acquire RS virus infection in the hospital, often via hospital personnel.[128]

Treatment

No specific antiviral drug is available. Mist and oxygen are useful for treatment of severe bronchiolitis.

Prevention

Killed virus vaccine resulted in more severe clinical illness when the infant was exposed to the naturally occurring illness.[132] Attenuated virus vaccines are under investigation.[122,136] The morbidity and mortality of RS virus infection in young infants is severe enough to need a vaccine.

RABIES VIRUS

Objectives

1. List the three domestic and the three wild animals that have the highest frequency of rabies.
2. Discuss the frequency of human rabies in the United States.
3. Indicate the frequency of animal bites and the number of persons given rabies postexposure prophylaxis each year.

Definitions

Rabies is a disease of biting animals, especially carnivores. Humans are a dead end for the virus, and usually do not transmit the disease further, although the saliva of a human with rabies contains the virus, and must be considered infectious.[150] Rabies was once called hydrophobia, and even recent case reports describe fear of drinking water (which results in choking and gagging) as a prominent symptom.

Pasteur showed the agent was present in the brains of rabid animals, and was the first to use preventive vaccination.[152] He used the term street virus in the way we now use wild virus; that is, street virus is the virus as found in nature. Pasteur used the term fixed virus to describe the laboratory strain of virus. Serial passage of the virus by intracerebral injection in rabbits leads to a shorter and shorter incubation period until the incubation period is fixed at about 14 days. Fixed virus is more virulent for dogs than street virus. Fixed virus also appears to be more virulent for humans than street virus, is an extremely hazardous virus for laboratory workers, and may be infectious by the airborne route.[140,142]

Frequency and Importance

Frequency in Animals. The risk of developing rabies after a bite depends in part on the species of the biting animal. The animals in the United States most likely to transmit rabies are skunks, bats, foxes. Skunk bites are especially dangerous, including those of pet descented skunks. Bats probably represent an uneradicable reservoir.[147] Rodents and lagomorphs (rabbits) seldom, if ever, transmit rabies to humans, and their bites are usually not treated with rabies vaccine.

In wild animals, there were about 2600 laboratory-confirmed cases per year in the United States, in the 1970s.[141] In rounded numbers, these were distributed per year as: skunks, 1500; bats, 500; foxes, 300; raccoons, 250; and others, 50. In domestic animals, there were about 400 laboratory-confirmed cases per year in the 1970s. In rounded numbers, there were distributed in 1977 as: cattle, 150; dogs, 120; cats, 100; and other farm animals, 30. Domestic animal rabies has been decreasing in frequency, particularly in dogs. In 1977, there were only 445 laboratory-confirmed cases of domestic animal rabies in the United States, compared to 1,119 cases in 1959, and 5,688 in 1953.[141]

Frequency in Humans. In the period of time from 1960 to 1970, there were one or two human rabies cases per year in the United States (Fig. 12-8), compared with 400 to 500 cases per year in Thailand.[147] Since 1972, there have been none to three cases per year in the United States.[140] There are probably at least a million incidents of animal bites annually in the United States, and an estimated 25,000 persons are treated each year with rabies vaccine.[151]

Clinical Patterns of Illness

Classic Rabies in Humans. There typically has been a bite by a clinically rabid animal. Often rabies vaccine has been given. Symptoms of nervousness and anxiety usually begin a month or two after the bite. A burning sensation or pain in the area of the bite is common. The disease typically progresses to an acute encephalitis, with anxiety, convulsions, and paralysis. Difficult and painful swallowing, with drooling, is common. Offering the patient water to drink may precipitate acute anxiety, leading to the name hydrophobia. Myocarditis may occur.[137,139]

Formerly it was assumed that once clinical findings began in a patient, the disease would always be fatal, although one patient with rabies in 1970 survived.[144] Another case in an immunized laboratory worker seemed to result from aerosol exposure, and was not fatal.[140]

Rabies in Dogs. Furious rabies refers to the excitability, restlessness and vicious manifestations in which the animal bites or snaps at everything. This is followed by the stage of dumb rabies, in which the animal is paralyzed. Typically, the dog has paralysis of the muscles of swallowing, and chokes. The owner may believe the dog is choking on a bone, and may be bitten when reaching into the dog's mouth. Most laboratory-confirmed rabies in dogs in the United States involve exposure to humans by a dog without aggressive behavior.[148]

Asymptomatic and subclinical rabies may occur in bats and skunks, which may be important reservoirs of the virus.[138,147]

Laboratory Approach

Histologic. The Negri body in the brain is usually considered pathognomonic for rabies. The fluorescent antibody staining of the brain has proved to be extremely useful to help see the Negri bodies quickly (Fig. 12-9). This is most useful in making the diagnosis in animals. Spinal cord can be used if the brain is damaged.[149] The virus can sometimes be detected before death by immunofluorescence staining of a facial skin biopsy, or an impression smear of the patient's cornea.

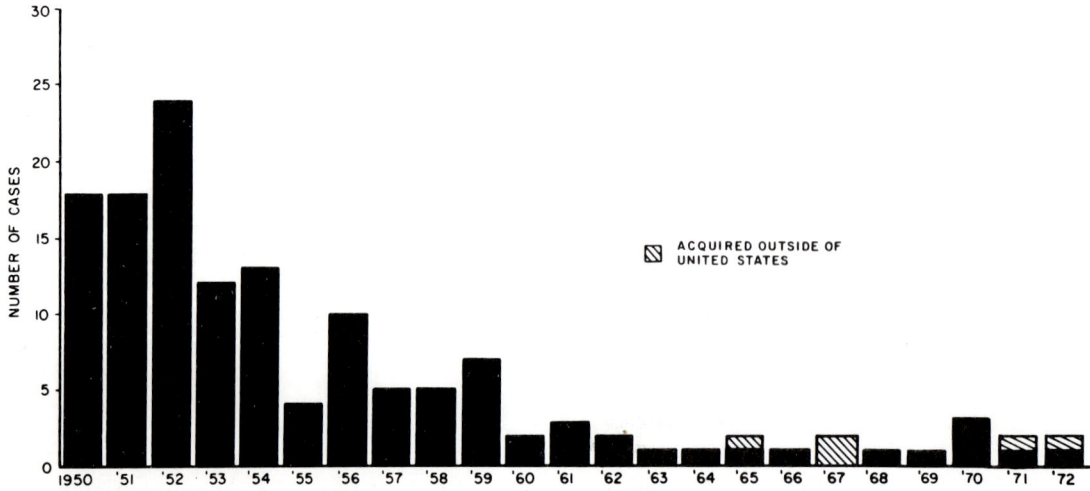

FIG. 12-8. *Reported human rabies frequency in the United States, 1950–1972.*

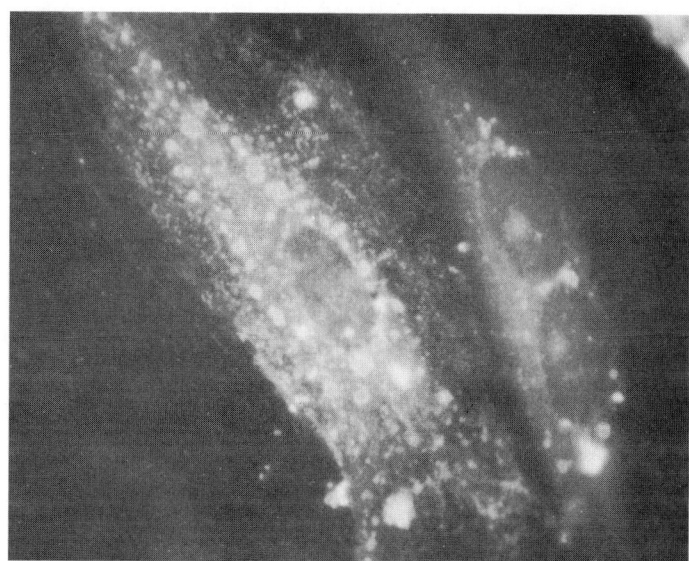

FIG. 12-9. Negri bodies seen in this photograph are white circular bodies found throughout the cell, outside of the muscles. This is a fluorescent antibody preparation in which the Negri bodies have a bright green fluorescence under the microscope, using a fluorescent light source. (Photo from Donald B. Nelson)

Viral Culture. The virus can be grown in cell cultures, such as human diploid cells. Usually no cytopathic effect is seen, but the virus can be detected by immunofluorescence using fluoroscein-labeled rabies antibody.

Animal Inoculation. Intracerebral inoculation of mice with brain material from the suspected animal can be done. This is slightly more sensitive than the standard staining techniques for Negri bodies, and comparable to the immunofluorescence method. It is useful if the brain has decomposed.

Serum Antibodies. Ordinarily, the patient dies of rabies before a second serum can be obtained. With use of supportive measures, such as mechanical ventilation with a respirator, a patient may survive long enough to develop antibodies. Occasionally, when no animal exposure has been recognized, serologic diagnosis might be useful.

Transport of Animal Specimens. The reference laboratory should be telephoned to find out their recommendations for transport of the animal specimen. Usually delivery by a private automobile is preferable to first class mail. The specimen should be properly packed, with a proper coolant, and labeling precautions.

Biologic Characteristics of Clinical Interest

Transmission. Aerosols can transmit the virus to experimental infections in foxes, and naturally occurring infections in man. Aerosol infections have apparently occurred in cave explorers and laboratory workers.[140,142,153] The virus also can be spread by the oral route. The possibility of immunization of wild animals by oral immunization using bait is under investigation.

Spread of Virus. The virus reaches the brain by way of peripheral nerves. Infection in experimental animals can be prevented by cutting the local nerve supply.

Interferon Stimulation. This may be an important mechanism for protection induced by the killed virus vaccine, although this is not clearly established.[153] This might account for the observation that it is essential to give the vaccine within 72 hours, and preferably within 24 hours of the bite, so that the virus can be eliminated before it spreads to the CNS from the wound. Treatment of monkeys with human interferon 24 hours after experimental inoculation of street rabies virus reduced the mortality rate.[146]

Treatment

Rabies has long been regarded as inevitably fatal in humans once the clinical symptoms of rabies have begun. Recently, long survivals and a fairly well documented cure have been reported.[144] In such cases, support of respiration with a mechanical ventilator, and human rabies immune globulin have been used.

Prevention

Rabies Control in Domestic Animals. The islands of Great Britain and Hawaii are rabies-free areas, which prevent introduction of the disease by requiring a 120-day quarantine for all imported carnivores.[142] Some large cities, such as New York City and Philadelphia, no longer recommend antirabies prophylaxis for humans bitten by stray or escaped dogs and cats, since rabies has not been observed in these animals in these cities for more than 30 years. However, available cats or dogs are placed under observation by the owner for 10 days. Immunization of domestic animals is very useful to decrease the risk to humans.[148]

Avoid Exposure to Wild Animals. Keeping a descented skunk as a pet is an unnecessary risk. Attempting to capture a live bat, raccoon, or fox is an unnecessary risk.

Pre-exposure Prophylaxis. Rabies vaccine can be given to animal handlers, veterinarians, and travelers to endemic areas, in advance of expected exposures.

Post-exposure Prophylaxis. Local wound care comes first, and consists of thorough cleaning with soap and water.[151] Then rabies vaccine, with or without human rabies immune globulin, might be used. The criteria for use of vaccine and human globulin are complicated. In general, both are used if the biting animal is a bat, skunk, or fox. In general, bites of rodents do not require vaccine or human globulin.

Human Rabies Immune Globulin (HRIG) can be prepared from the plasma of individuals (usually veterinarians) who have received preexposure rabies vaccine. This hyperimmune globulin is now available and is used instead of horse serum (rabies antiserum), which is now considered obsolete, as long as HRIG is available. Because HRIG is more potent than the previously used horse antirabies serum, 21 days instead of 14 days of vaccine are indicated.[145]

Duck Embryo Vaccine (DEV) has been used since about 1960 instead of the obsolete nervous system (Semple) vaccine. Repeated daily subcutaneous injections of the vaccine are given in order to stimulate the bitten persons to develop their own antibodies. However, about 10 percent of persons given DEV alone, and about 40 percent of persons given DEV and HBIG, did not have detectable rabies antibodies 90 days after exposure,[145] indicating a need for better vaccines.

Human diploid cell vaccine produced in human cell cultures produces a much higher titer of rabies antibodies compared to duck embryo vaccine.[143] In 1978, this vaccine became available from the Center for Disease Control in Atlanta, Georgia, for use in persons with a serious allergy to duck embryo vaccine, and is expected to be available eventually without limitations.

REFERENCES

Influenza

1. Beare, A.S., and Craig, J.W.: Virulence for man of a human influenza-A virus antigenically similar to "classical" swine viruses. Lancet, 2:4–6, 1976.
2. Bell, T.D., Chai, H., Berlow, B., and Daniels, G.: Immunization with killed influenza virus in children with chronic asthma. Chest, 73:140–145, 1978.
3. Berendt, R.F., Long, G.G., and Walker, J.S.: Influenza alone and in sequence with pneumonia due to *Streptococcus pneumoniae* in the squirrel monkey. J. Infect. Dis., 132:689–693, 1975.
4. Center for Disease Control: Influenza-respiratory disease surveillance. Report No. 98, 1972–1973. February, 1974.
5. Davenport, F.M., et al.: Pilot studies on recombinant cold-adapted live type A and B influenza virus vaccines. J. Infect. Dis., 136:17–25, 1977.
6. Dietzman, D.E., Schaller, J.G., Ray, C.G., and Reed, M.E.: Acute myositis associated with influenza B infection. Pediatrics, 57:255–258, 1976.
7. Dowdle, W.R.: Influenza anti-neuraminidase: the second best antibody. New Engl. J. Med., 286:1360–1361, 1972.
8. Gaydos, J.C., et al.: Swine influenza A at Fort Dix, New Jersey (January–February 1976). I. Case finding and clinical study of cases. J. Infect. Dis., 136 (Suppl.):S356–S362, 1977.
9. Greco, T.P., Askenase, P.W., and Kashgarian, M.: Postviral myositis: myxovirus-like structures in affected muscles. Ann. Int. Med., 86:193–194, 1977.
10. Gregg, M.B., Hinman, A.R., and Craven, R.B.: The Russian flu. Its history and implica-

tions for this year's influenza season. J.A.M.A., 240:2260–2263, 1978.

11. Gross, P.A., et al.: Influenza immunization in immunosuppressed children. J. Pediatr., 92:30–35, 1978.

12. Gross, P.A., and Ennis, F.A.: Influenza vaccine: split-product versus whole-virus types—how do they differ? (editorial). New Engl. J. Med., 296:567–568, 1977.

13. Gwaltney, J.W., Jr., DeSanctis, A.N., Metzgar, D.P., and Hendley, J.O.: Systemic reactions to influenza B vaccine. Am. J. Epidemiol., 105:252–260, 1977.

14. Jackson, G.G.: Sensitivity of influenza A virus to amantadine. J. Infect. Dis., 136:301–302, 1977.

15. Jackson, G.G., and Stanley, E.D.: Prevention and control of influenza by chemoprophylaxis and chemotherapy. Prospects from examination of recent experience. J.A.M.A., 235:2739–2742, 1976.

16. Kapila, R., et al.: A nosocomial outbreak of influenza A. Chest, 71:576–579, 1977.

17. Kilbourne, E.D.: Influenza: the vaccines. Hosp. Prac., 6:103–114, 1971.

18. ———: Influenza pandemics in perspective. J.A.M.A., 237:1225–1228, 1977.

19. Klein, R.C., Bobear, J.B., and Gohd, R.: Influenza, respiratory distress and clear chest films: report of four cases. South. Med. J., 78:5–6, 1978.

20. Laraya-Cuassay, L.R., et al.: Chronic pulmonary complications of early influenza virus infections in children. Am. Rev. Respir. Dis., 116:617–625, 1977.

21. Lindsay, M.I., Jr., et al.: Hong Kong influenza. Clinical, microbiologic, and pathologic features in 127 cases. J.A.M.A., 214:1825–1932, 1970.

22. Little, J.W., et al.: Amantadine effect on peripheral airway abnormalities in influenza. A study of 15 students with natural influenza A infection. Ann. Intern. Med., 85:177–182, 1976.

23. Maugh, T.H., II: Influenza: the last of the great plagues. Science, 180:1042–1044, 1973.

24. ———: Influenza (II): a persistent disease may yield to new vaccines. Science, 180:1159–1161, 1215, 1973.

25. Myer, H.M., Jr., Hopps, H.E., Parkman, P.D., and Ennis, F.A.: Review of existing vaccines for influenza. Am. J. Clin. Pathol., 70(Suppl):146–152, 1978.

26. Minor, T.E., Dick, E.C., Dick, C.R., and Inhorn, S.L.: Attenuated influenza A vaccine (Alice) in an adult population: vaccine related illness, serum and nasal antibody production, and intrafamily transmission. J. Clin. Microbiol., 2:403–409, 1975.

27. Moffet, H.L., et al.: Outbreak of influenza B in a children's home. J.A.M.A., 182:834–838, 1962.

28. Mogabgab, W.J.: The complications of influenza. Med. Clin. North Am., 47:1191–1199, 1963.

29. Morris, J.A., et al.: Immunity to influenza as related to antibody levels. New Engl. J. Med., 274:527–535, 1966.

30. Mostow, S.R., Flatauer, S., Paler, M., and Murphy, B.R.: Temperature-sensitive mutants of influenza virus. XIII. Evaluation of influenza A/Hong Kong/68 and A/Udorn/72ts and wild-type viruses in tracheal organ culture at permissive and restrictive temperatures. J. Infect. Dis., 136:1–6, 1977.

31. Murphy, B.R., Kasel, J.A., and Chanock, R.M.: Association of serum anti-neuraminidase antibody with resistance to influenza in man. New Engl. J. Med., 286:1329–1332, 1972.

32. Ogra, P.L., et al.: Clinical and immunologic evaluation of neuraminidase-specific influenza A virus vaccine in humans. J. Infect. Dis., 135:499–506, 1977.

33. Paisley, J.W., Bruhn, F.W., Lauer, B.A., and McIntosh, K.: Type A_2 influenza virus infections in children. Am. J. Dis. Child., 132:34–36, 1978.

34. Parrott, R.H., Kim, H.W., Vargosko, A.J., and Chanock, R.M.: Serious respiratory tract illnesses as a result of Asian influenza and influenza B infections in children. J. Pediatr., 61:205–213, 1962.

35. Partin, J.C., et al.: Isolation of influenza virus from liver and muscle biopsy specimens from a surviving case of Reye's syndrome. Lancet, 2:599–602, 1976.

36. Public Health Service Advisory Committee on Immunization Practices: Influenza Vaccine. Mort. Morb. Week. Rep., 27:285–286, 291–292, 351, 1978.

37. Richman, D.D., Murphy, B.R., Baron, S., and Uhlendorf, C.: Three strains of influenza A virus (H3 N2): interferon sensitivity in vitro and interferon production in volunteers. J. Clin. Microbiol., 3:223–226, 1976.

38. Sabin, A.S.: Correspondence. Amantadine and influenza: evaluation of conflicting reports. Jackson, G.G.: Reply. J. Infect. Dis., 138:557–568, 1978.

39. Scheinberg, M.A., et al.: Influenza: response of T—cell lymphopenia to thymosin. New Engl. J. Med., 294:1208–1211, 1976.

40. Schwarzmann, S.W., Adler, J.L., Sullivan, R.J., Jr., and Marine, W.M.: Bacterial pneumonia during the Hong Kong influenza epidemic of 1968–1969. Arch. Intern. Med., 127:1037–1041, 1971.

41. Shenouda, A., and Hatch, F.E.: Influenza A viral infection associated with acute renal failure. Am. J. Med., 61:697–702, 1976.

42. Smith, T.F., and Reichrath, L.: Comparative recovery of 1972–1973 influenza virus isolates in embryonated eggs and primary rhesus

monkey cell cultures after one freeze-thaw cycle. Am. J. Clin. Pathol., 61:579–584, 1974.

43. Webster, R.G., Kasel, J.A., Couch, R.B., and Laver, W.G.: Influenza virus subunit vaccines. II. Immunogenicity and original antigenic sin in humans. J. Infect. Dis., 134:48–58, 1976.

Parainfluenza

44. Appel, M.J.G., and Percy, D.H.: SV-5-like parainfluenza virus in dogs. J. Am. Vet. Med. Assoc., 156:1778–1781, 1970.

45. Craighead, J.E., and Brennan, B.J.: Cytopathic effects of parainfluenza virus type 3 in organ cultures of human respiratory tract tissue. Am. J. Pathol., 52:287–300, 1968.

46. Doane, F.W., et al.: Rapid laboratory diagnosis of paramyxovirus infections by electron microscopy. Lancet, 2:751–753, 1967.

47. Hall, C.B., Geiman, J.M., Breese, B.B., and Douglas, R.G., Jr.: Parainfluenza viral shedding in children: correlation of shedding with clinical manifestations. J. Pediatr., 91:194–198, 1977.

48. Kapikian, A.Z., et al.: An outbreak of parainfluenza 2 (croup-associated) virus infection. Association with acute undifferentiated febrile illness in children. J.A.M.A., 183:324–330, 1963.

49. ———: Inoculation of human volunteers with parainfluenza virus type 3. J.A.M.A., 178:537–541, 1961.

50. Konerding, K., and Moffet, H.L.: New episodes of fever in hospitalized children. Am. J. Dis. Child., 120:515–519, 1970.

51. Marks, M.I., Nagahama, H., and Eller, J.J.: Parainfluenza virus immunofluorescence. In vitro and clinical application of the direct method. Pediatrics, 48:73–78, 1971.

52. Parrott, R.H., et al.: III. Myxoviruses: Parainfluenza. Am. J. Publ. Health, 52:907–917, 1962.

53. Parrott, R.H., Vargosko, A.J., Kim, H.W., and Chanock, R.M.: Clinical syndromes among children. Am. Rev. Respir. Dis., 88:(Part 2), 73–88, 1963.

54. Wigley, F.M., Fruchtman, M.H., and Waldman, R.H.: Aerosol immunization of humans with inactivated parainfluenza type 2 vaccine. New Engl. J. Med., 283:1250–1253, 1970.

Mumps

55. Adair, C.V., Gauld, R.L., and Smadel, J.E.: Aseptic meningitis, a disease of diverse etiology: clinical and etiologic studies on 854 cases. Ann. Intern. Med., 39:675–704, 1953.

56. Anon: Steroid therapy in mumps (editorial). Br. Med. J., 2:146, 1959.

57. ———: Mumps vaccine: more information needed (editorial). New Engl. J. Med., 278:275–276, 1968.

58. Brickman, A., and Brunnell, P.A.: Susceptibility of medical students to mumps: comparison of serum neutralizing antibody and skin test. Pediatrics, 48:447–450, 1971, Pediatrics, 49:314–315, 1971.

59. Brunell, P.A., Brickman, A., O'Hare, D., and Steinberg, S.: Ineffectiveness of isolation of patients as a method of preventing the spread of mumps. Failure of the mumps skin test antigen to predict immune status. New Engl. J. Med., 279:1357–1361, 1968.

60. Candel, S., and Wheelock, M.C.: Serum amylase and serum lipase in mumps. Ann. Intern. Med., 25:88–96, 1946.

61. Center for Disease Control: Mumps—United States. Morb. Mort. Week. Rep., 27:379–380, 1978.

62. Ennis, F.A.: Immunity to mumps in an institutional epidemic. Correlation of insusceptibility to mumps with serum plaque neutralizing and hemagglutination–inhibition antibodies. J. Infect. Dis., 119:654–657, 1969.

63. Eylan, E., Zmucky, R., and Sheba, C.H.: Mumps virus and subacute thyroiditis. Evidence of a causal association. Lancet, 1:1062–1063, 1957.

64. Fiumara, N.J., McCroan, J.E., Stokes, J., Jr., and Katz, S.L.: Correspondence. Use of mumps vaccine. New Engl. J. Med., 278:681–684, 1968.

65. Foy, H.M., Cooney, M.K., Hall, C.E., Bor, E., and Maletzky, A.J.: Isolation of mumps virus from children with acute lower respiratory tract disease. Am. J. Epidemiol. 94:467–472, 1971.

66. Gellis, S.S., McGuinness, A.C., and Peters, M.: A study on the prevention of mumps orchitis by gamma globulin. Am. J. Med. Sci., 210:661–664, 1945.

67. Gold, H.E., Boxerbaum, B., and Leslie, H.J., Jr.: Mumps arthritis. Am. J. Dis. Child., 116:547–548, 1968.

68. Harris, R.W., et al.: Mumps in a Northeast metropolitan community. I. Epidemiology of clinical mumps. Am. J. Epidemiol., 88:224–233, 1968.

69. Henle, G., Henle, W., and Rosenberg, P.: Isolation of mumps virus from human beings with induced apparent or inapparent infection. J. Exp. Med., 88:223–232, 1948.

70. Horton, G.E.: Mumps myocarditis: case report with review of the literature. Ann. Intern. Med., 49:1228–1239, 1958.

71. Kilham, L., and Margolis, G.: Induction of congenital hydrocephalus in hamsters with attenuated and natural strains of mumps virus. J. Infect. Dis., 132:462–466, 1975.

72. Kocen, R.S., and Critchley, E.: Mumps epididymo—orchitis and its treatment with cortisone. Br. Med. J., 2:20–24, 1961.

73. Kolars, C.P., and Spink, W.W.: Thrombocytopenic purpura as a complication of mumps. J.A.M.A., 168:2213–2215, 1958.

74. Lennette, D.A., Emmons, R.W., and Lennette, E.H.: Rapid diagnosis of mumps virus in-

fections by immunofluorescence methods. J. Clin. Microbiol., 2:81–84, 1975.

75. Lennette, E. H., et al.: Serologic responses to para-influenza viruses in patients with mumps virus infection. J. Lab. Clin. Med., 61: 780–788, 1963.

76. Levitt, L.P., et al.: Central nervous system mumps. Neurology, 20:829–834, 1970.

77. Levitt, L.P., Mahoney, D.H., Jr., Casey, H.L., and Bond, J.O.: Mumps in a general population. Am. J. Dis. Child., 120:134–138, 1970.

78. Norrby, E., Grandien, M., Örvell, C.: New tests for characterization of mumps virus antibodies: hemolysis inhibition, single radial immunodiffusion with immobilized virions, and mixed hemadsorption. J. Clin. Microbiol., 5: 346–352, 1977.

79. Person, D.A., Smith, T.F., and Herrmann, E.C., Jr.: Experiences in laboratory diagnosis of mumps virus infections in routine medical practice. Mayo Clin. Proc., 46:544–548, 1971.

80. Philip, R.N., Reinhard, K.R., and Lachman, D.B.: Observations on a mumps epidemic in a "virgin" population. Am. J. Hyg., 69:91–111, 1959.

81. Public Health Service Advisory Committee on Immunization Procedures: Mumps vaccine. Morb. Mort. Week. Rep., 26,:393–394, 1977.

82. Reed, D., Brown, C., Merrick, R., Sever, J., and Feltz, E.: A mumps epidemic on St. George Island, Alaska. J.A.M.A., 199:967–971, 1967.

83. Russell, R.R., and Donald, J.C.: The neurological complications of mumps. Br. Med. J., 2:27–30, 1958.

84. St. Geme, J.W., Jr., et al.: Experimental gestational mumps virus infection and endocardial fibroelastosis. Pediatrics, 48:821–826, 1971.

85. Salt, W.B., II, and Schenker, S.: Amylase—its clinical significance: a review of the literature. Medicine, 55:269–289, 1976.

86. Smith, M.H.D.: Letters to editor. Mumps virus vaccine. Pediatrics, 42:907–909, 1969.

87. Sultz, H.A., Hart, B.A., Zielezny, M.: Is mumps virus an etiologic factor in juvenile diabetes mellitus? Preliminary report. J. Pediatr., 86:654–656, 1975.

88. Thomas, F.B., Perkins, R.L., and Saslaw, S.: Paralytic mumps infection in two sisters. Arch. Intern. Med., 121:45–49, 1968.

89. Warren, W.R.: Serum amylase and lipase in mumps. Am. J. Med. Sci., 230:161–168, 1955.

90. Wong, T.-W., Straus, F.H., III, and Warner, N.E.: Testicular biopsy in the study of male infertility. I. Testicular causes of infertility. Arch. Pathol., 95:151–159, 1973.

Measles

91. Babbott, F.L., Jr., and Gordon, J.E.: Modern measles. Am. J. Med. Sci., 228:334–361, 1954.

92. Baker, J.A.: Measles vaccine for protection of dogs against canine distemper. J. Am. Vet. Med. Assoc., 156:1743–1746, 1970.

93. Barkin, R.M.: Measles mortality. Analysis of the primary cause of death. Am. J. Dis. Child., 129:307–309, 1975.

94. Black, F.L.: Serologic epidemiology in measles. Yale J. Biol. Med., 32:44–50, 1959.

95. Center for Disease Control: Measles Surveillance. 1972 Summary, issued August, 1973.

96. Cherry, J.D., et al.: Urban measles in the vaccine era: a clinical, epidemiologic, and serologic study. J. Pediatr., 81:217–230, 1972.

97. Cherry, J.D., Feigin, R.D., Lobes, L.A., Jr., and Shackleford, P.G.: Atypical measles in children previously immunized with attenuated measles virus vaccines. Pediatrics, 50:712–717, 1972.

98. Davis, S.D., and Wedgwood, R.J.: Antibiotic prophylaxis in acute viral respiratory diseases. Am. J. Dis. Child., 109:544–554, 1965.

99. Enders, J.F., McCarthy, K., Mitus, A., and Cheatham, W.J.: Isolation of measles virus at autopsy in cases of giant cell pneumonia without rash. New Engl. J. Med., 261:875–881, 1959.

100. Fulginiti, V.A., Eller, J.J., Downie, A.W., and Kempe, C.H.: Altered reactivity to measles virus. Atypical measles in children previously immunized with inactivated measles virus vaccines. J.A.M.A., 202:1075–1080, 1967.

101. Fulton, R.E., and Middleton, P.J.: Immunofluorescence in diagnosis of measles infections in children. J. Pediatr., 86:17–22, 1975.

102. Gerson, K.L., and Haslam, R.H.A.: Subtle immunologic abnormalities in four boys with subacute sclerosing panencephalitis. New Engl. J. Med., 285:78–82, 1971.

103. Greenberg, M., Pellitteri, O., and Eisenstein, D.T.: Measles encephalitis. I. Prophylactic effect on gamma globulin. J. Pediatr., 46:642–647, 1955.

104. Henson, T.E., et al.: Measles antibody titers in multiple sclerosis patients, siblings, and controls. J.A.M.A., 211:1985–1988, 1970.

105. Jabbour, J.T., and Sever, J.L.: Serum measles antibody titers in patients with subacute sclerosing panencephalitis, compared with parents and siblings. J. Pediatr., 73:905–907, 1968.

106. Kiessling, W.R., Hall, W.W., Yung, L.L., and Ter Meulen, V.: Meales virus-specific immunoglobulin M response in subacute sclerosing panencephalitis. Lancet, 1:324–327, 1977.

107. Koffler, D.: Giant cell pneumonia. Arch. Pathol., 78:267–273, 1968.

108. Kohn, J.L., and Koiransky, H.: Successive roentgenograms of the chest of children during measles. Am. J. Dis. Child., 28:258–270, 1929.

109. LaBoccetta, A.C., and Tornay, A.S.: Measles encephalitis. Report of 61 cases. Am. J. Dis. Child., 107:247–255, 1964.

110. Link, H., Norrby, E., and Olsson, J-E.:

Immunoglobulins and measles antibodies in optic neuritis. New Engl. J. Med., 289:1103–1107, 1973.

111. Linnemann, C.C., Jr.: Measles vaccine: immunity, reinfection and revaccination. Am. J. Epidemiol., 97:365–371, 1973.

112. Murphy, J.V., and Yunis, E.J.: Encephalopathy following measles infection in children with chronic illness. J. Pediatr., 88:937–942, 1976.

113. Olding-Stenkuist, E., and Bjorvatn, B.: Rapid detection of measles virus in skin rashes by immunofluorescence. J. Infect. Dis., 134:463–469, 1976.

114. Payne, F.E., Baublis, J.V., and Itabashi, H.H.: Isolation of measles virus from cell cultures of brain from a patient with subacute sclerosing panencephalitis. New Engl. J. Med., 281:585–589, 1969.

115. Siegel, M.M., Walter, T.K., and Ablin, A.R.: Measles pneumonia in childhood leukemia. Pediatrics, 60:38–40, 1977.

116. Surgina, D.W.R., Bank, L.J., and Ackerman, A.B.: Role of measles virus in skin lesions and Koplik's spots. New Engl. J. Med., 283:1139–1142, 1970.

117. Weinstein, L., and Franklin, W.: The pneumonia of measles. Am. J. Med. Sci., 217:314–324, 1949.

Respiratory Syncytial Virus

118. Beem, M.: Repeated infections with respiratory syncytial virus. J. Immunol., 98:1115–1122, 1967.

119. Beem, M.O.: Acute respiratory illnesses in nursery school children: a longitudinal study of the occurrence of illness and respiratory viruses. Am. J. Epidemiol., 90:30–44, 1969.

120. Bruhn, F.W., Mokrohisky, S.T., and McIntosh, K.: Apnea associated with respiratory syncytial virus infection in young infants. J. Pediatr., 90:382–386, 1977.

121. Bruhn, F.W., and Yeager, A.S.: Respiratory syncytial virus in early infancy. Circulating antibody and the severity of infection. Am. J. Dis. Child., 131:145–148, 1977.

122. Buynak, E.B., Weibel, R.E., McLean, A.A., and Hilleman, M.R.: Live respiratory syncytial virus vaccine administered parenterally. Proc. Soc. Exp. Biol. Med., 157:636–642, 1978.

123. Center for Disease Control: Respiratory Syncytial Virus—Missouri. Morb. Mort. Week. Rep., 26:351, 1977.

124. Chanock, R.M., et al.: Respiratory syncytial virus. I. Virus recovery and other observations during 1960 outbreak of bronchiolitis, pneumonia, and minor respiratory illnesses in children. J.A.M.A. 176:647–667, 1961.

125. Giles, T.D., and Gohd, R.S.: Respiratory syncytial virus and heart disease. A report of two cases. J.A.M.A. 236:1128–1130, 1978.

126. Hall, C.B., et al.: Respiratory syncytial virus infections within families. New Engl. J. Med. 294:414–419, 1976.

127. Hall, C.B., and Douglas, R.G., Jr.: Clinically useful method for the isolation of respiratory syncytial virus. J. Infect. Dis., 131:1–5, 1975.

128. Hall, C.B., Geiman, J.M., Douglas, R.G., Jr., and Meagher, M.P.: Control of nosocomial respiratory syncytial virus infections. Pediatrics, 62:728–732, 1978.

129. Hall, W.J., Hall, C.B., and Speers, D.M.: Respiratory syncytial virus infection in adults. Clinical, virologic, and serial pulmonary function studies. Ann. Intern. Med., 88:203–205, 1978.

130. Johnson, K.M., Chanock, R.M., Rifkind, D., Kravetz, H.M., and Knight, V.: Respiratory syncytial virus. IV. Correlation of virus shedding, serologic response, and illness in adult volunteers. J.A.M.A. 176:663–667, 1961.

131. Kaul, A., et al.: Respiratory syncytial virus infection. Rapid diagnosis in children by use of indirect immunofluorescence. Am. J. Dis. Child., 132:1088–1090, 1978.

132. Kim, H.W., et al.: Respiratory syncytial virus disease in infants despite prior administration of antigenic inactivated vaccine. Am. J. Epidemiol., 89:422–434, 1969.

133. Henderson, F.W., Collier, A.M., Jr., and Denney, F.W.: Respiratory-syncytial-virus infections, reinfections, and immunity. A prospective longitudinal study in young children. New Engl. J. Med., 300:530–534, 1979.

134. Neligan, G.A., Steiner, H., Gardner, P.S., and McQuillin, J.: Respiratory syncytial virus infection of the newborn. Br. Med. J., 3:146–147, 1970.

135. Wohl, M.E.B., and Chernick, V.: State of the art. Bronchiolitis. Am. Rev. Respir. Dis., 118:759–781, 1978.

136. Wright, P.F., et al.: Evaluation of a live, attenuated respiratory syncytial virus vaccine in infants. J. Pediatr., 88:931–936, 1976.

Rabies

137. Anderson, J.A., Daley, F.T., Jr., and Kidd, J.C.: Human rabies after antiserum and vaccine postexposure treatment. Case report and review. Ann. Intern. Med., 64:1297–1302, 1966.

138. Bell, J.F., Sancho, M.I., Diaz, A.M., and Moore, G.J.: Nonfatal rabies in an enzootic area: results of a survey and evaluation of techniques. Am. J. Epidemiol., 95:190–198, 1970.

139. Bhatt, D.R., et al.: Human rabies. Diagnosis, complicating and management. Am. J. Dis. Child., 127:862–869, 1974.

140. Center for Disease Control: Follow-up on rabies—New York. Morb. Mort. Week. Rep., 26:249–250, 1977.

141. Center for Disease Control: Rabies Surveillance. Annual Summary. 1977. Issued September 1978.

142. Conomy, J.P., Leibovitz, A., McCombo, W., and Stinson, J.: Airborne rabies encephalitis: demonstration of rabies virus in the human central nervous system. Neurology, 27:67–69, 1977.

143. Hafkin, B., et al.: A comparison of a WI-38 vaccine and duck embryo vaccine for preexposure rabies prophylaxis. Am. J. Epidemiol., 107:439–443, 1978.

144. Hattwick, M.A.W., et al.: Recovery from rabies. A case report. Ann. Intern. Med., 76:931–942, 1972.

145. Hattwick, M.A.W., Corey, L., and Creech, W.B.: Clinical use of human globulin immune to rabies virus. J. Infect. Dis., 133(Suppl.): A266–A272, 1976.

146. Hilfenhaus, J., et al.: Administration of human interferon to rabies virus-infected monkeys after exposure. J. Infect. Dis., 135:846–849, 1977.

147. Kaplan, M.: Epidemiology of rabies. Nature, 221:421–425, 1969.

148. Kappus, K.D.: Canine rabies in the United States, 1971–1973: study of reported cases with reference to vaccination history. Am. J. Epidemiol., 103:242–249, 1976.

149. Lee, T.K., and Becker, M.E.: Validity of spinal cord examination as a substitute procedure for routine rabies diagnosis. Appl. Microbiol., 24:714–716, 1972.

150. Meyer, K.F.: Man contracting rabies from man. J.A.M.A. 165:158–159, 1957.

151. Public Health Service Advisory Committee on Immunization Practices: Rabies. Morb. Mort. Week. Rep., 25:403–406, 1976.

152. Vallery-Radot, R.: Life of Pasteur. Garden City, New York, Garden City Pub. Co.

153. Wiktor, T.J., Postic, B., Ho, M., and Koprowski, H.: Role of interferon induction in the protective activity of rabies vaccines. J. Infect. Dis., 126:408–418, 1972.

154. Winkler, W.G., Fashinell, T.R., Leffingwell, L., Howard, P., and Conomy, J.P.: Airborne rabies transmission in a laboratory worker. J.A.M.A. 226:1219–1221, 1973.

13
Other RNA Viruses

ENTEROVIRUSES

Objectives

1. Describe how enteroviruses are distinguished from rhinoviruses.
2. Describe the typical clinical picture of acute paralytic poliomyelitis and of nonparalytic polio.
3. Discuss what other diseases should be considered in patients whose illness resembles that of acute poliomyelitis.
4. Describe the diseases that can be produced by coxsackieviruses.
5. Describe the diseases that can be produced by echoviruses.
6. Describe how enteroviruses are recognized in the laboratory and how the diagnosis of a poliovirus infection can be confirmed in the laboratory.
7. Describe how the laboratory diagnosis of Coxsackie and ECHO viruses differ from the laboratory diagnosis of poliovirus.
8. Define the two groups of coxsackieviruses and how they are distinguished in the laboratory.
9. Describe how wild poliovirus is distinguished in the laboratory from attenuated poliovirus (vaccine virus).

Classification

Enteroviruses are members of a larger group of viruses called picornaviruses. In the word pico-RNA-virus, pico means small, and RNA in the name refers to the fact that picornaviruses contain ribonucleic acid (RNA) rather than deoxyribonucleic acid (DNA). The other viruses in the picornavirus group are the rhinoviruses, which are distinguished from enteroviruses by growing in the cooler environment of the nose and by being acid-labile, so they are found in nasal, but not rectal, cultures. Enteroviruses are acid-stable, survive gastric acidity, multiply in the intestine, and may be found in high titer in rectal cultures (Table 13-1).[17]

The enteroviruses are subdivided into three groups: polioviruses, echoviruses (ECHO viruses), and coxsackieviruses (Coxsackie viruses) (Table 13-1). There is a spectrum of virulence in these three groups which can be seen in their effect in experimental infection of animals, and which parallels their virulence in humans.

Polioviruses

Polioviruses are characterized by the ability to cause paralysis in chimpanzees when injected into the spinal cord. There are three serotypes: Type 1, Type 2, and Type 3, defined by the cytopathic neutralization test. If a virus isolated from a patient produces a cytopathic effect (CPE) in a cell culture, and the CPE can be neutralized (prevented) by one of the three poliovirus antisera, then the isolate is that type of poliovirus.

TABLE 13-1. CLASSIFICATION OF PICORNAVIRUSES

Picornaviruses: small (about 30 nanometers), RNA
    Rhinoviruses: acid-labile
    Enteroviruses: acid-stable
        Polioviruses: disease in primates
        Coxsackie viruses: disease in mice
        ECHO viruses: no disease in mice or primates

Coxsackieviruses

Coxsackieviruses are named for the town of Coxsackie, New York, the source of the first isolates. (Coxsackie is a Native American word, pronounced Cook' sock' ie by the people who live in the town. However, the word is usually pronounced Cock' sack' ie when used to describe the virus.) The major defining characteristic of coxsackievirus is pathogenicity for mice. There are two groups of coxsackieviruses. Group A coxsackieviruses produce predominately muscle damage in mice. Group B coxsackieviruses produce predominately brain damage in mice.

Types are defined by neutralization tests in cell cultures or by mouse protection tests. Group A contains about 25 types, and Group B contains six types.

All Group B viruses grow very well in cell cultures, just as polioviruses do. However, most Group A viruses do not grow in cell cultures. Therefore, in order to detect all types of Group A coxsackievirus infections, the patient's specimens must be inoculated into mice. Since this is a cumbersome and expensive procedure, it is reserved for special situations, such as research projects or severely ill patients.

Echoviruses

ECHO is an abbreviation for enteric (found in the enteric tract) cytopathic (producing pathologic changes in cells), human (found in humans), orphan (originally "viruses without a disease"). Echoviruses are enteroviruses that are not pathogenic for primates or mice, and are of lower pathogenicity for humans compared to coxsackieviruses. There are about 30 serotypes.

Viruses originally classified as echoviruses have been occasionally reclassified when more information has been obtained about their biologic characteristics. Echovirus type 28 was found to be acid-labile, and has been reclassified as rhinovirus type 1. Echovirus type 9 has been reclassified as coxsackievirus A-23, because of its pathogenicity for suckling mice.

Recently recognized enteroviruses now may be classified simply as enterovirus type 66, or whatever number is designated, rather than as an echovirus or a coxsackievirus.

Frequency and Importance

Enteroviruses, as implied by the name, can be recovered from the enteric tract. They can survive the acidity of the stomach and so can be cultured from rectal swabbings, as well as from throat swabbings.[17] However, enteroviruses are not a frequent cause of enteric disease, such as gastroenteritis. Coxsackie and ECHO viruses are the most frequent cause of viral meningitis in the United States. Approximately 5000 cases of aseptic meningitis syndrome, which is often viral, are reported annually in the United States.

Enteroviruses are a very rare cause of fatal or brain damaging encephalitis, and are a rare cause of fatal myocarditis. The role of enteroviruses in the etiology of idiopathic myocarditis is controversial.

Paralytic disease due to wild poliovirus or poliovaccine virus is rare in the United States, so a single case of suspected wild poliovirus infection stimulates a vigorous public health effort for laboratory confirmation and mass immunization, if confirmed. When paralytic disease due to wild poliovirus infection does occur in the United States, it usually occurs in sharply localized outbreaks. An outbreak occured in 1972 with paralytic disease in 11 students in a school in which two-thirds of the group had not had any prior immunization against poliomyelitis.[12,20] The outbreaks in 1979 among Amish remained localized to that group.

In the late 1970s, outbreaks of 60 to 100 cases occured in central America and the Netherlands. Several cases were imported from the Netherlands to Canada. In the United States, about two or three imported cases occur each year. However, in the late 1970s most cases in the United States (about five to ten cases per year) were the result of poliovaccine virus, either in individuals who were given the vaccine (recipients) or close contacts (discussed later in this section).

Poliovirus Spectrum of Illness

Infection with a poliovirus can produce a spectrum of severity of illness ranging from asymptomatic infection to permanent paralysis or death.[10] It is useful to review this spectrum even though poliovirus infection is now rare.

Asymptomatic Infection. No illness was probably the most common manifestation of poliovirus infection in the past, when this virus was prevalent. Such illness may have been very mild, but not totally asymptomatic.

Fever Without Localizing Signs. A brief episode of fever with no other findings was also a common manifestation of poliovirus infection. This was called abortive polio, or the minor illness of polio. Poliovirus was probably the major cause of summer febrile illnesses before poliovaccine was available. Now coxsackieviruses and echoviruses are probably the most common causes of fever without localizing signs in the summer.

Aseptic meningitis syndrome is typically manifested by fever and stiff neck, with spinal fluid findings of about 20 to 500 cells per cubic millimeter, with normal glucose and protein. Initially the white blood cells may be predominately neutrophiles, but within a day or 2 lymphocytes predominate. This pattern of poliovirus infection was originally called non-paralytic polio. After it was recognized that coxsackieviruses and echoviruses also caused this clinical pattern, the term non-paralytic polio was abandoned. The terms aseptic meningitis syndrome or nonpurulent meningitis are now used, since they are a more accurate description of this syndrome.

Poliovirus was probably the major cause of aseptic meningitis syndrome before poliovaccine became available. Now coxsackieviruses and echoviruses are the most frequent cause of the aseptic meningitis syndrome.

Paralytic Poliomyelitis Syndrome. If flaccid paralysis is present, or develops soon after the patient is found to have aseptic meningitis syndrome, the patient should be diagnosed as having paralytic poliomyelitis syndrome. The severity and extent of the paralysis sometimes progress. A patient with the aseptic meningitis syndrome (non-paralytic polio) may develop "paralytic polio" during the first week or so of the illness. Most illnesses resembling paralytic poliomyelitis are now caused by coxsackieviruses or echoviruses, but are likely to have less severe paralysis than is associated with polioviruses.

Bulbar Poliomyelitis. Cranial nerve paralysis and involvement of the respiratory center in the medulla can occur and can progress to death.

Coxsackievirus Spectrum of Illness

Syndromes in the Poliovirus Spectrum. Paralysis due to coxsackievirus is extremely rare. However, the other syndromes that were once commonly due to poliovirus are now frequently due to coxsackieviruses, especially fever without localizing signs and nonpurulent meningitis.

Coxsackie B Virus Syndromes. Pleurodynia is typically caused by a Coxsackie B virus. Pleurodynia is a syndrome characterized by severe pain in the chest, usually related to breathing (pleuritic pain), and usually associated with fever.[1] Pleurodynia may be mistaken for a pulmonary embolus or an acute myocardial infarction, although most individuals with pleurodynia are too young to have coronary artery disease.

Acute pericarditis in children or young adults, and infantile myocarditis are *frequently* due to Coxsackie B viruses. Transient benign arrhythmias in the newborn infant can be caused by Coxsackie B virus.

Pharyngitis is *occasionally* due to a Coxsackie B virus, but usually the patient has high fever and minimal pharyngitis.

Coxsackie A Virus Syndromes. Ulcerative pharyngitis (herpangina) is usually due to a Coxsackie A virus. A generalized rash, especially involving the palms and soles, with ulcers in the mouth or throat is called hand-foot-mouth syndrome, and is usually due to a Coxsackie A virus. Minor respiratory illnesses, such as rhinitis or bronchitis, are occasionally due to a Coxsackie A virus.

Echovirus Spectrum of Illness

Echoviruses are a common cause of aseptic meninigitis syndrome, fever without localizing signs, and fever–rash syndromes (exanthems). Illnesses in the coxsackievirus and poliovirus spectrum may occasionally be caused by echoviruses, but etiologic proof depends on recovery of the virus from brain or CSF, because an echovirus is recovered from the stool of about five percent of normal individuals. Neither echoviruses nor coxsackievirus appear to be recovered from infants with severe diarrhea much more frequently than from normal infants.[15] However, in young infants, echovirus can be a cause of severe diarrhea or even fatal disseminated disease.[16]

Laboratory Diagnosis

Virus Culture. Most enteroviruses are easily recovered using cell cultures. Coxsackie A viruses are the exception as they usually cannot be isolated unless the specimen is inoculated into suckling mice. All enteroviruses produce a very similar cytopathic effect, so the laboratory cannot be sure that an enterovirus is a poliovirus unless the virus is identified by a CPE neutralization test (Fig. 13-1). Enteroviruses produce a CPE on cell cultures at an average of 5 to 9 days after inoculation, and the laboratory can then give a preliminary diagnosis of an enterovirus isolate. The physician may be able to deduce on clinical grounds whether the enterovirus is most likely to be an echovirus, coxsackievirus, or poliovirus.

Recovery of an enterovirus from feces may be coincidental to the patient's illness, so interpretation of such an isolate depends on whether there exists a known relationship between the virus and the particular illness involved.

Serum Antibodies. It is not practical to do serologic studies to detect an enterovirus infection. There are no satisfactory group antigens for enteroviruses. Therefore, the laboratory cannot use a single antigen for serologic diagnosis of all echoviruses, or all coxsackieviruses, or all polioviruses, as can be done for adenoviruses. The patient's acute and convalescent sera would have to be tested against three polio types, about 30 ECHO types and about six coxsackie B types,

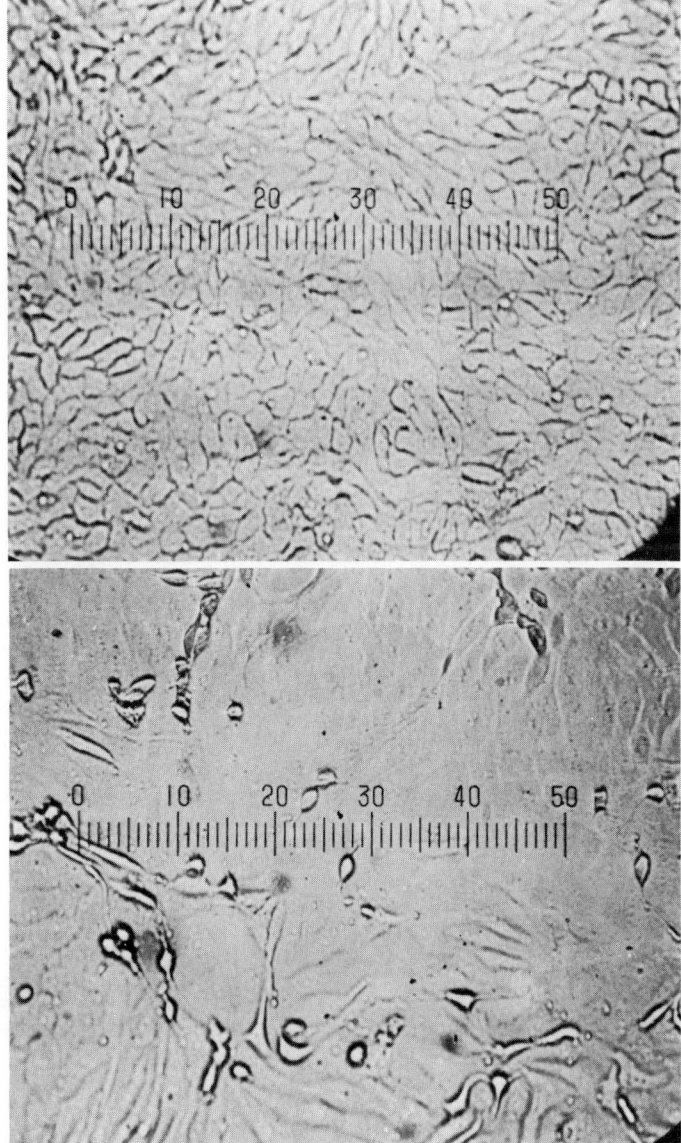

FIG. 13-1. Top, *An uninoculated cell culture.* Bottom, *ECHO virus type 1 cytopathic effect on rhesus monkey cell culture.*

which is not practical. In cases of paralysis, however, the possibility of wild poliovirus infection is serious enough to test the patient's sera for antibodies against the three types of poliovirus. However, laboratory diagnosis of echovirus and coxsackievirus infection depends on the culture of the virus from the patient's throat, stool, or spinal fluid.

Biologic Characteristics of Clinical Interest

Seasonal Illnesses. Coxsackievirus infections occur most frequently in summer and fall in the Northern Hemisphere. This was also true of poliovirus infections in the past. Echoviruses are more prevalent in summer and fall, but are found all year (Fig. 13-2).

Susceptibility of Immature Animals. Suckling mice, like young infants, are particularly susceptible to severe disease when infected by coxsackieviruses.

Growth in Cell Cultures. Enteroviruses (polioviruses) were the first viruses grown in cell cultures. The laboratory diagnosis of most virus diseases and production of virus vaccines depend primarily on the use of cell cultures.

Coxsackievirus Infections in Mice. Diabetes mellitus can occur in mice after experimental infection, and is associated with destruction of the islets of Langerhans.[7] Coxsackievirus infection in humans has been suggested as possibly an occasional cause of diabetes, but this remains unproved.

Myocarditis also can be produced in mice and is made worse by pretreatment with corticosteroids, by stress, or by a diet which elevates the mouse's cholesterol.[13]

Biologic Characteristics of Poliovaccine Virus

Poliovirus was the first virus for which a live attenuated vaccine was developed and extensively studied. Studies of the efficacy of live attenuated poliovaccine provided detailed information about a human viral infection. Prospective studies have been done to confirm many details which were previously studied by accidental recovery of the virus from sick patients. Many important principles were discovered or confirmed during the study of poliovaccine.

Interference. When poliovaccine is given, the three types of live attenuated poliovaccine are given together, and sometimes only one or two types infect the patient. The trivalent vaccine is usually given three times, because one type of poliovaccine can interfere with infection with another. The trivalent vaccine is usually given with 2 months between doses to lessen the possibility of interference, because excretion of the virus in the stool can last 4 to 8 weeks. Natural infections with other enteroviruses can also interfere with infection by the poliovaccine virus, so an extra fourth dose is usually given.[8]

Markers. When paralytic disease occurs, and a poliovirus is isolated, it is important to be able to distinguish wild poliovirus from the attenuated poliovirus vaccine used for immunization. The characteristics that can be used to distinguish these two viruses are called markers. Antigenic differences are often used to define polioviruses as vaccine-like, nonvaccine-like, or intermediate. A temperature marker also has been used, which is based on the effect of incubation temperature on viral growth. Wild poliovirus grows much better at 40 °C than attenuated vaccine virus.

Immunology of Newborns. When poliovaccine is given to breast-fed newborns, antibodies in colostrum (IgA antibodies) can prevent infection.[19] Thus, immunization should certainly not be given in the first few days of life when colostrum (high protein breast secretions), containing many antibodies, is present. Antibodies acquired by transplacental transfer (IgG antibodies) can also prevent infection with attenuated vaccine. Usually immunization with poliovaccine is begun at 2 months of age. This is another reason for

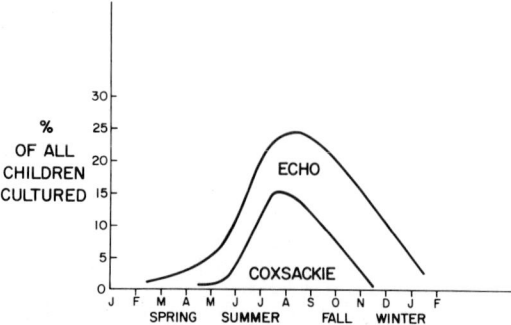

FIG. 13-2. *Seasonal frequency of viruses recovered from all viral cultures taken at Children's Memorial Hospital, Chicago, Illinois, 1964–1969.*

multiple feedings of the vaccine, because some transplacental antibodies still present may interfere with infection at the initial feedings.

Adventitious Agents. A Simian virus (SV-40) was present in some lots of vaccine between 1954 and 1961. This Simian virus resists inactivation by formalin and produces undifferentiated sarcomas several months after injection into newborn hamsters. However, it has not been related to tumors in humans (p. 247). No adverse effects of this virus contaminant have been observed in the vaccine recipients who were exposed.

Vaccine-Associated Disease. Some early attenuated poliovaccine strains were neurovirulent for monkeys, and the passage of poliovaccine virus through children was observed to increase its virulence for monkeys in some instances.[14] Attenuated poliovirus can produce viremia, and well-documented cases of paralysis related to poliovaccine have been reported, both in vaccine recipients and their contacts.[2,6,11] Even in 1962, the first year that attenuated poliovaccines were introduced into the United States, some cases of vaccine-associated paralysis were observed. In the 1970s, about five to ten cases of vaccine-associated paralysis were reported each year (Fig. 13-3).

The Center for Disease Control defines vaccine-associated poliomyelitis on the basis of the epidemiologic evidence of exposure to vaccine virus or wild virus, regardless of whether the virus recovered is vaccine-like, nonvaccine-like, or intermediate by its laboratory characteristics. In fact, a small percentage of pre-vaccine-era wild poliovirus strains have vaccine-like laboratory characteristics.

Imported cases and cases occurring in small outbreaks ("epidemics") are associated with viruses with the laboratory characteristics of wild poliovirus (Fig. 13-3).

Recipient cases have been defined as occurring 4 to 30 days after feeding of the vaccine, with onset of paralysis within 60 days after feeding, significant residual lower motor neuron paralysis, laboratory evidence of recent poliovirus infection, and with other diseases excluded. The virus recovered from vaccine recipients typically is vaccine-like in its laboratory characteristics. Contact-associated cases are counted separately from recipient-associated cases because a contact case may be infected with a poliovaccine virus that may have gained some virulence by passage through the person who actually was given the vaccine.[4] The virus recovered from vaccine contacts typically is vaccine-like according to the laboratory markers, but may have characteristics of both vaccine and wild virus.

Endemic (sporadic) cases without known vaccine contact are labeled in Fig. 13-3 with a question mark. The origin of the virus is presumed to be the poliovaccine, because the cases occur without known contact with wild virus. However, these virus isolates may have laboratory characteristics of wild virus, vaccine virus, or both. Poliovaccine virus is very prevalent in the United States, and has replaced the wild poliovirus in surveillance studies as the most prevalent virus in human sewage, and unrecognized exposures are probably frequent.

Immune deficiency has been detected in only a few of the patients who develop vaccine-associated poliomyelitis (Fig. 13-3).[21]

In summary, "vaccine-associated" poliomyelitis might be caused by infection with a vaccine virus with increased virulence, or might be secondary to an immune defect in the host. Rarely, it may be caused by infection with what appears to be wild poliovirus from an unrecognized exposure, as from an asymptomatic person recently arrived from a country where wild poliovirus is prevalent.

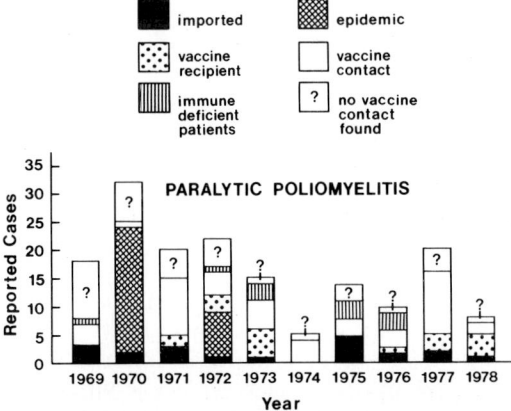

FIG. 13-3. *Paralytic poliomyelitis in the United States, 1969–1978.*[3,4] *This classification is based on the apparent source of the virus, using epidemiologic criteria. 1978 data may be revised later. In 1979, several small epidemics occurred among unvaccinated Amish people.*

Treatment

No specific chemotherapy is available for any enterovirus infection. Supportive care includes mechanical ventilation in paralytic poliomyelitis syndrome.

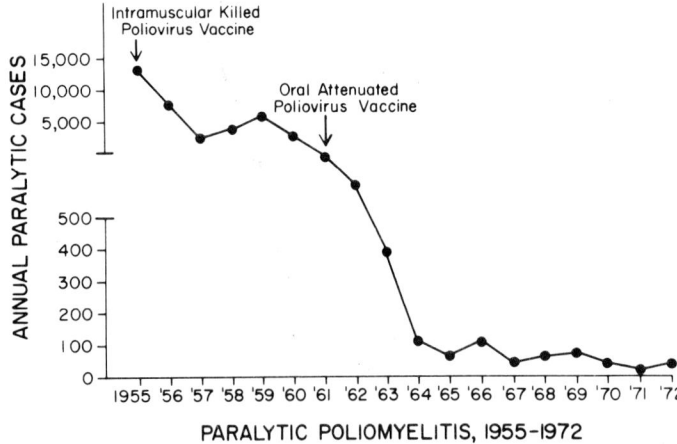

FIG. 13-4. *Paralytic poliomyelitis in the United States, 1955–1972. Compare with Fig. 13-3.*

Prevention

Poliovaccine has proved to be extraordinarily effective in eliminating poliomyelitis in the United States (Figs. 13-3 and 13-4). In 1977, live attenuated poliovaccine (oral poliovaccine) is recommended for routine immunization of infants and children less than 18 years of age, and adults travelling to countries where polio is prevalent.[5] It is excreted in the stool for up to 2 months and is contagious to close contacts via the fecal-oral route.

Killed (inactivated) poliovirus vaccine is recommended in a few specific situations, such as children or adults with immune deficiency diseases or immunosuppressed states.[5] The killed vaccine is also recommended for family contacts of individuals with immune defects, because of the contagion risk of the live vaccine.

RHINOVIRUSES

Objectives

1. Discuss the clinical significance of rhinoviruses' growth at 33°C and acid-lability.
2. Describe how rhinoviruses are detected by laboratory methods.
3. Discuss the problems of immunization against rhinoviruses.

Definitions and Types

Rhinoviruses were so named because of their importance in rhinitis and the common cold syndrome. Earlier names for these viruses were muriviruses, coryzaviruses, and common cold viruses.

Rhinoviruses produce a cytopathic effect which resembles that of enteroviruses, but, unlike enteroviruses, are inactivated by gastric acid. One of the first rhinoviruses discovered was originally classified as ECHO 28, on the basis of cytopathic effect. It was later reclassified as rhinovirus type 1A because of acid-lability.

Rhinoviruses are divided into H (human) strains, which grow well only in human cell cultures, and M (monkey) strains, which also grow well in monkey cell cultures. Over 100 serologic types of rhinovirus have been described.

Frequency and Importance

Rhinoviruses are the most frequent viral cause of the common cold syndrome. Rhinovirus rarely cause serious respiratory disease, but can precipitate asthmatic attacks. The major importance of rhinovirus illness is usually defined in terms of worker or student absenteeism.

Clinical Patterns of Illness

Rhinoviruses are a *common* cause of:

Common Cold Syndrome. Inoculation of susceptible volunteers with a rhinovirus results in an illness which begins about a day after inoculation, and lasts about 4 days.[22,30] The virus continues to be recoverable from the nasopharynx for about 1 to 2 weeks.[22,25,30]

In one well-described study of clinical illnesses, more than 90 percent of 38 volunteers had nasal discharge, nasal obstruction and inflamed nasal mucosa.[30] More than 50 percent had sneezing, cough, and malaise. Only about 25 percent had chilliness, sweats, myalgia, pharyngeal injection, or mild cervical adenopathy. Only 16 percent had an oral temperature over 100 ° F; only 5 percent had vomiting or diarrhea, and none had rales.[30]

Rhinovirus infections sometimes result in common colds with moderately severe symptoms, which precipitate asthmatic attacks in asthmatic children, and appear to be one of the most important viral precipitants of such asthmatic attacks.[28]

Rhinoviruses are a rare cause of serious respiratory illness.[31] However, a 61-year-old patient with pneumonia complicating a malignancy had a rhinovirus recovered from the lung at autopsy.[23]

Laboratory Diagnosis

Viral Culture. This is the most practical method of laboratory diagnosis. Laboratory confirmation that an individual's common cold is concurrent with isolation of a rhinovirus is of little importance to the patient, but has value in research on the common cold.

Serum Antibodies. Serologic diagnosis is not convenient because of the large number of virus types.

Biologic Characteristics of Clinical Interest

Temperature Preference. Rhinoviruses usually grow best in cell cultures at 33 ° C, a temperature about the same as that of the nose.

Acid-Lability. Rhinoviruses are destroyed by acidity. This distinguishes them from enteroviruses, and accounts for the observation that they are not recovered from rectal cultures.

Nasal Antibodies. Type-specific rhinovirus nasal antibody is associated with immunity, according to studies of rhinovirus vaccine. A person may have symptomatic reinfection with the same rhinovirus type several times before there is sufficient nasal antibody to prevent further reinfection.

Contagion. Rhinoviruses can be spread, under experimental conditions, by self-inoculation of the nose or eyes from virus on the fingers, which are contaminated by contact with skin or environmental surfaces.[27] This may be a more important mode of transmission than sneezing.[27] Kissing does not transmit rhinovirus effectively, apparently because inoculation of the pharyngeal mucosa is not particularly effective in initiating infection.[24] Thus, although kissing may be important in lowering resistance, it does not appear to be an important route of rhinovirus contagion. Rhinoviruses are difficult to recover from handkerchiefs compared to nonporous surfaces. Thus, in spite of folklore to the contrary, rhinoviruses might be spread by finger contact as well as coughing and sneezing aerosols.[26]

Treatment

Symptomatic treatment needs to be individualized. Nose drops, antihistamines, oral decongestants, expectorants, and patent medicines have not been shown to be superior to rest and simple diet. Vitamin C has been of no statistically significant value in controlled studies of experimental rhinovirus infections.[32,33] The economic loss ascribed to absenteeism caused by the rhinovirus common cold may actually be less than the wasted money spent for unstudied or worthless cold remedies, including antibiotics.[29]

Prevention

Vaccines. The need for rhinovirus vaccines is based primarily on the frequency, rather than the severity, of rhinovirus infections. The safety, efficacy, and duration of immunity for a rhinovirus vaccine have not been adequately studied. Prevention of the common cold by a vaccine composed of many rhinovirus types seems unlikely to be practical. With so many serotypes, it seems unlikely that the severity or risk of the common cold justifies the expense of the development or the risk of the use of a vaccine. For adults, and sometimes for school age children, the cold is often a socially acceptable excuse for missing work or school, whether the real reason is a hangover, a depression, an unpleasant work duty, or an important athletic event. The opposite view is that the common cold is a significant cause of work absenteeism, and does significant harm to the American economy.

Although rhinovirus vaccines do not appear to be practical, experiments with these vaccines followed by challenge with the virus are of interest in defining the mechanisms of rhinovirus infections. Rhinovirus vaccines are more protective if given intranasally than intramuscularly, because nasal antibodies are necessary for immunity and are produced by nasal inoculation, but not by intramuscular inoculation.

Host Factors. Exposure to cold does not influence resistance to rhinovirus infection.[25]

ARTHROPOD-BORNE ENCEPHALITIS VIRUSES (ARBOVIRUSES)

Objectives

1. Describe the typical clinical and spinal fluid findings of an acute encephalitis that might be caused by an arbovirus.
2. Discuss the epidemiology of arbovirus encephalitis in the United States, particularly with respect to vectors and animal reservoirs, and the arboviruses that can be found in your state.
3. Describe how to confirm the diagnosis of an arbovirus infection in the laboratory.

Definitions

Arbo is a contraction of arthropod-borne. In the past, arboviruses were called arborviruses. These viruses have an insect vector, usually a mosquito or a tick, and an animal reservoir. Most of the arboviruses are named for the geographic region where they were first recovered. Hence, a fever or an encephalitis with a geographic name, such as St. Louis encephalitis or California encephalitis, is likely to be due to an arthropod-borne virus. More exotic examples include West Nile fever, Semliki forest virus, and Bolivian hemorrhagic fever.

Classification

Arboviruses were originally classified together on the basis of their common denominator of an arthropod vector. However, as more information was obtained about the size, structure, composition, and antigenic relationships of arboviruses, this classification has been changing. Most arthropod-borne viruses that cause encephalitis are now classified as togaviruses. Toga refers to the outer lipoprotein envelope of the virus group, and is defined on the basis of the virus characteristics, rather than the vector. The togaviruses include most of the arboviruses, but also include rubella virus. California encephalitis virus is classified as a bunyavirus.

This section deals with the arboviruses that cause encephalitis in the United States (Table 13-2). Other viruses often included with arboviruses are discussed in Chapter 14. Colorado tick fever is an important arthropod-borne virus, but is an arenavirus (Chap. 14), and usually does not produce encephalitis. Lymphocitic choriomeningitis virus is an occasional cause of acute neurologic disease in the United States, but is an arenavirus, not a togavirus. Yellow fever virus and dengue virus are also arthropod-borne togaviruses uncommonly found in the United States and are discussed in a later chapter.

Frequency and Importance

The arboviruses found in the United States are important because of the encephalitis they may produce, with occasional brain damage or death. In the 22 years from 1955–1976 an average of about 300 cases of arbovirus encephalitis were reported each year, with about 15 deaths per year.[40] St. Louis encephalitis occasionally produces major epidemics; for example, 1815 cases in 1975, especially in Texas, Alabama, and Mississippi. Arbovirus encephalitis in the United States is usually sharply localized to areas where the reservoir and vector are present. When all the arboviruses are considered, six states have an average of at least ten cases of arbovirus encephalitis reported annually (Fig. 13-5).

TABLE 13-2. VIRUSES TRANSMITTED BY MOSQUITOES OR TICKS, INDIGENOUS OR OCCASIONALLY IMPORTED INTO THE UNITED STATES.

Virus	Vector
TOGAVIRUSES	
Dengue	Mosquito
Eastern equine encephalitis	Mosquito
Powassan	Tick
St. Louis encephalitis	Mosquito
Western equine encephalitis	Mosquito
Venezuelan equine encephalitis	Mosquito
Yellow fever	Mosquito
ARENAVIRUSES	
Colorado tick fever	Tick
Lassa fever	?
Lymphocytic choriomeningitis	?
BUNYAVIRUSES	
California encephalitis	Mosquito

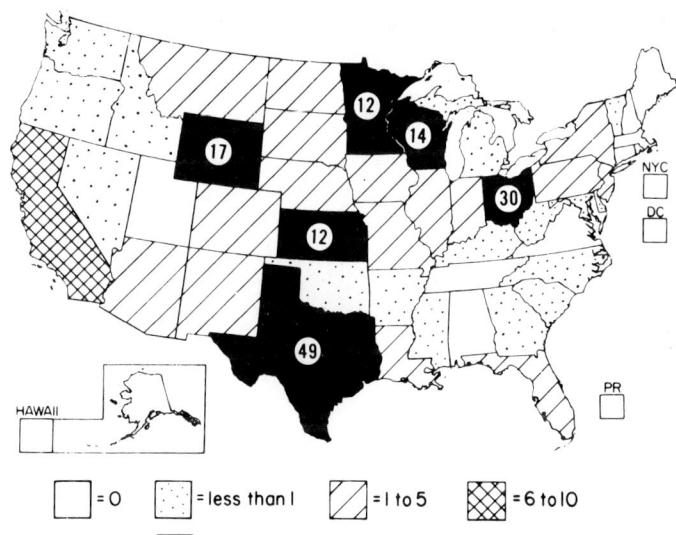

FIG. 13-5. *Arbovirus encephalitis: average annual reported cases, 1965–1971.*[35]

Clinical Patterns of Illness

In endemic areas of the United States, the arboviruses are an *occasional* cause of:

Acute encephalitis is characterized by clinical signs of brain involvement and spinal fluid pleocytosis. Typically the patient has fever and a change of consciousness, which may range from disorientation to coma. Other signs may include stiff neck, convulsions, weakness or paralysis or abnormal movements. However, a severe and nontransient change in consciousness is an essential feature.

The spinal fluid shows an increased number of white blood cells (more than 10 per cubic millimeter). The typical CSF cell count is 50 to 500 white blood cells, with a predominance of lymphocytes.

Nonpurulent meningitis resembles acute encephalitis, except that there is no clinical evidence of definite brain involvement. The patient typically has fever, stiff neck, and a CSF pleocytosis with less than 500 white blood cells per cubic mm, predominately lymphocytes.

Meningoencephalitis is a term used by some authorities for illnesses with minimal to moderate CSF lymphocytosis and both meningeal and encephalitis signs.[37] However, it is useful to try to separate the illnesses with definite brain involvement (encephalitis) from the illnesses without definite brain involvement (nonpurulent meningitis), because of the difference in prognosis between the two groups. Any of the arboviruses that usually cause acute encephalitis also sometimes cause nonpurulent meningitis.[43,44,46]

In the 1975 epidemic of St. Louis encephalitis in Mississippi, 58 percent of the patients with laboratory-documented infection had encephalitis, 15 percent had aseptic meningitis, 21 percent had fever without CNS findings, and six percent had other patterns of illness.[42]

Arthropod-Borne Encephalitis Viruses in the United States

In the United States, there are six arboviruses that can cause acute encephalitis. The states in which these viruses have been most frequent from 1968–1976 are shown in Table 13-3.

TABLE 13-3. TOTAL REPORTED CASES OF ARBOVIRUS ENCEPHALITIS FOR THE YEARS 1965–1972, 1975, AND 1976, FOR STATES WITH THE HIGHEST FREQUENCY.

St. Louis encephalitis virus: Ill., 466; Ohio, 426; Tex., 417; Ind., 406; Miss., 210; Tenn., 97; Kan., 95; Ala., 59; Mo., 58; Ky., 41; Colo., 38; Calif., 32; N.J., 30.
California encephalitis virus: Ohio, 293; Wisc., 148; Minn., 113; Ia., 25; Ind., 21.
Western equine encephalitis virus: Colo., 104; N.D., 63; Tex., 44; Calif., 38; S.D., 28; Kan., 26; Minn., 25; Mont., 21; Wyo., 16; Neb., 13.
Eastern equine encephalitis virus: N.J., 12; Fla., 6; Ga., 4; Mass., 2; Del., La., Pa., S.C., Va., each 1.
Venezuelan encephalitis virus: Tex., 17; Fla., 2; Ark., 1.
Powassan virus: Upstate N.Y., 3.

California Encephalitis Virus. This virus is found primarily in the midwestern states, particularly Ohio, Wisconsin, and Minnesota (Fig. 13-5).[36,37] Most of the neurologic illnesses caused by infection with this virus are encephalitic.[37] Meningeal signs are usually found, and Babinski's sign, tremor, paresis, chorea or blurred disc margins are occasionally found.[36] It is typically a disease of children and is rarely found in individuals older than 20 years of age.

St. Louis Encephalitis Virus (SLE). Severe outbreaks of encephalitis due to SLE virus occurred in Florida in 1962,[43] in Texas in 1966,[44] and Texas and Alabama in 1976. The majority of patients with encephalitis were usually older than 40. Headache was severe. Encephalitis was manifested by an altered level of consciousness, especially confusion. Tremors were noted in the majority of patients. Paralysis, convulsions, ataxia, or diplopia were occasionally observed. Urinary symptoms of urgency, frequency, incontinence or retention were the earliest manifestations of illness in about 25 percent of patients, and unexplained renal insufficiency were noted in some patients.

Eastern Equine Encephalitis Virus and Western Equine Encephalitis Virus. Eastern and Western equine encephalitis viruses are antigenically distinct. Recently they have been less frequent in the United States, compared to California and St. Louis encephalitis viruses.[38]

Horses are an important sentinel of disease. Most cases occur in rural areas. Children appear to have more severe disease than adults. Early in the illness, the CSF typically shows a predominance of neutrophils.

Venezuelan Equine Encephalitis Virus. Human cases occasionally occur in the United States, and 18 cases occurred in Texas in 1971.[39]

Powassan Virus. This arbovirus, named after a town in Canada, has been identified as a very rare cause of encephalitis in Canada and northern United States.[45]

Asymptomatic or Subclinical Infection

On the basis of antibody studies, it appears that subclinical infection is common during outbreaks of some arboviruses. Fever with headache and no other remarkable findings is occasionally the only manifestation of an arbovirus infection.[37,43]

Laboratory Diagnosis

Serum Antibodies. Most state reference laboratories will test paired sera from a patient with encephalitis for antibodies against the arboviruses found in the state, and this is the most practical method for diagnosis of an arbovirus infection. Rapid detection of California encephalitis infection is possible by counterimmunoelectrophoresis.

Virus Culture. Isolation of the virus is a possible means of laboratory diagnosis, but is less practical than serologic methods. Cytopathic effects are produced by some arboviruses in cell cultures, and hemadsorption can sometimes be used to demonstrate the presence of an arbovirus in the cells. However, intracerebral inoculation of mice is usually the most sensitive method for virus isolation, although usually is not readily available. Mouse inoculation is essential for virus isolation for some of the arboviruses; for example, California encephalitis virus.

Biologic Characteristics of Clinical Interest

Viremia. This occurs in the animal host, and is an essential feature for transmission of the virus by blood-sucking arthropods.

Seasonal Frequency. Since vectors are involved in the transmission of arboviruses, human illnesses occur only during months the vectors are found—usually from about May through October, depending on the climate.[40] Infected mosquitoes can lay infected eggs, and this transovarian transmission is the mechanism of survival of the California encephalitis virus over the winter.[41] Furthermore, venereal transmission of California encephalitis virus from male to female mosquitoes has been demonstrated.[47]

Relation to Vectors and Reservoirs. The frequency of human arbovirus disease is closely related to mosquito control, which is influenced by flooding and other conditions which allow an increase in mosquitoes. Lack of availability of the usual animal host for the mosquito also increases the frequency of biting of humans. Usual reservoirs include pheasants and small wild birds for the equine encephalitis and rabbits and small wild mammals for California encephalitis virus.[34] The horse and other equine animals are dead end hosts for most equine encephalitis viruses, probably because the viremia is too brief and low in titer to be infective to mosquitoes. Powassan is found in small animals such as skunks, whose ticks rarely attack humans.[34]

Treatment and Prevention

No specific viral chemotherapy is available. Corticosteroids or mannitol are sometimes used to decrease acute brain swelling in encephalitis, but are not of proved value. Exposure to mosquitoes should be avoided, if possible. Mosquito control is sometimes useful.

RUBELLA VIRUS

Objectives

1. Describe the classical picture of rubella-like illness in young children, and contrast it with the picture seen in young adults.
2. List the five major congenital malformations or abnormalities that may result from congenital rubella infection.
3. Describe the evidence that rubella virus infection may occur without a rash.
4. Discuss the methods used in the laboratory for the recovery of rubella virus.
5. Discuss the methods used for the serologic diagnosis of rubella virus infection, and describe how susceptibility to rubella can be determined.
6. Define reinfection and describe the evidence that reinfection with rubella virus can occur.
7. Describe what should be done if a woman in the first trimester of pregnancy believes that she has been exposed to rubella.

Definitions

Rubella is a name that has been used both for the virus and for the rash disease it causes. The rash disease is also called three-day measles and German measles—terms that medical professionals should teach lay people to avoid. It is better to use the term rubella-like illness unless laboratory evidence documents rubella virus infection.[64]

Frequency and Importance

In the 1960s, approximately 50,000 cases of rubella were reported annually, although these reports based on the clinical diagnosis are only gross estimates. Rubella vaccine was licensed in 1969, and the number of reported cases has gradually decreased, with about 25,000 cases reported in 1972, and about 20,000 cases in 1977, the lowest number in many years.[52] However, sporadic rubella outbreaks continue to be recognized in high schools and colleges, and the highest attack rate occurs in those 15 to 19 years of age. About 20 to 40 cases of congenital rubella syndrome are reported each year. Many cases undoubtedly go unrecognized and unreported, especially those with mental retardation not recognized until preschool years. An unknown number of fetuses with congenital rubella damage are aborted.

Rubella is important almost exclusively because of the congenital malformations which can be produced in the fetus when a pregnant woman is infected. Rubella virus infection in young children typically is a mild disease, not in itself worth preventing. In young adults, the illness is slightly more severe, as described below. However, a single newborn infant with multiple handicaps is a major public health problem for rehabilitation.

The 1963–1964 epidemic in the United States was unusually severe, and probably involved most of the then susceptible individuals. With widespread use of the vaccine, a future outbreak with such a high frequency of complications seems unlikely. However, the vaccine is still not widely used, and localized outbreaks will probably continue, especially in young adults.

Clinical Patterns of Illness

Rubella virus is a *possible* cause of:

Rubella-like Illness. Typical rubella virus infection in young children 1 to 10 years of age is usually a mild disease, with little or no fever, and a confluent maculopapular rash lasting about 3 days.[50] Generalized lymphadenopathy, particularly behind the ears (postauricular) and at the back of the head (occipital), is usually present. The spleen may be palpable.

In adolescents or young adults, fever may be more prominent, in the 101°F to 103°F range, and complications are more frequent. Complications include arthritis, which typically involves multiple joints, especially the fingers, wrist, knees, or ankles, often with some joint effusions.[54]

Thrombocytopenic purpura is uncommon, but may be seen as late as several weeks after the rash of rubella.[48] Encephalitis is a rare complication, but has been documented by laboratory studies.[74]

Modified measles-like illness has been observed in adolescents and young adults with laboratory-confirmed rubella infection. These include prodromal respiratory symptoms, cough, rhinitis, conjunctivitis, pharyngitis, palatal petechiae, and a rash lasting longer than 5 days.[59] Pain on motion of the eyes, chills and fever have also been observed in young adults.[54]

Generalized Lymphadenopathy. Rubella virus infection is occasionally manifested only by generalized lymphadenopathy, without a rash.[50]

In the newborn period, rubella virus is an *occasional* cause of:

Congenital Malformations. Congenital heart disease, particularly patent ductus arteriosus or pulmonic stenosis, congenital cataracts or glaucoma, and congenital deafness are the malformations most clearly associated with fetal infection with rubella virus (Figure 13-6).[53]

Neonatal Jaundice with Thrombocytopenia. Hepatosplenomegaly with jaundice, purpura or petechiae due to thrombocytopenia, and low birth weight for the period of gestation can be due to fetal infection with rubella virus.[48,53]

Learning Disabilities. On the basis of follow-up studies of children born of women with rubella virus infection in pregnancy, it appears that rubella virus must be regarded as a cause of delayed speech and mental retardation.[60]

Diabetes Mellitus. Infants with congenital rubella infection appear to have an increased risk of developing diabetes mellitus in early childhood, presumably because of pancreatic damage from the infection.[63]

Asymptomatic or Subclinical Infection

Rubella without rash can occur, as has been demonstrated experimentally by serial intramuscular injection or intranasal inoculation of filtered serum or respiratory secretions in children.[66] Rubella without a rash has also been

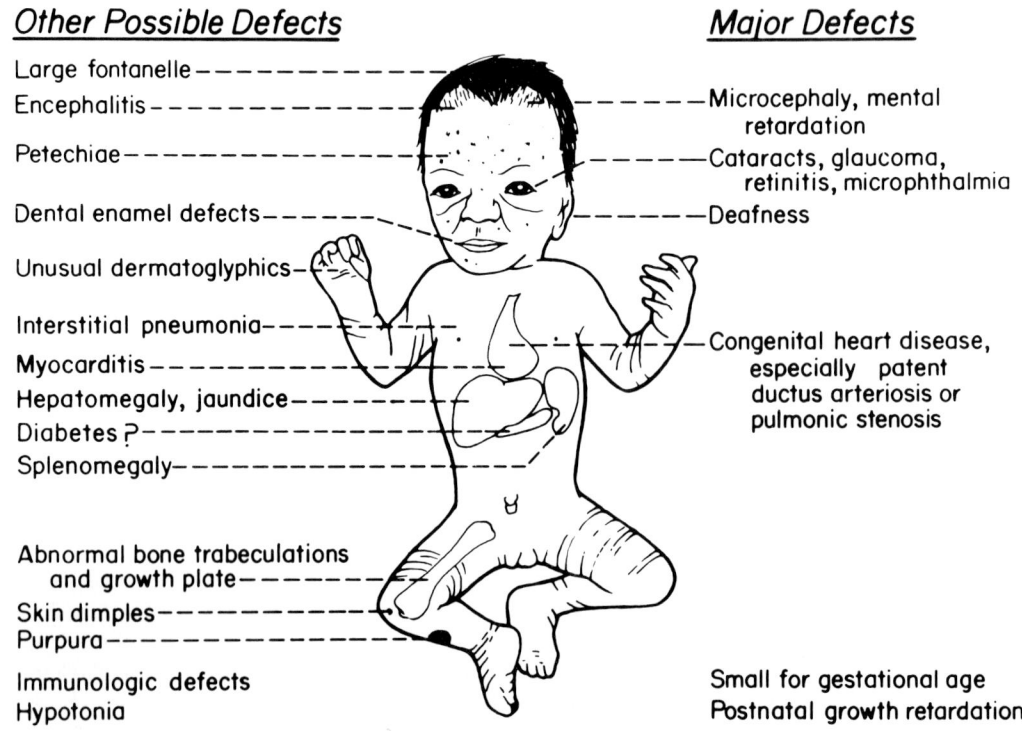

FIG. 13-6. *Congenital rubella syndrome.*

observed in laboratory-confirmed, naturally occurring cases.[49,50]

Other viruses or allergies can cause a rubella-like illness, and there has been confusion and controversy over the clinical diagnosis of rubella for more than 100 years.[55,64] Laboratory studies have shown that the clinical diagnosis of rubella by a physician and a patient's history of having had rubella are both unreliable. Therefore, it is essential to do specific serologic studies of a patient with the clinical diagnosis of rubella, if the prevention of congenital rubella infection is involved.

Many lay people believe that a person can have rubella more than once. Most experts have held the opposite view, that a second episode of clinical rubella would not be documented if studied by virus laboratory methods. However, reinfection with a second clinical illness has recently been reported, although this is presumably exceedingly rare.[75]

Laboratory Diagnosis

Serum Antibodies. The measurement of hemagglutination inhibition antibodies is the most useful and most available serologic test for rubella virus infection. It can be used to detect previous infection (immunity) and to prove a recent infection, using paired sera, or identification of rubella-specific IgM antibody using a single serum.[69] The demonstration of a serologic response using paired serum in the hemagglutination inhibition (HI) test is much more practical than isolation of the virus.

The ELISA method can be used to determine antibodies to rubella virus (Fig. 13-7) and may become used more widely as it is as sensitive and more convenient than the HI test (p. 63).[57,68]

Virus Culture. The virus is best detected by interference, its property of preventing the growth of a test virus, such as ECHO 11, in green monkey kidney cell cultures. Not all interfering agents are rubella. Identification of the virus depends on neutralization of the interference effect with specific antiserum.[70]

Biologic Characteristics of Clinical Interest

Reinfection. Naturally occurring infection with rubella virus should presumably produce lifetime immunity to significant clinical disease, just as naturally occurring mumps, measles, and chickenpox presumably do. However, reinfection (defined as a rise in antibody titer because of exposure to the virus) can occur with rubella virus without any clinical disease. Such a rise in antibody titer has been demonstrated to occur in experiments in chimpanzees and in naturally exposed children with rubella antibodies from past natural infection.[62] Naturally occurring reinfection with arthritis has been documented in an adult woman, with a significant rise in rubella neutralizing antibodies,[75] but such a clinical manifestation of reinfection is exceedingly unusual. An antibody titer rise, without viremia, also has been observed in individuals who have been exposed to natural rubella, after receiving rubella vaccine. Since viremia does not seem to occur in vaccinated persons, even though an antibody titer rise can occur, fetal infection is probably extremely unlikely.

Virus in the Rash. Rubella virus can be recovered from organ cultures of punch biopsies of the skin, both from the rash lesion and from normal appearing skin.[61] This indicates that the appearance of a rash in rubella infection probably implies viremia has occurred.

Contagion. Rubella virus can be recovered as early as 7 days before the rash and as long as 21 days after the rash.[58] Probably the contagious period does not extend this long for most individuals with the disease.

Immune persons are presumed to be unable to transmit the disease. The virus presumably spreads from person to person by respiratory droplets. The virus also can be recovered from the uterine cervical secretions during the illness, but

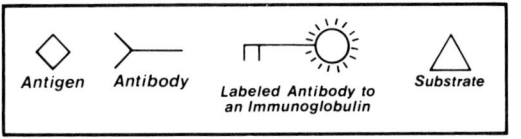

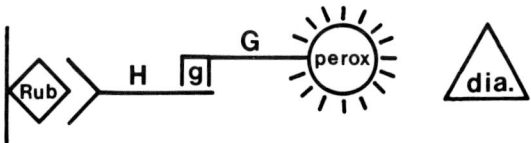

FIG. 13-7. *ELISA method for antibody detection. Rub=rubella antigen from cell cultures; H=human serum specimen to be tested; G=goat antiserum to human IgG, conjugated with peroxidase; dia.=diaminobenzidine and H_2O_2 to detect the peroxidase. g=IgG part of human serum being tested.*

the risk to the fetus of the ascending route of infection is unknown.[73]

Infants with congenital rubella syndrome are highly contagious and excrete the virus in throat, urine, or stools for a variable number of months in the first year of life, and can be a source of infection for susceptible individuals.[72]

Herd Immunity. This concept refers to the lack of spread of a disease through a large group if a large percentage of the group is immune. The concept assumes random mixing of the members of the group. However, actual human behavior typically involves membership in a large number of small family, social, or occupational groups, and the validity of the concept of herd immunity has recently been challenged.[56]

At one time, immunization of children was expected to protect women of childbearing age from exposure. Observations of the indirect protection afforded has indicated herd immunity does not occur in military groups[67] or in communities in which elementary school children are fully immunized.[65]

Treatment

No specific viral chemotherapy is available. Surgical procedures are useful in the treatment of post-rubella congenital heart disease and cataracts.

Prevention

Susceptibility Testing. Serum antibodies can be measured cheaply and accurately and susceptible women of childbearing age should receive vaccine.

Abortion. Termination of pregnancy is advisable if rubella virus infection occurs during early pregnancy. However, determination of susceptibility by measuring antibodies soon after puberty, and use of the vaccine, if necessary, is far better medical care than waiting until abortion becomes a consideration.

Vaccine. A single live attenuated vaccine is available. The vaccine is assumed to have some risk of producing some of the same congenital malformations as the wild virus infection, so that its use during pregnancy is contraindicated. The vaccine can produce arthritis, but presumably is less likely to do so than would the natural disease.

Use of rubella vaccine is unlikely to eliminate congenital rubella syndrome entirely, and its effect may never be adequately determined, because of lack of reporting of the reasons for abortion. However, immunization of children and nonpregnant susceptible women of childbearing age will certainly reduce the frequency to some extent, and so immunization is one of the important methods of prevention, and exceeds the risk of the vaccine.

There had been no nationwide outbreak of rubella in the United States through the end of 1979, whereas prior to the availability of rubella vaccine, an outbreak occurred about every 8 years. However, elimination of rubella in all age groups is an indirect goal, and a less important goal than the elimination of congenital rubella syndrome. All four modes of prevention of congenital rubella syndrome are important: immunization of children, immunization of non-pregnant susceptible females of childbearing age, abortion of pregnant women who develop rubella infection, and immunization of selected male medical professionals.

Vaccination of Male Medical Personnel. In 1978, the United States Public Health Service recommended that all persons (implying males) who come in contact with pregnant women or sick children (as in a hospital or clinic) be immune to rubella to avoid getting the disease and transmitting it to pregnant women.[71] This implies testing such males, and asking them to be immunized if they are susceptible. This approach to the voluntary altruistic immunization of male health care personnel is not unreasonable considering the safety of the vaccine, but should not decrease the emphasis on susceptible females of childbearing age as the primary targets of immunization.

REFERENCES

Enteroviruses

1. Bain, H.W., McLean, D.M., and Walker, S.J.: Epidemic pleurodynia (Bornholm Disease) due to Coxsackie B-5 virus. Pediatrics, 27:889–903, 1961.
2. Balduzzi, P. and Glasgow, L.A.: Paralytic poliomyelitis in a contact of a vaccinated child. New Engl. J. Med., 276:796–797, 1967.
3. Center for Disease Control: Morbidity Mor-

tality Weekly Rep., Vols. 21, Annual supplements, Summary 1972, and Summary 1964. p. 4

4. ———: Poliomyelitis Surveillance Summary 1974–1976. Issued October 1977. p. 1-8.

5. ———: Poliomyelitis prevention. Morbidity Mortality Weekly Rep., 26:329–330, 335–336, 1977.

6. Chang, T., Weinstein, L. and MacMahon, H.E.: Paralytic poliomyelitis in a child with hypogammaglobulinemia: probable implication of Type I vaccine strain. Pediatrics, 37:630–636, 1966.

7. Coleman, T.J., Gamble, D.J. and Taylor, K.W.: Diabetes in mice after Coxsackie B_4 virus infection. Br. Med. J., 2:25–27, 1973.

8. Feldman, R.A., Holguin, A.H. and Gelfand, H.M.: Oral poliovirus vaccination: a study suggesting enterovirus interference. Pediatrics, 33:526–533, 1964.

9. Gelfand, H.M.: Oral vaccine: associated paralytic poliomyelitis, 1962. J.A.M.A., 184:948–956, 1963.

10. Horstmann, D.: The clinical epidemiology of poliomyelitis. Ann. Intern. Med., 43:526–533, 1955.

11. Horstmann, D.M., et. al.: Viremia in infants vaccinated with oral poliovirus vaccine (Sabin). Am. J. Hyg., 79:47–63, 1964.

12. Kraus, G.: Details of poliomyelitis outbreak (letter). New Engl. J. Med., 288:1357–8, 1973.

13. Loria, R.M., Kibrick, S and Madge, G.E.: Infection of hypercholesterolemic mice with coxsackievirus B. J. Infect. Dis., 133:655–662, 1976.

14. Melnick, J.L., Benyesh-Melnick, M. and Brennan, J.C.: Studies on live poliovirus vaccine. Its neurotropic activity in monkeys and its increased neurovirulence after multiplication in vaccinated children. J.A.M.A., 171:1165–1172, 1959.

15. Moffet, H.L., Shulenberger, H.K. and Burkholder, E.R.: Epidemiology and etiology of severe infantile diarrhea. J. Pediatr., 72:1–14, 1968.

16. Morens, D.M: Enteroviral disease in early infancy. J. Pediatr., 92:374–377, 1978.

17. Syverton, J.: Enteroviruses. Pediatrics, 24:643–653, 1959.

18. Tindall, J.P. and Callaway, J.L.: Hand-foot-and-mouth disease—it's more common than you think. Am. J. Dis. Child., 124:372–375, 1972.

19. Warren, R.J., Lepow, M. Bartsch, G.E. and Robbins, F.C.: The relationship of maternal antibody, breast feeding, and age to the susceptibility of newborn infants to infection with attenuated polioviruses. Pediatrics, 34:4–13, 1964.

20. Weinstein, L.: Poliomyelitis—a persistent problem (editorial). New Engl. J. Med., 288:370–372, 1973.

21. Wright, P.F., et al.: Vaccine-associated poliomyelitis in a child with sex-linked agammaglobulinemia. J. Pediatr., 91:408–412, 1977.

Rhinoviruses

22. Cate, T.R., Couch, R.B., and Johnson, K.M.: Studies with rhinoviruses in volunteers: production of illness, effect of naturally acquired antibody, and demonstration of a protective effect not associated with serum antibody. J. Clin. Invest., 43:56–67, 1964.

23. Craighead, J.E., Meier, M., and Cooley, M.H.: Pulmonary infection due to rhinovirus type 13. New Engl. J. Med., 281:1403–1405, 1969.

24. D'Alessio, D. J., Peterson, J.A., Dick, C.A., and Dick, E.C.: Transmission of experimental rhinovirus colds in volunteer married couples. J. Infect. Dis., 133:28–36, 1976.

25. Douglas, R.G., Lindgren, K.M., and Couch, R.B.: Exposure to cold environment and rhinovirus common cold. Failure to demonstrate effect. New Engl. J. Med., 279:742–747, 1968.

26. Gwaltney, J.W., Jr., and Hendley, J.O.: Rhinovirus transmission. One if by air, two if by hand. Am. J. Epidemiol., 107:357–361, 1978.

27. Hendley, J.O., Wenzel, R.P., and Gwaltney, J.M., Jr.: Transmission of rhinovirus colds by self-inoculation. New Engl. J. Med., 288:1361–1364, 1973.

28. Minor, T.E., et. al.: Viruses as precipitants of asthmatic attacks in children. J.A.M.A., 227:292–298, 1974.

29. Moffet, H.L.: Common infections in ambulatory pediatrics. Ann. Intern. Med., 89(Part 2): 743–745, 1978.

30. Mufson, M.A., et al.: Effect of neutralizing antibody on experimental rhinovirus infection. J.A.M.A., 186:578–584, 1963.

31. Portnoy, B., Eckert, H.L., and Salvatore, M.A.: Rhinovirus infection in children with acute lower respiratory disease; evidence against etiologic importance. Pediatrics, 36:899–905, 1965.

32. Schwartz, A.R., et al.: Evaluation of the efficacy of ascorbic acid in prophylaxis of induced rhinovirus 44 infection in man. J. Infect. Dis., 128:500–505, 1973.

33. Walker, G.H., Bynoe, M.L., and Tyrrell, D.A.: Trial of ascorbic acid in prevention of colds. Br. Med. J., 1:603–606, 1967.

Arboviruses

34. Anon.: Arboviruses in the Northeastern United States. New Engl. J. Med., 279:380–381, 1968.

35. Center for Disease Control: Neurotrophic viral diseases surveillance. Annual encephalitis summaries. 1965–1971, 1976 p. 2; p. 6-8.

36. Chun, R.W.M., Thompson, W.H., Grabow, J.D., and Matthews, C.G.: California arbovirus encephalitis in children. Neurology, 18:369–375, 1968.

37. Cramblett, H.G., Stegmiller, H., and Spencer, C.: California encephalitis virus infections in children. J.A.M.A., *198:*108–112, 1966.

38. Ecklund, C.M.: Human encephalitis of the western equine type in Minnesota in 1941: clinical and epidemiological study of serologically positive cases. Am. J. Hyg., *43:*171–193, 1946.

39. Ehrenkranz, N.J., Sinclair, M.C., Buff, E., and Lyman, D.O.: The natural occurrence of Venequelan equine encephalitis in the United States. First case and epidemiologic investigations. New Engl. J. Med., *282:*298–302, 1970.

40. McGowan, J.E., Bryan, J.A., and Gregg, M.B.: Surveillance of arboviral encephalitis in the United States, 1955–1971. Am. J. Epidemiol., *97:*199–207, 1973.

41. Pantuwantana, S., et al.: Isolation of LaCrosse virus from field collected *Aedes triseriatus* larvae. Am. J. Trop. Med. Hyg., *23:*246–250, 1974.

42. Powell, K.E., and Blakley, D.L.: St. Louis encephalitis. The 1975 epidemic in Mississippi. J.A.M.A., *237:*2294–2298, 1977.

43. Quick, D.T., Thompson, J.M., and Bond, J.O.: The 1962 epidemic of St. Louis encephalitis in Florida. IV. Clinical features of cases occurring in the Tampa Bay area. Am. J. Epidemiol., *81:*415–427, 1965.

44. Ray, C.G., Sciple, G.W., Holden, P., and Chin, T.D.Y.: Acute, febrile CNS illnesses in an endemic area of Texas. Pub. Health Rep., *82:*785–793, 1967.

45. Smith, R., et al.: Powassan virus infection. A report of three human cases of encephalitis. Am. J. Dis. Child., *127:*691–693, 1974.

46. Southern, P.M., et al.: Clinical and laboratory features of epidemic St. Louis encephalitis. Ann. Intern. Med., *71:*681–689, 1969.

47. Thompson, W.H., and Beaty, B.J.: Venereal transmission of LaCrosse (California encephalitis) arbovirus in *Aedes triseriatus* mosquitoes. Science, *196:*530–531, 1976.

Rubella Virus

48. Bayer, W.L., et al.: Purpura in congenital and acquired rubella. New Engl. J. Med., *273:*1362–1366, 1965.

49. Bisno, A.L., Spence, L.P., Stewart, J.A., and Casey, H.L.: Rubella in Trinidad: seroepidemiologic studies of an institutional outbreak. Am. J. Epidemiol., *89:*74–81, 1969.

50. Brody, J.A., et al.: Rubella epidemic on St. Paul Island in the Pribilofs, 1963. I. Epidemiologic, clinical, and serologic findings. J.A.M.A., *191:*619–623, 1963.

51. Center for Disease Control: Rubella surveillance, January 1972–July 1973. Issued 1973.

52. ———: Rubella and congenital rubella, 1977–1978. Morb. Mort. Week. Rep., *27:*495–497, 1978.

53. Cooper, L.Z., et al.: Rubella. Clinical manifestations and management. Am. J. Dis. Child., *118:*18–29, 1969.

54. Finklea, J.F., Sandifer, S.H., and Moore, G.T., Jr.: Epidemic rubella at The Citadel. Am. J. Epidemiol., *87:*367–372, 1968.

55. Forbes, J.A.: Rubella: historical aspects. Am. J. Dis. Child., *118:*5–11, 1969.

56. Fox, J.P., et al.: Herd immunity: basic concept and relevance to public health immunization practices. Am. J. Epidemiol., *94:*179–189, 1971.

57. Gerna, G., and Chambers, R.W.: Rubella antibody assay by the immunoperoxidase technique: comparison with the hemagglutinin-inhibition test for determination of immune status. J. Infect. Dis., *133:*469–472, 1976.

58. Green, R.H., et al.: Studies of the natural history and prevention of rubella. Am. J. Dis. Child., *110:*348–365, 1965.

59. Gross, P.A., et al.: A rubella outbreak among adolescent boys. Am. J. Dis. Child., *119:*326–331, 1970.

60. Hardy, J.B., McCracken, G.H., Jr., Gilkeson, M.R., and Sever, J.L.: Adverse fetal outcome following maternal rubella after the first trimester of pregnancy. J.A.M.A., *207:*2414–2420, 1969.

61. Heggie, A.D.: Pathogenesis of the rubella exanthem: distribution of rubella virus in the skin during rubella with and without rash. J. Infect. Dis., *137:*74–77, 1978.

62. Horstmann, D.M., Pajot, T.G., and Leibhaber, H.: Epidemiology of rubella. Subclinical infection and occurrence of reinfection. Am. J. Dis. Child., *118:*133–136, 1969.

63. Johnson, G.M., and Tudor, R.B.: Diabetes mellitus and congenital rubella infection. Am. J. Dis. Child., *120:*453–455, 1970.

64. Kibrick, S.: Rubella and rubelliform rash. Bacteriol. Rev., *28:*452–457, 1964.

65. Klock, L.E., and Rachelefsky, G.S.: Failure of rubella herd immunity during an epidemic. New Engl. J. Med., *288:*69–72, 1973.

66. Krugman, S., Ward, R., Jacobs, K.G., and Lazar, M.: Studies on rubella immunization. 1. Demonstrations of rubella without rash. *151:*285–288, 1953.

67. Lehane, D.E., Newberg, N.R., and Beam, W.E., Jr.: Evaluation of rubella herd immunity during an epidemic. J.A.M.A., *213:*2236–2239, 1970.

68. Leinikki, P.O., Shekarchi, I., Dorsett, P., and Sever, J.L.: Enzyme-linked immunosorbent assay determination of specific rubella levels in micrograms of immunoglobulin G per milliliter

of serum in clinical samples. J. Clin. Microbiol., 8:419–423, 1978.

69. Millian, S.J., and Wegman, D.: Rubella serology: applications, limitations and interpretations. Am. J. Publ. Health, 62:171–175, 1972.

70. Parkman, P.D., Hopps, H.E., and Meyer, H.M.: Rubella virus. Isolation, characterization, and laboratory diagnosis. Am. J. Dis. Child., 118:68–77, 1969.

71. Public Health Service Advisory Committee on Immunization Practices: Rubella vaccine. Morb. Mort. Week. Rep., 27:451–459, 1978.

72. Schiff, G.M., and Dine, M.S.: Transmission of rubella from newborns. A controlled study among young adult women and report of an unusual case. Am. J. Dis. Child., 110:447–451, 1965.

73. Seppälä, M., and Vaheri, A.: Natural rubella infection of the female genital tract. Lancet, 1:46–47, 1974.

74. Sherman, F.E., Michaels, R.H., and Kerny, F.M.: Acute encephalopathy (encephalitis) complicating rubella. J.A.M.A., 192:675–681, 1965.

75. Wilkins, J., Leedom, J.M., Salvatore, M.A., and Portnoy, B.: Clinical rubella with arthritis resulting from reinfection. Ann. Intern. Med., 77:930–932, 1972.

14
Miscellaneous Viruses

HEPATITIS VIRUSES

Objectives

1. Describe the differences between hepatitis A and B viruses in terms of size, nucleic acid content, usual source, and incubation period. Define non-A, non-B hepatitis and describe the evidence for its existence.
2. Describe the antigenic components of hepatitis B virus, give the abbreviations, and describe the clinical significance of each.
3. Describe how hepatitis A and hepatitis B virus infection can be confirmed by laboratory tests.
4. Describe the frequent sources and routes of transmission of hepatitis A and B viruses. Describe methods of prevention of transmission.
5. Define hepatitis as a syndrome. Describe 4 clinical patterns of illness due to hepatitis A or B viruses. List 4 other causes of hepatitis.
6. Describe four high-risk exposure situations for hepatitis B.
7. Discuss the current indications for ordinary immune serum globulin (ISG) and hepatitis B hyperimmune globulin (HBIG) for prevention or treatment of hepatitis A and hepatitis B. Describe the sources and efficacy of these antibody preparations.

Definitions

The word hepatitis is used in two general senses, meaning a syndrome or a virus. Hepatitis as a syndrome refers to an illness with fever, jaundice, and a marked elevation of liver enzymes. The word hepatitis is also used to refer to hepatitis A virus or hepatitis B virus. If the word hepatitis is not further modified here, it usually refers to the syndrome hepatitis (which has many possible causes), rather than a particular hepatitis virus.

The two most frequent and important hepatitis viruses are hepatitis A virus (HAV) and hepatitis B virus (HBV). These were formerly called infectious hepatitis (IH) and serum hepatitis (SH) viruses, based on their usual mode of transmission by close contact or by serum products.

Hepatitis A virus infection is often associated with epidemiologic observations that indicate it is very contagious. Common source outbreaks from water and food, and spread between family members or within a group, are frequently observed.

Hepatitis B virus infections are characteristically associated with contact with blood or serum. Serum hepatitis got its name from the observation of hepatitis outbreaks that could be traced to a single source of serum or blood. Serum hepatitis virus was presumed to be different from infectious hepatitis virus because the illnesses seemed different in several ways. The illness of serum hepatitis had a much longer incubation period, and typically did not have an abrupt onset, with high fever, as did the illness of infectious hepatitis.

Another virus or viruses causing hepatitis in patients in whom HAV and HBV infection are excluded is now called non-A, non-B hepatitis. The term hepatitis C was once used, but has been abandoned because there may be more than one virus that causes non-A, non-B hepatitis.

Hepatitis A and B viruses differ in size, pathogenesis, and nucleic acid content (Table 14-1), but are discussed together because of years of clinical problems in distinguishing between them.

Types

Hepatitis A virus appears to be a single serotype, and infection appears to confer lifetime immunity.

Hepatitis B virus has numerous antigenic components, but like hepatitis A virus, appears to have a single serotype. Hepatitis B virus is quite different from hepatitis A in that chronic, persistent infection can occur, sometimes with progressive liver damage (see chronic active hepatitis, below).

Hepatitis B virus is identical with the particle described in electron micrographs as the Dane particle (named after D.S. Dane, who described it), and is composed of a core and a surface component (see Fig. 14-1). These particles are observed more frequently in the blood of patients with chronic hepatitis than in healthy hepatitis B carriers.[54,55]

Antigenic Components of Hepatitis B

Many antigenic components of hepatitis B virus have been identified.

Surface Antigen. This is abbreviated HB_sAg, but is conveniently called surface antigen in ordinary conversation (Fig. 14-1). It was formerly called Australia antigen or hepatitis associated antigen (HAA). (The abbreviation HAA was confusing because it referred to a hepatitis B antigen, not hepatitis A antigen.)

HB_sAg can be found in the blood during the acute phase of infection with HBV. In almost everyone, this antigen disappears within a month as the patient recovers. If HB_sAg persists in the blood after recovery, the patient is called an asymptomatic carrier.

Core Antigen. Abbreviated HB_cAg, this antigen is produced in liver nuclei and coated with surface antigen in the cytoplasm. Antibody to core antigen (anti-HB_c) can be detected in the serum of recently infected individuals (Fig. 14-1).

DNA Polymerase. This enzyme can be detected in the serum of recently infected individuals.[37]

E Antigen. Abbreviated HB_eAg, this antigen appears to be correlated with infectivity of the patient's serum.[62] The presence of e antigen is associated with large amounts of surface antigen, and hence, with a high probability of intact infective virus. The e antigen is not correlated with chronic or persistent hepatitis.[75] Anti-HB_e is associated with noninfectivity, but may be present in the chronic carrier state for surface antigen.[73] The e antigen has two subdeterminants: $HB_eAg/1$ and $HB_eAg/2$.

Hepatitis B Subtypes. Studies of transmission are aided by identification of lettered subtypes, such as ayr and adw_4. The a antigen is a group antigen found in all HBV isolates, and has several numbered subsubtypes ($a_1, a_2, a_3 \ldots$). Two sets of subtypes usually behave as alleles, so that one or the other only is found. These are y and d, and r and w. The w antigen has some numbered subtypes ($w_1, w_2, w_3, w_4 \ldots$).

Other antigenic components used in typing, with no clear clinical significance yet, are q, x, f, t, j, n, and g.

Subtype y is more common in drug addicts, and subtype d is more common in sporadic cases.[53]

TABLE 14-1. COMPARISON OF HEPATITIS A AND HEPATITIS B VIRUSES

	Hepatitis A Virus	Hepatitis B Virus
Older synonyms	Infectious hepatitis (IH)	Serum hepatitis (SH)
Nucleic acid	RNA	DNA
Diameter (nanometers)	47 nm	22 nm
Incubation period	1–2 months	> 2 months
Sources	Feces, contaminated food or water	Blood or blood products, rarely other body secretions
Chronicity of infection	Almost never	Occasionally

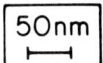
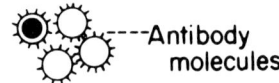
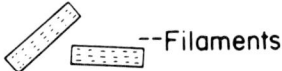
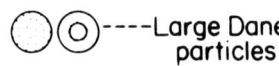

FIG. 14-1. *Hepatitis viruses, as seen by electron microscopy.*

Frequency and Importance

Hepatitis is a very common disease. In the United States, about 15,000 cases of hepatitis are reported annually as presumed hepatitis B, and about 35,000 to 40,000 cases of hepatitis are reported as presumed hepatitis A. An additional 8000 cases are reported as unspecified hepatitis. Cases are usually recognized and reported only if the patient is jaundiced (icteric), so the many anicteric cases are not included. In addition, as serologic studies for hepatitis B become more widely used, these figures will become more accurate. About 1000 deaths due to presumed viral hepatitis occur each year in the United States.[12]

Non-A, non-B hepatitis syndrome appeared to account for the majority of posttransfusion hepatitis in the United States in the late 1970s.[12]

Sources of Knowledge

It is useful to identify the bases of the present knowledge about hepatitis viruses.

Human Volunteer Experiments. In the 1940s and 1950s, prisoners or military personnel were used as volunteers in experiments with materials which produced jaundice when ingested or injected.[27,51]

Children have also been used as subjects of experimental studies.[39,40] These children were retarded and were entering an institution where infectious hepatitis was endemic, and the few individuals who would not get the naturally acquired disease were presumed to have been immune by previous infections.[38] These studies were justified as necessary in order to define the natural history of infectious hepatitis so that prevention could be improved, and did demonstrate the value of gamma globulin in modifying the disease.[40]

Single Source Outbreaks. Occasionally, a single exposure of many individuals has occurred, and has provided information about the incubation period and the spectrum of severity of hepatitis. An outbreak in a college football team exposed to contaminated water indicated that about 90 percent of those exposed were infected, with 32 percent having asymptomatic infection demonstrated only by elevated transaminase values.[50]

Outbreaks also have demonstrated a risk of hepatitis A from eating shellfish,[35] contaminated food (especially salads), and from exposure to chimpanzees.[16] Transfusion,[25] drug addiction,[46] and working in dialysis units[66] have been frequent sources of infection with hepatitis B virus.

Study of personnel working in dialysis units indicates that hepatitis A rarely if ever is spread by a parenteral mechanism,[68] presumably because the illness of hepatitis is often recognized and the period of viremia is brief.

Hepatitis B Surface Antigen in the Serum. The relationship of surface antigen to serum hepatitis was discovered in 1967. It was first called Australia antigen because it was first found in the serum of an Australian aborigine.[10] It is found in the serum of less than one percent of a normal general population—the asymptomatic carriers of the antigen. It appears to be the outer surface part of the virus of hepatitis B. HB_sAg can be obtained from human serum, purified and shown to be particles about 20 nanometers in diameter. However, it does not contain nucleic acid, and so is not the entire serum hepatitis virus.

Antibody against HB_sAg is abbreviated anti-HB_s. It can be produced by immunization of rabbits with human serum containing the antigen, and then absorption of the rabbit serum with human serum not containing the antigen. This rabbit antibody can be used to detect surface antigen in human serum, and can be conjugated with fluorescein to make a fluorescent antibody, which will combine with liver cells containing surface antigen.[48]

Surface antigen is found in the serum of the majority of patients with acute hepatitis following an exposure to blood products or needles. However, surface antigen is also found in about 30 percent of patients with acute hepatitis without a history of parenteral exposures.[10] This indicates that this agent can be transmitted without exposure to needles or blood products.

Illness Caused by Hepatitis A

Hepatitis A virus is a *frequent* cause of:

Acute hepatitis is best defined as a syndrome characterized by fever, enlarged or tender liver, jaundice, and evidence that the jaundice is of the hepatocellular type. Hepatocellular jaundice is defined by a markedly elevated serum transaminase, without evidence of hemolysis or obstruction as a more important source of the jaundice.

In a study of 25 United States servicemen with hepatitis presumed to be due to hepatitis A, 90 percent had anorexia, malaise, cigarette distaste, nausea, jaundice, hepatomegaly, and transaminase elevation.[13] The majority also had fever, weight loss, headache, splenomegaly, and albuminuria. Vomiting, itching, lymphadenopathy, and hematuria were also noted.

In one adult fed hepatitis A virus, disease developed about a month later.[11] Manifestations were malaise, fever to about 104°F, shaking chills, muscle aches, and sweating. The liver was slightly enlarged and tender. The maximum transaminase was 460 units, 2 days after the onset of symptoms. The maximum total bilirubin was 1.8 milligrams percent 6 days after the onset of illness.

In children fed hepatitis A virus, fever was the first sign of infection.[38] Elevation of transaminase preceded the jaundice by 5 to 10 days, and was at its peak when the jaundice was first noted. Vomiting and hepatomegaly coincided with the jaundice, but diarrhea was not prominent.

Asymptomatic or Subclinical Infection. Experimental studies have clearly established that hepatitis A virus infection can occur without jaundice, hepatomegaly, fever, or bile in the urine.[11,38] In asymptomatic experimental infections, transaminase elevation was invariably present, but other liver function studies were usually normal. Serum from such a patient produced hepatitis A when injected into volunteers.[38]

Illnesses Caused by Hepatitis B

Hepatitis B is a *frequent* cause of:

Subacute (Insidious) Hepatitis. Hepatitis caused by hepatitis B virus is often more subacute in onset than that caused by hepatitis A virus. The symptoms of nausea, vomiting, loss of appetite, and muscle aches are usually not prominent. There is often a history of exposure to injections or blood products. However, hepatitis due to HBV can usually not be distinguished from that due to HAV on clinical grounds alone.

Acute Polyarthritis. Hepatitis B virus occasionally produces an acute polyarthritis which can precede the jaundice by about a week, or can occur without any jaundice at all.[67] Typically there is painful swelling and effusion of many joints, which are not red or hot. Itching and a urticarial rash is often found. The serum transaminase is moderately elevated at the time of the arthritis and continues to rise for a week or so. Hepatitis B surface antigen is detectable during the illness.

Chronic active hepatitis is manifested by elevated liver enzymes (serum transaminases), and by histologic evidence of cellular infiltration of the liver. Chronic active hepatitis is often associated with the presence of hepatitis B surface antigen in the patient's serum in a high percentage of cases.[23] There is other evidence of chronic serum hepatitis virus infection in these cases, presumably related to a defect in immunity. Patients with chronic active hepatitis without surface antigenemia often have antibody to hepatitis B core antigen.[23] Thus, hepatitis B virus appears to be the most frequent, but not the only cause of chronic active hepatitis, which also can be produced by autoimmune mechanisms.[31]

Asymptomatic Infection and the Carrier State. Hepatitis B infection is often asymptomatic. In experimentally infected children, jaundice was unusual; but an elevated serum transaminase was noted about 2 months after the exposure, and lasted about 2 or 3 months.[39] Most children had a normal thymol turbidity, in contrast to children with experimental infection with hepatitis A.

In a prospective study of liver enzymes after blood transfusion, 10 of 56 patients developed transaminase elevations, without jaundice.[26] This study indicated that anicteric hepatitis is 100 times as common as icteric hepatitis after transfusion.

Mild or asymptomatic illness does not necessarily indicate a benign course, especially in the case of serum hepatitis virus infection. In a retrospective study of hepatitis B surface antigen in the serum of prisoner volunteers who had been inoculated with serum hepatitis virus, a persistent carrier state with residual liver dysfunction occurred more frequently in the volunteers who had mild or asymptomatic illness than in the volunteers who had more severe illnesses.[8]

Hepatitis B is the *usual* cause of:

Fulminant Hepatitis. This is defined as acute hepatitis severe enough to cause hepatic coma. Fulminant hepatitis is usually caused by hepatitis B virus. It is rarely caused by hepatitis A virus.[58]

Neonatal Hepatitis. Infants born to women who are chronic asymptomatic carriers of hepatitis B surface antigen have a risk of becoming infected, usually by ingestion of maternal blood at the time of delivery, especially if the mother is positive for e antigen.[43] Most infants who develop antigenemia do so about 2 to 12 weeks after delivery, but many infants later become negative for surface antigen.[19] Some women who are chronic carriers infect their infants repetitively in each pregnancy.[18,49]

The infection in newborn infants can be asymptomatic, fulminant, or chronic, with progressive cirrhosis of the liver.[18,64] Hepatitis B immune globulin administered to the baby within a few days after birth appears to prevent neonatal hepatitis B.

Acute hepatitis B in a pregnant woman is not more severe than in the non-pregnant woman, but is associated with an increased frequency of premature infants, some of whom become carriers.[29]

Non-A, Non-B Hepatitis

It was possible to define non-A, non-B hepatitis with certainty only after laboratory methods became available to confirm or exclude A or B virus infection.[2,3,9,20,34] Blood for transfusion was screened to exclude hepatitis B surface antigen-positive blood as soon as methods become available. However, hepatitis still occurred after transfusion with surface antigen-negative blood. Hepatitis A virus infection can now be excluded by antibody studies. Cytomegalovirus, Epstein-Barr virus, and toxoplasmosis, which also can be transmitted by transfusion and cause hepatitis, must also be excluded in order to make the diagnosis of non-A, non-B hepatitis.

Hepatitis C is a term formerly used for the presumed cause of this diagnosis by exclusion, but has been abandoned until details of a new agent are sufficient to characterize it adequately.

Post-transfusion hepatitis is now usually caused by non-A, non-B hepatitis virus or viruses. Non-A, non-B hepatitis can result in a chronic carrier state, chronic active hepatitis, or cirrhosis. Other diseases and other routes or transmission of these agents have not yet been fully characterized.

The agent or agents involved have been transmitted to chimpanzees and a 27 nanometer candidate virus has been observed on electron microscopy.

Prevention. Immune serum globulin given before a blood transfusion reduces the risk of subsequent non-A, non-B hepatitis.[34]

Laboratory Approach

Hepatitis B Surface Antigen. This antigen can be detected in the serum of most individuals with acute hepatitis B virus infection. It appears 2 to 8 weeks before the jaundice.[38] It is usually present for a short period (1 to 12 weeks), but in about 4 percent of patients, it persists for more than 3 weeks, and these patients usually develop chronic hepatitis.[52] Testing for this antigen is generally available in blood banks throughout the United States (Fig. 14-2).

Hepatitis B Serum Antibody. Antibody to HB_sAg can be detected by a number of methods. It is necessary to demonstrate a titer rise, since this antibody is detectable in about 20 percent of normal individuals.[41] The antibody appears after hepatitis B virus infection and persists indefinitely.[42]

A rise in antibody titer to core antigen is a new and accurate method for serologic diagnosis of hepatitis B infection.[32] These tests for antibody are usually available only in specialized laboratories.

Hepatitis A Serum Antibody. This laboratory test is not available for diagnosis of hepatitis A, except in research laboratories. Surveys of individuals of various ages and social classes in the United States indicate that the prevalence of hepatitis A antibody increases with age, poor personal hygiene, and lower social class.[17] The antibody is usually detected by using antigen present in feces and using immune electron microscopy. Newer methods utilize enzyme-linked (ELISA) techniques.[47]

Histologic Methods. Hepatitis A and hepatitis B viruses apparently cannot be distinguished from each other on the basis of histologic appearance of the liver, using conventional strains. However, HB_sAg can be detected in liver and other tissues using fluorescent antibody techniques.[48]

Immune Electron Microscopy. Hepatitis A virus has been observed in stool filtrates of volunteers who have been experimentally infected.[17,56] Treatment of the stool with serum obtained from volunteers convalescing from infectious hepatitis results in clumping of the virus particles, making them easier to locate (Fig. 14-1).[21]

Liver Function Studies. In asymptomatic individuals being screened for suitability as blood donors, a normal serum transaminase (less than 40 units) is closely correlated with absence of hepatitis B surface antigen.

Virus Culture. Icterogenic serum has been reported to produce a cytopathic effect in some cell cultures,[11] but most of these observations have been difficult to reproduce.

Biologic Characteristics of Clinical Interest

Comparative Incubation Periods. Hepatitis A infection has an incubation period of 2 to 8 weeks. Hepatitis B infection has an incubation period of 2 to 6 months. This difference is sometimes not distinct. For example, in children given hepatitis B virus by intramuscular injection, the incubation period was about 2 months.[39] Furthermore, if a sharp rise in serum transaminase level is taken as the end of the incubation period, the incubation periods of hepatitis A and B viruses overlap each other in experimental infections in the range of 40 to 50 days, making interpretation of that interval difficult.[38] However, if the onset of jaundice is taken as the end of the incubation period, as is likely in nonexperimental conditions, then there is a reasonably clear distinction between short incubation periods (less than 60 days), and long incubation periods (greater than 60 days).

The incubation period of hepatitis B virus disease also depends on the dose of the virus. In one study of volunteers, clinical hepatitis developed in about 2 months after undiluted serum, but took 3 to 4 months after a thousandfold dilution.[7]

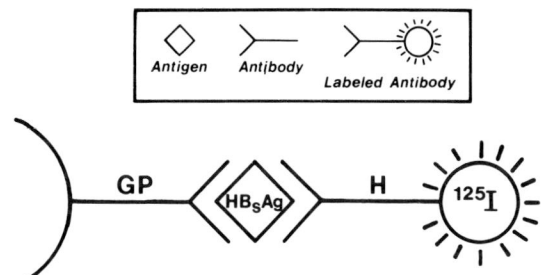

FIG. 14-2. *Detection of surface antigen (HB_s Ag) by a radioactive iodine isotope method. GP=guinea pig antibody against HB_s Ag. The patient's specimen containing HB_s Ag is captured by the guinea pig antibody, which is coated on a polystyrene bead. After rinsing, the ^{125}I labeled high-titer human antibody (H) against HB_s Ag is added, and rinsed. Bound ^{125}I, and therefore HB_s Ag, is detected and quantitated by counting the radioactivity.*

Transmission to Animals. In past years, many reports of transmission of a hepatitis virus to animals have been controversial. However, with availability of laboratory methods to detect hepatitis B antigens, it has been possible to prove animal infection. Chimpanzees appear to be a possible animal model for study of experimental infection of hepatitis B.[6] Chimpanzees and marmosets appear to be suitable animal models for Hepatitis A.

Hepatitis B and Other Diseases. Some chronic diseases with impaired immune mechanisms are associated with chronic hepatitis and an increased frequency of persistently detectable chronic carrier state for surface antigen.[10] These include Down's syndrome, lymphocyticleukemia, Hodgkin's disease, and lepromatous leprosy. Liver biopsy shows mild chronic hepatitis, and the transaminase is usually slightly elevated.

A statistical association of hepatitis B infection and hepatic carcinoma has been observed.[69]

Membranous nephropathy (a form of renal disease) has been observed in Japanese children born to women with hepatitis B surface antigenemia and is apparently transmitted from mother to child.[70]

Immune Complex Disease. Many manifestations of hepatitis B infection suggest that the virus and the antibody against it form complexes which can be deposited in various parts of the body. These manifestatations include polyarteritis nodosa, skin rashes with vaculitis, membranous glomerulonephritis, polymyositis, decreased serum complement, and cryoglobulinemia.[45]

Treatment

In mild to moderate hepatitis, observation is all that is needed.

In severe cases, supportive therapy includes restriction of dietary protein and use of oral neomycin to reduce proteinolysis by intestinal bacteria. Blood glucose should be monitored closely and hypoglycemia treated if it occurs. Drugs excreted by the liver should be avoided. Exchange blood transfusion and heterologous liver perfusion with pig's liver have been used in fulminating hepatitis to reduce the ammonia and other toxins in the blood.[61] Corticosteroids do not appear to be of value of fulminant hepatitis.[4] No specific chemotherapy is available.

In chronic active hepatitis, treatment is primarily symptomatic and supportive.

Complications

Complications of acute viral hepatitis in addition to massive necrosis of the liver include cardiac arrhythmias, hypoglycemia, fatal aplastic anemia, seizures, cryoglobulinemia, deafness and myositis.

Known hepatitis B carriers can be followed by liver function tests and biopsied if these tests suggest chronic active hepatitis.

Prevention of Exposure to Hepatitis A

Contagion of Hepatitis A. Studies feeding children hepatitis A virus indicated this virus was present in the feces for about a week after the onset of jaundice.[40] Another study in 6 children with naturally occurring disease indicated that the hepatitis A virus could not be detected after the liver enzymes had peaked and begun to decline.[57] This study concluded that precautions against fecal contamination were not usually necessary for patients hospitalized with type A hepatitis.[57]

Avoiding Hepatitis A Exposures. Prevention of fecal-oral exposures includes avoiding food or water from unsanitary sources. Uncontrolled exposure to patients with clinical hepatitis should be avoided.

Prevention of Exposure to Hepatitis B

Avoiding Exposures to Blood and Blood Products. Unnecessary transfusion of blood or blood products should be avoided. Neither gamma globulin (ISG) nor albumin can transmit HBV, but antihemophilic globulin can. Pooled plasma increases the risk by a factor of the number of donors in the pool and its use has been abandoned. Packed red blood cells have a low risk. Needles and syringes must be sterile and used for one individual only. Disposable needles and syringes should be used if available.

Screening of Donor Blood for Hepatitis B. Donors with a history of transmitting hepatitis have had a high frequency of HB_sAg.[71] However, in another study, ten donors had HB_sAg in their serum, but no recipient had developed hepatitis or had HB_sAg.[59] The discrepancies of the older blood transfusion data can best be explained by the observations that the hepatitis B surface antigen is not the complete infectious particle or that another virus or viruses (non-A, non-B hepatitis) were transmitted by blood transfusion.

Resistance to Sterilization. Hepatitis B virus is not completely inactivated by heating to 60°C for 10 hours.[63]

Hepatitis B Carriers

Avoiding Exposure to Hepatitis B Carriers. Hepatitis B virus can be contagious through close exposure to patients with serum hepatitis, without parenteral (needle) exposure.[22,28] This is consistent with early observations in experimental human infections that the virus can be transmitted by feeding.[38] Hepatitis B surface antigen has been recovered from the saliva, and transmission of the disease has apparently occurred from the bite of a mentally retarded child. However, most nonparenteral outbreaks have involved exposure to blood of a carrier, such as a bleeding patient. In addition, when the virus enters the mouth, it may need to enter through breaks in the oral mucosa in order to produce infection. The high frequency of sporadic hepatitis B disease in institutions for the retarded or distrubed or in underdeveloped areas presumably is related to poor sanitary practices, which allow oral ingestion of virus-containing blood or body secretions.

In two small incidents of exposure to naturally-occurring infections, HBV was not transmitted by sharing woodwind instruments or by sharing the dummies used to practice cardiopulmonary resuscitation.

Household contact appears to be sufficient exposure for increased risk of hepatitis B, especially with sexual contact, as described below.

Hepatitis B Antigen in Body Fluids. A chronic carrier of hepatitis B surface antigen may excrete the antigen in feces, urine, saliva, semen, nasal secretions, tears, or genital secretions.[72] Since the surface antigen is not the whole infective virus, its presence does not automatically mean these fluids contain virus. Further studies are being done to determine the contagiousness of these fluids using core antigen or e antigen as guides to infectivity. There is very little evidence as yet that hepatitis B virus has been spread independently of serum- or blood-contaminated body fluids. In one study, HB_sAg positive saliva produced disease in gibbons after injection but not by the oral or nasal routes.[5]

Frequency of Hepatitis B Carriers. Hospital personnel have a higher frequency of hepatitis B antibody because of their greater exposure to blood. Persons with a higher than average probability of being carriers include laboratory technicians who handle blood, workers in hemodialysis units, operating room nurses, pathologists, surgeons, and housekeepers in a hospital.[15] Persons requiring frequent transfusions, such as hemophiliacs, and parenteral drug addicts have a very high risk.

About one to two percent of adult patients admitted to a general hospital have surface antigen, and about 20 percent have antibody to surface antigen in their blood.[44] However, routine screening of hospital admissions does not seem justified because carriers do not seem to be enough of a risk to hospital personnel,[44] and it is an expensive way to detect asymptomatic liver disease.

Dentists and physicians have a significantly higher frequency of past hepatitis B infection than first-time blood donors.[65] In this study, the prevalence of surface antigen was 1.7 percent for dentists, 0.8 percent for physicians, and 0.3 percent for the blood donors.

In one study of personnel in a German medical center, the following carrier rates were found for hepatitis B surface antigen: housekeeping, eight percent; physicians and medical students, four to five percent; nurses, student nurses, medical technicians, two or three percent; student technicians and administration, less than one percent.[33] Persons in surgical departments had HB_sAg carrier rates about twice that of those in non-surgical departments.

Sexual Contact. Spouses or sexual contacts of patients with *acute* hepatitis B are at increased risk for hepatitis B, and are candidates for receiving hepatitis B immune globulin.[36] However, sexual contacts of chronic carriers are not considered candidates for HBIG. Evidence that the spread between sexual contacts is not by a venereal route has been obtained by study of prostitutes, nuns, and a control group, and indicated no significant difference in hepatitis B antibody in these groups.[1] Homosexual men have a higher frequency of past or asymptomatic chronic hepatitis B infection, compared to heterosexual male patients in a venereal disease clinic.

Screening for Hepatitis B Carriers. At the present time, it is not clear what occupational groups should be screened for hepatitis B surface antigenemia. The risk of transmitting the virus probably depends on the carriers exposing others to their blood. Of all occupations, the risk might be highest from dentists or dental hygienists who put their fingers in the patient's mouth. A hepatitis B outbreak has been traced to an oral surgeon, who was a surface antigen carrier, and possibly infected patients from minor cuts on his fingers.[60]

Restrictions on HB_sAg Carriers. The presence of e antigen in the serum of chronic surface antigen carriers indicates a greater contagiousness of the carrier. Chronic elevation of the serum transaminase correlates with the presence of e antigen and hence with high infectivity. Possibly these tests may prove useful in defining which carriers should be especially careful.

Chronic carriers of HB$_s$Ag should be educated about their risk to others to the degree that accurate and definite information is available. They should, of course, not be blood donors, and should be careful to avoid letting their blood come in contact with others. Restriction of occupational activities has not yet been recommended. Dentists and dental hygienists who are carriers may be contagious because they put their hands (which may have minor cuts) into patients' mouths. They can be advised to wear gloves. Women who are carriers can transmit the hepatitis B virus through breast milk or bleeding nipples, and it is recommended that these women not nurse their babies.

Antigen-positive hemodialysis patients should be segregated with respect to equipment, location, and contact with antigen-negative personnel and patients.

When a patient is found to be an asymptomatic carrier of surface antigen, e antigen and anti-e should be measured, if the tests are available. If the serum transaminase is elevated, this also may indicate greater infectivity, but also may indicate the presence of chronic active hepatitis (p. 237).

Hepatitis A Immunization

Immune Serum Globulin (ISG). Ordinary ISG is of value in modifying hepatitis A virus infection in family contacts or comparably closely exposed contacts.[12] The dose is 0.01 to 0.02 milliliter per pound.

ISG may often be indicated in persons intimately exposed in relatively unhygienic day care centers or nurseries for very young children.[76] However, it does not appear to be practical to try to prevent hepatitis in restaurant clientele after a restaurant source of a hepatitis A outbreak has been identified.[14] ISG may also be indicated when exposure is highly likely and cannot be prevented by adequate hygiene, as in new admissions to an institution for the mentally retarded. ISG is also recommended for travelers to countries where hepatitis A is common if the traveler cannot take ordinary hygienic precautions.

ISG is also given after a needle stick from a needle used in a patient with hepatitis and serum negative for hepatitis B surface antigen, in order to protect the person stuck from possible non-A, non-B hepatitis. ISG is also recommended for individuals closely exposed to recently infected non-human primates, particularly chimpanzees.

Vaccine. No vaccine is likely to be available until the virus can be grown in high titer in cell culture.

Hyperimmune Hepatitis A Globulin. This preparation theoretically could be produced by the same methods as HBIG, but at the present time is not commercially available.

Hepatitis B Immunization

ISG. ISG has been shown to be of value in preventing hepatitis B, under the same circumstances in which HBIG might have been given, if the ISG is known to have a titer of 1:32 to 1:64 of anti-HB$_s$.[74]

Hyperimmune Hepatitis B Immune Globulin (HBIG). This hyperimmune globulin is prepared from donor blood screened for a high titer of anti-HB$_s$ and became commercially available in the United States in 1977. It should not be given to an exposed person who is known to be positive for HB$_s$Ag[12], because of the risk of immune complex disease (antigen–antibody–complement complexes which may produce fever, rash, polyarthritis, or nephritis).

HBIG is currently recommended for HB$_s$Ag-negative individuals with exposure to blood known to be positive for HB$_s$Ag, by accidental ingestion (as by pipetting), splash to the eye, or by an accidental needle puncture.[12] HBIG is also recommended for the infant just born to a woman who is a chronic carrier of HB$_s$Ag. It also is indicated for the spouse of a patient who develops *acute* hepatitis B infection.

It is *not* currently recommended for surface antigen-negative personnel in dialysis units, spouses of chronic surface antigen carriers, or others with a chronic exposure to individuals with hepatitis B.

HBIG postponed but did not prevent hepatitis B in acutely exposed medical personnel in a recent large controlled trial, suggesting it may be less effective than originally believed.[24]

Vaccines. Hepatitis B vaccines are currently under investigation. They are prepared from viral antigens obtained from plasma of chronic carriers of hepatitis B, since the virus cannot be adequately grown in cell cultures. These vaccines are not really killed virus vaccines, but are purufied antigenic components, particularly hepatitis B surface antigen. Because of the high frequency of hepatitis B in male homosexuals, members of this group have volunteered to be subjects in vaccine trails.

YELLOW FEVER VIRUS

This virus is named for the jaundice and fever it typically produces. The demonstration in 1900 that yellow fever was transmitted by mosquitoes was an important landmark in the history of medicine. Yellow fever was an important disease in residents of the continental United States in colonial times, in American troops in Cuba during the Spanish-American war, and in men who first were digging the Panama Canal.

At the present time, all cases seen in the United States occur in individuals who have recently returned from endemic areas (Fig. 14-3). About 50 cases of jungle yellow fever are reported annually in the jungles of South America, in spite of a vaccine program there, because of the reservoir in non-human primates.

The typical manifestations of yellow fever virus infection are fever, jaundice, severe headache, severe myalgia, conjunctivitis, leukopenia, markedly elevated serum transaminase, and albuminuria.[78] In severe cases, there may be hemorrhagic manifestations, particularly bleeding from nose or gums. Hematemesis can occur. In mild cases, there may be no jaundice, with fever lasting about 3 days, headache, myalgia, conjunctivitis, and albuminuria.

Treatment and Prevention

Supportive therapy may be helpful for renal or liver insufficiency.

Exposure to mosquitoes should be avoided, if possible. Insect control is useful in some areas.

A live attenuated virus vaccine is effective in the prevention of yellow fever. It should be used by travelers to the yellow fever endemic zone (Fig. 14-3) in order to enter other countries. Some countries in the endemic zone, such as Colombia, do not require yellow fever vaccine for entry. However, another country, such as India, may require the immunization to enter if the traveler has recently been to the endemic zone.[77]

DENGUE VIRUS

The origin of the word "dengue" and the reasons it is used for this disease are uncertain. One meaning refers to prude or dandy, and may refer to the strutting walk sometimes associated with severe myalgia. In Colombia, South America, dengue is a native word meaning beaten up, severely punished, or excessively exercised, as in the training of military recruits. This meaning also refers to the "break-bone" severity of the myalgia.

There are four types of dengue virus, and infection with one type does not protect against infection with the other three types. The virus is

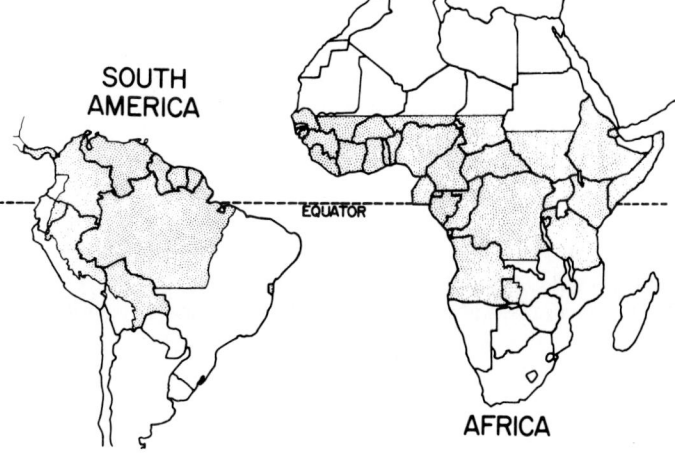

FIG. 14-3. *Yellow fever endemic zone is found in the equatorial regions of South America and Africa.*

classified as a flavivirus and is antigenically similar to the yellow fever virus.[81]

Dengue fever has occasionally occurred in large outbreaks in Puerto Rico, the Virgin Islands, and Carribbean countries, and formerly occurred in the Southeastern United States. It occasionally occurs in Hawaii, but it is now observed in mainland United States only in recent travelers to endemic areas, usually the Caribbean islands. About 50 cases of imported dengue were observed in the United States in 1977.

Dengue virus typically causes break-bone fever, with racking myalgia, arthralgia, retro-orbital pain, severe headache, conjunctivitis and leukopenia.[79,81] A rash typically occurs on the third or fourth day of fever. It is a biphasic illness, with a double-humped fever curve. Milder disease may be common.

Dengue hemorrhagic fever is a severe illness which occurs as an uncommon manifestation of dengue virus infection. It is apparently related to hypersensitivity from past dengue virus infections.[80,82]

A rise in titer of serum antibodies is usually the most practical way to confirm the diagnosis by laboratory methods. The virus can be recovered in cell cultures prepared from monkey kidneys or from mosquitoes, or by mouse inoculation.

Mosquitoes, once infected, continue to carry the virus for their lifetime of several months.[79] Monkeys are a satisfactory experimental animal for study of the disease.[80]

Treatment is supportive. Vaccines are currently investigational. Mosquito control by insecticides appears to be effective in controlling outbreaks of this disease.

COLORADO TICK FEVER VIRUS

This virus is found predominately in the Rocky Mountain states, where its wood tick vector is found, particularly in Idaho, Colorado, Nevada, Oregon, Wyoming, and Montana.[83] About 100 to 200 cases are confirmed by virus studies each year.

The virus is transmitted by a tick, and the history of a tick bite or brief exposure to a tick can usually be obtained.[84-87] The illness is biphasic, and begins about 4 to 6 days later, with aches, vomiting and fever. This first phase lasts about 2 or 3 days, and is followed by a remission of 1 or 2 days of feeling well, with no fever. Then fever, myalgia, backaches, and weakness recur for 2 or 3 days before the patient is finally well.[84] Weakness may persist for 1 to 3 months.[85] The severe muscle aches or backache are sometimes compared to the break-bone myalgia of dengue fever.

Abdominal pain, vomiting, arthralgia, and mild pharyngitis are not unusual, but splenomegaly, conjunctivitis, stiff neck, orchitis, pneumonitis, or a faint maculopapular rash are uncommon.[85] Leukopenia is almost always observed, with a shift to the left of the granulocytes, often with 10 percent to 20 percent band forms.[84,85] Occasionally, Colorado tick fever is associated with extensive bleeding. When a petechial rash is noted along with the history of a tick bite, the physician often suspects Rocky Mountain spotted fever.[87] Nonpurulent meningitis is another uncommon complication.

Persistent viremia, particularly in the erythrocytes, occurs for 4 weeks or longer in about half of the cases.[85]

Colorado tick fever virus is classified as an orbivirus because of its structure and double-stranded RNA. The infection can be confirmed in the laboratory by demonstrating a rise in antibody titer, or by detection of the virus in cell culture or in suckling mice.

Treatment is supportive. No vaccine is available. Ticks should be avoided.

CORONAVIRUSES

The coronavirus is named for its appearance on electron microscopy. The virus has petal-shaped projections resembling a corona of the sun. The virus is very difficult to grow in the laboratory, and knowledge about the disease it produces is incomplete. Isolation of the virus can be achieved in cell cultures for some coronavirus strains, but other strains can be recovered only in organ cultures of human tracheal tissue. Such strains are often labeled OC for organ culture, for example, OC38 and OC43.

Coronaviruses are an occasional cause of the common cold syndrome.[90] Experimental inoculation of human volunteers produces a common cold, perhaps of slightly shorter duration and with more nasal discharge than rhinovirus colds.[88]

In a study of hospitalized infants, coronaviruses appeared to be a cause of pneumonia, bronchitis, or bronchiolitis, ranking third in frequency after respiratory syncytial virus and parainfluenza viruses.[91] About 50 percent of coronavirus infections are asymptomatic. Reinfection is frequent and occurs in spite of the presence of serum neutralizing antibodies.

Detection of complement-fixing antibodies is the best way for laboratory confirmation of coronavirus infection.[89] The CF antibodies remain elevated for only a short time, so a single serum is useful.[89]

Other coronaviruses can produce naturally-occurring infections in animals, for example, avian infectious bronchitis virus (IBV) and mouse hepatitis virus (MHV).

LYMPHOCYTIC CHORIOMENINGITIS VIRUS

Lymphocytic choriomeningitis (LCM) virus was the first virus proved to cause nonpurulent meningitis, which was called lymphocytic choriomeningitis at that time. LCM virus infection has been demonstrated by serum antibody titer rise in about ten percent of patients with nonpurulent meningitis (aseptic meningitis syndrome).[94]

LCM virus can produce a variety of clinical patterns, but usually fever, severe myalgia, and headache are prominent. A recent hamster-associated outbreak in ten laboratory personnel indicated that infection with this virus usually does not produce nonpurulent meningitis.[92] In this outbreak, the individuals had fever, retro-orbital headache, and myalgia. Of the ten, seven had chest pain, three had confusion, two had arthritis, two had partial alopecia (loss of hair from the head), and three had orchitis. All four tested had leukopenia, and the lowest white blood count was 1200. All were exposed to infected hamsters, although mice have usually been regarded as the usual source of human infections.

In another pet-hamster associated outbreak, fever, severe myalgia, and pain on moving the eyes were common.[93] Arthritis, rash, lymphadenopathy, orchitis, or paralysis were rare, but it was not unusual for patients to describe the illness as the worst illness of their lives.[93] Prolonged and fatal LCM infections have been reported.

LCM virus is an occasional cause of nonpurulent meningitis. Typically, there is a predominance of lymphocytes. The spinal fluid glucose is occasionally lowered, which may lead the clinician to suspect tuberculous meningitis.

Asymptomatic or subclinical infection can occur with lymphocytic choriomeningitis virus, as about 10 percent to 20 percent of normal individuals tested have antibodies against this virus.[95]

Laboratory confirmation of infection is based on a rise in serum antibodies in paired sera. The virus can be recovered using animal inoculation, but this is usually not practical.

LCM virus is classified as an arenavirus on the basis of size, a common antigen, and sand-like granules in the virion (*arena* is Latin for sand).

LASSA VIRUS

Lassa virus is also an arenavirus. It is named for a village near the Sahara Desert in Nigeria, Africa. It is important because it is highly contagious, frequently produces fatal disease, and has been imported to the United States.[96] It has also produced death in a laboratory technician working with the virus in the United States.[97]

It is antigenically related to other rodent-associated viruses that produce human disease, including LCM (discussed above). Bush rats are apparently the reservoirs in Liberia, Sierra Leone, and Nigeria where Lassa virus is found. It is the most frequent cause of death in Sierra Leone.

The clinical disease in humans begins with fever, headache, myalgia, pharyngitis, cough, vomiting, or diarrhea. Leukopenia is often present. The disease can progress to produce pneumonia, pericarditis, hepatitis, and fatal hemorrhagic manifestations. The mortality is about 40 percent.[96]

The only treatment is supportive. Strict isolation is essential.

Other highly contagious viruses that can be imported from Africa include the Marburg virus of monkeys, and the Ebola virus, from Zaire and Sudan, discussed below.

MARBURG AND EBOLA VIRUSES

Marburg virus was a source of laboratory infection that was fatal to seven laboratory workers in Marburg and Frankfurt, Germany, in 1967.[99] It also has been called green monkey virus, after the African green monkey (vervet monkey), whose tissues were handled by laboratory workers who were infected.

Marburg virus is important because, like Lassa virus, it is contagious and has been shown to cause secondary cases in medical personnel exposed to sick patients.[100]

About 30 cases of illness were noted in Germany and Yugoslavia in 1967. Fever, headache, and myalgia occurred, followed by conjunctival infection, photophobia and a generalized rash.[100] Disseminated intravascular coagulation can occur terminally.

The virus is classified as a rhabdovirus, like rabies virus, because of its bullet-shaped appearance on electron micrography.

Ebola virus closely resembles Marburg virus. It has caused nonfatal infection in an investigator working with the virus in the laboratory.[98]

TUMOR VIRUSES

Definitions and Classifications

This section contains many definitions and observations of clinical interest, but recent reviews cover the basic science and clinical implications in more depth.[101,108,109,111]

Viruses associated with tumors in humans can be classified on the basis of their RNA or DNA content. Two viruses cause naturally occurring benign tumors in humans, the human papilloma virus and molluscum contagiosum virus. A few others have a statistical association with human malignancy. These include *Herpesvirus hominis*, Epstein–Barr virus, and hepatitis B virus (these are discussed in their own sections). Some human viruses can produce tumors when injected into animals, but this cannot be interpreted as evidence that the virus can cause malignancy in humans. Nevertheless, such animal models are important areas of research into the etiology of human cancer.

Cell Transformation. This occurs when a cell acquires the ability to multiply indefinitely.[108] It is sometimes called malignant or neoplastic transformation. It should be distinguished from DNA transformation, for example, in pneumococci (Chap. 1), and from lymphocyte transformation induced by mitogens.

Cell transformation is often accompanied by integration of the virus genome into the cell's DNA. Transformed animal cells often lose their normal chromosome patterns and are changed in cell morphology, rate of growth and antigenic composition. Sometimes they gain the ability to produce tumors when injected into animals.

Epstein–Barr virus has the ability to transform lymphocytes so that they can multiply indefinitely (provided nutrient fluids are provided and byproducts of metabolism are removed). This can be used to detect Epstein–Barr virus (Chap. 11).

A culture of cells, such as HEp-2, that can be subcultured continuously, consists of transformed cells. Primary cell cultures, such as rhesus monkey kidney cells, can be subcultured only a few times.

RNA Tumor Viruses

Retroviruses. Oncornaviruses are a subfamily of the larger family of retroviruses.[101] Retroviruses all contain reverse transcriptase.

Oncornaviruses. RNA tumor viruses include the oncornaviruses. Oncorna is a contraction of oncogenic and RNA. The oncogenic RNA viruses are similar to each other, and produce naturally occurring cancer in animals.[108,109] However, at present there is no proof that such viruses cause human neoplasms. Oncornaviruses have been called leukoviruses, because many can cause leukemia in animals. Examples include the Rous sarcoma virus, mouse leukemia virus, avian leukemia virus, and other tumor viruses of subhuman primates, cats, and rabbits.

Reverse transcriptase, an enzyme that catalyzes the flow of genetic information from RNA to DNA, is essential in tumor production by oncornaviruses.[111] However, whether RNA tumor viruses are involved in human malignancy remains to be determined.

Human Leukemia. This has long been suspected to be a viral disease,[110] but at present, there is no conclusive evidence that human leukemia is caused by a virus. Nevertheless, leukemia in human and other primates is being intensively studied for a possible oncornavirus etiology, especially Type C oncornaviruses.[105,107]

Human Type C Oncornaviruses. These viruses are members of the retrovirus family. Mammalian Type C viruses have an RNA genome, a reverse transcriptase, a particular EM appearance described below, and shared antigens.

The classification of Types A, B, and C oncornavirus is based on the size and shape of virus particles found in animal tumors.[105] Type C particles have large centrally located dense nucleoids. Type C particles appear to be formed by budding from the outer cell membrane.

Type C viruses can affect cells in two ways—by passing from one cell to another in the classical fashion, or by making a DNA copy of the RNA virus, using reverse transcriptase, and then being transmitted by cell division.

Type C particles have been observed in the blood leukocytes of humans with acute myelogenous leukemia.[103] Subhuman primates have oncogenic type C viruses (wooly monkey sarcoma virus and gibbon ape leukemia virus.)[103] These observations indicate the need for further investigation.

Type C viral antigen is also present in the antigen-antiviral-antibody complex found found in the kidney in disseminated lupus erythematosus described later in this chapter.

DNA Tumor Viruses

Papovaviruses. This is a family of viruses containing papilloma, polyoma, and vacuolating viruses as subgroups. Papova is an acronym derived from PApilloma, POlyoma, Vacuolating Agent. Papovaviruses are DNA viruses, many of which are oncogenic (can induce tumors when injected into animals). Many are capable of producing cell transformation as described above. There are two genera of papovaviruses: polyomaviruses and papillomaviruses.

Polyomaviruses. The polymaviruses are so named because they can produce many different types of tumors in animals.

Papilloma Viruses. The human papilloma virus is the cause of common warts, and it and molluscum contagiosum virus are the only DNA viruses known to cause human tumors. Several different types of human papilloma virus exist.[104]

Simian Virus 40 (SV-40) is a papilloma virus (Fig. 14-4). It was originally called vacuolating agent of monkeys since it produces a vacuolating cytopathic effect in monkey kidney cell cultures. This virus from monkey kidney cells contaminated many lots of both attenuated and killed poliovaccine for several years. No ill effects have been noted in the millions of individuals who ingested this virus or who were injected with killed poliovirus vaccine that contained some live SV-40.[102]

SV-40 has also been recovered from two pa-

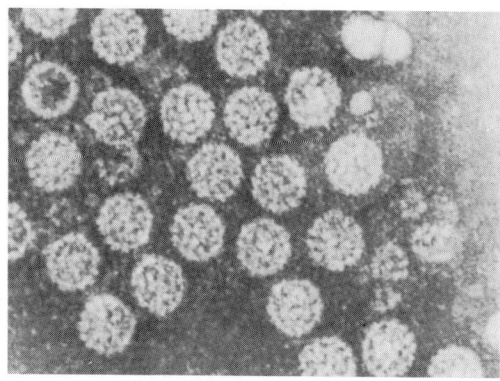

FIG. 14-4. *Simian virus 40. This virus is often found in monkey kidney cell cultures, and has been a contaminant of poliovaccines. (Photo from Dr. Albertina Alberts)*

tients with progressive multifocal leukoencephalopathy, but the significance is not clear. SV-40 and JC virus have some antigenic cross reactivity.

Simian Viruses Contaminating Vaccines. Both live and killed vaccines prepared for human use have a potential risk of containing simian viruses if prepared in monkey tissue. This has been used as an argument in favor of using human cell lines for vaccine preparation.[106]

Adenoviruses. Many types of human adenovirus produce tumors when injected into rodents (p. 178), but have not been associated with human tumors.

T Antigen. Tumor antigen (T antigen) can be detected in cells transformed by adenoviruses or polyomaviruses. Antibodies to T antigen can be detected in the serum of animals with adenovirus-induced tumors.

JC Papovavirus. This virus was first recovered from patients with progressive multifocal encephalopathy. It produces brain tumors when injected into owl monkeys, as discussed below in the section on slow viruses.

SLOW VIRUS DISEASES

Slow virus diseases are named for their long incubation period, which is measured in years, and for their slowly progressive course.[113] The original recognition of a slow virus disease was made with scrapie, a progressive neurologic disease of sheep, which makes them scrape themselves.

Slow virus diseases can be classified into two groups. The first group of slow virus diseases are produced by conventional viruses, which ordinarily produce acute illnesses, but for reasons apparently related to the host, occasionally produce the slow virus disease pattern.

Slow Virus Disease Due to Conventional Viruses

Measles virus, which rarely produces subacute sclerosing panencephalitis (Chap. 12) is an example of a conventional virus that can produce a slow virus disease.

JC virus (named for the patient's initials) is a papovavirus which has been recovered from the brain of patients with progressive multifocal leukoencephalopathy (Fig. 14-5). This disease has the course of a slow virus disease, and typically occurs in patients with defective cell-mediated immunity.[125,126] This appears to be another example of a conventional virus that is rarely associated with a slow virus disease. More than 50 percent of individuals over age ten have antibodies to JC virus, indicating that asymptomatic infection is common.[127]

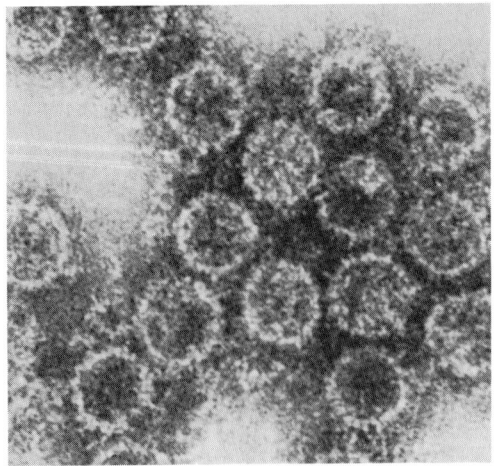

FIG. 14-5. *JC virus, a papovavirus recovered from a patient with progressive multifocal leukoencephalopathy. Spikelike projections between virus particles are antibody molecules in this immune electron microscopic preparation. Compare to diagram of EM preparation of hepatitis A virus (Fig. 14-1). (Photo from Dr. Albertina Alberts)*

SV-40, a tumor-producing simian virus, has rarely been recovered from patients with progressive multifocal encephalopathy.[118,129] Both SV-40 and JC papovavirus can induce tumors when injected into animals.

Aleutian mink disease is a chronic disease of mink, caused by a conventional virus. (It should be distinguished from transmissible mink encephalopathy, caused by an unconventional agent.) There is one reported case of a generalized fatal disease in a man who raised mink and was often bitten by them.[112] The patient had a recurrent illness, with generalized adenopathy, enlarged liver and spleen, and cavitary pneumonia. Autopsy findings were similar to findings in mink with Aleutian mink disease.

Visna virus is a retrovirus that produces a slowly progressive chronic neurologic disease in sheep. The spinal fluid findings resemble those of multiple sclerosis and subacute sclerosing leukoencephalopathy.[117]

Slow Virus Disease Due to Unconventional Agents

The second group of slow virus diseases are presumed to be of viral etiology because the disease is transmissible. However, these agents differ from conventional viruses by having no recognizable virions on electron micrographs, by producing no detectable immune response in the host, and by being unusually resistant to inactivation by physical or chemical agents, such as formalin. It is possible that these agents will not be defined as viruses after further characterization, but at the present time it is useful to refer to these agents as viruses.

Two diseases of animals and two diseases of humans are now classified in this second group of transmissible chronic neurologic diseases (Table 14-2). These diseases are scrapie in sheep; transmissible mink encephalopathy of mink and other animals (not the same as Aleutian mink disease); kuru, a disease of New Guinea cannibals; and Creutzfeldt–Jacob disease, which occurs in adults of any region or background.

Kuru was the first slow virus disease to be recognized in humans. This disease was observed in cannibals in New Guinea, and is apparently transmitted by cannibalism, particularly of the brain.[113] The disease is manifested by progressive ataxia and tremors, progressing to death. Kuru-like disease has been transmitted to chimpanzees by injection of brain suspension.[114]

Spongiform encephalopathy (Creutzfeldt–Jacob disease) is the second example of a chronic, progressive neurologic disease of humans that is transmissible, according to studies injecting human brain tissue into primates.[116] The patient develops a progressive intellectual deterioration called presenile dementia, with histologic changes in the brain that resemble those found in kuru.

Person-to-person transmission of Creutzfeldt–Jacob disease has occurred via a corneal transplant,[121] and via intracerebral stereotactic electrodes.[115] Isolation precautions are now recommended for patients with this disease.[115] The buffy coat of blood is infectious in experimental animal infections, indicating the disease may possibly be transmitted by blood transfusions.[122] Other modes of transmission that have been postulated included eating animal brain (as in ethnic specialty sausages), or sheep eyes (considered a delicacy in North Africa and the Middle East).

These two slow virus diseases of humans are exceedingly rare. However, they are important because study of the pathogenesis of these rare diseases may provide information important to the discovery of the etiology, pathogenesis, and possible specific treatment or prevention of similar chronic neurologic diseases which are much more common—especially multiple sclerosis.

Diseases of Unknown But Suspected Viral Etiology

Multiple sclerosis, which has high antibody titers to measles virus, is suspected of having a pathogenesis analogous to subacute sclerosing panencephalitis, but conclusive evidence is lacking.[119] Demyelination occurs in both of these human diseases, as well as in visna of sheep, a slow virus disease.

Other chronic neurologic diseases of suspected viral etiology include Guillain–Barré syndrome, amyotrophic lateral sclerosis, Alzheimer's disease, Pick's disease, paralysis agitans, and Parkinsonian dementia.[119]

Viroids. Recent studies suggest that scrapie maybe a viroid, the first viroid found in animals, although viroids are well-recognized pathogens in plants.[123] Viroids are very small molecules of

TABLE 14-2. CLASSIFICATION OF SLOW VIRUS DISEASES

Conventional Agents	Disease
Measles virus	Subacute sclerosing panencephalopathy
JC virus, SV-40 virus	Progressive multifocal leukoencephalopathy
Visna virus	Visna (sheep)
Aleutian Disease virus	Aleutian disease (mink)

Unconventional Agents	Disease
Unnamed	Creutzfeldt-Jacob disease
Unnamed	Kuru
	Scrapie (sheep)
	Transmissible mink encephalopathy

nucleic acid, less than one-tenth the size of a virus.

Multiple Sclerosis. According to current hypotheses, both genetic factors and an unidentified virus or viruses may be related to the etiology of multiple sclerosis. Patients with this disease have very high titers to measles virus, which is the virus most frequently suggested as a contributing cause. A viral agent that has been recovered from the bone marrow appears to be related to the paramyxovirus group—perhaps canine distemper virus or measles virus.[124] Paramyxovirus-like structures have been observed in electron micrographs of brains of multiple sclerosis patients.[128] However, these observations must be interpreted with caution, as such preliminary evidence obtained in the past has not been confirmed.

VIRUS-LIKE PARTICLES IN LUPUS

Disseminated lupus erythematosus is a chronic systemic disease of unknown cause. It is associated with circulating immune complexes containing DNA and antibody, which are deposited throughout the body, particularly in the skin and glomeruli.

The kidneys of patients with lupus contain a Type C oncornavirus antigen, demonstrable by fluorescent antibody staining.[134] Normal kidneys or kidneys from patients with non-lupus immune complex glomerulonephritis are negative by similar staining.

Lupus-like disease occurs naturally in a New Zealand laboratory mouse, in which Type C virus antigen has also been demonstrated in their immune complex glomerulonephritis.[134]

Virus-like tubular particles have been observed in electron micrographs of glomerular endothelium cytoplasm of patients with lupus, but not in normal patients.[130] These tubular structures have also been found in the lymphocytes of newborn infants of mothers with systemic lupus erythematosus.[133]

The significance of these virus-like particles remains unknown. However, a theoretical role of chronic virus infection in the pathogenesis of lupus has been proposed.[135] A variety of connective tissue disorders also have these cytoplasmic tubuloreticular structures.[131] In summary, lupus in humans appears to be a disease of unknown etiology, but a virus appears to be involved in the complex etiology.[133]

REOVIRUSES

REO is an abbreviation for Respiratory, Enteric, Orphan. Reoviruses have many similarities with the ECHO virus group, and reovirus type 1 was originally classified as echovirus type 10. Reoviruses differ from echoviruses by larger size, 60 to 70 nanometers, compared to the 30 nanometer size of echoviruses.

There are three reovirus types. Reovirus types 1 and 2 were originally classified as simian viruses. Reovirus type 3 was formerly called hepatoencephalitis virus.

The relative frequency and importance of reoviruses as a cause of human illness have not yet been clearly established, although mild febrile illnesses apparently occur as a result of reovirus infection.[137-139] The virus has been recovered from the lungs of a patient dying with severe pneumonia.[140]

Recent studies using immune electron microscopy have provided evidence that previously unrecognized reoviruses may be an important cause of acute idiopathic diarrhea of infancy. These viruses appear to be antigenically related to other reoviruses which can cause diarrhea in calves or mice (Nebraska calf diarrhea virus and epizootic diarrhea of infant mice virus).

The diarrhea-causing reovirus-like viruses are discussed later in this chapter under the heading of rotaviruses, which is their current name.

Biologic features of clinical interest include the observation that the RNA in reoviruses is double-stranded, in contrast to most RNA viruses, where it is single-stranded.[136] Orbiviruses, such as Colorado tick fever virus and other unnamed viruses observed in children with diarrhea, also have double-stranded RNA.

Naturally occurring reovirus infections have been observed in mice, dogs, cattle, and monkeys.

No specific treatment or vaccine is available.

ROTAVIRUSES

Rotavirus is the current name for a group of viruses that are an important cause of diarrhea. *Rota*, the Latin word for wheel, refers to the appearance of the virus on electron micrographs. These viruses were formerly called reovirus-like agents or orbiviruses (because of their size and double-stranded RNA) or duoviruses (because of their double-shelled capsid structure).[142]

Other poorly characterized viruses that can cause diarrhea include Norwalk virus and astroviruses, and caliciviruses.[46A]

Frequency and Importance

Rotaviruses are important because they appear to be the most frequent cause of diarrhea in young infants.[148] Fatalities are rare, and are usually secondary to shock from dehydration.[141]

Types

Several serotypes of rotavirus have been identified. Of these, type 2 appears to be the most important cause of infantile diarrhea in the United States.[151] Infection with one serotype does not provide resistance to infection by another serotype.[151]

Clinical Patterns of Illness

Rotaviruses are a *frequent* cause of:

Diarrhea in Infants. Vomiting and fever typically occur at the onset, followed by watery diarrhea.[148, 149] In young infants, who easily become dehydrated with diarrhea, hospitalization for intravenous fluids may be necessary.

Diarrhea in Adults. Adult contacts of infants with rotavirus diarrhea, especially hospital staff, often develop diarrhea.[146]

Other Diseases. Rotaviruses have been recovered from patients with diarrhea and rectal bleeding,[143] intussusception,[147] and Crohn's disease,[150] although a causal relationship has not been proved.

Laboratory Approach

Immune Electron Microscopy (IEM). Rotaviruses were first detected by IEM (Fig. 14-5), which is available in a few research laboratories.

ELISA. Rotaviruses can be readily detected by enzyme-linked immunoassay (see Fig. 3-9 and p. 63).

Serology. Nebraska calf diarrhea virus (NCDV) can be used as the antigen for a complement fixation test to detect rotavirus infection.[145] Human rotavirus from purified infected feces has also been used to make antigens for complement fixation. Radioimmunoassay and ELISA methods can be used for antigen or antibody detection. Neutralization of clumping of the virus on IEM can be used in research laboratories to demonstrate serum antibodies.

Characteristics of Clinical Interest

Importance of Breast Feeding. Antibodies to rotavirus have been demonstrated in colostrum and breast milk, especially in the first few days after delivery.[152]

Virus Size. Rotaviruses are approximately 75 nanometers in diameter on electron microscopy, somewhat smaller than reoviruses and orbiviruses, which are about 80 nanometers.[144,145] These viruses are much larger than the Norwalk virus, which is about 27 nanometers in diameter.

Peak Occurrence in Winter. In the United States and Japan, rotaviruses are a frequent cause of diarrhea in the winter, but are an uncommon cause of diarrhea in the summer.

Antigenic Cross-Reactions with Animal Rotaviruses. Some animal rotaviruses can be used for serologic diagnosis of human infections, as described above. Animal rotaviruses have some theoretical potential for a vaccine to be used for humans.

Treatment and Prevention

No specific treatment is available for rotavirus infection. Hospitalization and intravenous fluids are useful in young infants with rotavirus diarrhea. No preventive measures are available except isolation of hospitalized infants with diarrhea.

NORWALK VIRUS

The Norwalk virus was named for Norwalk, Ohio, where it was first recovered from elementary school children with gastroenteritis, manifested primarily by vomiting.[153] Originally called Norwalk agent, it is now known to be a virus on the basis of electron micrographs, and has been transmitted to humans by stool filtrates.[155] The disease produced consisted of predominately vomiting in some volunteers and predominately diarrhea in others.[155]

This virus is not the only infectious cause, or even the most frequent cause, of nonbacterial gastroenteritis. However, its discovery represented a major breakthrough in determining the etiology of this frequent syndrome, and was first seen by using the technique of immune electron microscopy. Later hepatitis A and later still rotaviruses were first seen by using immune electron microscopy to examine stool suspensions.

Types

At the present time, there appear to be two immunologically distinct 27 nanometer viruses, originally called the Norwalk agent and the Hawaii agent, from an outbreak of gastroenteritis in Hawaii.[156] A virus from a Montgomery County, Maryland outbreak (MC agent), appears not to produce illness in volunteers previously developing illness with Norwalk agent.

Clinical Patterns of Illness

Norwalk virus is a *frequent* cause of:

Epidemic Vomiting Disease. The Norwalk, Ohio, school outbreak was associated with vomiting in 92 percent and diarrhea in 38 percent of the sick children.[153] Inoculation of volunteers with a bacteria-free filtrate from stool from a boy (W) admitted during an outbreak in a boys' boarding school, was associated with profuse vomiting, straining, and retching, with mild diarrhea and mild fever.[154]

Several waterborne outbreaks in summer camps have been associated with recovery of a parvovirus and an antibody titer rise to Norwalk agent. The illnesses were associated with abdominal pain as well as nausea and vomiting. Headache and vomiting were less prominent.

Characteristics of Clinical Interest

Parvovirus Size. Norwalk virus is 27 nanometers in diameter on electron microscopy, the size of a parvovirus.

Laboratory Approach

Immune Electron Microscopy. IEM is the usual way to detect this virus, which has not yet been grown in cell cultures. This technique is available only in a few research laboratories.

Serology. Radioimmunoassay has been used in reference laboratories to demonstrate an antibody titer rise to Norwalk virus.

CAT SCRATCH DISEASE

Cat scratch disease is best defined as nonbacterial regional lymphadenopathy with a positive skin test with cat scratch antigen. Bacterial disease can be excluded by culture of aspirated pus or by histologic appearance. No infectious agent has been recovered in lymph nodes from patients with cat scratch disease, so most statements about the disease must lack conclusive proof.

Frequency and Importance

The actual frequency is not known, because laboratory confirmation is difficult, and the disease is benign and self-limited. Margileth observed about 100 cases in 10 years, and suggested it is relatively frequent.[162] It is important because it may be mistaken for a lymphatic neoplasm.

Clinical Patterns of Illness

The agent of cat scratch disease is an *occasional* cause of:

Regional Lymphadenopathy. The enlarged nodes are in the head or neck in about half the patients and in the axillary area in about half (Fig. 14-6).[162] Cat scratches are found in the majority of patients. Suppuration occurs in 10 to 25 percent of the cases.[158,162] Fever over 101° is noted in about 25 percent of patients, and lasts about a week. A rash or conjunctivitis or parotid enlargement is noted in about 3 to 5 percent of patients.

The enlarged node is noted about 2 weeks after the scratch, but may occur as late as 7 weeks or as early as 3 days after the scratch.[162]

Recurrence is rare. Rare complications include encephalopathy and thrombocytopenic purpura.

Oculoglandular Syndrome. Cat scratch disease is one of many causes of conjunctivitis with preauricular lymphadenopathy.

Laboratory Approach

Skin Test. The antigen is difficult to obtain. Reactions to various lots of antigen vary, suggesting strain variation.[161] The antigen is not available commercially, but may be available from individuals with a special interest in the disease. It is prepared from pus obtained from a node of a patient with the disease.[157] The interpretation of the test is also controversial.

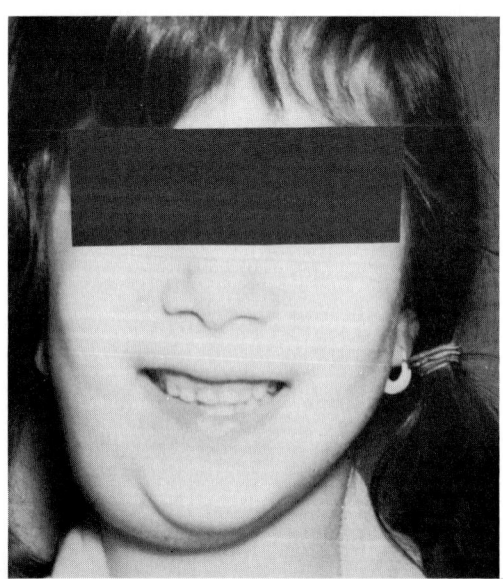

FIG. 14-6. *Cat scratch disease showing regional lymph node enlargement. (Photo from Dr. Andrew Margileth)*

Histology. Biopsy reveals characteristic, but nonspecific findings.[160] It is often done to exclude malignancy.

Serology. Complement fixing antigen has been made from pus, and antibody rise can be demonstrated. There appears to be cross reaction with chlamydia antigens (psittacosis or lymphogranuloma venereum agents).[161]

Culture. This should be done to exclude bacteria or mycobacteria, but the agent of cat scratch disease has not been isolated. Needle aspiration is simple and preferable to incision and drainage if the diagnosis is reasonably certain.

Leukocyte Lysis Test. The lysis of leukocytes in vitro by cat scratch disease antigen is a new diagnostic test which has not yet been fully evaluated.[163]

Biologic Characteristics of Clinical Interest

Other Animal Hosts. Several animals other than cats can transmit the disease, including dogs and monkeys.

Experimental Infections. Humans have been infected by experimental inoculation, but the nature of the agent is still unknown. It is presumed to be a virus, since usual techniques have failed to yield a mycobacteria or other culturable agent.

Treatment

No specific treatment is available. Steroids have been advocated, and appear to produce reduction in swelling, but probably should not be used without a confirmed diagnosis, and a compelling reason.[159]

Prevention

Unnecessary exposure to cat scratches should be avoided.

REFERENCES

Hepatitis Viruses

1. Adam, E., et al.: Type B hepatitis antigen and antibody among prostitutes and nuns: a study of possible venereal transmission. J. Infect. Dis., 129:317–321, 1974.
2. Alter, H.J., Purcell, R.H., Holland, P.V., and Popper, H.: Transmissible agent in non-A, non-B hepatitis. Lancet, 1:459–463, 1978.
3. Anon: non-A, non-B hepatitis (editorial). Lancet, 1:942–944, 1978.
4. Auslander, M.O., and Gitnick, G.L.: Vigorous medical management of acute fulminant hepatitis. Arch. Intern. Med., 137:599–601, 1977.
5. Bancroft, W.H., et al.: Transmission of hepatitis B virus to gibbons by exposure to human saliva containing hepatitis B surface antigen. J. Infect. Dis., 135:79–85, 1977.
6. Baker, L.F., et al.: Transmission of type B viral hepatitis to chimpanzees. J. Infect. Dis., 127:648–662, 1973.
7. Barker, L.F., and Murray, R.: Relationship of virus dose to incubation time of clinical hepatitis and time of appearance of hepatitis-associated antigen. Am. J. Med. Sci., 263:27–33, 1971.
8. ———: Acquisition of hepatitis-associated antigen. Clinical features in young adults. J.A.M.A., 216:1970–1976, 1971.
9. Blumberg, B.S.: Non-A, non-B hepatitis (editorial). Ann. Intern. Med., 87:111–115, 1977.
10. Blumberg, B.S., Sutnick, A.I., London, W.T., and Millman, I.: Australia antigen and hepatitis. New Engl. J. Med., 283:349–354, 1970.
11. Boggs, J.D., Melnick, J.L., Conrad, M.E., and Felsher, B.F.: Viral hepatitis. Clinical and tissue culture studies. J.A.M.A., 214:1041–1046, 1970.
12. Center for Disease Control: Immune globulins for protection against viral hepatitis. Morbidity Mortality Weekly Rep., 26:425–428, 441–442, 1977.
13. Conrad, M.E., Schwartz, F.D., and Young, A.A.: Infectious hepatitis—a generalized disease. A study of renal, gastrointestinal and hematologic abnormalities. Am. J. Med., 37:789–801, 1964.
14. Denes, A.E., et al.: Foodborne hepatitis A infection: a report of two urban restaurant-associated outbreaks. Am. J. Epidemiol., 105:156–162, 1977.
15. ———Hepatitis B infection in physicians. Results of a nationwide seroepidemiologic survey. J.A.M.A., 239:210–212, 1978.
16. Dienstag, J.L., et al.: Nonhuman primate-associated viral hepatitis A. Serologic evidence of hepatitis A virus infection. J.A.M.A., 236:462–464, 1976.
17. Dienstag, J.L., Szmuness, W., Stevens, C.E., and Purcell, R.H.: Hepatitis A virus infection: new insights from seroepidemiologic studies. J. Infect. Dis., 137:328–340, 1978.
18. Dosik, H., and Jhaveri, R.: Prevention of neonatal hepatitis B infection by high-dose hepatitis B immune globulin. New Engl. J. Med., 298:602–603, 1978.

19. Dupuy, J.M., et al.: Hepatitis B in children. II. Study of children born to chronic HB$_s$Ag mothers. J. Pediatr., 92:200–204, 1978.

20. Feinstone, S.M., et al.: Transfusion-associated hepatitis not due to viral hepatitis type A or B. New Engl. J. Med., 292:767–770, 1975.

21. Feinstone, S.M., Kapikian, A.Z., and Purcell, R.H.: Hepatitis A: detection by immune electron microscopy of a virus-like antigen associated with acute illness. Science, 182:1026–1028, 1973.

22. Garibaldi, R.A., et al.: Nonparenteral serum hepatitis. Report of an outbreak. J.A.M.A., 220:963–966, 1972.

23. Gerber, M.A., Zappi, T., Vernace, S.J., and Paronetto, F.: Antibodies to hepatitis B core antigen in hepatitis B surface antigen-positive and -negative chronic hepatitis. J. Infect. Dis., 135:1006–1009, 1977.

24. Grady, G.F., et al.: Hepatitis B immune globulin for accidental exposures among medical personnel: final report of a multicenter controlled trial. J. Infect. Dis., 138:625–638, 1978.

25. Grady, G.F., Chalmers, T.C., and Boston Inter-Hospital Liver Group: Risk of post-transfusion viral hepatitis. New Engl. J. Med., 271:337–342, 1964.

26. Hampers, C.L., Prager, D., and Senior, J.R.: Post-transfusion anicteric hepatitis. New Engl. J. Med., 271:747–753, 1964.

27. Havens, W.P., Jr.: Experiments in cross-immunity between infectious hepatitis and homologous serum jaundice. Proc. Soc. Exp. Biol. Med., 59:148–150, 1945.

28. Hersh, T., Melnick, J.L., Goyal, R.K., and Hollinger, F.B.: Nonparenteral transmission of viral hepatitis type B (Australia antigen-associated serum hepatitis). New Engl. J. Med., 285:1363–1364, 1971.

29. Hieber, J.P., Dalton, D., Shorey, J., and Combes, B.: Hepatitis and pregnancy. J. Pediatr., 91:545–549, 1977.

30. Hok, K.A., Nieman, R., Lackey, J.O., and Cabasso, V.J.: Australia antigen in a closed adult population monitored for serum glutamic oxalacetic transaminase. Appl. Microbiol., 20:6–10, 1970.

31. Hopf, U., Meyer zum Büschenfelde, K.-H., and Arnold, W.: Detection of liver-membrane autoantibody in HB$_s$Ag-negative chronic active hepatitis. New Engl. J. Med., 294:578–582, 1976.

32. Irwin, G.R., et al.: Serodiagnosis of hepatitis B infection by antibody to core antigen. J. Infect. Dis., 136:31–36, 1977.

33. Janzen, J., et al.: Epidemiology of hepatitis B surface antigen (HB$_s$Ag) and antibody to HB$_s$Ag in hospital personnel. J. Infect. Dis., 137:261–265, 1978.

34. Knodell, R.G., Conrad, M.E., and Ishak, K.G.: Development of chronic liver disease after acute non-A, non-B post-transfusion hepatitis. Gastroenterology, 72:902–909, 1977.

35. Koff, R.S., et al.: Viral hepatitis in a group of Boston hospitals. III. Importance of exposure to shellfish in a nonepidemic period. New Engl. J. Med., 276:703–710, 1967.

36. Koff, R.S., Slavin, M.M., Connelly, L.J.D., and Rosen, D.R.: Contagiousness of acute hepatitis B. Secondary attack rates in household contacts. Gastroenterology, 72:297–300, 1977.

37. Krugman, S., et al.: Viral hepatitis, type B. DNA polymerase activity and antibody to hepatitis core antigen. New Engl. J. Med., 290:1331–1335, 1974.

38. Krugman, S., and Giles, J.P.: Viral hepatitis. New light on an old disease. J.A.M.A., 212:1019–1029, 1970.

39. Krugman, S., Giles, J.P., and Hammond, J.: Infectious hepatitis. Evidence for two distinctive clinical, epidemiological, and immunological types of infection. J.A.M.A., 200:365–373, 1967.

40. Krugman, S., Ward, R., and Giles, J.P.: The natural history of infectious hepatitis. Am. J. Med., 32:717–728, 1962.

41. Lander, J.J., et al.: Antibody to hepatitis-associated antigen. J.A.M.A., 220:1079–1082, 1972.

42. Lander, J.L., Giles, J.P., Purcell, R.H., and Krugman, S.: Viral hepatitis, Type B (MS-2 strain). Detection of antibody after primary infection. New Engl. J. Med., 285:303–307, 1971.

43. Lee, A.K.Y., Ip, H.M.H., and Wong, V.C.W.: Mechanisms of maternal-fetal transmission of hepatitis virus. J. Infect. Dis., 138:668–671, 1978.

44. Linnemann, C.C., Jr., Hegg, M.E., Ramundo, N., and Schiff, G.M.: Screening hospital patients for hepatitis B surface antigen. Am. J. Clin. Pathol., 67:257–259, 1977.

45. London, W.T.: Hepatitis B virus and antigen–antibody complex disease (editorial). New Engl. J. Med., 296:1528–1529, 1977.

46. Louria, D.B., Hensle, T., and Rose, J.: The major medical complications of heroin addiction. Ann. Intern. Med., 67:1–22, 1967.

46A. Madeley, C.R.: Comparison of the features of astroviruses and caliciviruses seen in samples of feces by electron microscopy. J. Infect. Dis., 139:519–523, 1979.

47. Mathiesen, L.R., et al.: Enzyme-linked immunosorbent assay for detection of hepatitis A antigen in stool and antibody to hepatitis A antigen in sera: comparison with solid-phase radioimmunoassay, immune electron microscopy, and immune adherence hemagglutination assay. J. Clin. Microbiol., 7:184–193, 1978.

48. Millman, I., Zavatone, V., Gerstely, B.J.S., and Blumberg, B.S.: Australia antigen detected in the nuclei of liver cells of patients with viral hepatitis by the fluorescent antibody technique. Nature, 222:181–184, 1969.

49. Mollica, F., Musumeci, S., and Fischer, F.: Neonatal hepatitis in five children of a hepatitis B surface antigen carrier woman. J. Pediatr., 90:949–951, 1977.

50. Morse, L.J., et al.: The Holy Cross College football team outbreak. J.A.M.A., 219:706–708, 1972.

51. Neefe, J.R., Gellis, S.S., and Stokes, J.J.: Homologous serum hepatitis and infectious (epidemic) hepatitis: studies in volunteers bearing on immunological and other characteristics of etiological agents. Am. J. Med., 1:3–32, 1946.

52. Nielsen, J.O., Dietrichson, O., Elling, P., and Christoffersen, P.: Incidence and meaning of persistence of Australia antigen in patients with acute viral hepatitis: development of chronic hepatitis. New Engl. J. Med., 285:1157–1160, 1971.

53. Nielsen, J.O., and LeBouvier, G.L.: Copenhagen hepatitis acute program: Subtypes of Australia antigen among patients and healthy carriers in Copenhagen. A relation between the subtypes and the degree of liver damage in acute viral hepatitis. New Engl. J. Med., 288:1257–1261, 1973.

54. Nielsen, J.O., Nielsen, M.H., and Elling, P.: Differential distribution of Australia antigen-associated particles in patients with liver diseases and normal carriers. New Engl. J. Med., 288:484–487, 1973.

55. Popper, H., and Schaffner, F.: Hepatitis B antigen and prognosis in hepatitis. New Engl. J. Med., 228:518–519, 1973.

56. Purcell, R.H., Dienstag, J.L., Feinstone, S.M., and Kapikian, A.Z.: Relationship of hepatitis A antigen to viral hepatitis. Am. J. Med. Sci., 270:61–71, 1975.

57. Rakela, J., and Mosley, J.W.: Fecal excretion of hepatitis A virus in humans. J. Infect. Dis., 135:933–938, 1977.

58. Rakels, J., et al.: Hepatitis A virus infection in fulminant hepatitis and chronic active hepatitis. Gastroenterology, 74:879–882, 1978.

59. Reincke, V., et al.: A study of Australia antigen-positive blood donors and their recipients, with special reference to liver histology. New Engl. J. Med., 286:867–870, 1972.

60. Rimland, D., Parkin, W.E., Miller, G.B., Jr., and Schrack, W.D.: Hepatitis B outbreak traced to an oral surgeon. New Engl. J. Med., 296:953–958, 1977.

61. Sherlock, S., and Parbhoo, S.P.: The management of acute hepatic failure. Postgrad. Med. J., 47:493–498, 1971.

62. Shikata, T., et al.: Hepatitis B e antigen and infectivity of hepatitis B virus. J. Infect. Dis., 136:571–576, 1977.

63. ———: Incomplete inactivation of hepatitis B virus after heat treatment for 60°C for 10 hours. J. Infect. Dis., 138:242–244, 1978.

64. Shiraki, K., et al.: Hepatitis B surface antigen and chronic hepatitis in infants born to asymptomatic carrier mothers. Am. J. Dis. Child., 131:644–647, 1977.

65. Smith, J.L., et al.: Comparative risk of hepatitis B among physicians and dentists. J. Infect. Dis., 133:705–706, 1976.

66. Snydman, D.R., Bryan, J.A., Macon, E.J., and Gregg, M.B.: Hemodialysis-associated hepatitis: report of an epidemic with further evidence on mechanisms of transmission. Am. J. Epidemiol., 104:563–570, 1976.

67. Stevens, D.P., et al.: Anicteric hepatitis presenting as polyarthritis. J.A.M.A., 220:687–689, 1972.

68. Szmuness, W., et al.: Hepatitis type A and hemodialysis. A seroepidemiologic study in 15 U.S. centers. Ann. Intern. Med., 87:8–12, 1977.

69. ———: Prevalence of hepatitis B virus infection and hepatocellular carcinoma in Chinese-Americans. J. Infect. Dis., 137:822–829, 1978.

70. Takekoshi, Y., et al.: Strong association between membraneous nephropathy and hepatitis B surface antigenemia in Japanese children. Lancet, 2:1065–1068, 1978.

71. Taswell, H.F., Shorter, R., Poncelet, T.K., and Maxwell, N.G.: Hepatitis-associated antigen in blood donor populations. Relationship to post-transfusion hepatitis. J.A.M.A., 214:142–144, 1970.

72. Tiku, M.L., et al.: Distribution and characteristics of hepatitis B surface antigen in body fluids of institutionalized children and adults. J. Infect. Dis., 134:342–347, 1976.

73. Tiku, M.L., et al.: Hepatitis B e antigen and antibody activity in hepatitis B virus infection. J. Pediatr., 91:540–544, 1977.

74. Varma, R.R.: Hepatitis B surface antigen carrier state in neonates. Prophylaxis with large doses of conventional immune serum globulin. J.A.M.A., 236:2302–2304, 1976.

75. Werner, B.G., and Blumberg, B.S.: e antigen in hepatitis B virus infected dialysis patients: assessment of its prognostic value. Ann. Intern. Med., 89:310–314, 1978.

76. Williams, S.V., Huff, J.C., and Bryan, J.A.: Hepatitis A and facilities for preschool children. J. Infect. Dis., 131:491–495, 1975.

Yellow Fever Virus

77. Center for Disease Control: Health Information for International travel 1978. Morb. Mort. Week. Rep., 27(Suppl.):1–96, 1978.

78. Lebrun. A.: Jungle yellow fever and its control in Gemena, Belgian Congo. Am. J. Trop. Med., 12:398–407, 1963.

Dengue Virus

79. Ehrenkranz, N.J., et al.: Pandemic dengue in Caribbean countries and the southern United States—past, present and potential problems. New Engl. J. Med., *285:*1460–1469, 1971.

80. Halstead, S.B., Shotwell, H., and Casals, J.: Studies on the pathogenesis of dengue infection in monkeys. II. Clinical laboratory responses in heterologous infection. J. Infect. Dis., *128:*15–22, 1973.

81. Morens, D.M., Woodall, J. P., and López-Correa, R.H.: Dengue in American children of the Caribbean. J. Pediatr., *93:*1049–1051, 1978.

82. Russell, P.K., Chumdermpadestsuk, S., and Piyaratn, P.: A fatal case of dengue hemorrhagic fever in an American child. Pediatrics, *40:*804–807, 1967.

Colorado Tick Fever Virus

83. Eklund, C.M., Kohls, G.M., and Brennan, J.M.: Distribution of Colorado tick fever and virus-carrying ticks. J.A.M.A., *157:*335–337, 1955.

84. Florio, L., Stewart, M.O., and Mugrage, E.R.: The experimental transmission of Colorado tick fever. J. Exp. Med., *80:*165–188, 1946.

85. Goodpasture, H.C., et al.: Colorado tick fever: clinical, epidemiologic, and laboratory aspects of 228 cases in Colorado in 1973–1974. Ann. Int. Med., *88:*303–310, 1978.

86. Silver, H.K., Meiklejohn, G., and Kempe, C.H.: Colorado tick fever. Am. J. Dis. Child., *101:*30–36, 1961.

87. Spruance, S.L., and Bailey, A.: Colorado tick fever. A review of 115 laboratory confirmed cases. Arch. Intern. Med., *131:*288–293, 1973.

Coronaviruses

88. Bradburne, A.F., Bynoe, M.L., and Tyrrell, D.A.J.: Effects of a "new" human respiratory virus in volunteers. Br. Med. J., *3:*767–769, 1967.

89. Cavallaro, J.J., and Monto, A.S.: Community-wide outbreak of infection with a 229E-like virus in Tecumseh, Michigan. J. Infect. Dis., *122:*272–279, 1970.

90. McIntosh, K., et al.: Seroepidemiologic studies of coronavirus infection in adults and children. Am. J. Epidemiol., *91:*585–592, 1970.

91. ———: Coronavirus infection in acute lower respiratory disaese of infants. J. Infect. Dis., *130:*502–507, 1974.

Lymphocytic Choriomeningitis Virus

92. Baum, S.G., Lewis, A.M., Jr., Rowe, W.P., and Huebner, R.J.: Epidemic nonmeningitic lymphocytic-choriomeningitis-virus infection. An outbreak in a population of laboratory personnel. New Engl. J. Med., *274:*934–936, 1966.

93. Biggar, R.J., Woodall, J.P., Walter, P.D., and Haughie, G.E.: Lymphocytic choriomeningitis outbreak associated with pet hamsters. Fifty-seven cases from New York State. J.A.M.A., *232:*494–500, 1975.

94. Meyer, H.M., Jr., et al.: Central nervous system syndromes of viral etiology. Study of 713 cases. Am. J. Med., *29:*334–347, 1960.

95. Wooley, J.G., Stimpert, F.D., Kessel, J.F., and Armstrong, C.: A study of human sera antibodies capable of neutralizing the virus of lymphocytic choriomeningitis. Public Health Rep., *54:*938–944, 1939.

Lassa Virus

96. Leifer, E., Gocke, D.J., and Bourne, H.: Lassa fever, a new virus disease of man from West Africa. II. Report of a laboratory-acquired infection treated with plasma from a person recently recovered from the disease. Am. J. Trop. Med. Hyg., *19:*677–679, 1970.

97. Zweighaft, R.M., et al.: Lassa fever: response to an imported case. New Engl. J. Med., *297:*803–807, 1977.

Marburg and Ebola Viruses

98. Emond, R.T.D., Evans, B., Bowen, E.T.W., and Lloyd, G.: A case of Ebola virus infection. Br. Med. J., *2:*541–544, 1977.

99. Kissling, R.E., Murphy, F.A., and Henderson, B.E.: Marburg virus. Ann. N.Y. Acad. Sci., *174:*932–945, 1970.

100. Smith, C.E.G., Simpson, D.I.H., Bowen, E.T.W., and Zlotnik, I.: Fatal human disease from vervet monkeys. Lancet, *2:*1119–1121, 1967.

Tumor Viruses

101. Baltimore, D.: Viruses, polymerases, and cancer. Science, *192:*632–636, 1976.

102. Fraumeni, J.F., Stark, C.R., Gold, E., and Lepow, M. L.: Simian virus in polio vaccine; follow up of newborn recipients. Science, *167:*59–60, 1970.

103. Gallagher, R.E., and Gallo, R.C.: Type C RNA tumor virus isolated from cultured human acute myelogenous leukemia cells. Science, *187:*350–353, 1975.

104. Gissman, L., Pfister, H., and Zur Hausen, H.: Human papilloma viruses (HPV): characterization of four different isolates. Virology, *76:* 569–580, 1977.

105. Gross, L.: The role of C-type and other oncogenic virus particles in cancer and leukemia (editorial). New Engl. J. Med., *294:*724–725, 1976.

106. Hayflick, L.: A comparison of primary monkey kidney, heteroploid cell lines, and human diploid cell strains for human virus vaccine preparation. Am. Rev. Resp. Dis., 88:287–294, 1963.

107. Karpas, A., Wreghitt, T.G., and Nagington, J.: Transformation of normal bone marrow cells by a leukaemic cell line associated with a presumptive new human virus. Lancet, 2:1016–1019, 1978.

108. Marx, J.L.: RNA tumor viruses: getting a handle on transformation. Science, 199:161–164, 1978.

109. Maugh, T.H., II: RNA viruses: the age of innocence ends. Science, 183:1181–1185, 1974.

110. Newell, G.R., et al.: Evaluation of "viruslike" particles in the plasmas of 225 patients with leukemia and related diseases. New Engl. J. Med., 278:1185–1190, 1968.

111. Temin, H.M.: The DNA provirus hypothesis. The establishment and implications of RNA-directed DNA synthesis. Science, 192:1075–1080, 1976.

Slow Virus Diseases

112. Chapman, I., and Jimenez, F.A.: Aleutian mink disease in man. New Engl. J. Med., 269:1171–1174, 1963.

113. Gajdusek, D.C.: Slow virus infections of the nervous system. New Engl. J. Med., 276:392–400, 1967.

114. ———: Unconventional viruses and the origin and disappearance of kuru. Science, 197:943–960, 1977.

115. Gajdusek, D.C., et al.: Precautions in medical care of, and in handling materials from, patients with transmissible virus dementia. New Engl. J. Med., 297:1253–1258, 1977.

116. Gibbs, C.J., Jr., and Gajdusek, D.C.: Experimental subacute spongiform virus encephalopathies in primates and other laboratory animals. Science, 182:67–68, 1973.

117. Griffin, D.E., et al.: The cerebrospinal fluid in visna, a slow viral disease of sheep. Ann. Neurol., 4:212–218, 1978.

118. Holmberg, C.A., et al.: Isolation of simian virus 40 from rhesus monkeys (Macaca mulatta) with spontaneous progressive multifocal leukoencephalopathy. J. Infect. Dis., 136:593–396, 1977.

119. Horta-Barbosa, L., Fuccillo, D.A., and Sever, J.L.: Chronic viral infections of the central nervous system. J.A.M.A., 218:1185–1188, 1971.

120. London, W.T., et al.: Brain tumors in owl monkeys inoculated with a human polyomavirus (JC Virus). Science, 201:1246–1249, 1978.

121. Manuelidis, E.E., et al.: Experimental Creutzfeldt–Jacob disease transmitted via the eye with infected cornea. New Engl. J. Med., 296:1334–1337, 1977.

122. Manuelidis, E.E., Gorgacz, E.J., and Manuelidis, L.: Viremia in experimental Creutzfeldt–Jacob disease. Science, 200:1069–1071, 1978.

123. Marx,: The scrapie agent: is it a viroid? Science, 203:532, 1979.

124. Mitchell, D.N., et al.: Isolation of an infectious agent from bone-marrows of patients with multiple sclerosis. Lancet, 2:387–391, 1978.

125. Narayan, O., et al.: Etiology of progressive multifocal leukoencephalopathy. Identification of a papovavirus. New Engl. J. Med., 289:1278–1282, 1973.

126. Padgett, B.L., et al.: JC papovavirus in progressive multifocal leukoencephalopathy. J. Infect. Dis., 133:686–690, 1976.

127. Padgett, B.L., and Walker, D.L.: Prevalence of antibodies in human sera against JC virus, an isolate from a case of progressive multifocal leukoencephalopathy. J. Infect. Dis., 127:467–470, 1973.

128. Tanaka, R., Iwasaki, Y., and Koprowski, H.: Paramyxovirus-like structures in brains of multiple sclerosis patients. Arch. Neurol., 32:80–83, 1975.

129. Weiner, L.P., et al.: Isolation of virus related to SV 40 from patients with progressive multifocal leukoencephalopathy. New Engl. J. Med., 286:385–390, 1972.

Virus-Like Particles in Lupus

130. Grausz, H., et al.: Diagnostic import of virus-like particles in the glomerular endothelium of patients with systemic lupus erythematosus. New Engl. J. Med., 283:506–511, 1970.

131. Györkey, F., Sinkovics, J.G., Min, K.W., and Györkey, P.: A morphologic study on the occurrence and distribution of structures resembling viral nucleocepsids in collagen disease. Am. J. Med., 53:148–158, 1972.

132. Klippel, J.H., Grimley, P.M., and Decker, J.L.: Lymphocyte inclusions in newborns of mothers with systemic lupus erythematosus. New Engl. J. Med., 290:96–97, 1974.

133. Marx, J.L.: Autoimmune disease: new evidence about lupus. Science, 192:1089–1091, 1150, 1976.

134. Panem, S., et al.: C-type virus expression in systemic lupus erythematosus. New Engl. J. Med., 295:470–475, 1976.

135. Ziff, M.: Viruses and the connective tissue diseases. Ann. Intern. Med., 75:951–958, 1971.

Reoviruses

136. Fields, B.N.: Genetic manipulation of reovirus—a model for modification of disease? New Engl. J. Med., 287:1026–1033, 1972.
137. Lerner, A.M., Cherry, J.D., Klein, J.O., and Finland, M.: Infections with reoviruses. New Engl. J. Med., 267:947–952, 1962.
138. Rosen, L., Evans, H.E., and Spickard, A.: Reovirus infections in human volunteers. Am. J. Hyg., 77:29–37, 1963.
139. Stanley, N.F.: Reoviruses, Br. Med. Bull., 23:150–155, 1967.
140. Tillotson, J.R., and Lerner, A.M.: Reovirus type 3 associated with fatal pneumonia. New Engl. J. Med., 276:1060–1063, 1967.

Rotaviruses

141. Carlson, J.A.K., et al.: Fatal rotavirus gastroenteritis. An analysis of 21 cases. Am. J. Dis. Child., 132:477–479, 1978.
142. Davidson, G.P., et al.: Importance of a new virus in acute sporadic enteritis in children. Lancet, 1:242–252, 1975.
143. Delage, G., McLaughlin, B., and Berthiaume, L.: A clinical study of rotavirus gastroenteritis. J. Pediatr., 93:455–456, 1978.
144. Flewett, T.H., et al.: Relation between viruses from acute gastroenteritis of children and newborn calves. Lancet, 2:61–63, 1974.
145. Kapikian, A.Z., et al.: New complement-fixation test for the human reovirus-like agent of infantile gastroenteritis. Lancet, 1:1056–1061, 1975.
146. Kim, H.W., et al.: Human reovirus-like agent infection. Occurrence in adult contacts of pediatric patients with gastroenteritis. J.A.M.A., 238:404–407, 1977.
147. Konno, T., et al.: Human rotavirus infection in infants and young children with intussusception. J. Med. Virol., 2:265–269, 1978.
148. Rodriguez, W.J., et al.: Clinical features of acute gastroenteritis associated with human reovirus-like agents in infants and young children. J. Pediatr., 91:188–193, 1977.
149. Tallet, S., et al.: Clinical, laboratory, and epidemiologic features of a viral gastroenteritis in infants and children. Pediatrics, 60:217–222, 1977.
150. Whorwell, P.J., et al.: Isolation of reovirus-like agents from patients with Crohn's disease. Lancet, 1:1169–1171, 1977.
151. Yolken, R.H., et al.: Epidemiology of human rotavirus types 1 and 2 as studied by enzyme-linked immunosorbent assay. New Engl. J. Med., 299:1156–1161, 1978.
152. ———: Secretory antibody directed against rotavirus in human milk—measurement by means of enzyme-linked immunosorbent assay. J. Pediatr., 93:916–921, 1978.

Norwalk Virus

153. Center for Disease Control: Epidemic gastroenteritis, possible winter vomiting disease, in an elementary school—Norwalk, Ohio. Morb. Mort. Week. Rep., 17:434–435, 440, 1968.
154. Clarke, S.K.A., et al.: A virus from epidemic vomiting disease. Br. Med., J., 3:86–89, 1972.
155. Dolin, R., et al.: Transmission of acute infectious nonbacterial gastroenteritis to volunteers by oral administration of stool filtrates. J. Infect. Dis., 123:307–312, 1971.
156. Wyatt, R.G., et al.: Comparison of three agents of acute infectious nonbacterial gastroenteritis by cross-challenge in volunteers. J. Infect. Dis., 129:709–714, 1974.

Cat Scratch Disease

157. Carithers, H.A.: Cat scratch skin test antigen: purification by heating. Pediatrics, 60:928–929, 1977.
158. Carithers, H.A., Carithers, C.M., and Edwards, R.O., Jr.: Cat scratch disease: its natural history. J.A.M.A., 207:312–316, 1969.
159. Eckhart, W.F., and Levine, A.I.: Corticosteroid therapy of cat scratch disease. Arch. Intern. Med., 109:463–468, 1962.
160. Johnson, W.C., and Helwig, E.B.: Cat scratch disease. Histopathic changes in the skin. Arch. Dermatol., 100:148–154, 1969.
161. Kalter, S.S.: Cat scratch disease: complement fixation and skin test results. Ann. Intern. Med., 55:903–910, 1961.
162. Margileth, A.M.: Cat scratch disease: nonbacterial regional lymphadenitis. The study of 145 patients and a review of the literature. Pediatrics, 42:803–818, 1968.
163. Rice, J.E., and Hyde, R.M.: Rapid diagnostic method for cat scratch disease. J. Lab. Clin. Med., 71:166–170, 1968.

15
Selected Protozoa

ENTAMOEBA HISTOLYTICA

Objectives

1. Describe the typical clinical picture of acute amebiasis.
2. Describe how to detect amebic infections using laboratory techniques and discuss some of the technical difficulties encountered.
3. Explain why a patient with the acute illness of amebic dysentery is less contagious than a chronic carrier.

Species and Life Cycle

The term amebiasis is usually used to mean infection with *Entamoeba histolytica*. Histolytica refers to the invasive power of this species, which secretes a lytic substance which allows tissue invasion. Other species of amoeba, (for example, *Entamoeba coli* and *dientamoeba fragilis*), may be found as normal flora in human feces, but *Dientamoebae* can cause diarrhea.[14]

The trophozoite is motile and survives only briefly after defecation. It is destroyed by gastric acid and so is not contagious.[4] The cyst is the usual infective form. Few cysts are passed by the acutely ill patient with diarrhea.

Humans are the reservoir, but can infect primates, dogs, and cats.

Frequency and Importance

About 3000 cases of amebiasis are reported annually in the United States, with New York City reporting more cases than any single state (Fig. 15-1).[2] The disease is important because it responds well to treatment, whereas untreated patients can develop an uncomfortable, debilitating, or even fatal disease. About 30 to 50 cases of fatal amebiasis are reported annually.[2]

Male homosexuals are more frequently infected than the general population.[7]

Clinical Patterns of Illness

In the United States, *Entamoeba histolytica* is an *uncommon* cause of:

Acute Dysentery-like Diarrhea. Dysentery is defined as diarrhea with blood, pus, and mucous. Acute amebic dysentery usually occurs after a heavy exposure.[3] There is diarrhea with gross blood or blood-streaked stools. The illness typically lasts about a week, but recurrence is frequent. Eosinophilia is unusual, and is not striking when present. Fever usually is less than 101° F, if present. Weight loss is common.

Chronic or Recurrent Diarrhea or Abdominal Pain. Mild diarrhea may alternate with constipation. Weakness, weight loss, and anemia may occur.[3,5]

Complications

Complications are rare. In amebic abscess of the liver, fever may be prominent or absent.[1,11] The alkaline phosphatase is characteristically elevated. The white blood count is often less than 20,000, but may be normal. The right diaphragm

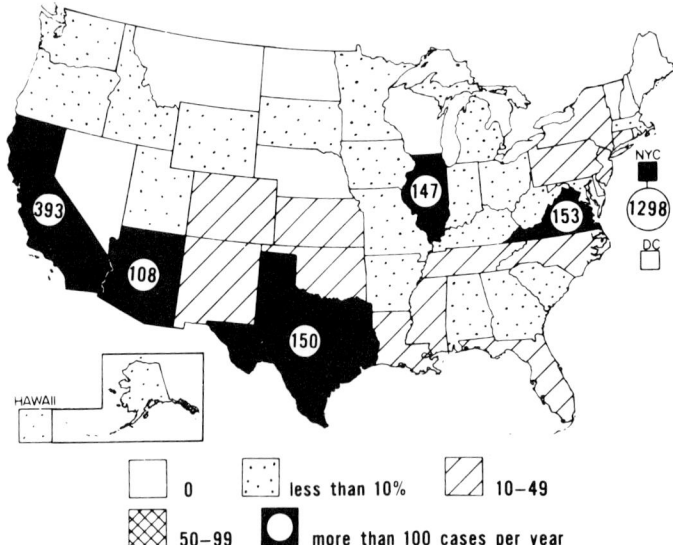

FIG. 15-1. *Amebiasis: average annual reported cases, 1970–1972.*[2]

may be elevated, and symptoms of pleural irritation may be present.

Brain abscess is rare.[6] The clinical findings are similar to brain abscess of bacterial etiology. The spinal fluid usually shows pleocytosis, elevated protein, and may be xanthochromic.

Fulminating colitis or toxic dilation of the colon is rare, but may resemble the clinical appearance of idiopathic ulcerative colitis.[1,13] Perforation of the colon is rare.

Asymptomatic or Subclinical Infection

Asymptomatic infection is common. These individuals go undetected, and are a major source of spread of infection to others.

Laboratory Diagnosis

Microscopic Examination of Stool.[4] Trophozoites are usually found only in patients with acute diarrhea. Immediate examination of a fresh specimen is necessary to identify the trophozoites. False-positive reports are usually due to mistaking fecal leukocytes or non-pathogenic amebae for the amebae.[4] Barium or mineral oil should not have been given for at least 3 days. Specimens obtained by aspiration proctoscopic examination are especially useful.

Concentration of the stool increases the probability of finding cysts. It is sometimes difficult to distinguish *Entamoeba coli* from *E. histolytica* cysts, and the clinician should recognize that many observers mistake one for the other. Overdiagnosis of amebiasis by stool examination can occur, especially if the laboratory workers are untrained.[4] The physician should take time to look at a positive specimen under the microscope, to be sure it is convincing. (Fig. 15-2).

Multiple examinations, perhaps one per week for 3 weeks, are better than three consecutive specimens to detect the parasite, in chronic or intermittent diarrhea. Polyvinyl alcohol (PVA) or a comparable preservative medium should be used to preserve the trophozoites so that examination need not be done immediately.

Serum Antibodies. Serologic tests were unreliable in the past because of the many antigens in the protozoa. Recently, adequate commercial serologic tests have been developed, and may be available in clinical laboratories.

Liver Scan. Radioactive isotopes are useful in recognizing hepatic amebic abscess.

Proctoscopy. This is useful to obtain specimens, and to identify ulceration. Rectal biopsy is useful in many cases. + immunofluorescence of section.

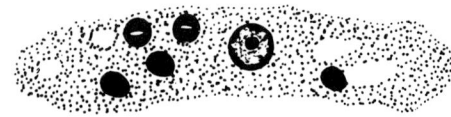

FIG. 15-2. E. histolytica *trophozoite, with ingested erythrocytes. The trophozoite exhibits a linear flowing motion, and is found in flecks of mucous. Trophozoites, not ova, are the forms found in acute amebic dysentary.*

Culture of the Stools. Culture for *E. histolytica* is possible, but is usually not available except in reference laboratories.

Biologic Characteristics of Clinical Interest

Contagion. Patients with acute illnesses are less contagious than asymptomatic carriers. The mature cyst is the contagious form. The patient with acute amebic dysentery may pass few or no cysts, and is much less contagious than the asymptomatic or chronic carrier who passes mature cysts.

Bacterial Synergism. Bacteria appear to enhance the virulence of amebae. Live coliform bacteria inoculated with amebae result in more virulent infections in hamsters, than amebae without bacteria.[12]

Adverse Effect of Corticosteroids. Treatment of another disease with corticosteroids may produce fulminating or even fatal amebiasis if the underlying amebiasis is not recognized, just as unrecognized tuberculosis may become disseminated during corticosteroid therapy.[10] If the patient has lived in the tropics or if latent amebiasis is a reasonable possibility, stool and serologic studies to exclude amebiasis should be done before non-emergency therapy with corticosteroids.[10]

Toxin Production. *E. histolytica* produces a heat-labile enterotoxin which is also cytotoxic to cell cultures.[9] This toxin is probably an important virulence factor in amebic colitis.

Treatment

Treatment is controversial and difficult. Metronidazole is effective in acute amebic dysentery, and appears to be the least toxic drug.[4] However, amebic liver abscess may occur in spite of treatment with this drug.[11] Side effects include acute confusion when taken with alcohol. Diiodohydroxyquin and iodochlorhydroxyquin are other chemotherapeutic agents which are effective, but can cause optic neuropathy.[4] Amebic liver abscess can be treated with choroquine or emetine, although the latter is potentially dangerous because of its toxicity.[4]

Diloxanide furoate appears to be the best drug in mild or asymptomatic amebiasis for eradicating cysts and eliminating symptoms.[4]

Prevention

Avoid Exposure. Water of uncertain purity can often be avoided or boiled. Raw fruits and vegetables should be washed thoroughly or avoided, when in endemic areas.[8]

Disinfection. Use of chlorinated tablets in drinking water is effective if sufficient time is allowed (up to 30 minutes).[8]

FREE-LIVING AMEBAS

Free-living amebas are not parasites of animals, but are found free-living in soil and water. Only rarely do free-living amebas cause disease in humans. The genera that have been associated with disease in humans are *Naegleria*, *Acanthamoeba*, and *Hartmanella*. In some reports some genus names are used interchangeably. Naegleriae have a flagellate stage whereas acanthamebae do not.

Disease due to free-living amebas is rare in the United States, with less than ten case reports of fatal disease before 1970. Retrospective study of autopsy materials indicates some cases have been missed at autopsy.[16,20,22] These organisms are important because of the possibility that anti-amebic chemotherapy might be life-saving in some patients if therapy were begun early enough.

The route of infection for naegleriae is thought to be through the nose, and most fatal cases have a history of swimming in lakes or rivers.[16,20] Outbreaks or second cases have been associated with common swimming areas in Virginia, California, and Czechoslovakia. Virulent naegleriae have been recovered from a cracked wall of an indoor swimming pool in Czechoslovakia where repeated outbreaks of amebic meningitis had occurred.[18] Free-living amebae (almost entirely *Hartmanella* (*Acanthamoebae*) species) have been recovered from the throats of normal individuals, and from the air.[19]

Most illnesses due to free-living amebas in the United States have been observed in Virginia, Florida, Texas, Georgia, and California.

Clinical Patterns of Illness

Primary Amebic Meningitis. The typical clinical pattern of illness is an acute meningoencephalitis or purulent meningitis, with headache, fever, and stiff neck.[16,22] Typically the spinal fluid shows 300 to several thousand white blood cells. Because the differential count of these cells is usually 80 percent to 90 percent segmented neurophils, and the spinal fluid glucose is low, a bacterial meningitis is usually suspected.

A subacute meningoencephalitis can be produced by acanthamoeba, whereas the acute meningoencephalitis is more likely to be due to naegleria.

Myocarditis can also be produced by these organisms,[16] and minor respiratory disease has been suspected as a manifestation of amebic infection.[22] Serious corneal infections by acanthamoebae have been reported.

Motile naegleria amebae have been observed in spinal fluid of patients with amebic meningoencephalitis, when the fluid has been examined carefully (Figure 15-3). The organism can be grown in tissue culture, where chemotherapy can be tested. Agar media also can be used if *E. coli* are added to enhance the growth of the ameba.[15] Amebic cysts can be found in the brain at autopsy.

Treatment

Treatment with anti-amebic drugs has been attempted in very few cases.[16] Amphotericin B, clotrimazole, and miconazole appear effective against naegleria in vitro.[17] Polymyxin B and pentamidine are somewhat effective against pathogenic acanthamoeba in vitro.[17] These drugs have not been adequately studied clinically. Recovery from acute amebic encephalitis after combined therapy with amphotericin B, miconazole, and rifampin has been reported.

GIARDIA LAMBLIA

Objectives

1. Describe the typical clinical pattern of giardiasis, including common sources of infection in the United States.
2. Describe the typical radiographic finding of giardiasis.
3. Describe the evidence that giardiasis produces diarrheal disease.

Definitions

Giardia lamblia is named for Giard, who studied the parasite, and Lambl, who first described it. It is a unicellular protozoan flagellate which has a trophozoite form and a cyst form.

Frequency and Importance

Giardiasis has only recently been widely accepted in the United States as a cause of disease. It is a relatively uncommon cause of diarrhea, but it is important because it can be mistaken for an intestinal malignancy, and therapy produces prompt relief of symptoms.

Most outbreaks have been associated with water. Travelers to foreign countries have sometimes been infected with giardia. Several cities in the United States have had outbreaks traced to contaminated water supplies.[31] Other outbreaks in the United States have been traced to drinking water from fecally contaminated mountain streams.

Giardiasis is increased in frequency in homosexual males,[30] suggesting sexual transmis-

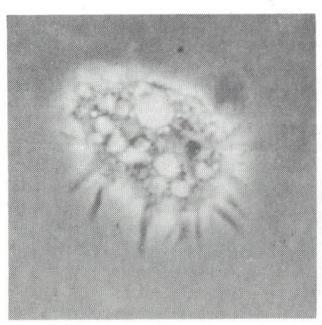
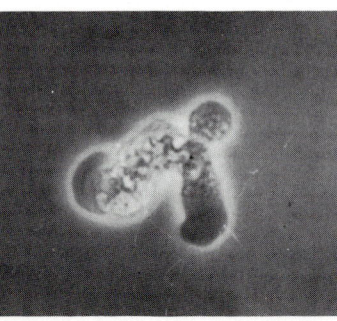

FIG. 15-3. *Trophozoites of free living amebae: (A) Hartmanella (Acanthamoeba) culbertsoni, (B) Naegleria aerobia. Trophozoites of Naegleria might be seen in spinal fluid in acute amebic meningoencephalitis. (Photo from Dr. Clyde G. Culbertson)*

sion in this group. Campers who drink untreated water from mountain streams are also at increased risk.

Clinical Patterns of Illness

In the United States, Giardia lamblia is an *uncommon* cause of:

Chronic Diarrhea. Malodorous, greasy, frothy stools, about five per day, for about 1 to 2 months, were observed in the Aspen, Colorado outbreak.[28] Weight loss, abdominal cramps, constipation alternating with diarrhea, flatulence, and chronic urticaria have also been described.[33] Eosinophilia is not found. In children, typical celiac syndrome can be produced, with foamy foul smelling stools, and abdominal distension.[25] Lactose intolerance can occur, secondary to the persistent diarrhea.

Profuse, watery diarrhea may be the presenting complaint.

Laboratory Diagnosis

Intestinal Roentgenograms. The intestine shows a characteristic edema of the duodenal and jejunal mucosa (Figure 15-4), which is not specific, and can also be produced by strongyloidiasis.[27]

Microscopic Examination of the Stool. Characteristic ova or trophozoite forms may be found in the stool. Giardia are best detected in the duodenum by aspiration or by passing a lead-weighted capsule on a string to which the giardia cling (Enterotest).

Biologic Characteristics of Clinical Interest

Evidence that Giardia Causes Disease. Outbreaks of chronic diarrheal disease have been observed, with recovery of the organism.[28] Statistical association of recovery of the parasite with diarrhea has been observed in a prospective study of fami-

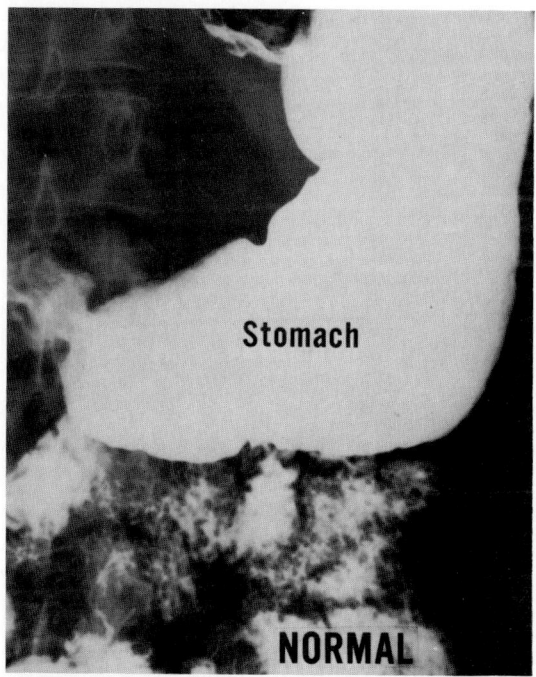
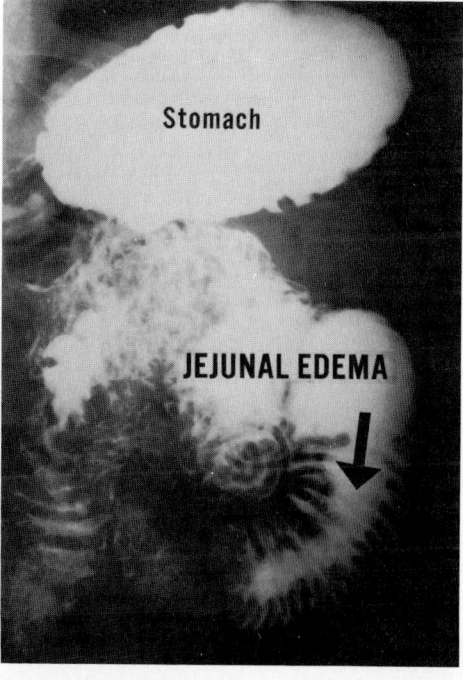

FIG. 15-4. *Barium examination of the stomach and intestine showing edema of the jejunum produced by giardiasis. The edematous jejunum is wider than normal and has deep edematous folds of mucosa (arrow). This patient acquired the disease during a skiing trip to Colorado. The diagnosis was confirmed by stool examination, but was first suspected by observation of this nonspecific jejunal edema. (Photo from Dr. Richard Logan)*

lies.³² Invasion of bowel mucosa by the protozoan has been demonstrated by intestinal biopsy, especially in patients with chronic malabsorption.²⁴ Response to chemotherapy with agents such as quinacrine is regularly demonstrated.²⁸,³³

Failure to Produce Experimental Disease. When volunteers were fed up to one million cysts, spontaneous disappearance of the organism occurred in most cases.²⁹ No significant clinical disease occurred in any volunteers, although a few volunteers had a few days of loose stools. This suggests that a very high parasite dose or special host factors are necessary to produce clinical disease.

Relation to Immunological Deficiency Disease. Chronic diarrhea associated with giardiasis, with response to therapy, occurs frequently in individuals with immunoglobulin deficiencies.²⁶

Person-to-Person Transmission. Although most outbreaks have occurred in association with contaminated water, person-to-person transmission has been documented in day care centers,²⁶ and among male homosexuals.

Prevention and Treatment

Uncertain water supplies should be avoided. Treating suspected water with iodine tablets or boiling for 10 minutes kills the cysts, but city chlorination procedures may not be effective.

Quinacrine is effective in stopping symptoms promptly.²⁸,³³,³⁴ Metronidazole is also effective, but is unlicensed for this use.³⁴ For children, furazolidone suspension is useful.³⁴

TOXOPLASMA GONDII

Objectives

1. Describe the classical clinical findings of congenital toxoplasmosis.
2. Describe the usual clinical picture of acquired toxoplasmosis.
3. Discuss the role of cats in the transmission of the disease.
4. Discuss the methods available for the laboratory confirmation of toxoplasma infection.
5. Discuss chemotherapy of toxoplasmosis.

Definitions

Toxoplasmosis is infection with the protozoan *Toxoplasma gondii*. The gondi is an African rodent which resembles a rat.

Frequency and Importance

Asymptomatic infection with toxoplasma is relatively frequent, as about 20 percent of the adult population have low titers of antibodies to the parasite.³⁹ Acquired clinical disease is rarely serious, except when it involves the eye, or occurs in the immunosuppressed patient. Congenital toxoplasmosis is an important cause of brain or eye damage in infancy, although it is very uncommon.

Clinical Patterns of Illness

Toxoplasma gondii is a *common* cause of:

Congenital chorioretinitis can occur as an isolated finding or in association with brain and liver disease in the newborn. Isolated chorioretinitis may be detected in the first few weeks of life by a routine fundiscopic exam, or much later when a defect in vision is noted (Fig. 15-5). Chorioretinitis due to toxoplasmosis also occurs in older children and adults. It is thought usually to be due to a reactivation of congenital disease, but acquired chorioretinitis has been clearly documented.⁴³

Heterophile-negative infectious mononucleosis syndrome has been best studied in accidental laboratory infections and in outbreaks due to

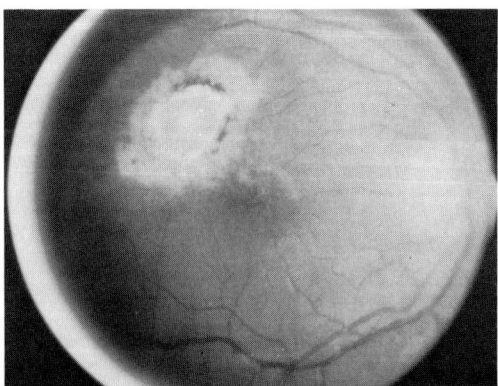

FIG. 15-5. *Chorioretinitis due to toxoplasmosis. As shown here, the pigmented lesion typically involves the macula and affects vision. (Photo from Dr. Thomas France)*

eating undercooked meat.[44,50] The incubation period is about 7 to 18 days. The illness resembles infectious mononucleosis, except for the absence of pharyngitis, but the heterophil is negative. Usual clinical findings include fever, headache, and generalized lymphadenopathy. Splenomegaly occurs in many patients; myalgia is prominent early and persists for weeks; and a transient red macular rash occurs in some individuals.[44] Lymphocytosis, sometimes with atypical lymphocytes, is usually present.

Neonatal Jaundice and Brain Damage. In utero infections are sometimes manifested during the newborn period by jaundice due to liver involvement.[38] Brain involvement often occurs, and is sometimes severe, with intracranial calcifications and microcephaly or hydrocephalus. Mental retardation also occurs, frequently with no other manifestations of disease, as has been demonstrated by IQ tests at a later age, in comparison to control infants.[35]

Toxoplasma gondii is a rare cause of myocarditis,[55] acquired chorioretinitis,[46] atypical pneumonia, and acute encephalitis.[50] Encephalitis or generalized disseminated toxoplasmosis may occur in patients with malignancy or patients treated with immunosuppressive drugs.[53] Concurrent infection with herpes simplex virus or cytomegalovirus is often found in such patients. Normal individuals rarely have disseminated toxoplasmosis.

Laboratory Approach

Serum Antibodies. The Sabin dye test is often used.[50] Acute and convalescent phase sera are necessary. Very high titers (greater than 1:1000) in the mother and infant are suggestive of congenital infection, but are not conclusive. The dye test depends on the use of living organisms, which normally are stained by the dye methylene blue, unless the organisms are damaged by the antibodies in the patient's serum.[39]

An indirect fluorescent antibody test has been developed which uses fluorescein-tagged antiserum specific for IgM.[49] This test is helpful, but not perfect, for detecting infections of the newborn, and recent infections in the adult. It utilizes formalin-killed toxoplasma organisms which are fixed on a slide (Fig. 15-6). The patient's serum is then added and any specific antibodies present adhere to the parasites. The patient's serum is then washed off, and fluorescent-labeled

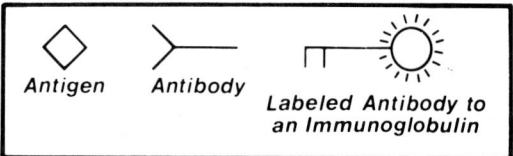
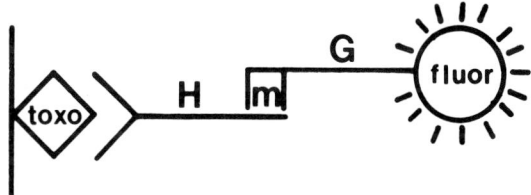

FIG. 15-6. *Indirect fluorescent antibodies against Toxoplasmosa gondii. toxo=toxoplasma organisms killed by formalin and fixed to a glass slide; H=human serum to be tested for the antibody; m=IgM in the human serum. G=goat antiserum prepared against human IgM and conjugated with fluorescein.*

antihuman IgM serum prepared in goats is added. If any IgM antibody has reacted with the toxoplasma, the fluorescent-labeled anti-IgM adheres to the toxoplasma–antibody complex, and the parasites have a green fluorescence under the proper wave length of light. It is called indirect fluorescent antibody (IFA) because the fluorescent-labeled antibody detects human antibody and not the toxoplasma antigen. This IFA test is also very useful in the detection of congenital syphilis (Chap. 9).

The complement fixation test is useful in some circumstances. The complement fixation test is usually positive in infants less than 2 years of age with severe congenital disease, but becomes negative in about 5 years.

An ELISA serologic test (Fig. 13-7) is under investigation and may be prove valuable.[54]

Culture of the Organism. This is sometimes possible by injecting mice with material from lymph nodes, brain and skeletal muscle.[48] The organism has been recovered in cell cultures, but the sensitivity of the method has not been adequately studied. Toxoplasmosis is a hazard to laboratory personnel, and attempts at culture are not recommended, except in laboratories with experienced personnel.

Histologic Diagnosis. The organisms can sometimes be found in biopsies of a lymph node or an autopsy. The trophozoite is seen microscopically as a crescent, with one end rounded and the other end pointed. Cysts are seen after the acute stages of the illness.

Skin Test. This is of little diagnostic value, because it does not become positive until several months to a year after infection.

Biologic Characteristics of Clinical Interest

Sources of Infection. Cats and members of the cat family appear to be the only animals in which the sexual phase of the toxoplasma life cycle can occur, and cat feces may be an important source of human infections.[37,41,45] In the past, these toxoplasma organisms had been mistaken for harmless *Isospora* species when found in cat feces. Either trophozoites or cysts from infected mice can be a source of infection for cats. Experimental toxoplasmosis can be produced in the chimpanzee by feeding the feces of experimentally infected cats.[37]

The cysts excreted in cat feces appear to remain infective in the soil for at least a year, and are a possible source of infection for herbivorous animals. Carnivorous animals and cannibalistic animals such as rats also perpetuate the infection by ingestion of infected animals. Inadquately cooked meat appears to be a source of the infection in humans.

Many mammals, rodents, and birds are naturally infected. The occurrence of toxoplasmosis in herbivorous animals such as sheep is hard to explain on the basis of exposure to cats or raw meat, and an alternative, unknown mechanism of spread may be involved.[40]

Growth in Cell Cultures. Monkey kidney cells will support the growth of toxoplasma, and have been used to study the effect of various therapeutic agents.

Treatment

Laboratory infections with toxoplasma are observed occasionally and have been treated with sulfadiazine and pyrimethamine (Daraprim), which have been shown to be effective in experimental infections in animals.[37,39,42] Folinic acid or yeast extract is usually used also to decrease the folic acid inhibition effect of the pyrimethamine in the human, without decreasing the effect on the organism. In one study of adults with chorioretinitis, pyrimethamine appeared to be more effective than a placebo in terms of clinical improvement.[47] In immunosuppressed patients with toxoplasmosis, sulfadiazine or pyrimethamine appeared to be effective.[52] In a study of subclinical congenital toxoplasmosis, infants treated with pyrimethamine and sulfadiazine in the first week of life had a higher IQ at 3 to 4 years of age than similar untreated infants.[35] At the present time, drug therapy is usually recommended in the above situations because of the risk to vision or intellectual function, even though efficacy is not clearly established. Treatment of uveitis or congenital toxoplasmosis would have no effect on damage already done by the infection.

Subscleral injection of clindamycin has some therapeutic value in chorioretinitis in experimentally infected rabbits.[51]

Prevention

Careful handwashing or wearing gloves has been recommended when changing a cat's litter pan, especially if the cat may catch infected mice. Meat should be thoroughly cooked. Pregnant women should avoid exposure to cat feces and undercooked meat. In France, where toxoplasmosis is much more frequent than in the United States, women have been followed during pregnancy for serologic evidence of acquired toxoplasmosis. In a controlled study, some of the pregnant women were treated with spiramycin when a rise in antibody titer occurred. The drug appeared to reduce the frequency of congenital infection but not of clinical disease in the newborn infant.[36] This therapy has not been studied in the United States because treatment of the pregnant woman with a folic acid inhibitor might be a risk to the fetus.

In general, once a woman has given birth to an infant with congenital toxoplasmosis, the risk of a second such congenital infection is exceedingly low, and the mother need not be given any chemotherapy.[36]

Cysts in meat can be killed by heating to 60 °C or freezing to −20 °C.[41]

MALARIA

Objectives

1. Discuss how malaria might occur in the United States in an individual who has not left the country.
2. Discuss the typical clinical picture of malaria.

3. Describe how the diagnosis of malaria can be confirmed by the laboratory.

4. Indicate the major areas of the world where chemoprophylaxis by visitors is indicated.

Definitions

The word malaria is derived from the Latin words for bad air. Malaria is best defined as an infection due to a parasite called a plasmodium. This parasite has part of its life cycle in the mosquito and part in humans (Fig. 15-7). It multiplies in human erythroid cells and typically produces chills and fever when the red blood cells rupture and release the parasite. The parasite also can develop in the liver and spleen in an exoerythrocytic or tissue form.

Malaria Fever Patterns. Fever occurring every other day is called tertian. Fever occurring every day is called quotidian. Quartan fever occurs every third day. The frequency of occurrence of fever is not useful to distinguish different species of malaria, since the patient often is infected with parasites which mature out of phase with each other, so that the fever occurs daily.

Species

There are four species of plasmodia that infect humans. *P. vivax* typically produces benign tertian malaria. It is called benign because it is rarely fatal, though it often relapses. Almost all fatal cases are due to *P. falciparum* which typically produces malignant tertian malaria. The severity of falciparum malaria can be explained by the involvement of all ages of erythrocytes by this species. *P. ovale* and *P. malariae* are rare in the United States. Other species, particularly plasmodia of monkeys, occasionally infect humans.[66]

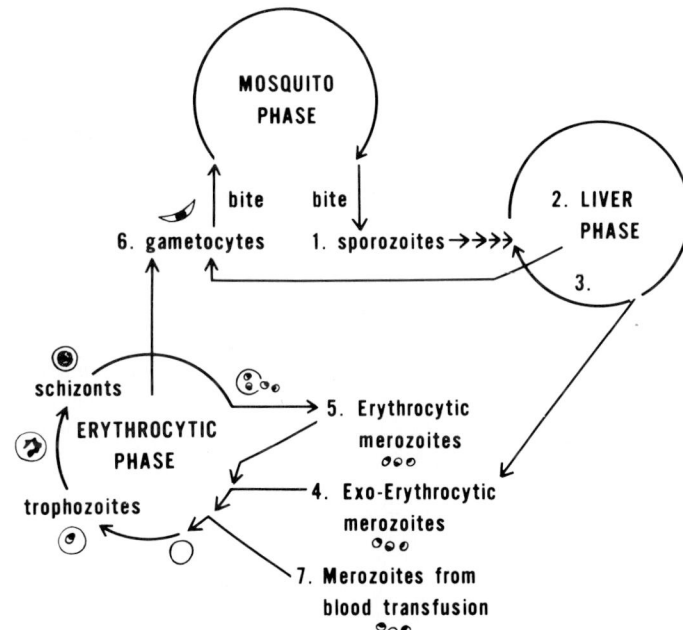

FIG. 15-7. *Malaria life cycle, simplified by omitting details of mosquito phase. 1. Mosquito bite injects sporozoites into plasma, which becomes noninfective within an hour, indicating entry of sporozoites into tissue (liver or spleen). 2. In the tissue or liver phase, sporozoites divide and develop into exoerythrocytic merozoites, which are released into plasma. 3. A secondary or persistent liver phase occurs in all species except P. falciparum and consists of invasion of liver cells by exorthyrocytic merozoites, which recycle and produce more merozoites. This phase may persist for years. 4. Merozoites released into plasma enter erythrocytes and develop into motile ameboid trophozoites and then into schizonts, which split and release erythrocytic merozoites into plasma with resulting chills and fever. 5. These erythrocytic merozoites can invade erythrocytes but cannot invade the liver and develop there. 6. Male and female gametocytes are produced after several cycles in the liver or erythrocytes and are the only forms which can develop in the mosquito. 7. Merozoites can also be introduced by blood transfusion, and can infect erythrocytes, but cannot invade the liver, so that a self-limiting disease of about 2 weeks' duration results.*

Frequency and Importance

Before 1962, there were less than 100 cases per year of malaria in the United States.[59] In the late 1960s, during the Viet Nam war, the number of cases per year of malaria with onset of symptoms in the United States increased, reached a peak of 4200 cases in 1970, but fell to 547 cases in 1977.[59] Most of these were imported cases, as described in the definitions given below.

Routine prophylaxis is not recommended for the usual traveler to malarious areas, unless the traveler plans to reside there for more than 1 or 2 weeks. Europe, Russia, Japan, Australia, the United States and Canada can be regarded as malaria-free areas.

Classification by Source

Malaria is classified in terms of its geographic origin.[59]

Imported Malaria. This is malaria acquired outside of a specific area. In 1968, all but 12 of 2610 cases of malaria in the United States were imported, most occurring in military personnel recently returned to the United States from southeast Asia.

Introduced Malaria. This is acquired by mosquito transmission from an imported case in an area where malaria is not a regular occurrence. In 1968, there were five cases of introduced malaria reported in the United States, usually in the vicinity of military bases, where anopheles mosquitoes were found.[59] In 1974, 12 introduced malaria cases occurred in California.[67] All introduced cases were *P. vivax*.

Two species of mosquitoes are found in the United States which can support the growth of human malaria parasites: *Anopheles freeborni* on the West coast and *A. quadrimaculatus*, on the East coast.

Induced Malaria (Transfusion Malaria). This is acquired by blood transfusions, platelet transfusions, or similar transferral of blood by means other than a mosquito. In 1968, there were seven induced cases of malaria in the United States. The donor typically had been to a tropical area and was asymptomatic. Malaria antibodies in such donors have usually been detectable by retrospective study, by use of an indirect fluorescent technique.

Indigenous Malaria. This is acquired by mosquito transmission in a geographic area where malaria is regularly found. Malaria was indigenous to most of the United States in the 1800s.

Relapsing Malaria. This is malaria acquired in the past, that relapses after an interval longer than the ordinary periodicity of the disease. Most military personnel with imported malaria in 1968 had vivax malaria, which has a relapse rate of 10 to 30 percent.[70]

Congenital Malaria.[68] In this case, clinical manifestations (especially fever) begin at 4 to 8 weeks of age.

Cryptic Malaria. This is malaria of unknown source.

Clinical Patterns of Illness

In endemic areas, *Plasmodium* species can cause:

Classical Acute Malaria. Intermittent episodes of bed-shaking chills, very high fever, drenching sweats, with headache, myalgia, and prostration is the classical pattern.[62] In falciparum malaria, the episodes may occur daily.[62] Splenomegaly is common, and occurs in about one-half of the patients. Hepatomegaly. anemia, leukopenia, thrombopenia, elevated serum bilirubin, and elevated SGOT levels each occur in about one-third of patients.

Malignant Malaria. Malignant or pernicious malaria is a form of malaria due to *P. falciparum*, which is dangerous and life threatening, primarily because of agglutination of red blood cells and production of thrombi in areas of vital importance, particularly the brain and liver.[56,62] The complications of malignant malaria include liver failure with gastrointestinal hemorrhage; cerebral involvement characterized by extremely high fever and convulsions, delirium and death; renal involvement, with hematuria, severe intravascular hemolysis, and extremely severe anemia, called blackwater fever; and progressive pulmonary edema.[62]

Splenic rupture is a rare complication of malaria due to any species.

Asymptomatic or Subclinical Infection

Subclinical or asymptomatic infection is commonly due to incomplete suppression by a prophylactic regimen. However, it may occur without prophylaxis.[62]

Laboratory Diagnosis

Blood Smear. The diagnosis is usually made by microscopic examination of a thick blood film. However, in acute illnesses, the parasitemia is often heavy enough so that the parasites can be seen in an ordinary Wright-stained or Giemsa-stained thin smear. Unless the technician has experience with malarial smears, a set of colored plates of the various forms of the parasite will be helpful.

Babesia is a genus of animal parasites that resemble plasmodia in blood smears. Occasionally humans are infected, usually by the bite of a tick, which is the usual vector for animals. The disease is endemic to Cape Cod, Nantucket and Martha's Vineyard in Massachusetts. Pentamidine is probably more effective therapy than antimalarial drugs such as chloroquinine.[65]

Serum Antibodies. Various serologic tests are available at various reference centers. Serologic diagnosis is most useful in the retrospective diagnosis of malaria in a blood donor. The fluorescent antibody technic appears to be the best serologic method.[66]

Culture. P. falciparum has recently been maintained in continuous culture of human erythrocytes.[69]

Nonspecific Laboratory Findings. The VDRL is positive in about one percent of patients.[61] Anemia, thrombopenia, leukopenia, and SGOT elevation may be present.[62] The anemia appears to be caused in part by a complement-mediated immune process, in addition to the destruction by parasites, and it may be persistent or sudden with scanty parasitemia.[70,4]

Biologic Characteristics of Clinical Interest

Life Cycle. The initial stage of infection in man has a primary tissue phase during which the sporozoites mature. After several weeks, a secondary tissue phase occurs, which is essential for the infection to become chronic or latent. Since P. falciparum has no secondary tissue phase, chronic infection with this species does not occur.

In transfusion malaria, no sporozoites are present in the blood. Therefore, chronic infection does not occur after transfusion, regardless of the infecting species.

Sickle Cell Disease. Humans with sickle cell trait, a genetic erythrocyte defect, survive better than normals in endemic areas of malaria.[66] Negroes are less frequently infected by P. vivax.[61]

Treatment

Chemotherapy. Quinine will control the clinical disease in all forms of malaria, but is rarely used except for resistant falciparum strains.[57,62] Pyrimethamine is often given in addition to quinine if the patient has come from a resistant falciparum area, such as parts of Brazil, Panama, Venezuela, and Asia,[57] and recently from Africa.[63]

Chloroquine is effective against sensitive P. falciparum. Chloroquine-resistant strains have emerged and are a serious clinical problem.[66] Primaquine is necessary to eliminate the tissue forms, as discussed in the Center for Disease Control guidelines for malaria prophylaxis.[58]

Prevention

Exposure to mosquitoes should be avoided as much as possible when in an endemic area. Chemoprophylaxis is often not considered or improperly done. Chemosuppression can be done with chloroquine phosphate, 500 mg once a week, for adults, beginning 2 weeks before and while in endemic areas, and for 4 to 6 weeks thereafter.[58] This is active only against the erythrocytic stages, so primaquine should be considered on an individual basis after the traveler returns to the United States.[58]

When travelling to areas where chloroquine-resistant falciparum malaria occurs, prophylaxis with a fixed combination of pyrimethamine and sulfadoxine is recommended.[58] However, this preparation is not licensed or available in the U.S., so it must be obtained when the traveller arrives in the endemic area.

Screening of blood donors should be done to exclude anyone visiting malaria areas within 2 years.[60]

Immunization against malaria is under investigation.[64] If infected mosquitoes are x-irradiated and then allowed to bite volunteers, immunity appears to be produced to the sporozoites (Fig. 15-7). The volunteer is resistant to challenge with sporozoite-infected mosquitoes, but not to challenge with the merozoites in a blood transfusion.

PNEUMOCYSTIS CARINII

This organism is named for the cystic appearance of the lung at autopsy, and for Carini, who submitted the original specimen for study.[76] The organism is believed to be a protozoan on the basis of electron microscopic demonstration of a life cycle, and on the basis of clinical response to an antiprotozoal drug.[76]

In 1967, the Parasitic Disease Drug Service, Center for Disease Control, Atlanta, Georgia, became the sole supplier in the United States of pentamidine, which is used to treat the disease.[81] Analysis of pathologically proved cases by this service indicates that there were an average of 65 cases a year, with infants less than a year of age with a primary immune deficiency disease constituting the most frequent group infected.[81]

The usual clinical illness produced by pneumocystis is a diffuse interstitial pneumonia.[80] Typically, the pneumonia occurs in a compromised host, especially infants with immune globulin deficiency or patients with leukemia in remission. Tachypnea, cyanosis, and hypoxemia due to an alveolar diffusion block are characteristically present. Often the fever is not high, and ascultation usually does not reveal significant abnormal findings. Marked leukocytosis or eosinophilia can occur. Focal pneumonia or normal chest roentgenogram (with tachypnea) may also occur.

Premature or debilitated infants are also especially susceptible, and epidemics have been observed in Europe.[80] Infants without prematurity or without immune deficiency rarely get pneumocystis pneumonia.

Asymptomatic infection can occur, as the organism has been observed as an incidental finding at autopsy.[80]

Laboratory diagnosis is made by histologic examination. An open lung biopsy can be done to obtain an adequate specimen for histologic evaluation. However, the organism has been observed in toluidine blue or methanamine silver-stained smears of pharyngeal secretions, sputum, gastric aspirates, percutaneous needle biopsy, or endobronchial brush biopsy.[71] Transtracheal aspirations are helpful to detect the organism. Fluorescent antibody methods may be available in some laboratories. The alveoli are filled with a foamy eosinophilic material, and a plasma cell infiltrate is common (Fig. 15-8). Dissemination to tissues other than the lung is extremely rare.[76]

No routine serologic test is available. However, the organism can be propagated for several passages in primary embryonic chick embryo lung,[78] and human fetal lung, so that a source of antigen is available for the development of serologic tests. Using serologic studies based on antigens propagated in these cultures, two-thirds of normal American children were found to have antibodies by 4 years of age.[79] Using fluorescent antibody methods, nearly 100 percent of children in the Netherlands were found to have serum antibodies by 2 years of age.[77] These studies indicate that infection with this organism is very common, but serious disease occurs almost exclusively in compromised hosts.

Human-to-human transmission is suggested by the outbreaks in nurseries, and outbreaks among compromised patients in close contact.[72]

Pentamidine isethionate was once the only effective therapy. However, trimethoprim–sulfisoxazole combination is as effective therapy and is less toxic than pentamidine.[74]

Prevention of pneumocystis pneumonia in immunosuppressed hosts is possible with chronic administration of trimethoprim–sulfamethoxazole combination.[73]

TRICHOMONAS VAGINALIS

Objectives

1. Describe the laboratory methods used for detection of *Trichomonas vaginalis*.
2. Describe the therapy of trichomonas vaginitis.

Clinical Information

Several species of *Trichomonas*, a flagellated protozoan, are found in humans. *T. vaginalis* is a frequent cause of vaginitis in adult women, and an occasional cause of urethritis in adult men.[82] Typically, there is a frothy, green discharge. Other *Trichomonas* species rarely cause extragenital disease, including mixed flora meningitis, and mixed flora empyema.[84]

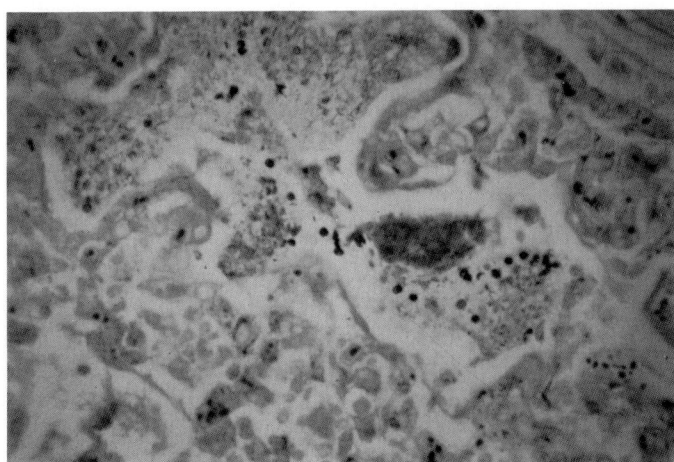

FIG. 15-8. Pneumocystis carinii *appear as dark oval bodies when stained with special silver stain. Foamy material is also seen throughout the several alveoli in this section of the lung. (Photo from Dr. Enid Gilbert)*

Laboratory diagnosis is best made by observation of the parasite on an unstained wet mount of vaginal or urethral secretions. Culture is possible. Treatment with a single dose of 2 gms. of metronidazole is effective,[83] although some strains are resistant. The consort should also be treated, and alcohol should be avoided for 48 hours. Metronidazole does not interfere with the ability to culture *N. gonorrheae*. Nitrimidazine also is effective, using a dose of 250 mg. twice a day for a week.[85]

REFERENCES

Entamoeba Histolytica

1. Adams, E.B., and MacLeod, I.N.: Invasive amebiasis: I. Amebic dysentery and its complications. II. Amebic liver abscess and its complications. Medicine, 56:315–334, 1977.
2. Center for Disease Control: Morbidity Mortality Annual Rep. 19–21, 26: Annual supplements. Summaries 1970–1972, 1977. pp. 4, 4, 4, 2.
3. Juniper, K., Jr.: Acute amebic colitis. Am. J. Med., 33:377–386, 1962.
4. Krogstad, D.J., Spencer, H.C., Jr., and Healy, G.R.: Current concepts in parasitology. Amebiasis. New Engl. J. Med., 298:262–265, 1978.
5. LeMaistre, C.A., et al.: Studies of a waterborne outbreak of amebiasis, South Bend, Indiana. I. Epidemiological aspects. Am. J. Hyg., 64:30–45, 1956.
6. Orbison, J.A., Reever, N., Leedham, C.L., Blumberg, J.M.: Amebic brain abscess. Medicine, 30:247–282, 1951.
7. Schmerin, M.J., Gelston, A., and Jones, T.C.: Amebiasis. An increasing problem among homosexuals in New York City. J.A.M.A., 238: 1386–1387, 1977.
8. Stamm, W.P.: Amoebic aphorisms. Lancet, 2:1355–1356, 1970.
9. Lushbaugh, W.B., et al.: Isolation of a cytotoxin–enterotoxin from *Entamoeba histolytica*. J. Infect. Dis., 139:9–17, 1979.
10. Stuiver, P.C., and Goud, T.J.L.M.: Corticosteroids and liver amoebiasis. Br. Med. J., 3:394–395, 1978.
11. Weber, D.M.: Amebic abscess of liver following metronidazole therapy. J.A.M.A., 216: 1339–1340, 1971.
12. Wittner, M., and Rosenbaum, R.M.: Role of bacteria in modifying virulence of *Entamoeba histolytica*. Am. J. Trop. Med., 19:755–761, 1970.
13. Wruble, L.D., Duckworth, J.K., Duke, D.D., and Rothchild, J.A.: Toxic dilatation of the colon in a case of amebiasis. New Engl. J. Med., 275:926–928, 1966.
14. Yang, J., and Scholten, T.: *Dientamoeba fragilis*: a review with notes on its epidemiology, pathogenicity, mode of transmission, and diagnosis. Am. J. Trop. Med. Hyg., 26:16–22, 1977.

Free-Living Amebas

15. Culbertson, C.G., Ensminger, P.W., and Overton, W.M.: Pathogenic *Naegleria* sp.—study of a strain isolated from human cerebrospinal fluid. J. Protozool., 15:353–363, 1968.
16. Duma, R.J., Ferrell, H.W., Nelson, E. C., Jones, M.M.: Primary amebic meningoencephalitis. New Engl. J. Med., 281:1315–1323, 1969.
17. Duma, R.J., and Finley, R.: In vitro susceptibility of pathogenic *Naegleria* and *Acanthamoeba* species to a variety of therapeutic agents. Antimicrob. Agents Chemother., 10:370–376, 1976.

18. Kadlec, V., Cerva, L., and Skvárová, J.: Virulent *Naegleria fowleri* in an indoor swimming pool. Science, 201:1025, 1978.
19. Kingston, D., and Warhurst, D.C.: Isolation of amoebae from the air. J. Med. Microbiol., 2:27–36, 1969.
20. Neva, F.A.: Amebic meningoencephalitis—a new disease? New Engl. J. Med., 282:450–452, 1970.
21. Robert, V.B., and Rorke, L.B.: Primary amebic encephalitis, probably from *Acanthamoeba*. Ann. Intern. Med., 79:174–179, 1973.
22. Symmers, W.ST.C.: Primary amoebic meningoencephalitis in Britain. Br. Med. J., 44:449–454, 1969.

Giardia Lamblia

23. Black, R.E., Dykes, A.C., Sinclair, S.P., and Wells, J.G.: Giardiasis in day care centers: evidence of person-to-person transmission. Pediatrics, 60:486–491, 1977.
24. Brandborg, L.L., et al.: Histological demonstration of mucosal invasion by *Giardia lamblia* in man. Gastroenterology, 52:143–150, 1967.
25. Burke, J.A.: Giardiasis in childhood. Am. J. Dis. Child., 129:1304–1310, 1975.
26. Hughes, W.S., Cereda, J.J., Hotzapple, P., and Brooks, F.P.: Primary hypogammaglobulinemia and malabsorption. Ann. Intern. Med., 74:903–910, 1971.
27. Marshak, R.H., Ruoff, M., and Linder, A.E.: Roentgen manifestations of giardiasis. Am. J. Roentgenol., 104:557–560, 1968.
28. Moore, G.T., et al.: Epidemic giardiasis at a ski resort. New Engl. J. Med., 281:402–407, 1969.
29. Rendtdorff, R.C.: Experimental transmission of human intestinal protozoan parasites. II. *Giardia lamblia* cysts given in capsules. Am. J. Hyg., 59:209–220, 1954.
30. Schmerin, M.J., Jones, T.C., and Klein, H.: Giardiasis: association with homosexuality. Ann. Intern. Med., 88:801–803, 1978.
31. Shaw, P.K., et al.: A community outbreak of giardiasis with evidence of transmission by a municipal water supply. Ann. Intern. Med., 87:426–432, 1977.
32. Wanner, R.G., Atchley, F.O., and Wasley, M.A.: Association of diarrhea with *Giardia lamblia* in families observed weekly for occurrence of enteric infections. Am. J. Trop. Med., 12:851–853, 1963.
33. Webster, B.H.: Human infection with *Giardia lamblia*: analysis of 32 cases. Am. J. Dig. Dis., 3:64–71, 1958.
34. Wolfe, M.S.: Giardiasis. New Engl. J. Med., 298:319–321, 1978.

Toxoplasma Gondii

35. Alford, C.A., Jr., Stagno, S., and Reynolds, D.W.: Congenital toxoplasmosis: clinical, laboratory, and therapeutic considerations, with special reference to subclinical disease. Bull. N.Y. Acad. Med., 50:160–181, 1974.
36. Desmonts, G., and Couvreur, J.: Congenital toxoplasmosis. A prospective study of 378 pregnancies. New Engl. J. Med., 290:1110–1116, 1974.
37. Draper, C.C., et al.: Experimental toxoplasmosis in chimpanzees. Br. Med. J., 2:375–378, 1971.
38. Eichenwald, H.F.: A study of congenital toxoplasmosis. With particular emphasis on clinical manifestations, sequelae, and therapy. In Siim, J.C. (ed.): Human Toxoplasmosis. Baltimore, Williams and Wilkins, 1959.
39. Feldman, H.A.: Toxoplasmosis. New Engl. J. Med., 279:1370–1375, 1431–1437, 1968.
40. Feldman, H.A.: Toxoplasmosis: an overview. Bull. N.Y. Acad. Med., 50:110–127, 1974.
41. Frenkel, J.K., Dubey, J.P., and Miller, N.L.: *Toxoplasma gondii* in cats: fecal stages identified as coccidian oocysts. Science, 167:893–896, 1970.
42. Frenkel, J.K., Weber, R.W., and Lunde, M.N.: Acute toxoplasmosis. J.A.M.A., 173:1471–1476, 1960.
43. Gump, D.W., and Holden, R.A.: Acquired chorioretinitis due to toxoplasmosis. Ann. Intern. Med., 90:58–60, 1979.
44. Kean, B.H., Kimball, A.C., and Christensen, W.N.: An epidemic of acute toxoplasmosis. J.A.M.A., 208:1002–1004, 1968.
45. Krick, J.A., and Remington, J.S.: Toxoplasmosis in the adult—an overview. New Engl. J. Med., 298:550–553, 1978.
46. Masur, H., Jones, N.C., Lempert, J.A., and Cherubini, T.D.: Outbreak of toxoplasmosis in a family and documentation of acquired retinochoroiditis. Am. J. Med., 64:396–402, 1978.
47. Perkins, E.S., Smith, C.H., and Schofield, P.B.: Treatment of uveitis with pyrimethamine (Daraprim). Br. J. Ophthalmol., 40:577–586, 1956.
48. Remington, J.S., and Cavanaugh, E.N.: Isolation of the encysted form of *Toxoplasma gondii* from human skeletal muscle and brain. New Engl. J. Med., 273:1308–1310, 1965.
49. Remington, J.S., and Desmonts, G.: Congenital toxoplasmosis: variability in the IgM-fluorescent antibody response and some pitfalls in diagnosis. J. Pediatr., 83:27–30, 1973.
50. Remington, J.S., Jacobs, L., and Kaufman, H.E.: Toxoplasmosis in the adult. New Engl. J. Med., 262:180–186, 237–241, 1960.
51. Tabbara, K.F., et al.: Clindamycin in chronic toxoplasmosis. Arch. Ophthalmol., 97:542–544, 1979.
52. Townsend, J.J., Wolinsky, J.S., Baringer, J.R., and Johnson, P.C.: Acquired toxoplasmosis.

A neglected cause of treatable nervous system disease. Arch. Neurol., 32:335–343, 1975.

53. Vietzke, W.M., Gelderman, A.H., Grimley, P.M., and Valsamis, M.P.: Toxoplasmosis complicating malignancy. Experience at the National Cancer Institute. Cancer, 21:816–827, 1968.

54. Walls, K.W., Bullock, S.L., and English, D.K.: Use of the enzyme-linked immunosorbent assay (ELISA) and its microadaption for the serodiagnosis of toxoplasmosis. J. Clin. Microbiol., 5:273–277, 1977.

55. Wertlake, P.T., and Winter, T.S.: Fatal toxoplasma myocarditis in an adult patient with acute lymphocytic leukemia. New Engl. J. Med., 273:438–440, 1965.

Malaria

56. Brooks, M.H., Kiel, F.W., Sheehy, T.W., and Barry, K.G.: Acute pulmonary edema in falciparum malaria. New Engl. J. Med., 279:732–736, 1968.

57. Butler, T., Warren, K.S., and Mahmoud, A.A.F.: Algorithms in the diagnosis and management of exotic diseases. XIII. Malaria. J. Infect. Dis., 133:721–726, 1976.

58. Center for Disease Control: Chemoprophylaxis of malaria. Morb. Mort. Week. Rep., 27(Suppl.):81–90, 1978.

59. ———. Malaria surveillance. Annual Summary 1977. Issued Sept., 1978. p. 3.

60. Chojmacki, R.E., Brazinsky, J.H., and Barret, O'N., Jr.: Transfusion-introduced falciparum malaria. New Engl. J. Med., 279:984–985, 1968.

61. Fisher, G.U., Gordon, M.P., Lobel, H.O., and Runcik, K.: Malaria in soldiers returning from Vietnam. Epidemiologic, therapeutic, and clinical studies. Am. J. Trop. Med., 19:27–39, 1970.

62. Heineman, H.S.: The clinical syndrome of malaria in the United States. A current review of diagnosis and treatment for American physicians. Arch. Intern. Med., 129:607–616, 1972.

63. Kean, B.H.: Chloroquine-resistant falciparum malaria from Africa. J.A.M.A., 241:395, 1979.

64. Miller, L.H.: Current prospects and problems for a malaria vaccine. J. Infect. Dis., 135:855–864, 1977.

65. Miller, L.H., Neva, F.A., and Gill, F.: Failure of chloroquine in human babesiosis (Babesia microti). Ann. Intern. Med., 88:200–202, 1978.

66. Neva, F.A.: Malaria—Recent progress and problems. New Engl. J. Med., 277:1241–1252, 1967.

67. Singel, M., Shaw, P.K., Lindsay, R.C., and Roberto, R.R.: An outbreak of introduced malaria in California possibly involving secondary transmission. Am. J. Trop. Med. Hyg., 26:1–9, 1977.

68. Thompson, D., Pegelow, C., Underman, A., and Powars, D: Congenital malaria: a rare cause of splenomegaly and anemia in an American infant. Pediatrics, 60:209–212, 1977.

69. Trager, W., and Jensen, J.B.: Human malaria parasites in continuous culture. Science, 193:673–675, 1976.

70. Waterhouse, B.E., and Riggenbach, R.D.: Malaria. Potential importance to civilian physicians. J.A.M.A., 202:683–685, 1967.

70A. Woodruff, A.W., Ansdell, V.E., and Pettitt, L.E.: Cause of anaemia in malaria. Lancet, 1:1005–1057, 1979.

Pneumocystis Carinii

71. Chan, H., et al.: Comparison of gastric contents to pulmonary aspirates for the cytologic diagnosis of Pneumocystis carinii pneumonia. J. Pediatr., 90:243–244, 1977.

72. Chusnid, M.J., and Heyrman, K.A.: An outbreak of Pneumocystis carinii pneumonia at a pediatric hospital. Pediatrics, 62:1031–1035, 1978.

73. Hughes, W.T., et al.: Successful chemoprophylaxis for Pneumocystis carinii pneumonitis. New Engl. J. Med., 297:1419–1426, 1977.

74. Hughes, W.T., et al.: Comparison of pentamidine isethionate and trimethoprim-sulamethoxazole in the treatment of Pneumocystis carinii pneumonia. J. Pediatr., 92:285–291, 1978.

75. Lau, W.K., Young, L.S., and Remington, J.S.: Pneumocystis carinii pneumonia. Diagnosis by examination of pulmonary secretions. J.A.M.A. 236:2399–2402, 1976.

76. LeGolvan, D.P., and Heidelberger, K.P.: Disseminated, granulomatous Pneumocystis carinii pneumonia. Arch. Pathol., 95:344–348, 1973.

77. Meuwissen, J.H.E.Th., et al.: Parasitologic and serologic observations of infection with pneumocystis in humans. J. Infect. Dis., 136:43–49, 1977.

78. Pifer, L.L., Hughes, W.T., and Murphy, M.J., Jr.: Propagation of Pneumocystis carinii in vitro. Pediatr. Res., 11:305–316, 1977.

79. Pifer, L.L., Hughes, W.T., Stagno, S., and Woods, D.: Pneumocystis carinii infection: evidence for high prevalence in normal and immunosuppressed children. Pediatrics, 61:35–41, 1978.

80. Sheldon, W.H.: Pulmonary Pneumocystis carinii infection. J. Pediatr., 61:780–791, 1962.

81. Walzer, P.D., et al.: Pneumocystis carinii pneumonia in the United States. Epidemiologic, diagnostic, and clinical features. Ann. Intern. Med., 80:83–93, 1974.

Trichomonas Vaginalis

82. Catterall, R.D.: Trichomonal infections of the genital tract Med. Clin. North Am. 56:1203–1209, 1972.

83. Fleury, F.J., et al.: Single dose of two grams of metronidazole for *Trichomonas vaginalis* infection. Am. J. Obstet. Gynecol., 128:320–322, 1977.

84. Masur, H., Hook, E., III, and Armstrong, D.: A *Trichomonas* species in a mixed microbial meningitis. J.A.M.A., 236:1978–1979, 1976.

85. Moffett, M., McGill, M.I., Schofield, C.B.S., and Masterton, G.: Nitrimidazine in the treatment of trichomoniasis. Br. M. Vener. Dis., 47:173–176, 1971.

Selected Nematodes

TRICHINELLA SPIRALIS

Objectives

1. Describe the life cycle of *Trichinella spiralis* and indicate the relationship of the life cycle to prevention and treatment of human disease.

Life Cycle

Trichinella spiralis is a roundworm found predominantly in swine. If humans ingest undercooked infested pork, the cysts around the larvae are digested by the gastric juice. The larvae invade the jejunal mucosa, where they mature in 3 to 5 days. The adult females are fertilized, imbed themselves in the intestinal mucosa, and produce larvae, which burrow into lymphatics and capillaries and travel via the blood to all areas of the body, but particularly the striated muscle.[4,7] The female produces larvae for a period of about 6 weeks. Encapsulation of the larvae after they reach muscle takes about 21 days.

Frequency and Importance

There has been a gradual decline in the frequency of trichinosis in the United States, from about 400 cases per year in the 1940s to about 100 to 200 cases per year in the 1970s (Figure 16-1).[2] Outbreaks still occur and still account for a large number of cases in some years. About two deaths are reported each year.

Clinical Patterns of Illness

Trichinella spiralis is a *common* cause of:

Marked Eosinophilia, with Fever and Myalgia. Eosinophilia, swollen eyes, muscle pain and tenderness, and fever are the classic findings.[4,7] The fever and muscle pains may lead the physician to diagnose influenza-like illness, but eosinophilia or periorbital swelling should suggest trichinosis. An erythematous or urticarial rash (hives) and splinter hemorrhages may occur. A retrospective history of ingestion of rare or raw meat (pork or possibly pork-contaminated) can usually be obtained.

Occasionally, ingestion of massive amounts of cysts produces more severe illness, with serious complications overshadowing the classical findings. Myocarditis may occur, with congestive heart failure. Encephalitis or focal paralysis may occur when many larvae reach the brain.[5] Respiratory muscle involvement can be severe enough to produce respiratory failure, requiring assisted ventilation.

Rare diseases that can produce myalgia, fever, and eosinophilia (but no eye swelling) include polyarteritis nodosa and dermatomyositis.

Asymptomatic Infection

Microscopic examination of diaphragms obtained at autopsy from 37 states in 1966–1968 indicated about four percent of the population had trichina larvae or cysts.[7] A lower percent was found in younger individuals. This is a lower frequency than was found in such surveys in the 1940s, when the frequency of infested diaphragms was about 10 to 20 percent.[7]

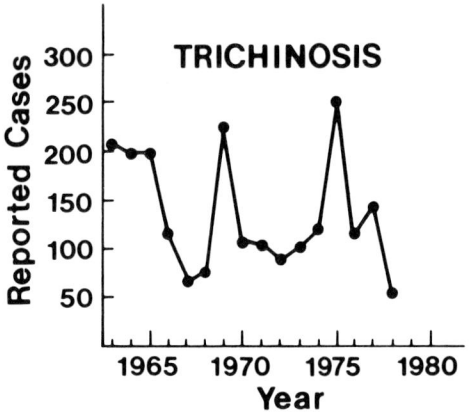

FIG. 16-1. Trichinosis: annual reported cases in the United States, 1963–1977.[2] Provisional data are reported for 1978.

Laboratory Diagnosis

Serologic Tests. These studies are complicated by the presence of the many antigens found in this parasite.[6] Many tests can be done, and selection should be left to the laboratory. Bentonite flocculation, latex agglutination, and counterimmunoelectrophoresis tests all appear useful. In the absence of reexposure, positive flocculation tests usually revert to negative within about 2 to 5 years,[3,4] so a positive flocculation test (1:5 or higher) implies recent infection. The test usually does not become positive until about 3 weeks after exposure, and is usually negative at the time of the onset of symptoms. The absence of eosinophilia correlates with an absence of a titer rise.[1]

Histology. Muscle biopsy is conclusive if positive for trichina cysts. However, encapsulation has usually not yet occurred at the peak of the symptoms.

Study of Suspected Source. Any suspected meat remaining can be examined for the presence of the parasite after digestion of the meat fibers with trypsin.

Skin Test. Commercially available antigen has been unreliable.[1]

Biologic Characteristics of Clinical Interest

Sources. Pork products, often consumed raw, account for almost all reported cases in the United States. Even pure ground beef can be contaminated by infected pork still in the grinder. Bear meat is an important source in Alaska, where bears are infected by eating raw meat. Venison occasionally has been incriminated. Rats, which are cannibalistic, may help to perpetuate the disease.

Conditions Necessary to Kill the Parasite. Cysts remain viable in muscle for many years. Freezing for 20 days, boiling of garbage for 30 minutes before feeding to swine, or cooking until no pink meat is present will destroy the cysts.

Treatment

Thiabendazole (Mintezol) is effective in experimental animals, and might be of value in human disease, except that the diagnosis is usually made after the larvae have reached the muscle.[4,7] Side effects of the drug include nausea or vomiting, but serious toxic effects are rare. To be effective in changing the course of the illness, treatment must be begun early enough to kill the female worms before they stop producing larvae, a period of about 6 weeks or less after ingestion. Thus, while often used, thiabendazole is usually of little value. Aspirin is used for relief of muscle aches. Corticosteroids may be indicated in cases with severe complications.

Long-term or permanent damage is exceedingly rare.[3]

Prevention

Only meat-eating animals can get the disease, so control of the diet of animals raised for sale for food can control the disease. An increasing number of states are prohibiting the feeding of garbage to swine, which is expected to decrease the frequency of human trichinosis. The disease can be prevented in humans by adequate cooking of pork or ground meat that may contain pork.

TOXOCARA SPECIES

Visceral larval migrans syndrome is characterized by eosinophilia, hepatomegaly, and pulmonary symptoms.[9,13,14,17] The usual cause in the United States is the dog roundworm larva, *Toxocara canis*, which has accidentally infected man. The cat roundworm, *Toxocara cati*, is an occasional cause.

Frequency and Importance

Toxocara infection is rare in the United States. It usually occurs in young children who may have poor hygiene and close exposure to dogs or cats. It is important because it may be mistaken for eosinophilic leukemia or a retinoblastoma, when the parasite lodges in the eye (Fig. 16-2).

Life Cycle

Typically, a child has ingested toxocara eggs, which hatch, enter the intestinal circulation, and are carried to various organs, especially the liver and lungs, but also to the kidneys, heart, striated muscles, and eyes. The larvae usually do not develop further, but often remain alive for months. Adult toxocara have been rarely recovered from the feces of humans.[16]

Parasites other than toxocara can also produce visceral larval migrans syndrome, especially outside the United States. These include animal filiariae, *Capillaria hepatica*, and *Ascaris suis*.[13,17]

Clinical Patterns of Illness

In the United States, *Toxocara canis* is a *frequent* cause of:

Extreme eosinophilia without symptoms is usually discovered when a white blood count is done for reasons unrelated to suspicion of the diagnosis. Hepatomegaly is typically present. Eosinophilia is in the range of 30 percent to 90 percent. It is first noted about a month after the ingestion of the toxocara, according to observations of experimental infection of two mentally defective infants, who were fed 200 *T. canis* eggs.[15] The eosinophilia may last at least 2 years.[9] Thus the extreme eosinophilia may be recognized months after the initial ingestion.

Exposure to a dog or cat, and a history of pica (abnormal ingestions, such as dirt) can usually be elicited. Hyperglobulinemia is usually present. Retinal involvement is rare. Myocarditis is rare.[11]

Toxocara canis is an *uncommon* cause of:

Mild Pneumonia or Asthma, With Extreme Eosinophilia. Chronic cough, often nocturnal, fever, and recurrent wheezing was noted in the majority of children with this syndrome in one large series from Puerto Rico and North Carolina.[13] A history of convulsions was reported

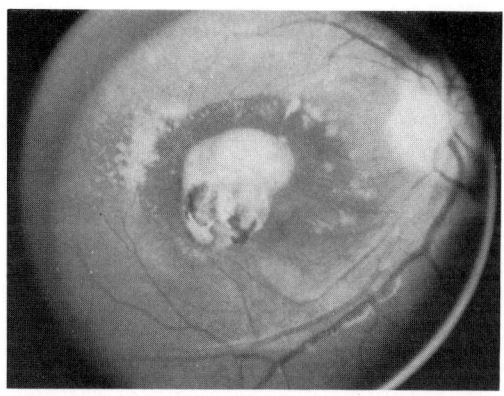

FIG. 16-2. *Toxocara canis involving the retina. The lesion resembles that of toxoplasmosis, but is less likely to involve the macula, perhaps because the larva tends to migrate away from highly vascular macula after it lodges there. (Photo from Dr. Thomas France)*

in 28 percent of the children in this group. Most of the children had an enlarged liver.

This syndrome of asthma and extreme eosinophilia was usually due to the human roundworms trichuris or ascaris, but was occasionally due to toxocara.

Ocular Infections. The pupil appears white and retinoblastoma may be suspected.[17]

Epilepsy. Children with idiopathic convulsive disorders had a higher frequency of serologic evidence of past toxocara infection compared to a control group, in one study, which suggests a possible effect of the parasite having migrated to the brain.[12]

Laboratory Diagnosis

Liver biopsy is definitive, but usually not necessary. Elevated serum globulins are usually present.[13] Serologic tests are sometimes available in a reference laboratory for detecting antibodies against Toxocara. A newly developed ELISA serologic test may prove useful.[10]

Treatment

Thiabendazole appears to be of value in human infections. It is associated with prompt return of appetite, improvement of adenopathy, liver enlargement, and lung findings within a week, and decrease in eosinophils over a period of several months.[8]

Diethylcarbamazine (Hetrazan) is of some value in experimental infections in mice.

Prevention

Deworming dogs is advisable.[14]

REFERENCES

Trichinella Spiralis

1. Barrett-Connor, E., Davis, D.F., Hamburger, R.N., and Kagan, I.: An epidemic of trichinosis after ingestion of wild pig in Hawaii. J. Infect. Dis., 133:473–477, 1976.
2. Center for Disease Control: Morbidity Mortality Weekly Rep. 19, 27: Annual supplements. Summaries, 1970, 1978. p.4, p.2.
3. Cox, P.M., Schultz, M.G., Kagan, I.G., and Preizler, J.: Trichinosis—five-year serologic and clinical follow up. Am. J. Epidemiol., 89:651–657, 1969.
4. Grove, D.I., Warren, K.S., and Mahmoud, A.A.F.: Algorithms in the diagnosis and management of exotic disease. VII. Trichinosis. J. Infect. Dis., 132:485–488, 1975.
5. Kramer, M.D., and Aita, J.F.: Trichinosis with central nervous system involvement. A case report and review of the literature. Neurology, 22:485–491, 1972.
6. Kagan, I.G.: Trichinosis: a review of biologic, serologic and immunologic aspects. J. Infect. Dis., 107:65–93, 1960.
7. Most, H.: Trichinosis—preventable—yet still with us. New Engl. J. Med., 298:1178–1180, 1978.

Toxocara Species

8. Aur, R.J.A., Pratt, C.B., and Johnson, W.W.: Thiabendazole in visceral larval migrans. Am. J. Dis. Child., 121:226–229, 1971.
9. Beaver, P.C., et. al.: Chronic eosinophilia due to visceral larva migrans. Report of three cases. Pediatrics, 9:7–19, 1952.
10. Cypess, R.H., Larva-specific antibodies in patients with visceral larval migrans. J. Infect. Dis., 135:633–640, 1977.
11. Friedman, S., and Hervade, A.R.: Severe myocarditis with recovery in a child with visceral larval migrans. J. Pediatr., 56:91–96, 1960.
12. Glickman, L.T., Cypess, R.H., Crumrine, P.K., and Gitlin, D.A.: Toxocara infection and epilepsy in children. J. Pediatr., 94:75–78, 1979.
13. Huntley, C.C., Costas, M.C., and Lyerly, A.: Visceral larval migrans syndrome: clinical characteristics and immunologic studies in 51 patients. Pediatrics, 36:523–536, 1965.
14. Schantz, P.M., and Glickman, L.T.: Toxocaral visceral larval migrans. New Engl. J. Med., 298:436–439, 1978.
15. Smith, M.H.D., and Beaver, P.C.: Persistence and distribution of toxocara larvae in the tissues of children and mice. Pediatrics, 12:491–497, 1961.
16. Von Reyn, C.F., Roberts, T.M., Owen, R., and Beaver, P.C.: Infection of an infant with an adult *Toxocara cati* (Nematoda). J. Pediatr., 93:247–249, 1978.
17. Zinkham, W.H.: Visceral larval migrans. A review and reassessment indicating two forms of clinical expression: visceral and other. Am. J. Dis. Child., 132:627–633, 1978.

Appendix
A Review of Essentials

In the foregoing text the author has endeavored to set forth the clinical essentials of microbiology. He believes that such a presentation, though short, will be an aid not only in learning but in clinical practice, and in the examinations that increasingly punctuate a medical career.

The following 104 questions were carefully drawn from the text. By answering them—even with the wrong answer—the student will learn central aspects of clinical microbiology; if he refers to the answer section following the questions, which gives the page in the text most pertinent to each question, this learning process will be enhanced even further.

Good luck!

Hugh Moffet

Review Questions

1. All of the following statements about *Staph. epidermidis* are correct EXCEPT which one:
   A. A common cause of false positive blood cultures
   B. A common cause of ventriculo-peritoneal shunt infections.
   C. Coagulase negative
   D. Normal flora of skin
   E. A common cause of food poisoning

2. All of the following statements about Group A streptococci are true EXCEPT which one:
   A. Beta-hemolytic on a sheep blood agar plate
   B. Usually bacitracin-sensitive
   C. Common cause of wound infections, cellulitis, and impetigo
   D. Tetracycline or a sulfa drug is an appropriate treatment
   E. Group can be identified by latex fixation or co-agglutination methods

3. Group B streptococci:
   A. Can be permanently eradicated from a woman's cervix by 10 days of oral penicillin.
   B. Is an important cause of septicemia and meningitis in newborn infants.
   C. Is found in only about one percent of cervix cultures from women of child-bearing age.
   D. Can produce septicemia, which is easily distinguished from neonatal respiratory distress syndrome, using clinical and radiologic criteria.
   E. Has only one type, so a monovalent vaccine will soon be practical.

4. The non-enterococcal Group D streptococci differ from enterococci by being:
   A. A cause of subacute infective endocarditis
   B. Usually susceptible to penicillin
   C. Sensitive to an optochin disk
   D. Normal flora of the urine
   E. Invariably beta-hemolytic on a sheep blood agar plate

5. Diphtheria:
   A. Can be diagnosed with accuracy by a Gram stain of the pharyngeal exudate.
   B. Can be treated with erythromycin while awaiting culture results to see if antitoxin is necessary.
   C. May require emergency treatment with a tracheostomy or tracheal intubation.
   D. Can be excluded by finding a positive Schick test.
   E. Can be readily diagnosed by testing the patient's serum for antibodies.

6. All of the following statements about diphtheria are true EXCEPT which one:
   A. Can be confused with severe pharyngeal infectious mononucleosis.
   B. Myocarditis with arrhythmias and congestive heart failure is a common mechanism of death.
   C. Can produce skin ulcers resembling impetigo.
   D. Can produce a slightly bloody nasal discharge if the membrane is in the nose.
   E. Frequently produces endotoxic shock and best treated with vigorous fluid replacement.

7. All of the following statements about *Listeria monocytogenes* are correct EXCEPT:
   A. Patients with disorders of cell-mediated immunity are especially susceptible to this microorganism.
   B. Is a gram positive rod sometimes mistaken for diphtheroids.
   C. Can produce purulent meningitis with a predominance of lymphocytes.
   D. Is associated with a good prognosis in early onset neonatal septicemia.
   E. Is associated with monocytosis in the peripheral blood of animals other than humans.

Questions 8–11:
   A. Pseudomembranous enterocolitis
   B. Muscle rigidity, best detected in unopposed muscles
   C. Outbreaks of meningitis

D. Gas gangrene
E. Cranial nerve paralysis

8. *Clostridium tetani*
9. *Clostridium botulinum*
10. *Clostridium perfringens*
11. *Clostridium difficile*

Questions 12–15:
A. Early muscle relaxant therapy
B. Sulfadiazine
C. Polyvalent antitoxin
D. Oral vancomycin
E. Heparin

12. *Clostridium difficile*
13. *Clostridium botulinum*
14. *Clostridium tetani*
15. *Nocardia asteroides*

Questions 16–18:
A. Spores imported in wool, hides, skins
B. Diarrheal food poisoning
C. Convulsions, rigidity, death from respiratory muscle paralysis
D. Anaerobic; mass (abscess) below jaw
E. Cranial nerve paralysis

16. *Bacillus anthracis*
17. *Actinomyces israelii*
18. *Bacillus cereus*

19. Which one of the following is the LEAST likely source of salmonella gastroenteritis:
A. Turtles
B. Dried (powdered) eggs
C. Feces of a human carrier
D. Respiratory droplets from a human carrier
E. Barbecued chicken

20. Antibiotic therapy usually prolongs the carrier state in:
A. Shigellosis
B. Salmonellosis
C. Both
D. Neither

21. *E. coli* can produce diarrhea by way of all of the following mechanisms EXCEPT:
A. Heat-stable enterotoxin
B. Heat-labile enterotoxin
C. Invasiveness
D. Adherence
E. Interference with synthesis of elongation factor

Questions 22–24:
A. Urinary infection with alkaline urine
B. Necrotizing pneumonia
C. Spread from person to person by aerosol
D. Red or orange pigmented colonies
E. Purulent conjunctivitis

22. *Klebsiella species*
23. *Proteus species*
24. *Serratia marcescens*

25. Which one of the following diseases is *not* frequently caused by *Pseudomonas aeruginosa*:
A. Burn infections
B. Swimmer's ear
C. Pustular rash from contaminated, heated pools
D. Ecthyma gangrenosum
E. Scalded skin syndrome

26. *Pseudomonas aeruginosa* is associated with all of the following EXCEPT:
A. Mucoid strains in patients with cystic fibrosis
B. Resistance to quarternary ammonium disinfectants
C. Production of water soluble blue-green pigment
D. Production of exotoxin
E. Good growth in anaerobic conditions

27. All of the following are causes of diarrhea in the United States EXCEPT:
A. *Campylobacter fetus*
B. *Vibrio cholerae*
C. *Vibrio parahaemolyticus*
D. *Yersinia enterocolitica*
E. *Herpesvirus hominis*

28. All of the following statements about *Vibrio cholerae* exotoxin are correct EXCEPT:
A. It is called choleragen.
B. It is heat-stable.
C. It is antigenically-related to an *E. coli* enterotoxin.
D. It increases chloride ion excreted in the small intestine.
E. Its effects are reversed by oral administration of glucose.

29. *H. influenzae* type b is a common cause of all of the following clinical syndromes in young children EXCEPT which one:
A. Laryngitis with epiglottitis
B. Facial cellulitis
C. Purulent meningitis
D. Septic arthritis
E. Purulent urethritis

30. *H. influenzae* type b capsular antigen can be detected by all EXCEPT which one of the following methods:
A. Slide agglutination of a suspected colony
B. Counterimmunoelectrophoresis of CSF
C. Latex agglutination of CSF
D. Coagglutination of CSF
E. Immune electron microscopy

31. All of the following statements about *Bordetella pertussis* are correct EXCEPT which one:
A. Sometimes produces fatal infection in young infants
B. Can be transmitted by infected hospital personnel
C. Treatment with pertussis hyperimmune globulin very effective
D. Eradicated by erthromycin in a few days of treatment, making the patient non-contagious
E. Can cause apneic spells or encephalopathy in young infants

32. Type b *H. influenzae* can be recovered from the nasopharynx of about _____ percent of healthy children:
A. 1
B. 3 to 5

C. 20 to 30
D. 50 to 60
E. 80 to 90

33. All of the following statements about ampicillin-resistant H. influenzae are correct EXCEPT which one:
   A. First detected in the United States in about 1974
   B. Is typically not associated with concurrent chloramphenicol resistance
   C. Is mediated by a beta-lactamase transmitted by a plasmid
   D. Can typically be successfully treated by a penicillinase-resistant penicillin such as nafcillin
   E. Can be detected in a H. influenzae colony by a laboratory test which takes only a few minutes

Questions 34–38:
   A. Domestic cat
   B. Mosquitoes
   C. Rabbits
   D. Beagles
   E. Imported raw goat's milk or cheese

34. Brucella mellitensis
34. Brucella canis
36. Francisella tularensis
37. California encephalitis virus
38. Toxoplasma gondii

Questions 39–43:
   A. Infected dog bite
   B. Fever, lymphadenopathy, chronic abscess of liver or spleen
   C. Suppurative inguinal lymphadenitis
   D. Oculoglandular syndrome
   E. Diarrhea, abdominal pain with mesenteric adenitis

39. Yersinia enterocolitica
40. Yersinia pestis
41. Pasteurella multocida
42. Francisella tularensis
43. Brucella abortus

44. Management of family members exposed to a patient with meningococcemia might include all EXCEPT which one of the following:
   A. Close observation and prompt evaluation of symptoms
   B. Administration of rifampin
   C. Administration of meningococcal vaccine
   D. Blood culture

45. Which one of the following is least likely to be due to gonococcal infection:
   A. Proctitis
   B. Tenosynovitis
   C. Skin pustules
   D. Periphepatitis
   E. Brain abscess

46. More severe disease due to the gonococcus can be related to all EXCEPT which one of the following:
   A. Strains which produce an IgA-cleaving proteinase
   B. Strains which have pili
   C. Strains which have been incubated with iron
   D. Patients with a C_7 complement deficiency
   E. Patients with a defect in cell-mediated immunity

47. All of the following statements about Legionnaires' disease bacillus (Legionella pneumophilia) is correct EXCEPT:
   A. It is contagious, especially between family members.
   B. It can produce progressive fatal pneumonia, especially in compromised hosts.
   C. It can produce a febrile illness with headache and muscle aches but without pneumonia.
   D. It can be overlooked by conventional Gram stains, but can be detected by special silver stains, of lung tissue.
   E. It is susceptible to erythromycin.

48. Anergy in tuberculosis is defined as:
   A. Confirmed tuberculosis with repeatedly negative tuberculin tests
   B. Confirmed tuberculosis with negative serum antibodies for tuberculosis
   C. A negative tuberculin test in a patient with a positive mumps skin test
   D. A positive tuberculin test with a normal chest x-ray
   E. A tuberculin chest that converts from negative to positive within 1 year

49. Preventive treatment, such as isoniazid for one year, is often recommended for certain individuals who have a positive tuberculin test without clinical or radiologic evidence of tuberculosis. In which one of the following situations is preventive therapy least appropriate:
   A. A 20 year old woman with a normal chest x-ray, whose tuberculin test has converted from negative to positive in the past year
   B. A 45 year old man with a normal chest x-ray who is found to have an intermediate tuberculin test that shows 8 mm of induration and 12 mm of erythema at 48 hours
   C. A 2 year old child with a normal chest x-ray, no known contact with tuberculosis, and an intermediate tuberculin test showing 15 mm. of induration
   D. A 12 year old boy who is found to have a positive tuberculin test
   E. A 30 year old man with a positive tuberculin test of unknown duration who is about to begin a course of immunosuppressive drugs.

50. Indicate which one of the follwing matches of antituberculous drug toxicities or side effects is incorrect:
   A. Rifampin—orange urine
   B. Streptomycin—VIIIth nerve damage
   C. Isoniazid—hepatitis
   D. Para-ammosalicylic acid—vomiting
   E. Ethambutol—pseudomembranous enterocolitis

51. Which one of the following sets of clinical findings is least likely to be associated with a non-tuberculous mycobacterial disease:
   A. Chronic cervical lymphadenitis in children
   B. Anesthetic hypopigmented plaques or nodules of the skin

C. Chronic cavitary pulmonary diseases
D. Chronic skin ulcerations after exposure to sea water or fish tanks
E. Chronic non-gonococcal urethritis

52. Which one of the following statements about histoplasmosis is *incorrect*:
   A. Found in the soil, especially in Ohio and Mississippi river valleys
   B. Often contagious to family contacts
   C. When disseminated, the yeast forms can often be seen in stained bone marrow smears
   D. Disseminated form characterized by hepatosplenomegaly, fever, leukopenia and thrombopenia
   E. Often contacted by inhalation of the mycelial form from areas contaminated by manure from chickens, pigeons, or other birds

53. Amphotericin B is often used for therapy of severe forms of all of the following diseases EXCEPT:
   A. Actinomycosis
   B. Histoplasmosis
   C. Candidiasis
   D. Cryptococcosis
   E. Aspergillosis

54. All of the following statements about coccidioidomycosis are correct EXCEPT:
   A. Found predominately in southern California and Arizona.
   B. Laboratory confirmation of infection can be done by testing paired sera for antibodies.
   C. Typical clinical manifestations include pleuritic chest pain, fever, hilar adenopathy, and erythema nodosum.
   D. Can be effectively treated by oral nystatin.
   E. Frequently associated with asymptomatic skin test conversion, using coccidioidin antigen.

55. Of the following pairs, select the one *incorrect* match:
   A. Blastomycosis—marked leukocytosis; naturally infects dogs
   B. Sporotrichosis—inoculation during gardening, with nodules in a line along lymphatics, with later ulceration
   C. *Candida albicans*—common cause of diaper rash, vaginitis, infections of hyperalimentation lines
   D. Aspergillosis—association with adrenal or parathyroid insufficiency
   E. *Cryptococcus neoformans*—predominance of lymphocytes, lowered glucose, elevated protein in the spinal fluid

56. A baby has just been born to a mother with a positive VDRL. The mother is not sure if she has been treated for syphilis. The best way to determine if the baby has been infected and has congenital syphilis is to:
   A. Obtain an IgM-specific FTA-ABS on the baby's serum.
   B. Obtain a VDRL on the baby's serum.
   C. Follow the baby for six weeks, watching for signs of congenital syphilis.
   D. Obtain a FTA-ABS on the baby's serum.
   E. Examine the baby's spinal fluid for spirochetes, using dark field microscopy.

57. Select the one *incorrect* match:
   A. *Leptospirosis*—exposure to animal urine
   B. *Borrellia recurrentis*—rat bite fever
   C. Syphilis—aorti aneurysms, delusions, loss of position sense
   D. *Leptospirosis*—fever, rash, hepatitis, nephritis, nonpurulent meningitis
   E. Syphilis—firm, painless ulcer on the glans penis

58. Select the one *incorrect* match:
   A. Mycoplasmas—atypical pneumonia
   B. Chlamydiae—neonatal conjunctivitis
   C. Richettsiae—fever and petechial rash
   D. Chlamydiae—non-gonococcal urethritis
   E. Ureaplasmas—trachoma

59. Which one of the following syndromes or diseases is *not* caused by a chlamydial species:
   A. Trachoma
   B. Severe pneumonia transmitted by psittacine birds
   C. Bilateral interstitial pneumonia in infants
   D. Suppurative inguinal adenitis, with occasional fistulae or strictures
   E. Bullous myringitis

60. Psittacosis can best be confirmed in the laboratory by:
   A. Culture on sheep blood agar plates
   B. Silver stain of throat swabbings
   C. Antibody studies on paired sera
   D. Special stain of a skin biopsy
   E. Culture of a rectal swabbing on rhesus monkey kidney cells

61. Which one of the following treatments would be *least* likely to be effective:
   A. A sulfa drug for psittacosis
   B. Tetracycline for Rocky Mountain spotted fever
   C. Tetracycline for non-gonococcal urethritis
   D. Erythromycin for mycoplasmal pneumonia
   E. Tetracycline for mycoplasmal pneumonia

62. Pick the one *incorrect* match:
   A. *Mycoplasma pneumoniae*—positive cold agglutinins
   B. *Mycoplasma pneumoniae*—recurrent pulmonary infections
   C. *Rickettsia rickettsii*—disseminated intravascular coagulation
   D. *Chlamydia trachomatis*—inclusion bodies which stain with iodine
   E. *Ureaplasma urealyticum*—lymphogranuloma venereum

63. Which one of the following descriptions is *least* likely to be a manifestation of *Herpesvirus hominis* infection:
   A. Dendritic keratitis
   B. Patent ductus arteriosus
   C. Gingivitis
   D. Ulcers of cervix
   E. Encephalitis

64. Varicella-zoster immune globulin is used for treatment or prevention of disease in which one of the following:
    A. A chickenpox-susceptible 4 year old child with leukemia and a family exposure to chickenpox 2 days ago
    B. A 24 year old immunosuppressed adult with disseminated zoster which began 4 days ago
    C. A 6 week old infant whose mother developed chickenpox 5 weeks after the birth of the infant
    D. A 64 year old man with zoster ophthalmicus
    E. A 6 year old normal child who has been exposed to chickenpox 2 days after receiving live attenuated poliovaccine

65. Which one of the following sets of clinical findings is LEAST likely to be caused by cytomegalovirus infection:
    A. Fever, splenomegaly, and atypical lymphocytosis
    B. Chronic interstitial pneumonia in a patient with a malignancy
    C. Jaundice, purpura, hepatosplenomegaly and microcephaly in a newborn infant
    D. Asymptomatic urinary excretion of the virus in the first week of life
    E. Patent ductus arteriosus, cataracts, and deafness

66. All of the following statements about infectious mononucleosis or EB virus infection are correct EXCEPT which one:
    A. A positive heterophile test or Mono Spot slide test is unusual in children less than 8 years old with EB virus infection.
    B. EB virus produces lymphoblastic transformation of leukocyte cultures of normal individuals.
    C. The incubation period is about 6 to 8 weeks.
    D. Most individuals in the United States have no serologic evidence of past EB virus infection until about 18 years of age.
    E. EB virus is regularly found in Burkitts' lymphoma cells.

67. All of the following statements about adenoviruses are correct EXCEPT which one:
    A. Some types produce sarcomas when injected into hamsters.
    B. Can produce progressive and fatal pneumonia in healthy individuals.
    C. A frequent cause of exudative pharyngitis in young children.
    D. Can cause ulcerative keratitis with a typical dendritic (branching) lesion.
    E. Can cause atypical pneumonia or influenza-like illness, especially in military populations.

68. Smallpox virus (and other pox viruses) can be distinguished from chickenpox virus (and other herpesviruses) by:
    A. Electron microscopy
    B. Cytopathic effect in cell cultures
    C. Inoculation of chimpanzees
    D. Hemagglutination of chick erythrocytes
    E. Inhibition of growth by RNA inhibitors

69. Complications of smallpox vaccination include all of the following except:
    A. Infection of atopic dermatitis by the virus
    B. Infection of the eye by the virus
    C. Progressive necrotizing viral infection at the inoculation site, in immunocompromised hosts
    D. Encephalitis
    E. Paralytic poliomyelitis syndrome

70. Select the one *incorrect* match:
    A. Monkeypox virus—resembles the illness of smallpox, when it infects humans
    B. Molluscum contagiosum—transmitted by eating raw shellfish
    C. Orf virus—produces nodular skin ulcerations in sheep or goats, occasionally transmitted to the hand of a human
    D. Vaccinia virus—dried crusts may contain living virus for many months
    E. Chickenpox virus—cannot be distinguished from zoster virus by electron microscopy or serologic tests

71. Classic influenza-like illness in a school-age child is best described by which one of the following sets of clinical findings:
    A. Persistent vomiting, with retching, but no fever
    B. Moderate fever, with moderate to severe diarrhea and dehydration
    C. Fever, cough, headache, sore throat, fatigue, weakness and myalgia
    D. Fever, stiff neck, confusion, and sometimes convulsions
    E. Fever, cough with purulent sputum, pleuritic chest pain, and generalized inspiratory rales or crackles

72. Complications of influenza virus infection include all EXCEPT which one of the following:
    A. Fulminating pneumonia
    B. Acute renal failure
    C. Severe laryngitis
    D. Acute encephalitis or encephalopathy
    E. Endocarditis

73. Influenza virus infection can be detected by all EXCEPT which one of the following laboratory techniques:
    A. Detection of the virus by hemadsorption after growth in monkey kidney cell cultures
    B. Detection of viral particles by immune electron microscopy of stool specimens
    C. Detection of a rise in antibody titer in paired sera
    D. Detection of the virus in lung tissue using fluorescent antibody staining
    E. Detection of the virus by hemagglutination, after growth of the virus in the amniotic sac of chicken eggs.

74. Indicate the one incorrectly matched pair of words or phrases referring to influenza virus:
    A. Hemagglutinin—involved with attachment of the virus to the host's cell surface
    B. Neuraminidase—involved with release of virus from the cell
    C. Antigenic drift—minor changes in the hemagglutinin
    D. Recombination—exchange of RNA units between different influenza A viruses
    E. Interference—blocking of the host's receptor sites by an enterovirus

75. Which one of the following is *not* a method of surveillance for influenza virus activity:
    A. School absenteeism
    B. Excess pneumonia mortality above threshhold rates
    C. Surveys of influenza-like illness at sentinel reporting station.
    D. Sales of antibiotics in military pharmacies
    E. Regular reporting of influenza-like illness from U.S. air force overseas installations

76. Select the one *incorrect* definition of a kind of influenza vaccine:
    A. H_3N_2 vaccine—one which contains subtype of virus with the hemagglutinin and neuraminidase antigen assigned these numbers
    B. Subunit vaccine—a purified preparation composed of disrupted viral components
    C. ts vaccine—a live virus vaccine which grows much better at nasal temperature than at lung temperature
    D. Bivalent vaccine—one which contains 2 different strains of vaccine virus
    E. "Alice" strain vaccine—a live, attenuated vaccine given by the oral route

77. Pick the one *incorrect* statement about parainfluenza virus:
    A. Detected by hemadsorption
    B. Can cause parotitis
    C. Important cause of laryngitis in children
    D. Preventable by use of live, attenuated virus vaccine
    E. A paramyxovirus, like measles and respiratory syncytial virus

78. All of the following statements about mumps virus infection are correct, EXCEPT which one:
    A. Sometimes complicated by aseptic meningitis syndrome
    B. Preventable by use of a live, attenuated viral vaccine
    C. Orchitis a common complication in males, often resulting in sterility
    D. Detected in cell cultures by hemadsorption
    E. Commercial hyperimmune globulin of little or no value for prevention of complications

79. Which one of the following is *not* a complication of measles virus infection:
    A. Giant cell pneumonia
    B. Severe laryngitis in children
    C. Acute encephalitis with subsequent mental retardation
    D. Subacute sclerosing panencephalitis
    E. Communicating hydrocephalus

80. All of the following sets of findings may be present with measles virus infection EXCEPT which one:
    A. Multinucleated giant cells in nasal secretions and sputum
    B. Dissemination of unrecognized tuberculosis
    C. Measles virus present in the skin rash and in Koplik spots
    D. Orchitis, arthritis, and chorioretinitis
    E. Absence of rash in a child with leukemia

81. Measles virus infection can be made milder by which one of the following:
    A. Presence of transplacentally-acquired antibodies at 15 months of age
    B. Administration of immune serum globulin (ISG) within a few days after exposure
    C. Administration of an antibiotic within a few days after exposure
    D. Administration of immune serum globulin (ISG) as soon as the rash appears
    E. Administration of killed measles vaccine within a few days after exposure

82. All of the following clinical patterns have been associated with respiratory syncytial virus infection EXCEPT which one:
    A. Severe bronchiolitis in young infants
    B. Apneic episodes in very young infants
    C. Adult respiratory distress syndrome
    D. Common cold syndrome in adults
    E. Exacerbation of asthma in young adults

83. All of the following statements about respiratory syncytial virus are correct EXCEPT which one:
    A. Freezing specimens before culture significantly decreases virus recovery.
    B. Reinfection with the virus is not unusual.
    C. Tends to produce outbreaks of disease.
    D. Older children or adults a frequent source for infections in infants.
    E. Readily preventable by use of a killed virus vaccine.

84. All of the following statements about rabies are correct EXCEPT which one:
    A. About 20 human cases in the United States each year.
    B. Bats, skunks, and foxes an important reservoir.
    C. The virus can travel from the bite area to the brain by way of peripheral nerves.
    D. Nervous tissue (Semple) vaccine or equine antiserum is now avoided in the United States.
    E. Rapid diagnosis in a suspected human case can be made by fluorescent antibody stain of an impression smear of the patient's cornea.

85. All of the following are useful ways of preventing human rabies EXCEPT which one:
    A. Use of human diploid cell vaccine after exposure
    B. Use of human immune rabies globulin after exposure
    C. Immunization of domestic animals with a live attenuated rabies virus vaccine
    D. Imposing quarantines on importation of animals into rabies-free islands
    E. Certifying and inspecting pet stores which sell descented skunks as pets

86. Enteroviruses are characterized by:
    A. Acid-lability
    B. Frequent production of diarrhea in humans
    C. Large size (120 nanometers)
    D. Frequent cause of the aseptic meningitis syndrome
    E. Frequent cause of atypical pneumonia

87. Select the one *incorrect* match:
    A. Coxsackie viruses—disease in mice
    B. ECHO viruses—no disease in mice or primates
    C. Poliovirus—disease in chimpanzees
    D. Group A coxsackie viruses—herpangina
    E. ECHO viruses—pleurodynia, myocarditis

88. All of the following statements about enteroviruses are correct EXCEPT which one?
    A. Many types of Group A coxsackieviruses require mouse inoculation for recovery of the virus.
    B. Echoviruses, polioviruses, and coxsackie B viruses produce similar cytopathic effects and require neutralization with type-specific antiserum for identification.
    C. ECHO virus infections can be detected by finding an antibody titer rise to the antigen that all echoviruses share.
    D. Enteroviruses often can be detected in throat washings during the first few days of the infection.
    E. ECHO virus are detected in rectal swabbings of about 5 percent of asymptomatic infants.

89. All of the statements about poliovirus or poliovaccine virus are correct EXCEPT which one:
    A. Wild poliovirus can often be distinguished from poliovaccine virus by laboratory methods such as temperature sensitivity and antigenic analysis.
    B. Most individuals who develop paralysis due to poliovaccine virus do not have any immunologic defect.
    C. Colostrum usually has a high content of poliovirus antibody and often interferes with poliovaccine infection in the first few days of life.
    D. Natural infection with other enteroviruses can interfere with infection with poliovaccine virus.
    E. Only about 1 or 2 cases of poliovaccine-associated paralysis occur annually in the U.S.

90. All of the following statements about rhinoviruses are correct EXCEPT which one:
    A. A practical vaccine containing the 5 to 10 serotypes which produce the most severe illness is likely to be available within a few years.
    B. Rhinoviruses grow best in cell cultures incubated at 35°C.
    C. Rhinoviruses are destroyed by gastric acid.
    D. Rhinoviruses can be readily transmitted by finger to nose or eyes in human experiments.
    E. Serologic methods are not clinically practical for the detection of a rhinovirus infection.

91. Select the one *incorrect* match:
    A. St. Louis encephalitis virus—transmitted by mosquitoes
    B. California encephalitis virus—highly contagious between humans
    C. Eastern equine encephalitis virus—viremia occurs in the animal host
    D. Arthropod-borne encephalitis viruses—diagnosis best confirmed in the laboratory by testing paired sera for antibodies
    E. Venezuelan equine encephalitis virus—has sometimes been found to cause disease acquired in the United States

92. Typical rubella-like illness in a grade school child would be characterized by:
    A. Cough, high fever, swollen red eyes, and a confluent rash that lasts at least 5 days
    B. Migratory arthritis, hives-like rash, and occasional myocarditis
    C. Fever, neck pain on flexion, decreased blood pressure, and a petechial rash
    D. Mild fever and respiratory symptoms with a papular rash and generalized lymphadenopathy
    E. Sore throat with tonsillar exudate, tender lymph nodes in the neck, with a generalized fine sandpaper-like rash that is most prominent in skin creases

93. All of the following statements about rubella virus infection are correct EXCEPT which one?
    A. The infection can occur without a rash.
    B. Infection in a pregnant woman can produce a number of fetal defects.
    C. A past history of rubella is not reliable enough to judge the immunity of a woman of child-bearing age.
    D. A rubella hemagglutination-inhibition (HI) titer is a reliable guide to immunity.
    E. Viral culture is a faster and more reliable way to detect present infection than testing acute and convalescent sera for antibodies.

94. ALL EXCEPT which one of the following methods are currently used to try to prevent the occurrence of congenital rubella syndrome?
    A. Immunization of susceptible women of child-bearing age
    B. Immunization of children
    C. Abortion of women who develop rubella during the first trimester of pregnancy
    D. Immunization of male medical personnel
    E. Use of human hyperimmune rubella globulin for women who develop rubella during pregnancy

95. All of the following statements about congenital rubella syndrome are correct EXCEPT which one?
    A. Patent ductus arteriosus or pulmonic stenosis is the most frequent cardiac defect.
    B. The infants have an increased risk of later developing diabetes mellitus.
    C. The newborn infant with this syndrome is not contagious.
    D. The newborn infant is typically small for gestational age and has postnatal growth retardation.
    E. Learning disabilities and mental retardation can occur as part of the syndrome.

96. All of the following statements about hepatitis A virus are correct EXCEPT which one?
    A. Formerly called infectious hepatitis
    B. Incubation period 1 to 2 months
    C. Sources include other patients with the disease, water, and food
    D. Hepatitis A virus has several serotypes and repeated infections can occur.
    E. Hepatitis A rarely produces a fulminant or a chronic hepatitis.

97. All of the following statements about hepatitis B virus are correct EXCEPT which one?
    A. Formerly called serum hepatitis
    B. Incubation period longer than 2 months
    C. Sources have included blood transfusions and injections
    D. Important antigenic components include surface antigen and e antigen
    E. Typically has an acute onset with high fever, but complete recovery without chronic infection

98. Most post-transfusion hepatitis in the United States is now due to:
    A. Hepatitis A
    B. Hepatitis B
    C. Non A, Non B hepatitis
    D. EB virus
    E. Cytomegalovirus

99. Select the one *incorrect* match:
    A. Yellow fever virus—preventable by a live attenuated vaccine
    B. Dengue virus—mosquito-borne febrile illness in Caribbean area, Virgin Islands, and Hawaii
    C. Colorado tick fever virus—transmitted by wood tick
    D. Lymphocyte choriomeningitis virus—reservoir in mice and hamsters
    E. Coronavirus—rare cause of hemorrhagic fever in North Africa

100. Select the one *incorrect* match:
    A. Spongiform encephalopathy (Creutzfeldt-Jacob disease)—transmissible by corneal transplantation
    B. Rotaviruses—the major cause of severe infantile diarrhea in the United States
    C. Norwalk virus—the major cause of epidemic vomiting disease in the United States
    D. Cat scratch disease—a reovirus which causes suppurative lymphadenitis
    E. JC virus—a papovavirus found in patients with progressive multifocal leukoencephalopathy

101. All of the following statements about amebiasis are correct EXCEPT which one:
    A. Patients with acute amebic dysentary are not very contagious.
    B. Eosinophilia is unusual in amebiasis.
    C. Serologic tests may be helpful in the diagnosis of amebic abscess.
    D. Some enteric bacteria appear to exert a synergistic effect with amebae in the production of disease.
    E. Corticosteroids are effective treatment of mild amebiasis.

102. Select the one *incorrect* match:
    A. *Naegleriae* species—purulent meningitis with low spinal fluid glucose
    B. *Giardia lamblia*—chronic diarrhea with edamatous jejunal mucosa
    C. *Toxoplasma gondii*—congenital chorioretinitis
    D. *Pneumocystis carinii*—interstitial pneumonia with oxygen diffusion block in a patient with leukemia in remission
    E. Malaria—reliable prophylaxis with chloroquine

103. Select the one *incorrect* match:
    A. *Trichomonas vaginalis*—therapy with metronidazole
    B. *Trichinella spiralis*—puffy eyes, eosinophilia, muscle pain
    C. *Toxocara canis*—extreme eosinophilia, treatment with thiabendozole
    D. *Pneumocystis carinii*—treatable and preventable with trimethoprim/sulfamethoxazole
    E. *Toxocara canis*—chorioretinitis, microcephaly, intracerebral calcification

104. The antiviral drug useful for prevention of influenza virus infection is:
    A. Ara-A
    B. Amantadine
    C. Beta-methasone
    D. Cyclohexamide
    E. Idoxuredine

ANSWERS TO REVIEW QUESTIONS

1. E	p. 2	20. B	p. 45, 48	39. E	p. 81	58. E	p. 149
2. D	p. 18	21. E	p. 52–54	40. C	p. 81	59. E	p. 157
3. B	p. 20	22. B	p. 56	41. A	p. 83	60. C	p. 152
4. B	p. 21	23. A	p. 57	42. D	p. 84	61. A	p. 149
5. C	p. 28–30	24. D	p. 57	43. B	p. 85	62. E	p. 151
6. E	p. 28–30	25. E	p. 58	44. D	p. 94	63. B	p. 163
7. D	p. 31	26. E	p. 60	45. E	p. 96	64. A	p. 169
8. B	p. 32	27. E	p. 62	46. E	p. 98	65. E	p. 228
9. E	p. 33	28. B	p. 64	47. A	p. 105–106	66. D	p. 172
10. D	p. 33	29. E	p. 74	48. A	p. 113	67. D	p. 163
11. A	p. 34	30. E	p. 75	49. B	p. 115	68. A	p. 181
12. D	p. 34	31. C	p. 79	50. E	p. 115–116	69. E	p. 180
13. C	p. 35	32. B	p. 76	51. E	p. 117–118	70. B	p. 183
14. A	p. 35	33. D	p. 77	52. B	p. 125	71. C	p. 189
15. B	p. 37	34. E	p. 85	53. A	p. 36	72. E	p. 189–190
16. A	p. 37	35. D	p. 85	54. D	p. 128	73. B	p. 190–191
17. D	p. 36	36. C	p. 84	55. D	p. 132	74. E	p. 220
18. B	p. 38	37. B	p. 224	56. A	p. 144	75. D	p. 193
19. D	p. 45	38. A	p. 266	57. B	p. 142	76. E	p. 195

77.	D	p. 197	**84.** A	p. 208	**91.** E	p. 226	**98.** C	p. 238
78.	C	p. 199	**85.** E	p. 210	**92.** D	p. 227	**99.** E	p. 244
79.	E	p. 203–204	**86.** A	p. 216	**93.** E	p. 229	**100.** D	p. 252
80.	D	p. 203–204	**87.** E	p. 218	**94.** E	p. 230	**101.** E	p. 261
81.	E	p. 204–205	**88.** C	p. 219	**95.** C	p. 230	**102.** E	p. 269
82.	C	p. 205	**89.** E	p. 221	**96.** D	p. 235	**103.** E	p. 265
83.	E	p. 207	**90.** A	p. 223	**97.** E	p. 237	**104.** B	p. 196

Index

Numerals in italics indicate a figure; "t" following a page number indicates a table.

Abdominal abscesses, 55, 82
Abdominal infections, intra-. *See also* Gastroenteritis
   enteric bacteria as cause of, 42
Abortion, in cases of rubella during early pregnancy, 230
Abscesses, abdominal, 55, 82
   due to bacteroides, 65
Absidia, 135
Acanthamoeba, 261, 262
Achromobacter, 61
Acinetobacter, 61
Acinetobacter anitratus, 60, 61
Acinetobacter lwoffi, 60, 61
Actinobacillus actinomycetemcomitans, 106
Actinomyces, 35–36, 106
Actinomyces bovis, 36
Actinomyces israelii, 36
Actinomyces viscosus, 36
Actinomycosis, 36
Acute febrile respiratory illness, 176
Acute respiratory disease (ARD), 176
Adenitis, cervical, as form of actinomycosis, 36
    due to mumps virus, 199
    due to mycobacteria, 118
   inguinal, due to *C. trachomatis*, 151
   mesenteric, due to adenoviruses, 177
    due to *Yersinia*, 81
Adenoid degeneration agents, 176
Adenoidal-pharyngeal-conjunctival viruses, 178
Adenoviruses, 176–179, *177*, *178*, 247
   B. pertussis associated with, 80
Adherence, in Group B streptococci, 20
   in pertussis, 80; in gonococci, 98
   in streptococci, 18, 20
Adhesiveness, plasmid-controlled, 54
Adrenal insufficiency, in histoplasmosis, 124–125
Aerobacter, 56
Aerococci, 3
Aeromonas hydrophilia, 106
Aeromonas shigelloides, 106
Agammaglobulinemia, acquired, after Epstein-Barr virus infection, 174
Agglutination *See* Latex agglutination
Agglutinins, cold, in pneumonias, 158
Alastrum, 180
Alcaligenes, 61
Aleutian mink disease, 248, 249t.
Alkalescens dispar, 47
Allescheria species, 136
Amantadine, in influenza, 196
Amebas, free-living, 261–262, *262*
Amebiasis, 259–261, *260*
Ampicillin, resistance to, 77
Amylase elevation, serum, in mumps virus, 200–201

Anemia, progressive hemolytic, due to *Histoplasma capsulatum*, 123
Anergy, 113–114
Animal bites, infected with *P. multocida*, 83
Animal inoculation tests, in diagnosis of botulism, 34
Animals. *See also* specific names
   amebiasis and, 259
   anthrax and, 38
   arboviruses and, 224, 224t., 226–227
   aspergillosis and, 135
   cat scratch disease and, 252–253
   chlamydia and, 149–153
   cocksackieviruses and, 217
   corona viruses and, 245
   cryptococcus due to, 134
   diseases of, resembling measles, 204
     transmittable to humans, 81
   hepatitis and, 236, 239
   histoplasmosis due to, 126
   in LCM, Lassa, and Marburg viruses, 245
   leptospirosis and, 147
   mycoplasma and, 158
   nematodes and, 275–278
   Pasteurella species and, 83
   plague and, 82
   rabies and, 207–208
   reoviruses and rotaviruses and, 250–251
   rickettsia and, 153–155
   salmonella and, 45
   slow viruses and, 248–249, 249t.
   staphylococci and, 6
   toxoplasmosis and, 264–266
   tularemia and, 84
Anthrax, 37–38, *37*
Antibiotics, colitis induced by, 34
   resistance to. *See also* Resistance transferable, 50
   tolerance of, 5
Antibody(ies), detection of. *See* ELISA
   heterophile, of infectious mononucleosis, 175
   transplacental (maternal), 20
Antibody technique, fluorescent. *See* Fluorescent antibody technique
Antibody titer rise, 14, *14*
Antigen(s), Australia, 236. *See also* Antigen, surface
   in body fluids, 75, 93, 97
   in pneumococcal infections, 9
   core, 235
   detection of. *See* Counterimmunoelectrophoresis
   e, 235
   surface, 235, 236–237, 238, *239*
   tumor (T), 247
   type C virus, 247, 250
Antigenic drift, influenza virus, 191, 191t., 192

289

Antistreptolysin O (ASO), 14–15, *14*
Antistreptolysin O test, 15
Apnea, due to respiratory syncytial virus, 205
Arboviruses, 224–227, 224t., *225*, 225t.
Arenaviruses, 224, 245
Arthritis, 2, 16, 74
    due to *N. gonorrheae*, 96
    due to *N. meningitidis*, 91
Arthropod-borne viruses, 224–227, 224t., *225*, 225t.
Ascaris suis, 277
Aspergillomas, 134
Aspergillosis, 134–135
Aspergillus species, 134–135
Asthma, influenza and, 189
    due to rhinoviruses, 222
    due to toxocara, 277
Astroviruses, 250
Atelectasis, due to *B. pertussis*, 79
Attenuation, of streptococci, 18
Australia antigen, 236. *See also* Antigen, surface
Autolysis, pneumococcus and, 9

Babesia, 269
Bacillary dysentery. *See* Shigellosis
Bacillus, 37–38
Bacillus anthracis, 37
Bacillus cereus, 37, 38
Bacillus subtilis, 37
Bacitracin sensitivity, 12–13, *13*
Bacteremia, due to bacteroides, 65
    clostridial, 33
    enterobacter, in shigella, 49
    due to *H. influenzae*, 74
    occult pneumococcal, 8
    transient, 6
Bacteriocins, 47, 56, 60
Bacteroides, 64–66
Bacteroides corrodens, 107
Bacteroides melaninogenicus, 66
Bacteriophage, 27
Basidiobolus, 135
Battey bacillus, 117
BCG immunization, 117
Bedsoniae, 149
Bentonite flocculation test, 276
Beta-lactamases, 4–5, 37, 98
    TEM type, 5
Beta-lactamase test, in *H. influenzae*, 77
Bifidobacteria, 108
Biological false positives (BFP), 144
Bites, animal, infected with *P. multocida*, 83
    human, infected with bacteroides, 65
Blastomyces dermatiditis, 128–129
Blastomycosis, 128–129, *129*
Blood transfusion, cytomegalovirus and, 171
    Epstein-Barr virus and, 175–176
    hepatitis virus due to, 238
    malaria due to, 268
    syphilis due to, 143
Bordetella bronchiseptica, 61–62
Bordetella parapertussis, 78
Bordetella pertussis, 78–80
Borrelia species, 141–142
Botryomycosis, 136
Botulism, 33, 32, *32*, 35
Brain, abscess of, due to actinomyces or nocardia, 36
    damage to, neonatal, due to toxoplasmosis, 265
    effects of influenza on, 190
Branhamella catarrhalis, 106
Breast feeding, *E. coli* and, 55
    rotaviruses and, 251
Brill-Zinsser disease, 155
Bronchiectasis, due to adenoviruses, 176
Bronchiolitis, due to adenoviruses, 176
    due to respiratory syncytial virus, 205, 206–207
Bronchitis, due to *B. pertussis*, 79
    due to *H. influenzae*, 74
Brucella, 84–86

Brucella abortus, 85
Brucella canis, 85, 86
Brucella melitensis, 85
Brucella suis, 85
Brucellosis, 84–86, *85*, 85t.
Bubo, 82, 151, 152
Bunyaviruses, 224
Burkitt's lymphoma, 172, 173, 175, *175*
Burn infections, pseudomonas in, 58

Calciviruses, 250
Calymmatobacterium (Donovania) granulomatis, 107
CAMP test, 19
Campylobacter species, 62, 63, 64
Candida albicans, 131–133, *132*
Candida tropicalis, 131
Candidiasis, 131–133
    chronic mucocutaneous, 131, 132, 133
Cannula sepsis, due to *Candida albicans*, 131
Capillaria hepatica, 277
Cardiobacterium hominis, 107
Carriers, convalescent, 16, 45
    of hepatitis virus, 237–242
    of meningococci, 91–92, 94–95
    of salmonella, 45
    in shigellosis, 48
    of staphylococcal strains, 4
    of streptococci, 16, 17, 20
    transient, 16
    of typhoid, 45–46
Cat scratch disease, 252–253, *252*
Catalase test, 11
Cats, toxoplasmosis due to, 264–266
Cell transformation. *See* Transformation
Cephalosporinases, 4
Cephalosporins, sensitivity to, 56
Cephalosporium, 136
Cervical adenitis, 16, 36
    due to mumps virus, 199
    due to mycobacteria, 118
Cervicitis, due to *C. trachomatis*, 151
    due to mycoplasmas, 157
Cervix, carcinoma of, due to herpes simplex virus, 164
    lesions of, due to herpes; 163; chlamydiae, 151
Chancre, syphilitic, 142, *143*
Chancroid, 73, 142, *143*
Chickenpox, 162, 166–169, *167*
    compared with smallpox, 181, *181*
    congenital, 167
    neonatal, 167
Chickenpox-like illness, 166
Chlamydia, 149–153
Chlamydia psittaci, 149, 150
Chlamydia trachomatis, 149–151, *151*
Chlamydospores, 131
Chloramphenicol resistance, 77
Cholera, 62–64
Cholera toxin, detection of, 63, *63*
Choleragen, 64
Chorioretinitis, congenital, due to toxoplasmosis, 264, *264*
Chromobacterium species, 107
Chromoblastomycosis, 136
Citrobacter, 58
Cladosporium, 136
Clavulanic acid, 5
Clostridial infections, necrotizing, 35
    postoperative, 33, 35
Clostridium, 31–35
    Propionibacterium compared to, 34
Clostridium botulinum, 32, 34
Clostridium difficile, 32, 34
Clostridium perfringens, 32–35, 38
Clostridium sordelli, 34
Clostridium tetani, 32, 34
Coagglutination method, for beta-hemolytic, streptococci, 13, 19, *19*
    for *E. coli* enterotoxin, 53
    for gonococci, 97

Coagglutination method *(continued)*
    for *H. influenzae*, 76
    for influenza, 191
    for meningococci, 90, 93
    in salmonella diagnosis, 44
Coagulase tests, for staphylococci, 3, 4
Cocci, gram-negative, 90–99
    gram-positive, 1–22
Coccidioides immitis, 126–128
Coccidioidin skin test, *126*, 127
Coccidioidomycosis, 126–128
Cold agglutinins, in pneumonias, 158
Coliform, 41
Colitis, antibiotic-induced, 34
    without high fever, due to *Shigella* species, 48
Colonization, plasmid-controlled, 54
Colonization factor antigen (CFA), 54
Colorado tick fever virus, 244
Common cold syndrome, due to coronaviruses, 244
    due to parainfluenza virus, 196
    due to respiratory syncytial virus, 205
    due to rhinoviruses, 222–223
Complement deficiency, in gonorrhea patients, 98
Complement fixation text, 177
Congenital chickenpox, 167
Congenital chorioretinitis, due to toxoplasmosis, 264, *264*
Congenital infection syndrome, chronic, 170
    due to congenital cytomegalovirus infection, 170
Congenital malformations, due to herpes simplex virus, 163
Congenital rubella syndrome, 227, 228, *228*, 230
Conjugation, 5, 10, 50
Conjunctival viruses, adenoidal-pharyngeal-, 178
Conjunctivitis, due to adenovirus, 176, *177*, 178
    inclusion, in newborn, due to *C. trachomatis*, 149, 150, 151, *151*
    purulent, due to gonorrhea, 96, *96*, 99
        due to vaccinia virus, 181
    sporadic, due to *C. trachomatis*, 151
Convulsions, as complication of shigellosis, 49
Cording, 114
Corneal ulceration, due to herpes simplex virus, 163
Coronaviruses, 244–245
Corynebacteria, 27–30
Corynebacterium acnes, 34, 108. See also *Propionibacterium acnes*
Corynebacterium diphtheriae, 27–30
Corynebacterium vaginale, 27
Coryneform bacteria, 27
Counterimmunoelectrophoresis (CIE), for candidiasis, 132
    for clostridia, 35
    in diphtheria, 29
    in *H. influenzae*, 76, *76*
    for meningococci, 93
    in mycoplasmal pneumonia, 158
    for staphylococcal infections, 4, 9
    for streptococci, 19
Coxiella burnetti, 153
Coxsackie viruses, *198*, 216, 217, 218, 220, *220*
Creutzfeldt-Jacob disease, 249
Cryptococcal meningitis, 133–134
Cryptococcosis, 133
Cryptococcus neoformans, 133–134
Cunninghamella, 135
Curvularis, 136
Cystitis, hemorrhagic, due to adenoviruses, 177
Cytomegalovirus, 169–172, *171*
Cytomegalovirus cytopathic effect (CPE), 170, *171*
Cytopathic effect (CPE), in enteroviruses, 216, 219, *219*
    in rhinoviruses, 222

Dawson's encephalitis, 203
Deafness, as complication of mumps virus, infection, 199
Dehydration, of infants, due to diarrhea, 54, *55*
Dengue virus, 243–244
Dermatitis, diaper, 131
Dermatophilus, 136
Diabetes mellitus, due to mumps virus, 200
    due to rubella virus, 228

Diaper dermatitis, 131
Diarrhea. *See also* Diarrheal food poisoning
    acute dysentery-like, due to amebiasis, 259, 260
    acute salmonella, 43, 44, 46
    due to amebiasis, 259
    chronic, due to Giardia lamblia, 263
    in cholera, 62–63
    dysentery-like, 63
        in shigellosis, 47
    epidemic, of newborn, 52
    fulminating fatal, due to *Shigella* species, 48
    due to reoviruses, 250
    due to rotaviruses, 250–251
    severe infantile, 177
    Shigella and *Salmonella* as causes of, 41, 44, 46, 47, 48
    sporadic, in infants, 52
    traveler's, 52
Diarrheagenic *E. coli*. *See Escherichia coli*, diarrheagenic
Diarrheal food poisoning, due to *Bacillus cereus*, 37, 38
    clostridial, 33
    due to *E. coli*, 52
    salmonella as cause of, 43–45, 46
    staphylococcal, 2, 38
    due to vibrios, 63
Dientamoeba fragilis, 259
Dimorphic fungi, 122
Diphasic morphology, 125
Diphtheria, 28–30, *28*
    cutaneous, 28, 29
Diphtheria bacillus, 27–29
Diphtheroids, 27, 30
    anaerobic, 108
Diplococcus pneumoniae. See Pneumococci
DNA. *See also* Plasmids
    in transferable antibiotic resistance, 50
DNA transfer, in pneumococci, 10
DNA tumor viruses, 246, 247
DNA viruses, 162–183
Drug resistance. *See also* Resistance transferable, 45, 50
Duck embryo vaccine (DEV), 210
Duoviruses, 250
Dysentery. *See also* Diarrhea
    amebic, 259
    bacillary, 47

Ear, infections of, 7, 16, 74, 91
    swimmer's, 58
Eaton agent, 156
Eberthella typhosa, 43
Ebola virus, 246
Echoviruses, *198*, 216, 217, 218, *219*, 220, *220*
    differentiated from reoviruses, 250
Eczema, due to herpes simplex virus, 180–181
    vaccinia-infected, 180–181
Edwardsiella, 58
Eikenella corrodens, 107
Electrolytes, in diarrheagenic *E. coli*, 54
Electron microscopy, of adenovirus, 177, *178*
    for chickenpox, 168
    for Epstein-Barr virus, 174, *174*
    immune, in hepatitis A, 239
        for Norwalk virus, 252
        for reoviruses, 250
        for rotaviruses, 251
    of JC virus, *248*
    of parainfluenza, 197, *198*
    for smallpox, 181, *181*
ELISA (enzyme-linked immunosorbent assay) technique, for chlamydiae, 152
    for cholera, 53, 63, *63*
    for *E. coli* and *V. cholerae*, 53, 63, *63*
    general principles of, 63, *63*
    for herpes simplex, 164, *165*
    in influenza, 76, 191
    for klebsiella, 57
    in Legionnaire's disease, 107
    for meningococci, 93
    in mycoplasma infections, 158

ELISA *(continued)*
    for parainfluenza, 197
    for pneumococci, 9
    for rotavirus, 251
    in rubella, 229, *229*
    in toxoplasmosis, 265
Elongation factor (EF-2), 29
Empyema, 2
Encephalitis, acute, due to herpes simplex virus, 163
    due to arboviruses, 224–226, *225*, 225t.
        California, 225t., 226
        Eastern equine and Western equine, 225t., 226
        Powassan virus, 225t., 226
        St. Louis, 225t., 226
        Venezuelan, 225t., 226
    in chickenpox, 167
    due to measles virus, 202, 203
    due to mumps virus, 199–200
Encephalopathy, due to measles virus, 204
    spongiform, 249
Endocarditis, *Staphylococcus aureus* and, 6
    subacute bacterial, 22
Endocrine disease, candidiasis associated with, 132
Endospores, in coccidioidomycosis, 126
Endothelial lesion, in rickettsia, 155
Endotoxin, in meningococci, 93
    in shigella, 49
Entamoeba coli, 259, 260
Entamoeba histolytica, 259–261
Enteric bacteria, 41–43, 55–56
Enteric fever pattern, 44
Enteric pathogen, 41
Enteritis necroticans, 34
Enterobacter, 41, 49, 56, 57
Enterobacter agglomerans, 107
Enterobacter hafnia, 58
Enterobacteriaceae, 41, 42t.
Enterococci, 3, 21
Enterocolitis, pseudomembraneous, due to clostridia, 32
    staphylococcal, 2
Enteropathogenic *E. coli*. *See Escherichia coli*, enteropathic
Enterotoxins, in *Bacillus cereus*, 38
    in amebiasis, 261
    in cholera, 64
    in *E. coli*, 52–53, 54
    staphylococcal, 2, 5–6
Enteroviruses, 216–222
Enzyme-linked immunosorbent assay technique. *See* ELISA
Eosin-methylene-blue (EMB), 52
Eosinophilia, in amebiasis, 259
    aspergillosis associated with, 135
    in coccidioidomycosis, 127
    in giardiasis, 263
    mycoplasmas associated with, 158
    in pertussis, 79
    in toxocaria, 277
    in trichinosis, 275
Epidermolyotic toxin, 2
    staphylococcal, 2, 5, 6
Epididymitis, acute "idiopathic," due to *C. trachomatis*, 150
Epilepsy, due to toxocara, 277
Episome. *See* Plasmids
Epstein-Barr virus, 172–176, *174*, *175*, 246
Erwinia species, 107
Erysipelothrix rhusiopathiae, 107
Erythema multiforme, in histoplasmosis, 124
Erythema nodosum, in histoplasmosis, 124
Erythrogenic toxin, 17
Escherichia coli, 50–56, *51*
    in cross reaction with *H. influenzae*, 77, 78
    diarrheagenic, 50–55
    as lactose fermenter, 41
    shigella and, 47, 48, 50
Escherichia freundii, 58
Esophagitis, due to *Candida albicans*, 131
Eukaryotes, 122
Exotoxins, in clostridia, 34–35
    Corynebacteria diphtheriae associated with, 29

Exotoxins *(continued)*
    in shigella, 49
Eyelids, in infectious mononucleosis, 172, *173*
Eyes. *See also* Conjunctival viruses; Conjuctivitis
    H. influenzae as cause of infections of, 74
    herpes simplex virus involving, 163
    herpes zoster involving, 167, *168*
    infections of, due to toxocara, 277

Face, herpes simplex virus affecting, 162
Fermenters, non-lactose and lactose, 41, 42
Fibroelastosis, endocardial, as complication of mumps, 200
Flavobacterium, 61
Flora, normal, of bowel, *E. coli* in, 55
    enterococci in, 22
    Klebsiella in, 56
    Pseudomonas aeruginosa in, 59
    of colon, skin, and genital tract, clostridia in, 35
    of human feces, amoeba in, 259
    of mouth, mycobacteria in, 118
        female genital tract and, bacteroides in, 65
        or pharynx, mycoplasma in, 156
        or vagina and bowel, *C. albicans* in, 132
    of nasopharynx, pneumococci in, 9
    of nose and throat, corynebacteria in, 27
        H. influenzae in, 76
    of pharynx, streptococci in, 17
    of skin, corynebacteria in, 27
        staphylococci in, 2, 4
        nose and, staphylococci in, 4
    of throat, *Klebsiella* in, 56
Fluids and electrolytes, in diarrheagenic *E. coli*, 54
Fluorescence, in bacteroides, 66
    in pseudomonas, 59
    in tinea, 122
    for tuberculosis, 113
Fluorescent antibody (FA) technique, in bacteroides, 65
    in cholera, 63
    in cytomegalovirus, 171
    in Epstein-Barr virus, 173
    general principles of, 63, 265
    in herpes simplex, 164
    in histoplasmosis, 125
    indirect (IFA), in toxoplasmosis, 265, *265*
    in influenza, 191
    in Legionnaire's disease, 105
    in measles, 204
    in parainfluenza virus, 197
    in pertussis, 79
    in plague, 82
    in pneumocystis, 270
    in rabies, 208, 209, *209*
    in Rocky Mountain spotted fever, 154
    in streptococci detection, 11, 13, *13*, 15
    in syphilis, 144
    in toxoplasmosis, 265
Fluorescent treponemal antibody-absorbed (FTA-ABS), 144
Fonsecaea, 136
Food poisoning, 2, 16, 33, 38, 45
Foreign bodies, infections associated with, 2
Francisella tularensis, 83–84
Frei test, 152
Fungi, 122–136
    classification of 122, 123t.
    dimorphic, 122
    mycelial, 122
Fusarium, 136
Fusobacterium, 65, 108

Gangrene, clostridia and, 32, 33
    gas, 35
    synergism in, 22
Gastroenteritis, acute, due to shigella, 47–48
    due to Norwalk virus, 251
Genital herpes, 163
Genital infections, 16, 65, 73, 96, 142, 150, 163
Genital mycoplasmas, 155–159

Gardnerella — see *haemophilus*

INDEX 293

Genitourinary, see genital or urinary
Gentamicin resistance, 5
Geotrichium, 131
German measles. *See* Rubella
Giant cells, formation of, in measles virus, 203, *203,* 204
Giardia lamblia, 262–264
Giardiasis, 262–264, *263*
Gingivostomatitis, 162–163
Glomerulonephritis, 11, 13, 17
Gonococcal bacteremia, disseminated, 96, 99
Gonorrhea, 95–99
Gram-negative cocci, 90–99
Gram-negative rods, 27–38
  enteric, 41–66
  small, 73–86
Gram-positive cocci, 1–22
Granuloma inguinale, 107
Group II-D, 107
Guillain-Barré syndrome, 195
Guthrie test, 57

Haemophilus aphrophilus, 106
Haemophilus ducreyi, 73, 78, 142, *143*
Haemophilus hemoglobinophilus, 76
Haemophilus hemolyticus, 14, 76
Haemophilus influenzae, 76–78, 93
Haemophilus parainfluenzae, 77
Haemophilus vaginalis, 27 — or G. vaginalis
Hafnia, 58
Halophilic vibrios, 62, 64
Hand-foot-mouth syndrome, 218
Hansen's bacillus, 111
Hartmanella, 261, *262*
Hawaii agent, 251
HB-1, 107
Heart-reactive antibodies, 17
Heat-labile toxin (LT), 52–53
Heat-stable toxin (ST), 52, 53
Hemadsorption (HA), in influenza virus, 191, *192*
  in mumps virus, 201
Hemagglutination, in influenza virus, 191
  in measles virus, 204
Hemagglutination inhibition (HI) test, 190, 229
Hemagglutinins, in influenza virus, 191–192, *192*
Hemolysis, 11–12, *12,* 14
  alpha, 11
  alpha-prime, 12, 14
  beta, 11–12
  gamma, 12
Hepatitis, 234–243
  acute, 237
  chronic active, 237
  due to leptospira, 146
  non-A, non-B, 238
  post-transfusion, 238
  subacute, 237
Hepatitis viruses, 234–243
  A, 234–235, 235t., 236, *236,* 237–240
  B, 234–238, 235t., *236,* 239–242
  C, 238
  fulminant, 238
  neonatal, 238
  non-A, non-B, 238
  serum, 234
Hepatoencephalitis virus, 250
Hepatosplenomegaly, fever with, due to *Histoplasma capsulatum,* 123, *124*
Herd immunity, 230
Herellea vaginicola. See Acinetobacter anitratus, 60, 61
Herpes genitalis or progenitalis, 163
Herpes simplex keratitis, 163, *163*
Herpes simplex virus, 162–165
Herpes zoster. *See* Zoster
Herpes zoster virus, 162
Herpesvirus hominis, 162–165, *168*
Herpesvirus varicellae, 162, 166, *168*
Herxheimer. *See* Jarisch-Herxheimer
Heteroimmunization, 77, 78

Heterophile antibody, of infectious mononucleosis, 175, 264–265
Hexoses, 60
Histoplasma capsulatum, 122–126, 136
Histoplasmin skin test, in histoplasmosis, 125
Histoplasmosis, 122–126, *123, 124*
Human bites, infected with bacteroides, 65
Human diploid cell vaccine, 210
Human rabies immune globulin (HRIG), 210
Hyaluronidase, 17
Hydrocephalus, as complication of mumps virus infection, 200
Hydrogen sulfide production, by bacteria, 42
Hyperbaric oxygen, in treatment of gas gangrene, 35
Hyperimmune hepatitis B immune globulin (HBIG), 242
Hyperlucent lung, unilateral, 177

IgA proteinase, 94, 98
IgM-specific antibody, 144
Ileal loop test, for enterotoxin, 52–53, *53*
Immune complexes, gonococcal, 98
  in hepatitis B, 240
Immune electron microscopy. *See* Electron microscopy, immune
Immune serum globulin (ISG), 242
Immunity, herd, 230
Immunologic diseases, *Candida* infections, associated with, 132
Indirect fluorescent antibody test (IFA), in toxoplasmosis, 265, *265*
Infants. *See also* Neonatal; Newborn
  botulism in, 33
  dehydration of, due to diarrhea, 54, *55*
  pneumonia in, 150
Infectious mononucleosis, 172–173, *173*
  heterophil-negative, due to toxopasmosis, 264–265
Infectious mononucleosis-like syndrome, 170
Influenza, 188–196, *198. See also Haemophilus influenzae*
  antigenic drift associated with, 191, 191t., 192
  hospital-acquired, 195–196
Influenza-like illness, due to adenoviruses, 176
  classic, 189
  due to *coccidioides immitis,* 127
  as defined by Center for Disease Control, 193
  due to Legionnaires' disease bacillus, 105
  due to *M. pneumoniae,* 157
Inguinal adenitis, due to *C. trachomatis,* 151
Interference, in poliovaccine, 220
  in rubella, 229
  in staphylococci, 6–7
Interferon, stimulation of, in rabies, 209
Intra-abdominal infections, enteric bacteria as cause of, 42
Intracellular infection, in brucellosis, 86
Intrinsic resistance, 5
Intussusception, 177
Invasiveness, of *E. coli,* 53
Isospora species, 266

Jarisch-Herxheimer reaction, 86, 142, 145
Jaundice, in hepatitis, 236, 237
  neonatal, in cytomegalovirus, 170
  in rubella, 228
  due to toxoplasmosis, 265
JC virus, 247, 248, *248,* 249t.

K-1 antigen, *E. coli* and, 56
Kaposi's varicelliform eruption, 181
Kawasaki disease, 146
Keratitis, herpes simplex, 163, *163*
Kerion, 122
Klebsiella, 41, 56–57
Klebsiella ozaenae, 56
Klebsiella pneumoniae, 56, 57
Klebsiella rhinoschleromatis, 56
Kuru, 248–249

L forms, 156
Lactobacilli, 108

Laryngitis, with epiglottitis, due to *H. influenzae*, 74
  in influenza, 189
  membranous, due to *C. diphtheriae*, 28
  due to parainfluenza virus, 196
Lassa virus, 245
Latency, of cytomegalovirus, 171
  in herpes simplex virus, 164–165
  in varicella-zoster virus, 168–169
Latex agglutination, in *H. influenzae*, 76
  in histoplasmosis, 125
Latex fixation test, for cryptococcus, 133, *134*
  for pneumococci, 9
  for streptococci, 13

Learning disabilities, due to rubella virus, 227, 228
Legionnaires' disease bacillus, 103, *104*, 105–106
Legionella pneumophilia, 103, *104*
Leprosy, 118, 119
Leptospira, 141, 144, 145–147
Leptospira interrogans, 146
Leptospirosis, 146–147
Leptotrichia, 65
Leukemia, study of, as viral disease, 246
Leukocyte lysis test, 253
Leukoencephalopathy, in JC virus, 248, *248*
Leukoviruses, 246
Limulus lysate test, 93
Listeria monocytogenes, 30–31
  bacteria similar to, 107
Listerosis, 30–31
  neonatal, 31
Liver, abscesses of, in brucellosis, 85
  enlarged, in histoplasmosis, 123, *124*
  in hepatitis, 237, 239
Locus minoris resistentiae, 6
Lung, unilateral hyperlucent, 177
Lupus, virus-like particles in, 249–250
Lygranum, in diagnosis of psittacosis, 152
Lymphadenitis, suppurative, in tularemia, 84
  due to *Yersinia pestis*, 81
Lymphadenopathy, generalized, in rubella virus, 228
  hilar, in tuberculosis 111–112
  regional, 81, 84, 252, *252*
Lymphangitis, nodular, in sporotrichosis, 130
Lymphocytic choriomeningitis virus (LCM), 245
Lymphogranuloma venereum, 149, 150, 152
Lymphoma, Burkitt's 172, 173, 175, *175*

Maduromycosis, 135
Malaria, 266–270, *267*
Malignant pustule, definition of, 37
Mannitol fermentation, 4
Marburg virus, 245–246
Markers, to distinguish polioviruses, 220
Measles, *198*, 202–205, *202*, *203*
  atypical, 203
  definitions of, 202
  "red" and "German," 202. *See also* Rubella
  as slow virus, 249, 249t.
Measles-like illness, mild, 202–203
  modified, 228
  severe, 202
Melioidosis, 59
Meningitis, aseptic. *See* Meningitis, nonpurulent
  candida, 131
  cryptococcal, 133–134
  enteric bacteria as cause of, 42
  with hemorrhagic skin lesions, 2
  neonatal, due to streptococci, 20
  nonpurulent, due to adenoviruses, 177
    due to arboviruses, 225
    due to histoplasmosis, 124
    due to leptospira, 146
    with low glucose, due to *Cryptococcus neoformans*, 133
    tuberculous, 112, 116
    due to lymphocytic choriomeningitis virus, 245
    due to mumps virus, 199
    due to poliovirus, 218

Meningitis, nonpurulent *(continued)*
  primary amebic, due to free-living amebas, 261–262
  due to *Pseudomonas aeruginosa*, 59
  purulent, due to *H. influenzae*, 74, 75
    due to *Listeria monocytogenes*, 31
    due to *Neisseria meningitidis*, 90, 91, 93
    due to pneumococci, 7
Meningococcus, 90–95, *91*, *92*
Meningococcus, 90–95
Meningoencephalitis, 225
Mental retardation, due to rebella virus, 227, 228
  due to cytomegalovirus, 170
  due to toxoplasmosis, 265
Mesenteric adenitis, due to adenoviruses, 177
  due to *Yersinia*, 81
Methicillin resistance, 5
Micrococcacus, 1, *1*, 3
Mima polymorpha. *See Acinetobacter lwoffi*, 97
Mink disease, Aleutian, 248, 249t.
Miyagawanella, 149
Molluscum contagiosum, 183, 247
Moniliasis, 131
Monkeypox, 182
Mononucleosis. *See* Infectious mononucleosis
Monosporium apiospermum, 135
Moraxella, 61
Moraxella vaginicola, 61
Mortierrella, 135
Mosquito, arboviruses due to, 224, 224t., 226–227
  dengue due to, 244
  malaria due to, 267, 268, 269
  yellow fever due to, 243
Mouth. *See* Flora, normal; Stomatitis
Mucocutaneous lymph node syndrome, 146
Mucor, 135
Mucormycosis, 135
Multiple sclerosis, associated with measles virus, 204
  suspected viral etiology of, 249
Mumps virus, 197–201, *198*, *200*
Myalgia, in dengue fever, 243, 244
Mycelial form of fungi, 122
Mycetoma, 122
Mycobacteria, 111–119
Mycobacterium bovis, 113, 117
Mycobacterium leprae, 111
Mycobacterium tuberculosis, 111–117
Mycoplasmas, 155–159
  genital, 156
  pharyngeal, 156
Mycoplasma hominis, 156, 157
Mycoplasma mycoides, 156
Mycoplasma pneumoniae, 155–158, *158*
Mycoplasma salivarium, 156
Mycotic microorganisms, 123t.
Myocarditis, due to coxsackievirus, 218
Myringitis, bullous, due to *M. pneumoniae*, 157

Naegleria, 261, 262, *262*
Nail infections, due to *C. albicans*, 131
Nasopharyngeal carcinoma, Epstein-Barr associated with, 175, *175*
Nebraska calf diarrhea virus (NCDV), 250, 251
Negri body, in rabies, 208, *209*
Neisseria, 90, 93
Neisseria catarrhalis, 106
Neisseria gonorrheae, 95–99
Neisseria lactamica, 92, 93
Neisseria meningitidis, 90–93
Nematodes, 275–278
Neonatal brain damage, due to rubella, 227, 228
  Toxoplasmosis, 265
Neonatal chickenpox, 167
Neonatal hepatitis, 238
Neonatal jaundice, in cytomegalovirus, 170
  due to rubella virus, 228
  due to toxoplasmosis, 265
Neonatal listerosis, 31
Neonatal meningitis, 20

Neonatal sepsis, suspected, due to herpes simplex virus, 163
Neonatal septicemia. *See* Septicemia, neonatal
Neonate. *See* Infants; Neonatal entries; Newborn
Neuramindase, from influenza virus, 192, *192*
Newborns. *See also* Congenital entries; Infants; Neonatal entries
   E. coli involving, 51–55, *55*
   gonococcal conjunctivitis in, 99
   inclusion conjunctivitis in, 149, 150, 151, *151*
   Listeria monocytogenes involving, 31
   poliovaccine for, 220–221
   streptococci involving, 20
Nocardia, 36
Nocardia asteroides, 36
Nonfermenters, 41
   gram-negative rods as, 60–62
Non-motile (NM), *E. coli* serotype labeled as, 51
Normal flora. *See* Flora, normal
Norwalk virus, 251–252

Oculoglandular syndrome, 84, 252
Omniserum, Quellung reaction with, 8
Oncornaviruses, 246
   human type C, 246–247
Oophoritis, due to mumps virus, 199
Optochin disc, 8, *9*
Orbiviruses, 244, 250
Orchitis, due to mumps virus, 199
Orf virus, 183
Ornithosis, 149
Orthomyxoviruses, *198*
Osteomyelitis, 2, 16, 59, 118
   tuberculous, 112
Otitis externa, due to pseudomomas, 58
Otitis media, purulent, 7
Oxidase test, for *Neisseria* species, 93
Oxidizers, 41, 60
Oxygen, hyperbaric, in treatment of gas gangrene, 35

Pancreatitis, due to mumps virus, 200
Panencephalitis, subacute sclerosing (SSPE), 203
Papanicolaou smear, abnormal, due to chlamydial cervicitis, 151
Papilloma viruses, 247
Papovaviruses, 247
   JC, 247, 248, 249t.
Parapertussis, Bordetella, 78
Paracolon, 41
Parainfluenza virus, 196–197, *198*
Paralytic disease, due to poliovirus, 217, 218
Paramyxoviruses, *198*
Parapertussis, bordetella, 78
Paratyphoid fever, 44
Parotitis, due to mumps virus, 198–201
   due to parainfluenza virus, 196–197
Parrot fever. *See* Psittacosis
Passaging, of streptococci, 18
Pasteurella, 83
Pasteurella multocida, 83
Pasteurella pestis. *See Yersinia pestis*
Pelvic inflammatory disease (PID), due to *N. gonorrheae*, 96, 99
Penicillinase, 4
Penicillinase-resistant penicillin, 6
Penicillium species, 136
Peptococci, 1, *1*, 108
Peptostreptococci, 108
Pericarditis, acute, due to coxsackie-virus, 218
Perinatal infections. *See also* Neonatal entries; Newborns
   Listeria monocytogenes as cause of, 31
Peritonitis, primary, 8
Pertussis, Bordetella, 78–80
Pertussis-like illness, 78, 79
   due to adenoviruses, 176–177
Petechial rash. *See* Rash
Petriellidium, 136
Phage typing, of *Staph. aureus*, 2, *2*
Pharyngeal-conjunctival viruses, adenoidal-, 178
Pharyngeal mycoplasms, 155–159

Pharyngitis, exudative, 16, 157
   due to mumps virus, 199
   membraneous, due to *C. diphtheriae*, 28
   nonstreptococcal, due to adenoviruses, 176
   streptococcal, 11, 13, 15, 16–17, 18
   ulcerative, due to coxsackievirus, 218
      due to herpes simplex virus, 163
Pharyngo-conjunctival fever, due to adenoviruses, 176
Phenylketonuria, 57
Phialophora, 136
Photochromogens, 117
Phycomycosetes, 135
Phycomycosis, 135
Picornaviruses, *198*, 216, 216t.
PIE syndrome. *See* Pulmonary infiltrate with eosinophilia
Pilus, in conjugation, 50
   on gonococci, 98
Pinta, 141
Plague, 81–83, *81*
Planococcus, 1, *1*, 3, *3*
Plasmids, 5
   E. coli and, 54
   gonococci and, 98
   H. influenzae and, 77
   salmonella and, 45
   shigella and, 50, *50*
Plasmid-controlled colonization, 54
Plasmid-mediated resistance, 18
Plasmodia falciparum, 267, 268, 269
Plasmodia malariae, 267
Plasmodia ovale, 267
Plasmodia vivax, 267, 268
Pleisomonas shigelloides, 106
Pleurodynia, 218
Pleuropneumonia-like organisms (PPLO), 155, 156
Pneumatocele, 2, 3
Pneumococci, 3, *3*, 7–22
Pneumocystis carnii, 270, *271*
Pneumonia, acute, due to anthrax bacillus (wool-sorter's disease), 37
   due to *Aspergillus* species, 134
   due to *Histoplasma capsulatum*, 124
   due to adenoviruses, 176
   atypical, due to adenovirus, 176
      in coccidioidomycosis, 126
      due to mycoplasmas, 157
   bilateral interstitial, in infants, 150
   Broad Street, 103
   chlamydial, 150, 152
   diffuse interstitial, in pneumocystis, 270
   focal chronic, due to *Actinomyces* species, 36
   due to *H. influenzae*, 74
   due to influenza virus, 189–190, 193
   due to *Klebsiella*, 56
   due to measles virus, 203, *203*
   due to *mycobacterium tuberculosis*, 111–112, *112*
   mycoplasmal, 156–158
   due to *N. meningitidis*, 91
   perihilar, due to *B. pertussis*, 79, *79*
   pneumococcal, 7
   progressive, in Legionnaires' disease, 104, 105
   progressive fatal, due to adenovirus, 176
   due to pseudomonas, 58–59
   due to respiratory syncytial virus, 205
   segmental, 7, *8*
   severe interstitial, in chickenpox, 166
   staphylococcal, 2, *3*
   due to toxocara, 277
Poliomyelitis, 218, *221*
   vaccine-associated, 221
Poliovaccine virus, 220–222, *221*, *222*
Polioviruses, 216, 217–218
Polyarthritis, rheumatic, 17; hepatitis B, 237
Polyomaviruses, 247
Polyribose phosphate (PRR), 73
Pontiac fever, 103
Postoperative infections, clostridial, 33, 35
Postpartum fever, due to *M. hominis*, 157

Postperfusion syndrome, 170
Powassan virus, 225t., 226, 227
Poxvirus officinale, 179
Precipitin tests, 11
Pregnancy. *See also* Congenital entries
  rubella during, 227, 228, 230
  syphilis during, 143
  toxoplasmosis and, 266
Proctitis, gonococcal, due to gonorrhea, 96
Prokaryotes, 122
Propionibacteria, 34
Propionibacterium acnes, 27, 108
Prostatitis, due to herpes simplex virus, 163
Protein, M, 18
Proteus, 57
Proteus mirabilis, 57
Protoplasts, 156
Prototheca, 136
Protozoa, 259–271
Providencia, 57
Pseudomonas, 58–60
Pseudomonas aeruginosa, 58–60
Pseudomonas mallei, 59
Pseudomonas pseudomallei, 59
Psittacosis, 149, 150, 152
Pulmonary alveolar proteinosis, 36
Pulmonary infiltrate with eosinophilia (PIE syndrome), 134
Pure protein derivative (PPD), 118
Purpuric rash, in meningococcemia, 91, *92*
Pustule, malignant, definition of, 37
Pyelonephritis, 157
Pyogenic infections, *Staph. aureus* as cause of, 2

Q fever, 153, 154
Quellung reaction, 7
  in *H. influenzae* diagnosis, 75
  with omniserum, 8

R factor, *See* Plasmids, 5
Rabies virus, 207–210, *208, 209*
Rapid plasma reagin (RPR) test, 144
Radioimmunoassay (RIA), general principles of, 63, *239*
Rash, due to *coccidioides immitis,* 127
  in measles, 202
  in mycoplasmal infections, 157
  petechial, in meningococcemia, 91, *92*
    in murine typhus, 153
    due to *R. rickettsia,* 154
  purpuric, in meningococcemia, 91, *92*
  in rubella, 227, 229
  in rubella-like illness, 227
  swimmer's, 59
  in syphilis, 142
  in zoster, 166
Rat(s), in plague, 81, 82
Rat bite fever, 142
Reagin screen test (RST), 144
Recombination, in influenza viruses, 192–193
Reinfection, in mumps, 201
  with respiratory syncytial virus, 207
Relapsing fever, 141
Renal failure, acute, in influenza, 190
Reoviruses, 250
Reproductive failure, due to *Ureaplasma urealyticum,* 157
Reservoirs. *See also* Animals; names of animals and insects
  of amebiasis, 259
  of anthrax, 38
  of arboviruses, 224, 224t., 226–227
  of cat scratch disease, 252–253
  of Lassa virus, 245
  of LCM virus, 245
  of Marburg virus, 245
  of plague, 82
  of rabies, 207–208
  of Rocky Mountain spotted fever, 155
  of salmonella, 45
  of staphylococci, 6
  of tularemia, 84

Resistance, to ampicillin, 77
  to antibiotics, due to R factor, 45
    transferable, in shigella, 50
  chloramphenicol, 77
  gentamicin, 5
  intrinsic, 5
  methicillin, 5
  to penicillin, 6, 10
    penicillinase, 6
  plasmid-mediated, 18
  transferable drug, 45, 50
Resistance transfer factor, 50, *50*
Respiratory disease. *See also* Pneumonia
  acute (ARD), due to adenoviruses, 176
  mild, due to meningococci, 92
  minor, influenza virus as cause of, 189
Respiratory Distress Syndrome (RDS), 20
Respiratory syncytial virus, *198,* 205–207, *206*
Retroviruses, 246
Reverse transcriptase, 246
Reye's syndrome, 167
Rheumatic fever, acute, 11, 12, 14–19
Rhinitis, bleeding, due to *C. diphtheriae,* 28–29
  due to rhinoviruses, 222
Rhinoviruses, *198,* 222–224
Rhizopus, 135
Rhodocrous, 136
Richmond's classification, 4–5
Rickettsia, 153–155
  Proteus species and, 57
Rickettsia rickettsi, 153
Rickettsialpox, 153, *153*
RNA, in influenza virus, 192–193, *192*
RNA viruses, helical, 188–210
  other, 216–230
Rocky Mountain spotted fever (RMSF), 153–155, *153, 154*
Rods, gram-negative, enteric, 41–66
  small, 73–86
  gram-positive, 27–38
Rotaviruses, 250–251
Rubella, 202, 227–230
Rubella-like illness, 227–228, *229*
Rubeola. *See* Measles

Sabin dye test, in toxoplasmosis, 265
St. Elizabeth's Hospital outbreak, 105
Salmonella, 43–46
Salmonella choleraesuis, 43, 45
Salmonella derby, 45
Salmonella enteritidis, 43
Salmonella paratyphi, 43, 44
Salmonella typhi, 43, 45
Salmonella typhimurium, 44, 45, 46
Salmonellosis, 43
Salpingitis, acute, due to *C. trachomatis,* 150–151
  due to *N. gonorrheae,* 96, 99
Sarcina, 1, *1*, 3
Satellite phenomenon, 76
Scalded-skin syndrome, 5, *6*
Scarlet fever, 17, 18
Scarlet fever-like illness, 146
Schick test, 30
Scotochromogens, 117
Scrapie, 248, 249
Septicemia, *E. coli* in 55, 56
  enteric bacteria as cause of, 42
  fulminating pneumococcal, 8
  neonatal, *E. coli* in, 55
    due to *Listeria monocytogenes,* 31
    due to streptococci, 20
  primary (meningococcemia), 91
  due to pseudomonas, 59
  due to *Yersinia pestis,* 82
Sereny test, 53
Serologic tests for syphilis (STS), 144
Serratia, 57
Serratia marcescens, 57
"Shaggy heart," 79

Sheep, slow virus of, 248, 249, 249t.
Sheep blood, inhibitory effect of, 76
Sheep blood agar, 14
Shigella, 41–42, 47–50
Shigella dispar, 47, 48
Shigella dysentereae, 47, 48, 49
Shigella flexneri, 47, 48, 49
Shigella sonnei, 47, 48, 49
Shigellosis, 47–50
Shingles, 166
Sickle cell disease, malaria and, 269
Simian virus 40 (SV-40), 221, 247, *247*, 248, 248, 249t., 250
Sinusitis, 7
Skin, infections of, due to *C. albicans*, 131
 cutaneous diphtheria, 29
 scalded-skin syndrome, 5, 6
 staphylococcal, 1, 2
 streptococcal, 11, 16
 lesions of, of herpes simplex, 163
  in histoplasmosis, 125
  pustular-ulcerative, due to vaccinia virus, 181
 ulcers of, chronic, mycobacteria in, 118
  with nodular lymphangitis, in sporotrichosis, 130
Skin tests, for blastomycosis, 129
 candida, 132
 in cat scratch disease, 252
 coccidioidin and spherulin, *126*, 127
 in cryptococcosis, 134
 histoplasmin, 125
 in lymphogranuloma venereum, 152
 mumps, 200
 for mycobacteria, 113–114
 in tularemia, 84
Slow virus diseases, 247–249, 249t.
Smallpox, 179–182
 classical (variola major), 180
 compared with chickenpox, 181, *181*
 fulminating hemorrhagic, 180
 inapparent, 180
 variola minor form of, 180
Smallpox vaccine, to prevent herpes labialis, 165
Spheroplasts, 156
Spherulin skin tests, *126*, 127
Spirillum species, 142
Spirochetes, 141–147
Spleen, abscesses of, in brucellosis, 85
 enlarged, in histoplasmosis, 123, *124*
Spongiform encephalopathy, 249
Sporothrix schenckii, 129–131
Sportrichium, 130
Sporotrichosis, 129–131, *130*
Staph. albus. See Staph. epidermidis
Staph. aureus, 1–7
 enterocolitis attributed to, 34
 hemolysis size produced by hemolytic, *12*
 streptococci and, 19, 22
Staph. epidermidis, 1, 2, 4
Staph. saprophyticus, 2
Staphylococcus, 1–7
 differentiated from streptococci, 11
Stomatitis, due to *C. albicans*, 131
Streptobacillus moniliformis, 108, 142
Streptococcosis, 16
Streptococcus(i), 3, *3*, 11–22
 anaerobic, 22
 Group A, 11–19
 Group B, 19–21
 Group D, 21
  non-enterococcal, 21–22
 Other than beta-hemolytic, 21–22
 Staph. aureus and, 19, 22
 viridans group of, 21
Streptococcus agalactiae, 19. *See also* Streptococcus, Group B
Streptococcus bovis, 22
Streptococcus equinus, 22
Streptococcus faecalis, 21
Streptococcus mutans, 21
Streptococcus pneumoniae. See Pneumococci

Streptococcus salivarus, 21
Streptococcus sanguis, 21
Streptolysins, 14–15
Streptozyme test, 15
Subacute sclerosing panencephalitis (SSPE), 202, 203–204
Surveillance methods, for influenza, 193–194
Swarming, of *Proteus* species, 57
Swimmer's ear, 58
Swimmer's rash, 59
Syncytial cytopathic effect (CPE), 206
Synergism, in amebiasis, 261
 in gangrene, 22
Syphilis, 141–145
 congenital, 143
 in pregnancy, 143
 primary, 142
 secondary, 142
 serologic tests for (STS), 144
 tertiary, 142–143
 transfusion, 143

T antigens (tumor antigens), in adenoviruses, 178
T strains, 156
Teichoic acid, 4
TEM type of beta lactamase, 5
Tenosynovitis, due to *N. gonorrheae*, 96
Tests, animal inoculation, in botulism, 34
 antistreptolysin O, 15
 Bentonite flocculation, 276
 beta-lactamase, in *H. influenzae*, 77
 CAMP, 19
 catalase, 11
 coagulase, for staphylococci, 3, 4
 complement fixation, 177
 ELISA. *See* ELISA
 fluorescent antibody. *See* Fluorescent antibody technique
 fluorescent treponemal antibody-absorbed (FTA-ABS), 144
 Frei, 152
 Guthrie, 57
 hemagglutination inhibition (HI), 190, 229
 IgM-specific antibody, 144
 ileal loop, for enterotoxin, 52–53, *53*
 latex fixation, 9, 13, 133, *134*
 leukocyte lysis, 253
 limulus lysate, 93
 oxidase, for *Neisseria* species, 93
 precipitin, 11
 rapid plasma reagin (RPR), 144
 reagin screen (RST), 144
 Sabin dye, in toxoplasmosis, 265
 satellite, 76
 Schick, 30
 Sereny, 53
 serologic, for syphilis (STS), 144
 skin. *See* Skin tests
 for streptococcal grouping, 12–13, 12t., *13*
 streptozyme, 15
 triple sugar iron (TSI) agar, 42
 tuberculin, 113–114
 Tzanck, 164
 VDRL, 144
 Weil-Felix, for rickettsial infection, 154
Tetanus, 32, 35
Thermophilia, in aspergillosis, 135
Three-day measles. *See* Rubella
Thrombophlebitis, pelvic, due to bacteroides, 65
Thrush, due to *C. albicans*, 131
Tick fever virus, Colorado, 244
Tick typhus. *See* Rocky Mountain spotted fever
Ticks, viruses transmitted by, 153–155, 224, 224t., 244
Tinea, 122, 131
Todd units, 15
Togaviruses, 224
Tolerance, antibiotic, 5
Torula histolytica, 133
Torulopsis glabrata, 133
Torulosis, 133
Toxic shock syndrome, staphylococcal, 5

Toxins. *See also* Endotoxins; Enterotoxins; Exotoxins
  in amebiasis, 261
  epidermolytic, 2, 5, 6
  erythrogenic, 17
  heat-labile (LT), 52–53
  heat-stable (ST), 52, 53
Toxocara, 276–278
Toxocara canis, 276, 277, 277
Toxocara cati, 276
Toxoplasma gondii, 264–266, 264, 265
Toxoplasmosis, 264–266, 264, 265
Trachoma, 149, 150, 151
Transduction, 4, 5, 10
Transfer factor, for leprosy, 119
  in tuberculosis, 116
Transformation, DNA, 246
  in Epstein-Barr virus, 174
  neoplastic, 246
  in pneumococci, 10
Transfusion. *See* Blood transfusion
Transplacental (maternal) antibody, 20
Treponema, 141
Treponema microdentium, 141, 144
Treponema pallidum, 141–145
Trichinella spiralis, 275–276
Trichinosis, 275–276, 276
Trichomonas, 271
Trichomonas vaginalis, 270–271
Triple sugar iron (TSI) agar test, 42
Tuberculin tests, 113–114
Tuberculosis, 111–117
  bovine, 113, 117
Tularemia, 83–84, 84
Tumor antigen, 247
Tumor viruses, 246–247
  DNA, 247
Tumors, due to adenoviruses, 178
  Epstein-Barr virus associated with, 172, 173, 175, 175
Typhoid fever, 43, 43. 44
Typhus, 153–154
  murine, 153
  recrudescent, 155
  scrub, 153–154
  tick, 153

Ulceroglandular syndrome, in tularemia, 84
Ureaplasma ureolyticum, 156
Ureaplasmas, 156
Urethritis, gonococcal, 96
  gonorrheal, 150
  nongonococcal, 150, 157
  due to *Trichomonas* species, 270
Urinary tract infections, *E. coli* in, 55, 56
  due to enteric bacteria, 42
  due to enterococci, 22
  due to *H. influenzae*, 74

Urinary tract infections *(continued)*
  due to *P. aeruginosa*, 59
Urine cytology, for cytomegalovirus infection, 170, 171
Uterine infections, due to bacteroides, 65
  clostridial, 33

V factor, in *H. influenzae*, 75, 75
Vaccinia-infected eczema, 180–181
Vaccinia viruses, 179–182
Vaginitis, candida, 131
  corynebacteria as cause of, 27
  due to mycoplasmas, 157
  purulent, due to *N. gonorrheae*, 96
  due to *Trichomonas vaginalis*, 270
Varicella, 166–169
  progressive disseminated, 166–167
Varicella bullosa, 167
Varicella-zoster virus, 164, 164, 166–169
Variola major, 180
Variola minor, 180
Variola sine eruptione, 180
VDRL test, 144
Veillonella genus, 108
Venereal disease, 95–99, 141–145, 150
Vibrio(s), 62–64
  halophilic, 62, 64
Vibrio alginolyticus, 62
Vibrio cholerae, 53, 56, 62, 63, 63, 64
Vibrio parahaemolyticus, 62, 63, 64
Viroids, 249
Visna virus, 248, 249t.
Vomiting disease, epidemic, 251–252

Warts, 247
Weil-Felix reaction, in rickettsial infection, 57, 154, 155
Weil's disease, 146
Whooping cough. *See* Pertussis
Wool-sorter's disease, 37
Wound infections, *E. coli* in, 56
  Staphylococcal, 2

X factor, *H. influenzae* and, 75, 75

Yaws, 141
Yellow fever virus, 243, 243
Yersinia, 81
Yersinia enterocolitica, 81
Yersinia pestis, 81
Yersinia pseudotuberculosis, 81

Zoonoses, 81. *See also* Animals
Zoster, 162, 166–169
  disseminated, 167
Zoster-like illness, 166
Zoster opthalmicus, 168
Zygomycosis, 135